DESIGN OF EXPERIMENTS

STATISTICS: Textbooks and Monographs

1. The Generalized Jackknife Statistic, *H. L. Gray and W. R. Schucany*
2. Multivariate Analysis, *Anant M. Kshirsagar*
3. Statistics and Society, *Walter T. Federer*
4. Multivariate Analysis: A Selected and Abstracted Bibliography, 1957–1972, *Kocherlakota Subrahmaniam and Kathleen Subrahmaniam*
5. Design of Experiments: A Realistic Approach, *Virgil L. Anderson and Robert A. McLean*
6. Statistical and Mathematical Aspects of Pollution Problems, *John W. Pratt*
7. Introduction to Probability and Statistics (in two parts), Part I: Probability; Part II: Statistics, *Narayan C. Giri*
8. Statistical Theory of the Analysis of Experimental Designs, *J. Ogawa*
9. Statistical Techniques in Simulation (in two parts), *Jack P. C. Kleijnen*
10. Data Quality Control and Editing, *Joseph I. Naus*
11. Cost of Living Index Numbers: Practice, Precision, and Theory, *Kali S. Banerjee*
12. Weighing Designs: For Chemistry, Medicine, Economics, Operations Research, Statistics, *Kali S. Banerjee*
13. The Search for Oil: Some Statistical Methods and Techniques, *edited by D. B. Owen*
14. Sample Size Choice: Charts for Experiments with Linear Models, *Robert E. Odeh and Martin Fox*
15. Statistical Methods for Engineers and Scientists, *Robert M. Bethea, Benjamin S. Duran, and Thomas L. Boullion*
16. Statistical Quality Control Methods, *Irving W. Burr*
17. On the History of Statistics and Probability, *edited by D. B. Owen*
18. Econometrics, *Peter Schmidt*

19. Sufficient Statistics: Selected Contributions, *Vasant S. Huzurbazar (edited by Anant M. Kshirsagar)*
20. Handbook of Statistical Distributions, *Jagdish K. Patel, C. H. Kapadia, and D. B. Owen*
21. Case Studies in Sample Design, *A. C. Rosander*
22. Pocket Book of Statistical Tables, *compiled by R. E. Odeh, D. B. Owen, Z. W. Birnbaum, and L. Fisher*
23. The Information in Contingency Tables, *D. V. Gokhale and Solomon Kullback*
24. Statistical Analysis of Reliability and Life-Testing Models: Theory and Methods, *Lee J. Bain*
25. Elementary Statistical Quality Control, *Irving W. Burr*
26. An Introduction to Probability and Statistics Using BASIC, *Richard A. Groeneveld*
27. Basic Applied Statistics, *B. L. Raktoe and J. J. Hubert*
28. A Primer in Probability, *Kathleen Subrahmaniam*
29. Random Processes: A First Look, *R. Syski*
30. Regression Methods: A Tool for Data Analysis, *Rudolf J. Freund and Paul D. Minton*
31. Randomization Tests, *Eugene S. Edgington*
32. Tables for Normal Tolerance Limits, Sampling Plans and Screening, *Robert E. Odeh and D. B. Owen*
33. Statistical Computing, *William J. Kennedy, Jr., and James E. Gentle*
34. Regression Analysis and Its Application: A Data-Oriented Approach, *Richard F. Gunst and Robert L. Mason*
35. Scientific Strategies to Save Your Life, *I. D. J. Bross*
36. Statistics in the Pharmaceutical Industry, *edited by C. Ralph Buncher and Jia-Yeong Tsay*
37. Sampling from a Finite Population, *J. Hajek*
38. Statistical Modeling Techniques, *S. S. Shapiro*
39. Statistical Theory and Inference in Research, *T. A. Bancroft and C.-P. Han*
40. Handbook of the Normal Distribution, *Jagdish K. Patel and Campbell B. Read*
41. Recent Advances in Regression Methods, *Hrishikesh D. Vinod and Aman Ullah*
42. Acceptance Sampling in Quality Control, *Edward G. Schilling*
43. The Randomized Clinical Trial and Therapeutic Decisions, *edited by Niels Tygstrup, John M Lachin, and Erik Juhl*
44. Regression Analysis of Survival Data in Cancer Chemotherapy, *Walter H. Carter, Jr., Galen L. Wampler, and Donald M. Stablein*
45. A Course in Linear Models, *Anant M. Kshirsagar*
46. Clinical Trials: Issues and Approaches, *edited by Stanley H. Shapiro and Thomas H. Louis*
47. Statistical Analysis of DNA Sequence Data, *edited by B. S. Weir*
48. Nonlinear Regression Modeling: A Unified Practical Approach, *David A. Ratkowsky*
49. Attribute Sampling Plans, Tables of Tests and Confidence Limits for Proportions, *Robert E. Odeh and D. B. Owen*
50. Experimental Design, Statistical Models, and Genetic Statistics, *edited by Klaus Hinkelmann*
51. Statistical Methods for Cancer Studies, *edited by Richard G. Cornell*
52. Practical Statistical Sampling for Auditors, *Arthur J. Wilburn*
53. Statistical Methods for Cancer Studies, *edited by Edward J. Wegman and James G. Smith*

54. Self-Organizing Methods in Modeling: GMDH Type Algorithms, *edited by Stanley J. Farlow*
55. Applied Factorial and Fractional Designs, *Robert A. McLean and Virgil L. Anderson*
56. Design of Experiments: Ranking and Selection, *edited by Thomas J. Santner and Ajit C. Tamhane*
57. Statistical Methods for Engineers and Scientists: Second Edition, Revised and Expanded, *Robert M. Bethea, Benjamin S. Duran, and Thomas L. Boullion*
58. Ensemble Modeling: Inference from Small-Scale Properties to Large-Scale Systems, *Alan E. Gelfand and Crayton C. Walker*
59. Computer Modeling for Business and Industry, *Bruce L. Bowerman and Richard T. O'Connell*
60. Bayesian Analysis of Linear Models, *Lyle D. Broemeling*
61. Methodological Issues for Health Care Surveys, *Brenda Cox and Steven Cohen*
62. Applied Regression Analysis and Experimental Design, *Richard J. Brook and Gregory C. Arnold*
63. Statpal: A Statistical Package for Microcomputers—PC-DOS Version for the IBM PC and Compatibles, *Bruce J. Chalmer and David G. Whitmore*
64. Statpal: A Statistical Package for Microcomputers—Apple Version for the II, II+, and IIe, *David G. Whitmore and Bruce J. Chalmer*
65. Nonparametric Statistical Inference: Second Edition, Revised and Expanded, *Jean Dickinson Gibbons*
66. Design and Analysis of Experiments, *Roger G. Petersen*
67. Statistical Methods for Pharmaceutical Research Planning, *Sten W. Bergman and John C. Gittins*
68. Goodness-of-Fit Techniques, *edited by Ralph B. D'Agostino and Michael A. Stephens*
69. Statistical Methods in Discrimination Litigation, *edited by D. H. Kaye and Mikel Aickin*
70. Truncated and Censored Samples from Normal Populations, *Helmut Schneider*
71. Robust Inference, *M. L. Tiku, W. Y. Tan, and N. Balakrishnan*
72. Statistical Image Processing and Graphics, *edited by Edward J. Wegman and Douglas J. DePriest*
73. Assignment Methods in Combinatorial Data Analysis, *Lawrence J. Hubert*
74. Econometrics and Structural Change, *Lyle D. Broemeling and Hiroki Tsurumi*
75. Multivariate Interpretation of Clinical Laboratory Data, *Adelin Albert and Eugene K. Harris*
76. Statistical Tools for Simulation Practitioners, *Jack P. C. Kleijnen*
77. Randomization Tests: Second Edition, *Eugene S. Edgington*
78. A Folio of Distributions: A Collection of Theoretical Quantile-Quantile Plots, *Edward B. Fowlkes*
79. Applied Categorical Data Analysis, *Daniel H. Freeman, Jr.*
80. Seemingly Unrelated Regression Equations Models: Estimation and Inference, *Virendra K. Srivastava and David E. A. Giles*
81. Response Surfaces: Designs and Analyses, *Andre I. Khuri and John A. Cornell*
82. Nonlinear Parameter Estimation: An Integrated System in BASIC, *John C. Nash and Mary Walker-Smith*
83. Cancer Modeling, *edited by James R. Thompson and Barry W. Brown*
84. Mixture Models: Inference and Applications to Clustering, *Geoffrey J. McLachlan and Kaye E. Basford*

85. Randomized Response: Theory and Techniques, *Arijit Chaudhuri and Rahul Mukerjee*
86. Biopharmaceutical Statistics for Drug Development, *edited by Karl E. Peace*
87. Parts per Million Values for Estimating Quality Levels, *Robert E. Odeh and D. B. Owen*
88. Lognormal Distributions: Theory and Applications, *edited by Edwin L. Crow and Kunio Shimizu*
89. Properties of Estimators for the Gamma Distribution, *K. O. Bowman and L. R. Shenton*
90. Spline Smoothing and Nonparametric Regression, *Randall L. Eubank*
91. Linear Least Squares Computations, *R. W. Farebrother*
92. Exploring Statistics, *Damaraju Raghavarao*
93. Applied Time Series Analysis for Business and Economic Forecasting, *Sufi M. Nazem*
94. Bayesian Analysis of Time Series and Dynamic Models, *edited by James C. Spall*
95. The Inverse Gaussian Distribution: Theory, Methodology, and Applications, *Raj S. Chhikara and J. Leroy Folks*
96. Parameter Estimation in Reliability and Life Span Models, *A. Clifford Cohen and Betty Jones Whitten*
97. Pooled Cross-Sectional and Time Series Data Analysis, *Terry E. Dielman*
98. Random Processes: A First Look, Second Edition, Revised and Expanded, *R. Syski*
99. Generalized Poisson Distributions: Properties and Applications, *P. C. Consul*
100. Nonlinear L_p-Norm Estimation, *Rene Gonin and Arthur H. Money*
101. Model Discrimination for Nonlinear Regression Models, *Dale S. Borowiak*
102. Applied Regression Analysis in Econometrics, *Howard E. Doran*
103. Continued Fractions in Statistical Applications, *K. O. Bowman and L. R. Shenton*
104. Statistical Methodology in the Pharmaceutical Sciences, *Donald A. Berry*
105. Experimental Design in Biotechnology, *Perry D. Haaland*
106. Statistical Issues in Drug Research and Development, *edited by Karl E. Peace*
107. Handbook of Nonlinear Regression Models, *David A. Ratkowsky*
108. Robust Regression: Analysis and Applications, *edited by Kenneth D. Lawrence and Jeffrey L. Arthur*
109. Statistical Design and Analysis of Industrial Experiments, *edited by Subir Ghosh*
110. *U*-Statistics: Theory and Practice, *A. J. Lee*
111. A Primer in Probability: Second Edition, Revised and Expanded, *Kathleen Subrahmaniam*
112. Data Quality Control: Theory and Pragmatics, *edited by Gunar E. Liepins and V. R. R. Uppuluri*
113. Engineering Quality by Design: Interpreting the Taguchi Approach, *Thomas B. Barker*
114. Survivorship Analysis for Clinical Studies, *Eugene K. Harris and Adelin Albert*
115. Statistical Analysis of Reliability and Life-Testing Models: Second Edition, *Lee J. Bain and Max Engelhardt*
116. Stochastic Models of Carcinogenesis, *Wai-Yuan Tan*
117. Statistics and Society: Data Collection and Interpretation: Second Edition, Revised and Expanded, *Walter T. Federer*
118. Handbook of Sequential Analysis, *B. K. Ghosh and P. K. Sen*
119. Truncated and Censored Samples: Theory and Applications, *A. Clifford Cohen*
120. Survey Sampling Principles, *E. K. Foreman*

121. Applied Engineering Statistics, *Robert M. Bethea and R. Russell Rhinehart*
122. Sample Size Choice: Charts for Experiments with Linear Models: Second Edition, *Robert E. Odeh and Martin Fox*
123. Handbook of the Logistic Distribution, *edited by N. Balakrishnan*
124. Fundamentals of Biostatistical Inference, *Chap T. Le*
125. Correspondence Analysis Handbook, *J.-P. Benzécri*
126. Quadratic Forms in Random Variables: Theory and Applications, *A. M. Mathai and Serge B. Provost*
127. Confidence Intervals on Variance Components, *Richard K. Burdick and Franklin A. Graybill*
128. Biopharmaceutical Sequential Statistical Applications, *edited by Karl E. Peace*
129. Item Response Theory: Parameter Estimation Techniques, *Frank B. Baker*
130. Survey Sampling: Theory and Methods, *Arijit Chaudhuri and Horst Stenger*
131. Nonparametric Statistical Inference: Third Edition, Revised and Expanded, *Jean Dickinson Gibbons and Subhabrata Chakraborti*
132. Bivariate Discrete Distribution, *Subrahmaniam Kocherlakota and Kathleen Kocherlakota*
133. Design and Analysis of Bioavailability and Bioequivalence Studies, *Shein-Chung Chow and Jen-pei Liu*
134. Multiple Comparisons, Selection, and Applications in Biometry, *edited by Fred M. Hoppe*
135. Cross-Over Experiments: Design, Analysis, and Application, *David A. Ratkowsky, Marc A. Evans, and J. Richard Alldredge*
136. Introduction to Probability and Statistics: Second Edition, Revised and Expanded, *Narayan C. Giri*
137. Applied Analysis of Variance in Behavioral Science, *edited by Lynne K. Edwards*
138. Drug Safety Assessment in Clinical Trials, *edited by Gene S. Gilbert*
139. Design of Experiments: A No-Name Approach, *Thomas J. Lorenzen and Virgil L. Anderson*

Additional Volumes in Preparation

Statistics in the Pharmaceutical Industry: Second Edition, Revised and Expanded, *edited by C. Ralph Buncher and Jia-Yeong Tsay*

Advanced Linear Models: Theory and Applications, *Song-Gui Wang and Shein-Chung Chow*

Multistage Selection and Ranking Procedures: Second-Order Asymptotics, *Nitis Mukhopadhyay and Tumulesh K. S. Solanky*

DESIGN OF EXPERIMENTS

A No-Name Approach

THOMAS J. LORENZEN

Mathematics Department
General Motors Research
Warren, Michigan

VIRGIL L. ANDERSON

Department of Statistics
Purdue University
West Lafayette, Indiana

MARCEL DEKKER, INC. NEW YORK • BASEL

Library of Congress Cataloging-in-Publication Data

Lorenzen, Thomas J.
Design of experiments: a no-name approach / Thomas J. Lorenzen, Virgil L. Anderson.
p. cm. -- (Statistics, textbooks and monographs; v. 138)
Includes bibliographical references and index.
ISBN 0-8247-9077-4
1. Experimental design. I. Anderson, Virgil L. (Virgil Lee). II. Title. III. Series.
QA279.L67 1993
001.4'34--dc20 93-2095
CIP

The publisher offers discounts on this book when ordered in bulk quantities. For more information, write to Special Sales/Professional Marketing at the address below.

This book is printed on acid-free paper.

MARCEL DEKKER, INC.
270 Madison Avenue, New York, New York 10016

Current printing (last digit):
10 9 8 7 6 5

PRINTED IN THE UNITED STATES OF AMERICA

To Sue and Avis

Preface

What is a "no-name" approach to the statistical design of experiments?

The "no-name" approach is a simpler yet more complete way of specifying and evaluating designs. Historically, designs were created for specific purposes and given specific names such as randomized complete block design, split-plot design, Latin square design, and so on, and so on. The emphasis in many textbooks is on presenting these named designs. When students are faced with designing experiments, they often find that none of the named designs exactly fits their needs. You often hear experiments described something like the following: "I have a split-plot design with a Latin Square on these factors and a randomized complete block on those factors."

The approach in this book is entirely different. Four fundamental concepts are given in Chapters **3**, **4**, **5**, and **7**. For ***any*** design requirements, these four fundamental concepts are used to write a unique mathematical model describing the experiment. From this unique mathematical model, the design combinations and run order are determined, the design can be evaluated to see if it meets the needs of the experimenter, and, after the experiment is run, the data can be analyzed. There is no need to be restricted to the class of "named" designs, all possible designs are covered.

The use of mathematical models to analyze data is standard practice since the advent of computer programs. Some books even introduce mathematical models to describe the "named" designs. This textbook is different because the mathematical models uniquely describe an experiment and the theory is completely developed around the mathematical model. This allows total flexibility in the design of the experiment. Experimenters are not limited to the "named" designs they have learned.

One of the authors postulated the uniqueness of the mathematical model (*Biometrics*, 1970, 255-268) while the other later proved the postulate (*Comm. in Stat.—Theory and Methods*, 1984, 2601-2623). This book is a result of the development of the theory since those papers.

In the first draft of this book names for designs were not even mentioned. The reviewers felt this was a mistake since much discussion centers around existing "named" designs. We agree and have included many "named" designs throughout this book. They can often be found in the problems at the end of each chapter since they are applications of the concept presented in the chapter.

This book has been written for an undergraduate senior or a first year graduate student with at least one course in basic statistics. We have had good success with General Motors engineers reading preliminary drafts on their own. We have also included enough of the theory behind the procedures and enough difficult problems that the book could serve as a basis for a more theoretical course. Since we have tried to appeal to a somewhat broader audience, we have included two icons throughout the book, a "nut and bolt" icon: and a "light bulb" icon: .

In the text, worked examples are indicated with a nut and bolt. At the end of each chapter, problems marked with a nut and bolt are considered essential to understanding the material. Solutions to these problems are provided to help those who study this material without the aid of an instructor.

Paragraphs marked with a light bulb are meant for theoretical statisticians only and can be skipped without loss of understanding of the material. They serve only as a guide for a more theoretical coverage of this material. Problems marked with a light bulb represent extensions beyond the basic material covered in the chapter. Many of these problems can be tackled by the probing non-statistician as well as the theoretical statistician.

This book has nine chapters. The first chapter presents the scientific method for designing an experiment, from the recognition stage to the implementation and summary stage. Chapter **2** presents the terminology, assumptions, and tests on the assumptions. Chapter **3** is the longest chapter. It presents the factorial concept, shows how to set up the unique mathematical model, and shows everything that can be done with the mathematical model. Chapters **4** and **5** present the nesting concept and restrictions on randomization. Chapter **6** is a return to Chapter **1**, having gained the knowledge of Chapters **2–5**. We have observed that students often get so involved with the technical details that they forget the fundamentals. This chapter helps the student see the whole picture more clearly with one thoroughly developed example. Chapters **7** and **8** present the fractionation concept, for two level factors and three and mixed level factors respectively. Finally, Chapter **9** covers some response surface ideas.

Thomas J. Lorenzen
Virgil L. Anderson

Contents

The glory and the nothing of a name.

GEORGE NOEL GORDON, LORD BYRON
ON SIR WINSTON CHURCHILL'S GRAVE

1

A Scientific Approach to the Design of Experiments

1.0 INTRODUCTION

Experiments are performed by people in nearly all walks of life. The basic reason for running an experiment is to find out something that is not known. By their very nature, experiments are designed to draw inferences about an entire population based on a few observations.

If experiments were perfectly repeatable and the important factors were perfectly separable, it would be easy to analyze and interpret the results. However, experiments are often run so that the effect of one factor is unknowingly confounded with the effect of a factor not considered in the experiment. Even with the best of experimental control, results vary from trial to trial. These reasons, and many more, add to the difficulty in analyzing the data from an experiment.

The role of statistics in experimental design has been to separate the observed differences into those caused by various factors and those due to random fluctuation. The classical method used to separate these differences is *analysis of variance*, or ANOVA. In general, the method consists of looking at the total variation in the data, breaking it into its various "accountable for" components, and running statistical tests in an attempt to find out which components influence the experiment.

The specific ANOVA heavily depends on the design of the experiment. If two different experiments have the same factors but are laid out and run differently, the variation components will be broken apart differently. The information available on the effects of some factors will increase, others will decrease, and still others may disappear completely. As a result of this drawback, specific designs have been developed to accomplish different tasks with different efficiencies.

A number of designs in common use today have grown out of the agricultural field. Generally, a design is associated with a specific name and the resulting ANOVA performed according to rules specific to the design. One disadvantage of this method is a lack of "feeling" about the design and an inability to extend the design to any other circumstance. To counteract this disadvantage, mathematical models have been developed that "describe" the design and the corresponding analysis. Hopefully, through the use of these models, the experimenter (or student, statistician, *etc.*) can visualize the layout and perform the ANOVA to draw conclusions.

Unfortunately, it has been shown *in textbooks* that designs can have the same mathematical model even if the layout is entirely different. (See Anderson, 1970, for an example.) This means the mathematical model does not adequately describe the layout of the experiment. Despite this drawback, these mathematical models are in wide use today. Quite frequently, the name associated with a given design is supposed to key the experimenter to the exact layout and eventually to the analysis without the experimenter ever knowing the meaning of the design or even the appropriate mathematical model.

The authors of this text do not believe it is correct to associate a layout with a name and then analyze the data in a fashion specific to that name. In the first place, there is no reason to force an experiment into a design that may have been constructed for other purposes. At best, we place undue restrictions on the experiment and may even lose valuable information. Even if information is not lost, we may be able to gain more information on the effect of factors of interest by using some non-named design.

Secondly, and more importantly, names are not needed! The problem with the classical approach is simply a deficiency in the mathematical model. With the inclusion of a few additional terms, the model serves as an understandable intermediary step, both describing the layout and serving as the basis for a rigorous analysis of the data. Instead of presenting many specific models, a few concepts will be presented. If these concepts are understood, **all of the named designs as well as any other design** can be comprehended.

Chapter **2** considers one factor designs, introducing the ANOVA terminology, listing the assumptions of the mathematical model, and giving a few statistical techniques. Chapter **3** presents the factorial concept and gives general methods for analyzing the collected data. Because it deals with both design and analysis, it is the longest chapter. Chapter **4** presents the nesting concept. Chapter **5** presents the concept of restrictions on randomizations . Chapter **6** reviews Chapter **1** from a more informed perspective. Experience has indicated that the material in Chapter **1** is not thoroughly understood until the mathematical framework in Chapters **2** through **5** has been presented. Chapter **7** presents the concept of fractionation for two level experiments. This concept applies for both fixed and random factors, whether the experiment is completely randomized or there are restrictions on the randomization. Chapter **8** extends the fractionation concept from two level to prime level experiments, presents methods useful for non-prime level and mixed level experiments, gives some orthogonal main effect plans, and discusses some non-orthogonal plans. Chapter **9** presents other ideas useful when designing for a specific form of a regression model.

1.1 THE SCIENTIFIC APPROACH

Thus far we have focused on the statistician's role in the design of experiments. Seldom do statisticians actually run the experiment. That part is almost always the responsibility of the experimenter involved. If experiments are to be run, and they most assuredly will be run, a scientific approach to designing them is needed. The following is an ordered list of requirements for scientific experimentation. The significance of each requirement will be discussed in detail in the remaining sections of this chapter.

- *Recognition that a problem exists.*
- *Formulation of the problem.*
- *Specifying the variable to be measured.*
- *Agreeing on the factors and levels to be used in the experiment.*
- *Definition of the inference space for the problem.*
- *Random selection of the experimental units.*
- *Layout of the design.*
- *Development of the mathematical model.*
- *Preliminary evaluation of the design.*
- *Redesigning the experiment.*
- *Collecting the data.*
- *Analyzing the data.*
- *Conclusions.*
- *Implementation.*

1.2 RECOGNITION THAT A PROBLEM EXISTS

In many industrial plants the accountants are able to show which areas are having problems by arraying the costs in various departments (sometimes called a *Pareto diagram*). Problems are indicated when the relative costs in these areas either have changed recently or are large with respect to some criteria.

In other plants the foreman may recognize trouble and bring it to the attention of a superior. Or a new technique may be considered to replace an existing technique. Or customers may be complaining about a specific defect. In any event, at least one area always seems to need attention, and alert management recognizes that a problem exists.

In research departments of universities or industries, people are continually encountering problems that cannot be solved analytically but can be attacked successfully using scientific experimentation. The recognition phase in these cases is usually short.

Many companies have a mission to continually improve products and processes. The problem here is to identify the causes of variation and experiment with different techniques to reduce variation.

1.3 FORMULATION OF THE PROBLEM

Cross-functional teams are the most successful means of formulating the problem after management agrees that a problem exists. The word *action* is important because the members of the cross-functional team for attacking a problem must be the ones who know the technical parts involved and be willing to state their thoughts. A person who knows the requirements for scientific experimentation and keeps the thoughts flowing is the best moderator for such a group. There must never be reticence by any participants, even if one's ideas are in conflict with those of a supervisor who may be present. Hence, a moderator from outside the plant who stimulates thought and prevents conflicts is usually preferred to a knowledgeable in-plant participant.

Usually this discussion period lasts from two to eight hours with all team members having a chance to express their views. Frequently forty to fifty possible causes of the problem will be put forth. The next step for the moderator is to get the participants to reduce these to a reasonable number of causes. Hopefully, the list will be reduced to eight or ten; preferably to four or five. Rarely is the ultimate list only two or three. Of course, the greatest difficulty in the actual formulation of the problem is to get the team members to agree on the ultimate causes to be tried in the forthcoming experiment.

When the number of possible causes is large and there is not enough knowledge about the process to narrow the list down, a two-staged approach must be used. The purpose of the first stage is to find the most important factors in the list. The second stage will study these factors in more detail.

To illustrate the formulation of a problem, consider an investigation related to the fabrication of men's synthetic felt hats. The manufacturer had experienced extreme difficulty in producing these hats so the flocking appeared on the molded rubber base in a uniform fashion—simulating real felt hats. In order to approach this problem, a cross-functional team was formed consisting of a development engineer, a manufacturing foreman, a chief operator, a sales representative, and a statistician. The statistician's job was to obtain all possible causes of imperfect hats. Factors which were possible causes were: thickness of the foam rubber base, pressure of molding, time of molding, viscosity of the latex used to glue the flocking to the molded rubber base, age of the latex, nozzle size of several different spray guns, direction of spraying, condition of the flocking, speed of drying, and location within the drying furnace.

After considerable discussion, the cross-functional team decided that the most serious problems were probably connected with the nozzle size and the pressure under which the latex was sprayed. In arriving at these various factors the team essentially forced a review of the entire production process. This in itself led to a better understanding of the production process and an eventual solution to the problem.

1.4 SPECIFYING THE VARIABLE TO BE MEASURED

One of the most difficult things in certain types of industrial experimentation is the specification of the variable to be measured (usually referred to as the *dependent variable* or simply y). Usually it is quite obvious what the dependent variable should be but the means of measuring it is sometimes quite difficult. In ideal cases, the value of the dependent variable is measured by some inspection tool. In other cases, it is a value that is almost impossible to measure and has to be graded by one or more inspectors. In a new class of problems defined by Taguchi (1987), the goal is to minimize the effect of production variation and customer use on the performance of a product. In this class of problems, y is a measure of variation over different production and customer use settings.

An ideal dependent variable is continuous, easily and accurately measurable, and directly related to the customer's perception of quality. Selection of the proper dependent variable(s) is the responsibility of the cross-functional team, not the statistician. However, the statistician has the responsibility to see that proper effort is put into finding the ideal dependent variable(s).

As one can imagine, the measuring of product quality in our synthetic hat example was quite difficult. It was decided that three dependent variables should be used: the hungry appearance of the flocking, the starchy appearance of the flocking, and the appearance of the brim. During the course of the investigation, the three dependent variables were found to be essentially independent of each other and consequently could be treated as three separate analyses. The standards for these variables were again arrived at through team action and resulted in a visual display board that gave the inspectors a realistic way to grade each dependent variable.

1.5 AGREEING ON THE FACTORS AND LEVELS TO BE USED IN THE EXPERIMENT

Like the dependent variable discussed in the previous section, the factors to be used in the experiment must either be measurable or distinguishable. In addition, the factors must be controllable, at least within a laboratory setting. This is because the design will tell exactly which combinations of the various factors should be run and in which order. This is determined by the needs of the experimenter.

Likewise, the number of levels of each factor will be determined by the needs of the experimenter. Sometimes the levels of a factor are predetermined and cannot be changed. Other times, the number of levels is optional and determined by trading off the experimental effort with the information to be gained by running the experiment. This trade-off will become clear as more is learned throughout the text.

Everything that affects the value of the response variable is a potential factor to be used in an experiment. This includes things that can be rigorously controlled, like the temperature of an oven or the amount of fertilizer added to a farm field, as well as things that simply cannot be controlled, like day to day fluctuations or differences in soil content from plot to plot. The levels of the latter factors cannot be set to a specified value. Rather each level represents a particular day effect or a particular plot effect and is generally considered to be random. More will be said about the different types of factors throughout the text.

It doesn't take much imagination to realize there are thousands upon thousands of potential factors in every experiment. It is up to the team of experts to come up with a list of the most reasonable factors, those likely to have the most effect on the response variable. Of course, if the team *knew* the factors there would be no reason to run an experiment.

In arriving at the most reasonable causes of nonconforming hats in the Section **1.3**, the team action essentially required a critical review of all production process factors. In this review the chief operator brought out the standard operating levels of the nozzle size as well as the pressure under which the latex was sprayed. Talk with the chief operator revealed that the latex pressure varied considerably due to the viscosity of the latex and from this information the pressure levels were eventually obtained. Additional inquiry ascertained that the manufacturing area had two different nozzle sizes that had been used interchangeably; consequently, the two sizes became the basis for this factor's levels.

The authors feel that, through team action of this type, one is almost always able to find realistic levels for all major factors. Occasionally, considerable effort is needed to find out just how shoddy one's manufacturing operations really are, and this example shows that actual production operations will allow their processes to operate at various levels as long as output is obtained. The determination of optimal levels is, of course, the desired end of this experimental investigation.

There are two specific tools that can help the team search for factors: the cause-and-effect (or fishbone) diagram and functional specification analysis.

In a cause-and-effect diagram, the dependent variable is written on the right with a thick arrow pointing from the left to the right. Each major factor is identified and has a branch arrow pointing to the thick main arrow. Factors influencing the major factors have thinner arrows pointing to the branch arrows. And so on. A cause-and-effect diagram for the felt hat example is given in Figure 1.1.

Functional specification analysis is similar to the cause-and-effect diagram but is better suited to design and product engineers. The response variable is written as some unknown function $f(\cdot)$ of the major factors. Then each major factor is written as another unknown function of other factors. And so on. An advantage of functional specifications over cause-and-effect diagrams is that interactions (to be defined later) are easily identified as factors appearing both in $f(\cdot)$ and in other functions of the major factors.

For the felt hat problem, we would write (Quality of Flocking) = f(Foam Base, Latex Glue, Drying, Molding, Spray Flocking). Then we would continue by writing (Foam Base) = g(thickness, supplier), (Latex Glue) = h(viscosity, age), and so on. One continues until all of the relations between the variables have been specified.

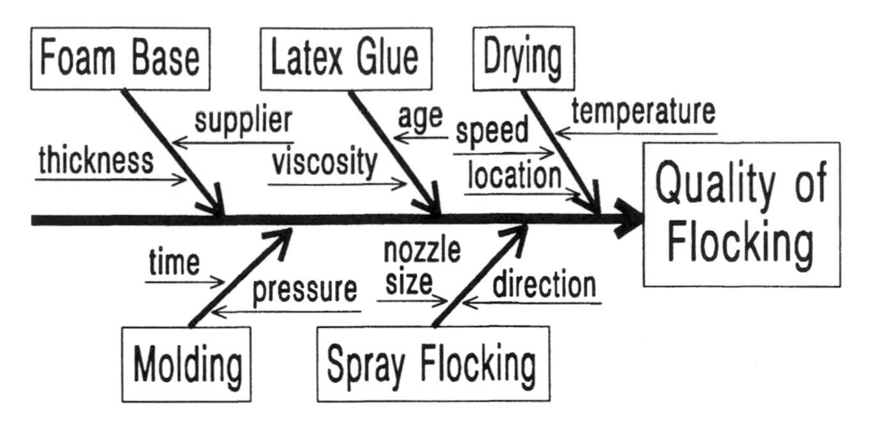

Figure 1.1 Cause-and-effect diagram for felt hat production.

1.6 DEFINITION OF THE INFERENCE SPACE FOR THE PROBLEM

After defining the problem, the experimenter must make decisions on the limits of the inference to be made from the results. Those limits within which the results will apply may be called the *inference space*. The inference space is indelibly linked to the factor selection as we have defined factors. All "factors" *not* selected for use in an experiment, either because one was not aware of the factors presence or because one was physically unable to set or change levels, must be either fixed for the entire experiment or allowed to vary naturally. If a factor not selected for the experiment is fixed at a given level, the experimental results will only apply for that specific level. In other words, the inference space will be restricted to that level only.

On the other hand, if a factor not selected for the experiment is allowed to vary naturally *and* complete randomization is used, its effect will become part of the error term used to test the factors studied in the experiment. This will make it more difficult to determine if a factor considered in the experiment has a significant effect or not.

One of the most difficult challenges faced by experimenters is to make the proper trade-offs between factors included in the experiment, factors whose levels are fixed during the experiment, and factors allowed to vary naturally. If too many factors are included in the experiment, the experimental effort becomes unmanageable. If too many factors are fixed in the experiment, whether knowingly or unknowingly, the inference space becomes too small to be meaningful. Finally, if too many factors are left to vary of their own free will, the error term will become too big and everything will become insignificant in the experiment.

For example, a chemist may want the results of the experiment to apply to all laboratories in the world. If this is the case, a factor called *laboratory* should be included in the experiment and the experiment should be conducted in such a way that a sufficiently large random sampling of all possible laboratories should be included. If the entire experiment was conducted in his lab only, the inference space would be his lab only. If the experiment was spread out randomly among all possible labs, laboratory to laboratory differences may make the variability in the results so large that it swamps the effects of all other factors.

In genetic studies using Drosophila (fruit flies), the variation between bottles is usually larger than the variation within bottles and the inference space is usually over bottles. Hence, a factor called bottles must be considered and the experiment must include several bottles so the results contain between bottle variation as well as within bottle variation. The inference space can then be expanded from a single bottle to all bottles. The general ideas involving correct error terms (developed later) have direct association with the inference space.

Great care should be placed at the initial stages of formulating the experiment to assure the wide applicability desired by the experimenter. (See Cox, 1958, Section 1.2, pp. 9–11.) Deming (1960) strongly emphasizes the difference between the circumstances under which the experiment is run and the circumstances under which the conclusions are to be applied and points out the dangers associated with inferring from one set of circumstances to the other. Our approach to this dilemma is to consider several different designs, carefully indicate the inference space for each design, note the detectability in each design (to be defined later), and let the team of experts decide which is the best trade-off.

1.7 RANDOM SELECTION OF THE EXPERIMENTAL UNITS

Throughout this book the term *experimental unit* will refer to the type of experimental material that receives the application of the various factors and is representative of the desired inference space. For example, if one is interested in investigating the effects of various types of fertilizers on potted plants, the potted plant would be the experimental unit and the *factor* would be the type of fertilizer. The *factor levels* (also called *treatment levels*) are the various types of fertilizer applied to the potted plants. In this case, extreme care should be taken before extending the inference to a field of plants as the field was not the experimental unit.

Similarly, if one is interested in investigating the effect of a certain type of teaching device, the experimental unit would be the individual student. In agronomy experiments, one often uses a plot of ground as an experimental unit. In this case the size and location of the plot may affect the growing habits of the particular plant under consideration, and again one has to be careful.

Once the experimental unit has been selected, the experimenter usually has many units from which to select. Consequently, it becomes necessary to randomly select a sufficient number of experimental units to be utilized in the experiment. Random selection is necessary at this point in order to protect against any bias in our experiment which is the result of some unknown factor having had prior influence on the experimental units. The number of experimental units randomly selected depends on the size of the experiment and prior knowledge of the standard error of the means.

In certain investigations the available experimental units will not be of the same homogeneous material. The experimenter will suspect that the responses will differ due to this nonhomogeneity as well as the effect of the factors applied to the experimental units. In agronomy experiments, this nonhomogeneity could be soil conditions. In Section **1.6** we discussed a genetic study using Drosophila. If the individual fruit fly is the experimental unit, then these fruit flies will be stratified according to their respective bottle. This type of nonhomogeneity will be referred to as *blocking* in the forthcoming chapters and must be carefully noted. Blocking affects the inference space and can generally be treated as a factor in the experiment. In the Drosophila experiment, the blocking factor could be called *bottle*. By explicitly recognizing bottles as a factor, the inference space will be across bottles and the bottle to bottle variation will be removed from the overall error term.

1.8 LAYOUT OF THE DESIGN

Having selected the experimental units, the experimenter has to decide the exact manner in which to run the experiment. This will be called the *layout of the design*.

To be assured that the probability statements made are as nearly correct as possible, laws of chance must enter into the experiment. We ideally want the units to be influenced only by the applied treatment combinations. If factors other than those specifically controlled by the experiment are present, we want to minimize their influence on the tests. This includes factors unknown to the experimenter as well as those not controlled in the experiment. Examples include systematic trends in the experimental units, factors that change over time, and factors that change over location. Factors that can be controlled by the experimenter should be kept at a fixed level, with the inference applying to this fixed level only.

The best way to minimize the influence of unknown factors is to *randomly assign units to treatment combinations and randomize the run order or location of the treatment combinations.* For example, in an electroplating experiment, magnetic material was electroplated to two different substrates, copper and brass, using three different magnetic charges. The dependent variable was the thickness of the plating. The layout was the order in which the plating was done. If all the material with copper substrates was plated prior to any of the brass substrates, there may be a thicker plate due to a build-up of resistance in the electroplating process or due to environmental changes over time, not because of a difference in the metal in the substrates. To ensure against unknown peculiarities (such as environmental conditions), the order of plating the material in combination with the charge should be completely randomized.

As a result of complete randomization, the estimates of parameters (such as average thickness of plating) should be unbiased and the tests of significance of the effects of treatments should be valid. The idea originated with Fisher (1926) and is further discussed by White (1951), Greenburg (1951), Kempthorne (1955 and 1977) and Lorenzen, (1984).

Unfortunately, complete randomization is not always possible. Sometimes it is not economically feasible to completely randomize the experiment. Sometimes, like the fruit fly example previously given, the experimental units are naturally stratified. In these cases we can still perform some sort of randomization. Perhaps, as in the fruit fly example, we can randomize between bottles and separately within each bottle. No matter what, we randomize somewhere in the design.

In this book we refer to the layout as the *design of the experiment.* This is the method in which treatments are placed on the experimental units, the order in which the treatment combinations are run, or the location of the treatment combinations.

Whenever the layout calls for restrictions on the randomization, for whatever reason, the restrictions must be noted and accounted for in the mathematical model and in the evaluation. This was the case for the synthetic hat experiment. Here the team decided that it would not be economically feasible to randomly interchange spray gun nozzles. Consequently, when a certain nozzle was set up, several treatment combinations were run by randomly adjusting the various gun pressures. This restriction on randomization gave rise to what will be called a *restriction error.*

This must be acknowledged and noted in the appropriate mathematical model (Anderson, 1970). Recognizing and noting restrictions such as the one given above will play an important role in this book. It is the sole topic of Chapter 5.

1.9 DEVELOPMENT OF THE MATHEMATICAL MODEL

After deciding on a possible layout for the experiment, the statistician must develop the appropriate mathematical model. Note that this is done in the preliminary stages of the experimental process and particularly before any data is collected. This mathematical model will serve as an intermediary step, both describing the layout and serving as the basis for the analysis of the collected data.

The model contains information on all of the factors and all of the levels of the factors used in the experiment. The factors will be designated by capital letters and the levels designated by the subscripts associated with the factors. Whenever a restriction on randomization occurs, it must be noted in the model. Such a restriction will be denoted by a Greek letter having the appropriate subscripts and placed in the model in its appropriate position. Rather than go into great detail about the mathematical model, we refer the reader to the appropriate chapters of this book. Since the model is one of the important contributions of this book, we will spend considerable time and effort clarifying the relationships between the layout, the model, and the analysis of the data.

1.10 PRELIMINARY EVALUATION OF THE DESIGN

Once the mathematical model is obtained, the statistician can use any of several existing techniques to derive the appropriate analysis of variance or ANOVA table. This table will consist of the factor labels, their interactions, the restriction errors, degrees of freedom, sums of squares, mean squares, expected mean squares, detectability, and F values. The ANOVA table will indicate all of the various factors whose effects can be tested as well as the appropriate test statistics. In addition, if the effect of some factor or interaction cannot be tested, the ANOVA table will indicate approximate and/or conservative tests. All of this information is known before the first observation is taken.

Upon reviewing this table, the statistician must note which effects of factors are directly testable, the amount of information available to make these tests, and the assumptions necessary to make conservative tests. He must then present these findings to the team to see if they are reasonable from a practical point of view. If the team decides that some effect should contain more information, some assumption may not hold, or some effect must have a direct test, the design will have to be modified. Even if the assumptions are met, there may be a better way to run the experiment. It is the statistician's responsibility to find the best trade-off between the ease of running the experiment and the information to be gained by running the experiment.

1.11 REDESIGNING THE EXPERIMENT

At this point, since no data has been collected, there is no cost to redesign the experiment. An alternate design can be proposed and the redesigned experiment compared to the original experiment for information available on the most important terms. This is accomplished by checking for the availability of tests and comparing the detectability for the terms in the two ANOVA tables. After considering several designs and comparing the resulting ANOVA tables, it will become clear to the experimenter and statistician where compromises have to be made for economic reasons. By placing a certain type of restriction on the randomization, it may be possible to obtain a conservative test and greatly reduce the cost of the experiment. One must recall, however, that there is a cost incurred for making an incorrect conclusion that is in addition to the experimental cost. This must be considered when trying to decide which design is best from an overall standpoint.

Again, we emphasize that several designs should be compared prior to the collection of the data. Only then can we be assured that we have arrived at the best possible design for the given set of conditions. The statistician must feel free to influence the number and levels of factors used in the experiment when it leads to a more efficient, general, or economical design.

1.12 COLLECTING THE DATA

After searching enough designs to be assured that the best possible design has been found, the experimenter can actually run the experiment (as given by the experimental layout, of course) and collect the data.

To the inexperienced investigator this particular topic may seem like a rather simple and straightforward procedure. However, extreme care must also be taken at this point. In some cases, carrying out the experiment may require an organized team with many observers. Team planning meetings may become quite extensive and detailed instructions may have to be prepared for operators, supervisors, and technical observers. The whole success of this scientific investigation depends upon the validity of all the data obtained. Even if the experiment is not too complicated, the authors have found that one of the most successful means of collecting data is for the statistician to make a specific data entry form which would be filled out by the inspector and further verified by the person in charge of running the experiment. The data form itself must be easy to use, understandable, and, in general, foolproof with respect to any misinterpretations. For a complicated experiment, it is sometimes wise to pretest the experimental plan and the data collection form. By this we mean to process a few experimental units that are designed specifically for pretesting the inspection and data recording procedures. Another possibility is to have the experimenter formulate data that could exist and analyze this dummy data as if it were actual data.

Considerable thought should be given to the collection of the data. For example, if the data is going to be entered into a particular computer program, it is well known that typing errors can be considerably reduced by having an efficient data form. In collecting extreme amounts of data, possibly the data can be obtained on magnetic tapes, floppy disks, data storage devices, or some type of mark sense cards. In general, the larger the experiment, the more care should be taken in the collection and coding of the data.

1.13 ANALYZING THE DATA

The actual analysis of the data will be primarily the statistician's responsibility. As pointed out earlier, this analysis is highly dependent on the design. Modern computing facilities have many so-called canned programs which will analyze the data from most of the classical designs. The input to these programs consists of the model, the factors, the number of levels from each factor, whether each factor is fixed or random (terms to be defined later), and finally, the data itself. Caution must always be exerted with these programs. As of the writing of this text, all such programs expect classical models. Some make assumptions about the nature of the factors causing improper tests and, possibly, false conclusions. The classical models cannot completely describe the layout of all experiments. Thus, the programs cannot always compute the proper test statistics. In general, canned programs cannot be modified to accept new models and, in particular, they cannot be modified to accept restriction errors. It is up to the statistician to recognize restriction errors, derive the proper tests, and modify the output of these programs accordingly. As will be pointed out throughout this text, it is easy to make these modifications.

If a modern computer is not available, one must resort to a hand calculator or similar device. Formulas for these calculations will be illustrated throughout this text and can be easily found in references such as Hicks (1982), Ostle and Malone (1988), and Steel and Torrie (1980). As with the canned programs, it will be necessary to insert restriction errors in the proper locations. When calculations are performed by hand, results should always be double-checked for accuracy as it is very easy to make a mistake somewhere along the line.

Once the basic calculations have been made the statistician should review the data for possible outliers and make graphs of the cell means in order to display the results of the experiment. The authors have found that graphical presentation of the data, complete with confidence intervals, is the best way to show management the important aspects of the problem.

1.14 CONCLUSIONS

Once the statistician has completed the analysis of the data it is again time for various members of the cross-functional team to meet and formalize their conclusions. In almost every case the experimenter will be able to physically interpret the statistical findings that have been brought to light by this experiment. Then, based on the magnitude of these findings and the cost of implementing any new procedures which might be required, the team must make recommendations to management. These recommendations are usually in some form of written and/or oral presentation to interested members of management.

1.15 IMPLEMENTATION

The implementation of the recommendations developed by the team is, of course, a management decision. The purpose of scientific investigation is to develop a realistic type of experiment, assure proper running of the experiment, and find dependable results. These findings have to be presented in such a fashion that management can make an intelligent decision about implementation.

Once management decides to make the process change recommended by the team, the team has another tremendous task facing it. This task is to assure management that the experimental procedure will be implemented as well as possible into the production process. It is not always a simple matter to implement a procedure developed in a pilot study into a full-fledged production process. In no case should a new procedure be put into production without further evaluation by some type of statistical study. It is the responsibility of the team to assure management that revisions to the process have indeed improved the production operation. Many times a prototype of the actual production set up is devised and tried out on a small-scale production basis. This gives an indication of how the results of the experiment may work in production without using actual production equipment as a testing device.

If the results of the prototype are successful, then the results of the experiment may be put into production. Many times changes are indicated by the prototype before implementation is completed.

1.16 SUMMARY

The remaining chapters will be concerned mainly with the design and analysis of the experiment. In this chapter we have discussed some aspects of the design, in addition to other problems faced by the experimenter. Here we summarize the aspects most closely associated with the design of the experiment.

A. Inference Space

Before the experiment is designed, the experimenter must decide how widely the results will apply. This will determine whether factors will be labeled fixed or random and have a direct bearing on the tests performed. The experimenter must compare the usefulness of the experiment with the costs involved in running the experiment. This demands knowledge of the inferences to be made prior to running the experiment. This stage appears to be the most neglected since many people do not take the time to think the problem through to its conclusions.

B. Randomization

This stage is necessary in order to construct probability statements about the results. Randomization allows for valid tests and unbiased estimates after the inference space has been decided upon. Restrictions in randomization generally lead to some sort of loss of information and must be noted in the model. The effect of such restrictions must be carefully pointed out to the experimenter. Often times, restrictions are purposely designed into the experiment because, with only a little loss of information, great cost savings can be achieved.

C. Redesigning the Experiment

Once the inference space is decided upon and the possible restrictions on randomizations noted, the statistician can determine the ANOVA table and the amount of information (if any) available for testing the effect of various factors and interactions. It is important at this point to redesign the experiment and compare ANOVA tables. Most often, a more efficient design exists. A few iterations usually produces the best possible design.

Both replication and fractionation (ideas developed in later chapters) should be carefully considered when redesigning the experiment.

The remainder of this text utilizes the above concepts in different layouts of factors on the experimental units. From a given layout, the appropriate mathematical model is developed. From the mathematical model, the corresponding ANOVA table is derived. This tells the experimenter what information will be available prior to the collection of any data. It is then a simple matter to redesign the experiment to obtain information not available from the initial design. A few iterations leads to the best possible design.

After the best possible design has been determined, the data can be collected. The resulting analysis will then be clear and conclusions can be made with confidence.

PROBLEMS

1.1 *Why must the experimenter be extremely careful about implementing the results of a laboratory experiment to a production process?*

1.2 *Labeling four units 1 through 4, explain how a random number table is used to randomly assign the units to two treatments, A and B, in such a way that each treatment is applied to two of the units.*

1.3 *In Problem* **1.2**, *label the treatments 1 and 2 and explain how a random number table could be used to randomly assign the treatments to the four units.*

1.4 *There are six ways in which the units can be assigned to the treatments (or vice-versa). Using the methods described in* **1.2** *and* **1.3**, *compute the probabilities of each of the six ways and compare. This is why you must randomly assign units to treatments, not the reverse.*

1.5 *Explain how you would randomly determine the order of running an experiment consisting of twelve units randomly assigned to four treatments.*

1.6 *An experiment was conducted for several different years on a certain section of highway. Describe the inference space for this experiment. If the results of the experiment were to apply to the entire highway, what would have to be done differently?*

1.7 *Describe the inference space for a laboratory experiment conducted on rats. What must be done to extend the inference space to humans in their normal settings? In general, statistical results cannot be used to make the jump from laboratory experiments on rats to the general public. Only the experimenter can make this jump, and it cannot be made on the basis of statistical evidence.*

1.8 *An experiment concerning the baking of cakes was run in the kitchen of a house. The number of eggs used was varied, the baking time and temperature were varied, and the amount of butter was varied. (a) What are some possible dependent variables and how would they be measured? (b) What are the factors for this experiment? (c) What are some factors not in this experiment that could have been in this experiment? (d) What are typical levels for each factor? (e) What is the inference space for this experiment? (f) How can this inference space be expanded?*

BIBLIOGRAPHY FOR CHAPTER 1

Anderson, V. L. (1970). "Restriction errors for linear models (an aid to develop models for designed experiments)," *Biometrics*, **26**, 255–268.

Anderson, V. L. and McLean, R. A. (1974). *Design of Experiments A Realistic Approach*, New York: Marcel Dekker. (Chapter 3)

Anderson, V. L. and McLean, R. A. (1974). "Restriction errors: another dimension in teaching experimental statistics," *The American Statistician*, **28**, 145–152.

Box, G. E. P., Hunter, W. G. and Hunter, J. S. (1978). *Statistics for Experimenters An Introduction to Design, Data Analysis, and Model Building*, New York: John Wiley & Sons. (Chapter 1)

Cox, D. R. (1958). *Planning of Experiments*, New York: John Wiley & Sons. (Chapters 1, 4, 5, 8)

Deming, W. E. (1960). *Sample Design in Business Research*, New York: John Wiley & Sons. (Chapter 1)

Finney, D. J. (1975). *An Introduction to the Theory of Experimental Design*, Chicago, Illinois: University of Chicago Press. (Chapters 1, 8, 9)

Fisher, R. A. (1926). "The arrangements of field experiments," *Journal of Ministry of Agriculture*, **33**, 503–513.

Fisher, R. A. (1973). *The Design of Experiments*, 14th ed., New York: Hafner. (Chapters I–IV)

Ghosh, S., editor. (1990). *Statistical Design and Analysis of Industrial Experiments*, New York: Marcel Dekker.

Greenburg, B. G. (1951). "Why randomize?," *Biometrics*, **7**, 309–322.

Hicks, C. R. (1982). *Fundamental Concepts in the Design of Experiments*, 3rd ed., New York: Holt, Rinehart and Winston. (Chapter 1)

Ishikawa, K. (1976). *Guide to Quality Control*, Tokyo: Asian Productivity Organization. (Chapter 3)

John, P. W. M. (1971). *Statistical Design and Analysis of Experiments*, New York: The Macmillan Co. (Chapter 1)

Kempthorne, O. (1955). "The randomization theory of experimental inference," *Journal of the American Statistical Association*", **50**, 946–967.

Kempthorne, O. (1977). "Why Randomize?," *Journal of Statistical Planning and Inference*, **1**, 1–25.

Lorenzen, T. J. (1984). "Randomization and blocking in the design of experiments," *Communications in Statistics—Theory and Methods*, **13**, 2601–2623.

Montgomery, D. C. (1991). *Design and Analysis of Experiments*, 3rd ed., New York, John Wiley & Sons. (Chapter 1)

Ostle, B. and Malone, L. C. (1988). *Statistics in Research*, 4th ed., Ames, Iowa: Iowa State University Press.

Sheffé, H. (1959). *The Analysis of Variance*, New York: Chapman and Hall. (Chapter 4)

Steel, R. G. D. and Torrie, J. H. (1980). *Principles and Procedures of Statistics: A Biometrical Approach*, 2nd ed.,New York: McGraw–Hill. (Chapter 6)

Taguchi, G. (1987). *System of Experimental Design*, Dearborn, MI: American Supplier Institute, Inc. (Chapter 4)

Yates, F. (1964). "Sir Ronald Fisher and the design of experiments," *Biometrics*, **21**, 307–321.

SOLUTIONS TO SELECTED PROBLEMS

1.1 *The inference space for a laboratory experiment is the laboratory, not the production process. If the experimenter feels the laboratory closely approximates the production process, the experimenter can make this leap of faith. The statistician cannot make such a leap.*

1.6 *The inference space is years on that particular section of highway. If the results are to apply to the entire highway, several randomly selected sections of highway must be used.*

1.8 *(a) Possible dependent variables include the height of the cake as measured in inches, the springiness as measured by depression of a finger, and taste as measured by a panel of friends. (b) There are four factors, number of eggs, baking time, baking temperature, and amount of butter. (c) Amount of liquid, amount of flour, type of flour, amount of baking powder or soda, time beating, type of pan baked in, which rack in the oven, and so on. (d) Number of eggs could be 1, 2, or 3. Baking time could be 30, 45, or 60* *minutes. Baking temperature could be 300 or 350°. Amount of butter could be 1/2 or 1 cup. Of course, a different number of levels and different values are also correct, depending on the recipe and the alternatives considered. (e) The inference space is to the type of cake baked, the kitchen used, the time of year, and the brand of the ingredients used. (f) Use different types of cakes, different kitchens, different times of the year, and different brands of ingredients. Alternatively, one could use non-statistical evidence such as the opinions of experts or laws of physics.*

2

One Factor Designs

2.0 INTRODUCTION

This chapter introduces the terminology, lists the assumptions, and gives tests for the assumptions used in the analysis of the data from designed experiments. We will assume that the reader has a little statistical background, perhaps a regression course, but encourage the dedicated reader with no background to continue on. After this chapter the going gets a little easier.

Notation will be introduced on an "as needed" basis. Enough detail will be given to indicate formal statistical tests where appropriate, but proofs of results are generally omitted. Paragraphs that require more statistical background will be marked with a "light bulb" and can be skipped by the reader with little background without loss of continuity. The statistician should read these paragraphs to understand why procedures are given and be able to handle the difficult and non-standard cases that periodically arise. All procedures will be illustrated with examples that are marked with a "nut and bolt".

In particular, this chapter will focus on experiments containing only one factor. The treatment levels of the factor to be applied to the experimental units are pre-specified and the inference of the experiment is only to these fixed levels. The experimental units are randomly selected from some large pool of all possible experimental units with the inference being to all such experimental units. Specifically, we shall indicate the hypothesis to be tested, give a test for this hypothesis, list the assumptions necessary for the validity of this test, and give tests of these assumptions. If the effect of the factor is significant, we give a method for comparing individual means to determine the possible cause of the significance. A method for determining the type II or β error (*i.e.*, the probability of accepting the null hypothesis given it is actually false) will be given along with the size of a difference that can be detected with 90% probability and the sample size necessary to detect a difference of a specified value.

2.1 LAYOUT OF THE EXPERIMENT

The *layout* of the experiment is the exact fashion in which the experiment is carried out. Label the levels of the factor of interest using the subscript i. The subscript i ranges from 1 to I, indicating there are a total of I different treatment levels of interest. Each level of the factor of interest will be applied to J different experimental units, for a total of IJ experimental units used in the experiment. Whenever the same number of experimental units are used for each level of the factor of interest, the experiment is considered *balanced*. Balanced designs will be the focus of this text.

Label the experimental units with the index j, j ranging from 1 to J. To indicate that there are different experimental units for each treatment level i, use the notation $j(i)$. The subscript $j(i)$ should be read j *depending on* i, j *nested within* i, or simply j *within* i to indicate that different experimental units are used for each treatment level. The concept of *nesting* will be covered in detail in Chapter **4**.

Let y_{ij} be the measured response for the jth replicate of treatment level i. This is the dependent variable. We wish to determine if the treatment averages differ or not. To assure the validity of the statistical test, we must randomly select the IJ experimental units, randomly assign the units to the treatment levels, and randomly run the experiment. This is called *complete randomization*. More will be said about the effects of failure to completely randomize in Chapter **5**. Randomization plays an important role in experimental design and will be carefully considered in this text.

To randomly select the IJ experimental units from a pool of N potential units, label the potential units 1 through N. Then use a random number table such as the one given in Appendix 1 to select IJ numbers with values between 1 and N. These are the units to be used in the experiment. If the units are coming off an assembly line, care must be taken to include units from different times, shifts, and operators to assure wide applicability of the results. (See Sections **1.6** and **1.7** on inference space and selection of experimental units.) To use the random number table, start at some random location using one column if N is a one digit number, two columns if N is a two digit number, three columns if N is a three digit number, *etc.* Read down the column(s), selecting the unit with that numbered label provided that number is between 1 and N and the unit has not been previously selected. Continue the process with the next column(s) until all IJ units have been selected.

Next we must randomly assign the IJ experimental units to the treatment levels. Relabel the units 1 through IJ in any fashion. Place the treatment levels down a column with level 1 appearing J times, level 2 appearing J times, *etc.*, to level I appearing J times. These can also be in any order. Use the random number table to select a number from 1 to IJ and assign that unit to the first treatment level in the column. Pick a different random number from 1 to IJ and assign that unit to the second treatment level in the column. Continue in this fashion until all units have been assigned to the treatment levels.

Finally, to randomize the run order of the experiment, again use the random number table to select a number from 1 to IJ. This unit, with its associated treatment level will be run first. Continue in this fashion until the experiment is completed. All of this should be done prior to the start of the experiment and laid out on a worksheet to be used by the experimenter. (See Section **1.12**.)

While this seems like a lot of unnecessary work, it is the only way to protect against biases or other unknown factors influencing the experimental results. It is the statistician's responsibility to use every means available to assure validity of the experimental results and randomization is one good method. We recommend the statistician give the experimenter a worksheet with the experiment already randomized to assure the experiment is properly carried out.

The process described above is called *complete randomization*. Randomization plays an important role in the design of experiments.

2.2 MATHEMATICAL MODEL AND ASSUMPTIONS

Assume that the observations y_{ij} come from the following mathematical model

$$y_{ij} = \mu + A_i + \varepsilon_{j(i)} \tag{2.2.1}$$

where μ is the overall mean of the process, A_i is the differential effect due to the ith treatment level of the fixed factor A_i, and $\varepsilon_{j(i)}$ is a random error component. We will assume that

$$\sum_{i=1}^{I} A_i = 0 \quad \text{and} \tag{2.2.2}$$

$$\varepsilon_{j(i)} \sim \text{Normal}(0, \sigma^2). \tag{2.2.3}$$

The assumption $\sum_{i=1}^{I} A_i = 0$ can be made without loss of generality because if $\sum_{i=1}^{I} A_i$ were equal to some value C, we could replace μ with

$\mu + C$ and $\sum_{i=1}^{I} A_i$ would then be equal to 0. Likewise, we can assume $\varepsilon_{j(i)}$ has mean 0. Note that the $\varepsilon_{j(i)}$ are assumed independent and identically distributed; in particular, that the variances of the error terms are homogeneous.

Our interest is in testing the equality of the mean response for each treatment level. Since we have removed the overall mean in (2.2.1), the mean responses will be equivalent whenever $A_1 = A_2 = \ldots = A_I = 0$. Alternately, the mean responses are equivalent whenever $\sum_{i=1}^{I} A_i^2 = 0$ or whenever $\sum_{i=1}^{I} A_i^2/(I-1) = 0$. This last form is of particular interest since, as will be seen, this is the exact quantity tested in our analysis of variance calculations. For shorthand notation we will use the following convention:

$$\Phi(A) = \sum_{i=1}^{I} A_i^2/(I-1). \tag{2.2.4}$$

Formally, we will be testing the hypothesis $\Phi(A) = 0$ or, equivalently, we will be testing the hypothesis that A has no effect on the response variable.

The natural estimate of μ is the mean of all the observations and, since A_i is defined as the differential effect of treatment level i from the mean, the natural estimate for A_i is the mean of the data collected for the ith treatment level minus the overall mean. Using the shorthand dot notation

$$\begin{aligned} \bar{y}_{i\cdot} &= \sum_{j=1}^{J} y_{ij}/(J) \qquad \text{and} \\ \bar{y}_{\cdot\cdot} &= \sum_{i=1}^{I}\sum_{j=1}^{J} y_{ij}/(IJ)\ , \end{aligned} \tag{2.2.5}$$

the natural estimate of A_i would be $\hat{A}_i = \bar{y}_{i\cdot} - \bar{y}_{\cdot\cdot}$ so the natural estimate of $\Phi(A)$ would be $\sum_{i=1}^{I}(\bar{y}_{i\cdot} - \bar{y}_{\cdot\cdot})^2/(I-1)$. We divide by $I-1$ instead of I since there is one linear restriction: $\sum_{i=1}^{I} \bar{y}_{i\cdot} = I\bar{y}_{\cdot\cdot}$.

Under the null hypothesis $\Phi(A) = 0$, it can be shown that $\sum_{i=1}^{I}(\bar{y}_{i\cdot} - \bar{y}_{\cdot\cdot})^2/(I-1)$ estimates σ^2/J so that $J\sum_{i=1}^{I}(\bar{y}_{i\cdot} - \bar{y}_{\cdot\cdot})^2/(I-1)$ estimates σ^2. Likewise, it can be shown that another estimate of σ^2 is given by the *pooled within treatment variation*, $\sum_{i=1}^{I}\sum_{j=1}^{J}(y_{ij} - \bar{y}_{i\cdot})^2/[I(J-1)]$. The estimates of σ^2 are orthogonal so, under the normality assumption, are independent and distributed as χ^2 random variables. The ratio of the estimates will have an F distribution. This is the motivation for the analysis of variance, ANOVA, table given next.

2.3 THE ANOVA TABLE FOR ONE FACTOR

For testing the effects of a single factor A_i, the computations can be summarized in Table 2.3.1 .

TABLE 2.3.1

ANOVA FOR ONE FACTOR DESIGN

Source	df	SS	MS	F
A_i	$I-1$	$J\sum_{i=1}^{I}(\bar{y}_{i\cdot}-\bar{y}_{\cdot\cdot})^2$	$SS(A)/(I-1)$	$MS(A)/MS(\varepsilon)$
$\varepsilon_{j(i)}$	$I(J-1)$	$\sum_{i=1}^{I}\sum_{j=1}^{J}(y_{ij}-\bar{y}_{i\cdot})^2$	$SS(\varepsilon)/[I(J-1)]$	
Total	$IJ-1$	$\sum_{i=1}^{I}\sum_{j=1}^{J}(y_{ij}-\bar{y}_{\cdot\cdot})^2$		

In this table, df stands for degrees of freedom, SS for sums of squares, MS for mean squares, and F for the calculated F value. We reject the null hypothesis of no treatment effect if the calculated F exceeds a critical F having $I-1$ and $I(J-1)$ degrees of freedom. Critical F values can be found in the tables in Appendix 2.

A simple way to remember this table is as follows. The total degrees of freedom is $IJ-1$, the total number of observations minus one for estimating the grand mean. The degrees of freedom for factor A_i equals $I-1$, the total number of treatment levels for A_i minus one for subtracting the overall mean. The degrees of freedom for error can be found by subtraction [df(A) + df(ε) = df($Total$)] or by direct calculation ($J-1$ degrees of freedom within each treatment level times the I different treatment levels).

The total sum of squares is the sum over all the observations of the square of each individual deviation from the overall mean. That is, take each individual observation, subtract the grand mean, square the result, and then sum over all observations.

The sum of squares for A_i is the sum over all the means of the square of the estimate of the ith effect weighted by the number of observations in each mean ($= J\sum_{i=1}^{I}\hat{A}_i^2$). Paralleling the explanation for the total sum of squares, the sum of squares for A_i is again the sum over all of the observations but, since the subscript j does not appear in the summand, the sum over j is replaced by J.

The sum of squares for $\varepsilon_{j(i)}$ is found by summing the square of the deviations of each observation from its treatment mean. Again, the sum is over all of the observations. We sum the square of the deviations from the treatment means as this is the natural estimate of the "error" in the observation.

As is demonstrated below, SS(ε) can also be found by subtracting SS(A) from SS($Total$). This demonstration shows that SS(A) and SS(ε) are orthogonal and, along with the normality assumption, shows that SS(A) and SS(ε) are independent.

$$\begin{aligned}
\text{SS}(Total) &= \sum_{i=1}^{I}\sum_{j=1}^{J}(y_{ij} - \bar{y}_{\cdot\cdot})^2 \\
&= \sum_{i=1}^{I}\sum_{j=1}^{J}(y_{ij} - \bar{y}_{i\cdot} + \bar{y}_{i\cdot} - \bar{y}_{\cdot\cdot})^2 \\
&= \sum_{i=1}^{I}\sum_{j=1}^{J}(y_{ij} - \bar{y}_{i\cdot})^2 + 2\left[\sum_{i=1}^{I}(\bar{y}_{i\cdot} - \bar{y}_{\cdot\cdot})\left[\sum_{j=1}^{J}(y_{ij} - \bar{y}_{i\cdot})\right]\right] \\
&\qquad + J\sum_{i=1}^{I}(\bar{y}_{i\cdot} - \bar{y}_{\cdot\cdot})^2 \\
&= \text{SS}(\varepsilon) + 0 + \text{SS}(A)
\end{aligned}$$

since $\sum_{j=1}^{J}(y_{ij} - \bar{y}_{i\cdot}) = \sum_{j=1}^{J} y_{ij} - J\bar{y}_{i\cdot} = 0$ for all i.

With the advent and popularity of desktop computing and statistical software packages, the importance of knowing the formulas for sums of squares has been greatly diminished. Such sums of squares are rarely, if ever, computed by hand. However, the formulas are still necessary, and it is important to know they are founded on a logical basis. Formulas will be given in this chapter and the next. The reader should not dwell on the formulas or on the computational aspects of ANOVA. Time is much better spent on learning the particular software package that will be used to analyze the data.

The mean square column is formed by dividing the sum of squares column by the degrees of freedom column and the F by forming the indicated ratio.

If computations are to be performed on a hand calculator, different formulas should be used. These formulas avoid the task of subtracting the overall mean from each individual observation or treatment mean and then squaring and summing. Many hand calculators automatically sum the data in a fashion useful for these calculations.

Define the following quantities

$$\mathrm{T}_{i\cdot} = \sum_{j=1}^{J} y_{ij} \tag{2.3.1}$$

$$\mathrm{T}_{\cdot\cdot} = \sum_{i=1}^{I}\sum_{j=1}^{J} y_{ij} \tag{2.3.2}$$

$$CT = \frac{\mathrm{T}_{\cdot\cdot}^2}{IJ} \tag{2.3.3}$$

where the letter T stands for the total over the dotted subscript(s) and CT denotes the correction term for the mean. Expanding the summations for SS(A) and SS($Total$) given in Table 2.3.1 yields the following.

$$\mathrm{SS}(Total) = \sum_{i=1}^{I}\sum_{j=1}^{J}(y_{ij} - \bar{y}_{\cdot\cdot})^2 = \sum_{i=1}^{I}\sum_{j=1}^{J} y_{ij}^2 - CT \tag{2.3.4}$$

$$\mathrm{SS}(A) = J\sum_{i=1}^{I}(y_{i\cdot} - y_{\cdot\cdot})^2 = \sum_{i=1}^{I} \mathrm{T}_{i\cdot}^2/J - CT \tag{2.3.5}$$

We compute SS(ε) by subtraction, SS(ε) = SS($Total$) - SS(A). As in Table 2.3.1, the MS column is found by dividing the SS column by the df column and the calculated F value is the indicated ratio.

Intuitively, CT is the correction term for the grand mean obtained by summing all of the observations, squaring the result, and dividing by IJ, the number of terms in the sum. SS(A) is obtained by summing all of the observations in each level of A_i (yielding $\mathrm{T}_{i\cdot}$), squaring each result, dividing by the number of terms in each sum—J in this case, summing over i, and correcting for the mean by subtracting CT. Likewise, SS($Total$) is obtained by summing the square of each observation (dividing by one if you wish since there is one observation in each square) and correcting for the mean by subtracting CT.

The equivalent ANOVA table is given below. This table is not recommended for computer calculations as roundoff errors can quickly accumulate. When calculating by hand, such roundoff errors become obvious because you will subtract two numbers that are the same in most of the decimal places. The problem is eliminated by subtracting a constant from each data point and rescaling the result. (F values are both scale and location invariant which means you will get the same results if you subtract an arbitrary number from the data or multiply by an arbitrary number.) Again, reject the null hypothesis of no treatment effect whenever the calculated F exceeds the critical F found in Appendix 2 using $I-1$ and $I(J-1)$ degrees of freedom.

TABLE 2.3.2

ANOVA FOR ONE FACTOR—HAND CALCULATIONS

Source	df	SS	MS	F
A_i	$I-1$	$\sum_{i=1}^{I} T_{i\cdot}^2/J - CT$	$SS(A)/(I-1)$	$MS(A)/MS(\varepsilon)$
$\varepsilon_{j(i)}$	$I(J-1)$	$SS(Total) - SS(A)$	$SS(\varepsilon)/[I(J-1)]$	
Total	$IJ-1$	$\sum_{i=1}^{I}\sum_{j=1}^{J} y_{ij}^2 - CT$		$CT = T_{\cdot\cdot}^2/(IJ)$

A research worker in a pharmaceutical firm was interested in finding out whether there were any differences in the disintegration times of four types of caplets. Label the factor caplets C_i, with the subscript i ranging from 1 to $I=4$ denoting the four types of caplets. It was decided that $J=6$ caplets of each type should be tested. Both the manufacturing and the testing procedures were carried out in a completely randomized fashion. The model describing this experiment is

$$y_{ij} = \mu + C_i + \varepsilon_{j(i)} \qquad i=1,\ldots,I \qquad j=1,\ldots,J \tag{2.3.6}$$

where y_{ij} is the disintegration time of the jth caplet within the ith type, μ is the overall mean disintegration time, C_i is the differential effect due to the ith type of caplet, and $\varepsilon_{j(i)}$ is the random error of the jth caplet within the ith type, assumed independent and identically distributed, *iid*, as Normal$(0,\sigma^2)$ with σ^2 unknown.

For this experiment, the 24 caplets were randomly tested with results given in Table 2.3.3.

We compute $\sum_{i=1}^{I}\sum_{j=1}^{J} y_{ij}^2 = 1141$, $\sum_{i=1}^{I} T_{i\cdot}^2 = 6441$, and $T_{\cdot\cdot} = 157$.

TABLE 2.3.3

DISINTEGRATION TIMES (SECONDS) FOR CAPLETS

Type of Caplet (i)	Observations (j)	Totals $T_{i\cdot}$	Averages $\bar{y}_{i\cdot}$
1	6,2,5,4,6,7	30	5.000
2	3,7,6,4,8,6	34	5.667
3	10,4,6,6,7,8	41	6.833
4	10,8,11,7,7,9	52	8.667

Therefore,

$$CT = T_{\cdot\cdot}^2/IJ = (157)^2/[(4)(6)] = 1027.0417$$

$$SS(C) = \sum_{i=1}^{I} T_{i\cdot}^2/J - CT = 6441/6 - 1027.0417 = 46.4583$$

$$SS(Total) = \sum_{i=1}^{I}\sum_{j=1}^{J} y_{ij}^2 - CT = 1141 - 1027.0417 = 113.9583$$

$$SS(\varepsilon) = SS(Total) - SS(C) = 113.9583 - 46.4583 = 67.5$$

The appropriate ANOVA is given in Table 2.3.4. The critical F value of 3.10, having 3 and 20 degrees of freedom and using an α of .05, can be found in Appendix 2. Since the calculated F of 4.59 exceeds 3.10, we reject the null hypothesis of no difference and conclude there is a significant difference in dissolving time due to types of caplets. The next logical step is to figure out what may have caused the difference. This is accomplished through a comparison of means.

TABLE 2.3.4

ANOVA FOR CAPLET DATA OF TABLE 2.3.3

Source	df	SS	MS	F_{calc}	$F_{3,20}(.05)$
C_i	3	46.4583	15.4861	4.59*	3.10
$\varepsilon_{j(i)}$	20	67.5	3.375		
Total	23	113.9583	* significant at the .05 level		

2.4 COMPARISON OF MEANS

The computation of confidence intervals around individual means, around the difference of two means, multiple comparison of means, and, for equally-spaced quantitative variables, polynomial trends will be given in this section. Multiple comparisons and polynomial trends should only be used if the effect of the factor of interest is significant. Multiple comparisons determine which of the mean levels differ while polynomial trends remove the linear, quadratic, *etc.*, trend of the means.

An estimate of σ in model (2.3.6) is given by

$$\hat{\sigma} = \sqrt{\mathrm{MS}(\varepsilon)} \tag{2.4.1}$$

having $\mathrm{df}(\varepsilon) = I(J-1)$ degrees of freedom. A $100(1-\alpha)\%$ confidence interval on a single mean is given by

$$\bar{y}_{i\cdot} \pm \mathrm{t}_{\alpha/2}\sqrt{\mathrm{MS}(\varepsilon)/J} \tag{2.4.2}$$

where $\mathrm{t}_{\alpha/2}$ is the $\alpha/2$ critical point of a t random variate having $\mathrm{df}(\varepsilon)$ degrees of freedom. The critical values are tabulated in Appendix 3. In particular, a 95% confidence interval on the mean disintegration time for Type 4 is given by

$$\bar{y}_{4\cdot} \pm \mathrm{t}_{.025}\sqrt{\mathrm{MS}(\varepsilon)/J} \;=\; 8.67 \pm 2.086\sqrt{3.375/6} \;=\; [7.10, 10.24].$$

A confidence interval on the difference of two *preselected* means is given by

$$\bar{y}_{i\cdot} - \bar{y}_{j\cdot} \pm \mathrm{t}_{\alpha/2}\sqrt{2\mathrm{MS}(\varepsilon)/J} \tag{2.4.3}$$

where again $\mathrm{t}_{\alpha/2}$ has $\mathrm{df}(\varepsilon)$ degrees of freedom and is found in Appendix 3. Care must always be taken to specify the two means of interest before collecting the data. If knowledge of the rankings is used to post-select the compared means, the confidence interval is invalidated. In particular, a 95% confidence interval on the difference in mean disintegration time between Types 2 and 3 is given by

$$\bar{y}_{2\cdot} - \bar{y}_{3\cdot} \pm \mathrm{t}_{.025}\sqrt{2\mathrm{MS}(\varepsilon)/J} \;=\; -1.167 \pm 2.213 \;=\; [-3.38, 1.05].$$

To be able to make all possible comparisons of means, a multiple comparisons procedure must be used. *This should only be used if there is a significant difference in the means and the experimenter is interested in finding out what may have caused the difference.* If the test on a factor effect is insignificant, we can conclude there is no difference in the mean levels and a multiple comparisons procedure should not be used.

Although there are several methods for multiple comparisons, including the least significance difference, Newman-Keuls sequential test, Scheffé's method, Tukey's honestly significant difference, and the Waller-Duncan test, the most popular is Duncan's multiple range test. This test is readily available in many computer packages and easily calculated by hand. To perform the test on I means, rank the means in descending order with the means in one column and the labels in an adjacent column. Start by computing the least significant ranges for comparing means spanning $k = 2, 3, \ldots, I$ values as follows

$$\mathrm{R}_k = \mathrm{q}_\alpha(k, \mathrm{df})\hat{\sigma}_{\bar{y}_{i.}} \tag{2.4.4}$$

where $\mathrm{q}_\alpha(k, \mathrm{df})$ can be found in Appendix 4, k is the number of means spanned in the comparison, df is the degrees of freedom used to estimate $\hat{\sigma}_{\bar{y}_{i.}}$, and $\hat{\sigma}_{\bar{y}_{i.}}$ is the estimated standard error of the mean of interest. For the one factor design, $\hat{\sigma}_{\bar{y}_{i.}} = \sqrt{\mathrm{MS}(\varepsilon)/J}$, as indicated in (2.4.2) above. $\mathrm{MS}(\varepsilon)$ is used since it tests $\mathrm{MS}(C)$ and we divide by J since there are J terms in each mean. As designs get more complicated in future chapters, $\mathrm{MS}(\varepsilon)$ will be replaced by more complicated expressions but the thought process will remain the same: use the square root of the MS testing the effect of interest divided by the number of observations in each level of the factor.

After computing R_k, put the means in descending order and perform pairwise subtractions starting with the most means spanned, I in this case, and working toward the least means spanned, two, corresponding to adjacent values. If the compared means are not significantly different, *i.e.*, less than R_k, then connect the two means by a vertical bar. If two means are already connected by a vertical bar, they are not significantly different and need not be tested again. This is why we start with spanning I means and work our way down to spanning two means; it reduces the computational effort.

The interpretation is as follows: If two means are connected by *any* vertical bar, then these two means *are not* significantly different. If *no* single bar connects the two means, then those two means *are* significantly different.

The computations are illustrated with the caplet data of Table 2.3.3. Using an α of .05, we get

$$\begin{aligned}
\mathrm{R}_4 &= \mathrm{q}_{.05}(4, 20)\hat{\sigma}_{\bar{y}_{i.}} = (3.18)(0.75) = 2.385 \\
\mathrm{R}_3 &= \mathrm{q}_{.05}(3, 20)\hat{\sigma}_{\bar{y}_{i.}} = (3.10)(0.75) = 2.325 \\
\mathrm{R}_2 &= \mathrm{q}_{.05}(2, 20)\hat{\sigma}_{\bar{y}_{i.}} = (2.95)(0.75) = 2.2125
\end{aligned}$$

We connect the four means if the difference between the largest and the smallest value is less than 2.385. We connect three means, *i.e.*, the first and the third mean or the second and the fourth mean, if the difference is less than 2.325. Finally, we connect two means, *i.e.*, adjacent means in the table, if their difference is less than 2.2125. The final result is given below.

Caplet Type	Mean
4	8.667
3	6.833
2	5.667
1	5.000

We conclude that Type 4 has a significantly longer disintegration time than Types 1 or 2, and that all other pairs are indistinguishable. Types 3 and 4 are indistinguishable since they are connected by the leftmost bar while Types 3, 2, and 1 are indistinguishable because they are connected by the rightmost bar. A visual comparison of the means is given in Figure 2.1.

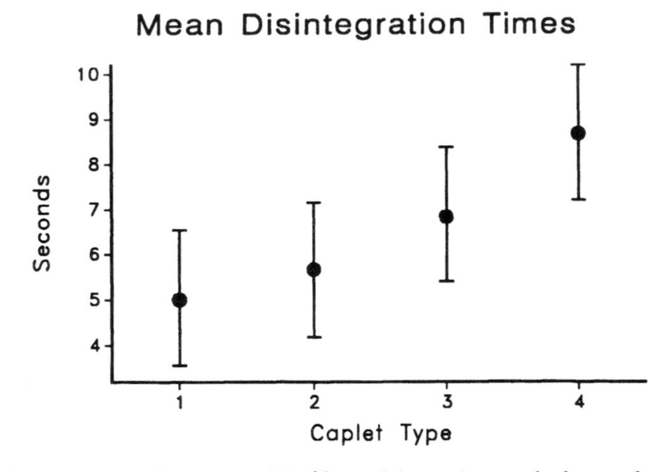

Figure 2.1 Means and 95% confidence intervals for caplet types.

For this particular example, the factor of interest is *qualitative* since the levels 1, 2, 3, and 4 are simply labels that connote no particular order. If the factor of interest was both *equally-spaced* and *quantitative*, polynomial trends may be of interest. Let us suppose that the four types of caplets were manufactured by adding different amounts of a bonding agent. Particularly, we will let C_1, C_2, C_3, and C_4 correspond to 10, 20, 30, and 40 mg of the bonding agent. The computation of polynomial trends can then be illustrated.

Since there are $n = 4$ mean values, it is possible to extract a linear, quadratic, and cubic (*i.e.*, one less than the number of means available) trend in the mean values. The goal is to find the lowest order polynomial that adequately describes the trend. To do this, we sequentially test higher and higher order polynomial trends and stop when there is no longer a significant lack of fit (Bozivich, Bancroft, and Hartley, 1956). Orthogonal polynomials are used to simplify calculations and are illustrated below.

Step one is to put the levels in their natural order and put their means, $\bar{y}_{i\cdot}$ based on m observations, in the adjacent column. Step two is to look up the orthogonal coefficients for the order of interest, z_i, in Appendix 5 and place these in a third column. The sum of squares for the order of interest is given by $m(\sum_{i=1}^{I} \bar{y}_{i\cdot} z_i)^2 / (\sum_{i=1}^{I} z_i^2)$, which can be calculated using a fourth column. For convenience, $\sum_{i=1}^{I} z_i^2$ is tabulated in the same appendix.

To determine the sum of squares due to the linear trend, we have

Bonding Agent (mg)	$\bar{y}_{i\cdot}(m = 6)$	z_{linear}	$\bar{y}_{i\cdot} z_{linear}$
10	5.000	−3	−15.000
20	5.667	−1	−5.667
30	6.833	1	6.833
40	8.667	3	26.000
			12.167

which yields $SS(C_{linear}) = 6(12.167)^2/20 = 44.4083$ and Table 2.4.1. You will notice that the *linear* and *lack of fit* terms are indented in the ANOVA table. This is a common convention indicating that these terms, as well as the df, SS, MS, and F values, are simply refinements of the C_i row, not separate entities. SS(*lack of fit*) is found by subtracting SS(*linear*) from SS(C_i). Degrees of freedom are also found by subtraction.

We conclude there is a significant linear trend and there is not a significant lack of fit. Therefore, a linear trend adequately describes the data and no further computations are needed. This is visually consistent with the mean plots given in Figure 2.1.

If there had been a significant lack of fit after removing the linear trend, we would repeat the above calculations for the quadratic trend. From Appendix 5, the coefficients are 1, −1, −1, and 1. The calculations, which are unnecessary since the lack of fit was insignificant, would yield

TABLE 2.4.1

ANOVA FOR CAPLET DATA OF TABLE 2.3.3

Source	df	SS	MS	F_{calc}	$F_{crit}(.05)$
C_i	3	46.4583	15.4861	4.59*	3.10
linear	1	44.4083	44.4083	13.16**	4.35
lack of fit	2	2.0500	1.0250	0.30	3.49
$\varepsilon_{j(i)}$	20	67.5000	3.3750		
Total	23	113.9583			

*significant at the .05 level
**significant at the .01 level

SS(*quadratic*) = 2.04. The computation of the sum of squares for the quadratic piece does not affect the sum of squares for the linear piece and the sum of squares for lack of fit would be found by subtracting both sum of squares from the sum of squares for C_i. If Appendix 5 were not used or the levels were not equally-spaced, SS(*linear*) would be affected by SS(*quadratic*) and would have to be recalculated with every added term. This is the advantage of using orthogonal components.

When the factor of interest is *quantitative* but not *equally-spaced*, simple tables do not exist. Under such circumstances, computer algorithms such as those given in Robson (1959) and Tadikamalla (1974) must be used. These algorithms will not be illustrated but are available in most statistical packages.

2.5 TESTING ANOVA ASSUMPTIONS AND TRANSFORMATIONS

The ANOVA model for one factor was given in (2.2.1) and has the basic assumptions given in (2.2.3): the error term is *iid* Normal$(0, \sigma^2)$. This section presents tests on the assumptions. First we will give tests on the assumption of equal variances, one of the i's in the *iid*. These are known as *tests on the homogeneity assumption*. If the homogeneity assumption is rejected, *i.e.*, the data is heterogeneous, we suggest some transformations that may help remove the heterogeneity. We will end this section with tests on the normality assumption. Generally speaking, transformations that remove the heterogeneity will also help remove non-normality. Both the homogeneity and the normality tests will apply to all ANOVA models, whether there is one factor of interest or thirty-five factors of interest.

Historically, the assumption thought to be most critical was the homogeneity of variance. However, Box (1954) demonstrated that the F test in ANOVA was most robust for α while working with a fixed model having equal sample sizes. He showed that for relatively large (one variance up to nine times larger than another) departures from homogeneity, the α level may only change from .05 to about .06. This is not considered to be of any practical importance. (It should be pointed out that the only time an α level increased drastically was when the sample size was negatively correlated with the size of the variance.)

When there are large departures from homogeneity, it is felt that the data should be transformed to produce more meaningful results. However, one must take care in the interpretation of the results after transforming since transforming also changes the form of the mathematical model. To our knowledge, no one has come up with an α level on homogeneity tests that protects against too much heterogeneity. A set of working rules that seems to be effective for the practitioner is as follows:

1. If the homogeneity test is accepted at $\alpha = .01$, do not transform.
2. If the homogeneity test is rejected at $\alpha = .001$, transform.
3. If the result of the homogeneity test is between $\alpha = .01$ and $\alpha = .001$, try very hard to find out the theoretical distribution from the investigator. If there is a practical reason to transform and the transformed variable makes sense, go ahead and transform. Otherwise, we recommend not transforming.

With the availability of transgeneration options in statistical programs for digital computers, there is a tendency to try various transformations, make the homogeneity test in each case, and select the transformation which yields the most favorable result. While this practice is not all bad, there is a danger in losing "touch" with the meaning of the variable being analyzed. Hence, use as much theory about the variable as possible and select a transformation that makes sense from a physical point of view.

We will use an example in the manufacture of tablets to demonstrate the tests for homogeneity of variance and the uses of transformations. Four types of magnesium trisilicate tablets,

(A)	magnesium stearate	16 mesh granule size
(B)	talc powder	16 mesh granule size
(C)	liquid petrolatum	16 mesh granule size
(D)	magnesium stearate	20 mesh granule size

were randomized in manufacture and testing order. Ten tablets of each

TABLE 2.5.1

DISINTEGRATION TIME (SECONDS) FOR 4 TABLET TYPES

	Types of Tablets							
	A		B		C		D	
	20	42	8	12	50	124	151	178
	28	25	10	24	67	72	125	151
	36	24	12	10	90	78	180	152
	16	31	16	19	103	70	149	161
	25	33	9	10	90	76	175	118
Mean	28		13		82		154	
Variance	59.56		26.22		430.89		436.22	

type were measured for disintegration time with the results given in Table 2.5.1.

Visual inspection of this data shows that the variances increase as the means increase. Thus, tests of homogeneity should be applied to this data. It is our experience that when one finds a relationship of this type, one usually finds a lack of normality. Transformations which tend to improve homogeneity also tend to improve lack of normality. This is discussed in Cochran (1947).

We will discuss three tests for homogeneity in the following subsections. The Bartlett (1937) test is an old, well-established test that has been programmed into many statistical packages. However, this test is a bit sensitive to nonnormality. If the tails of the distribution are too long, it tends to show significance too often. Using our working rules, we suggest using $\alpha = .001$ if it is thought the distribution of y has long tails. This will help prevent unnecessary transformations.

Another test, suggested by Bartlett and Kendall (1946), utilizes $\ln s^2$ as a variable. They suggest running an ANOVA on $\ln s^2$ in order to find out if the variances are heterogeneous and, if so, which treatments cause the heterogeneity. The basis for this test is the approximate normality of $\ln s^2$ when the sample size $n \geq 5$. To obtain even two observations of $\ln s^2$ per cell, however, there must be at least ten observations of y per cell. This ln transformation is often used to study the relation between variability and the experimental factors; especially in product design when one wishes to find the combination of levels of factors that minimizes product variability and therefore assures uniformity of operation of the product.

A third test of homogeneity that was best in a simulation study by Conover, Johnson, and Johnson (1981) is the modified Levene test, suggested by Levene (1960) and modified by Brown and Forsythe (1974). Simply run an ANOVA on $z_{ij} = \mid y_{ij} - \tilde{y}_i \mid$, where $\tilde{y}_i$ is the median of the data in the ith cell. This test is least sensitive to long tailed distributions but it is doubtful if z_{ij} is normally distributed.

2.5.1 Bartlett's Test (Equal Subclass Numbers)

This sub-section will illustrate Bartlett's test for equal subclass numbers. To test for the equality of k variances using Bartlett's test, where each variance is estimated with df degrees of freedom, we compute

$$\begin{aligned} M &= (df)[k \ln \overline{s^2} - \sum_{i=1}^{k} \ln s_i^2] \\ C &= 1 + (k+1)/[3(k)(df)] \\ \chi^2_{k-1} &= M/C \quad . \end{aligned} \tag{2.5.1}$$

The statistic χ^2_{k-1} follows a Chi-Squared distribution with $k-1$ degrees of freedom. Critical values are given in Appendix 6. We reject the hypothesis of homogeneity of variance if the calculated statistic exceeds the critical value found in Appendix 6. Rejection of homogeneity leads to transformation of the data.

The calculations for the tablet data of Table 2.5.1 are illustrated in Table 2.5.2. Using an α of .001 and 3 degrees of freedom, we find a critical value of 16.27. Since 20.60 exceeds the critical value, we reject the homogeneity of variances hypothesis at the .001 level and conclude the variances are indeed different.

2.5.2 Bartlett and Kendall ln s^2 ANOVA

The data from Table 2.5.1 was randomly assigned to two groups of five observations for each of the four tablet types. Table 2.5.3 shows the corresponding ANOVA for ln s^2. Note that it does not matter if we use ln (= $\log_e$) or $\log_{10}$ for this test as the two log functions are scale related and ANOVA is scale invariant.

Since $F_{calc} > F(.01)$ but $F_{calc} < F(.001)$, we are in the range where theoretical knowledge of the data must guide our actions. If it makes sense to transform, go ahead and apply an appropriate transformation. Otherwise, do not transform.

TABLE 2.5.2

BARTLETT'S TEST FOR HOMOGENEITY OF VARIANCES

i	Tablet type	s^2	$\ln s^2$
1	A	59.56	4.0870
2	B	26.22	3.2665
3	C	430.89	6.0659
4	D	436.22	6.0781
	Total	952.89	19.4975

$$\overline{s^2} = \frac{952.89}{4} = 238.22 \qquad \ln \overline{s^2} = 5.4732$$

$$M = (9)[4(5.4732) - 19.4975] = 21.558$$

$$C = 1 + (5)/[3(4)(9)] = 1.0463$$

$$\chi_3^2 = 21.558/1.0463 = 20.60$$

2.5.3 Modified Levene Test of Homogeneity

The medians for tablet types A, B, C, and D are given by 26.5, 11.0, 77.0, and 151.5 respectively. We subtract these medians from each value in Table 2.5.1 and take absolute values. A standard ANOVA is run on this new data. The data and corresponding ANOVA are given in Table 2.5.4 .

The modified Levene test does not reject at the .01 level. Therefore, according to our guidelines, we should not transform the data to eliminate homogeneity. However, since the three tests give conflicting advice, it is probably best to leave this decision to theory. If it makes theoretical sense to transform, then go ahead and transform the data.

2.5.4 Transformation of y

The results of the previous three sections show that the hypothesis of homogeneity of variances was rejected at the .001 level using Bartlett's test, rejected at the .01 level for the $\ln s^2$ test and not rejected at the .01 level for the modified Levene test. As can be seen in Figure 2.2, the sample means are approximately proportional to the sample variance. The experimenter indicated that this type of data usually appeared to have the sample mean proportional to the sample variance, so we decided the data

TABLE 2.5.3

LN s^2 ANOVA FOR HOMOGENEITY OF VARIANCES

Type of Tablet							
A		B		C		D	
s^2	ln s^2	s^2	ln s^2	s^2	ln s^2	s^2	ln s^2
59.0	4.0775	10.0	2.3026	449.5	6.1082	493.0	6.2005
52.5	3.9608	39.0	3.6636	510.0	6.2344	478.5	6.1707

ANOVA on ln s^2

Source	df	SS	MS	F_{calc}	$F_{3,4}(.01)$	$F_{3,4}(.001)$
Types of Tablets	3	15.4097	5.1366	21.83*	16.69	56.2
Error	4	0.9414	0.2353			
Total	7	16.3511	*significant at the .01 level			

should be transformed. When the mean is proportional to the variance, a square root transformation is usually employed. That is, each observation is replaced by the square root of the observation and the analysis is repeated. The results of performing Bartlett's test on the transformed data, $\sqrt{y}$, are given in Table 2.5.5. It is seen that the hypothesis of homogeneity of variances is now accepted even at the .05 level. The result is also substantiated by the ln s^2 analysis and the modified Levene test. At this point, one is ready to carry out the ANOVA on the $\sqrt{y}$ data.

As previously mentioned, transformations are utilized to obtain a response variable that will conform to the basic assumptions of the analysis of variance. The most familiar paper on this subject is Bartlett (1947). This paper contains a list of several often used transformations associated with various relationships between the mean and the variance of the population.

As was indicated above, a square root transformation should be used when the mean and variance of a population are proportional. Of course this assumes that all of the data are positive in value. When some values are negative, adding a constant before taking the square root is often an adequate solution.

TABLE 2.5.4

MODIFIED LEVENE TEST OF HOMOGENEITY OF VARIANCE

Absolute Deviation from the Median $\mid y_{ij} - \tilde{y}_i \mid$							
A		B		C		D	
6.5	15.5	3	1	27	47	0.5	26.5
1.5	1.5	1	13	10	5	26.5	0.5
9.5	2.5	1	1	13	1	28.5	0.5
10.5	4.5	5	8	26	7	2.5	9.5
1.5	6.5	2	1	13	1	23.5	33.5

ANOVA on $\mid y_{ij} - \tilde{y}_i \mid$

Source	df	SS	MS	F_{calc}	$F_{3,36}(.01)$	$F_{3,36}(.001)$
Types of Tablets	3	1089.90	363.30	3.35	4.39	6.77
Error	36	3901.00	108.36			
Total	39	4990.90				

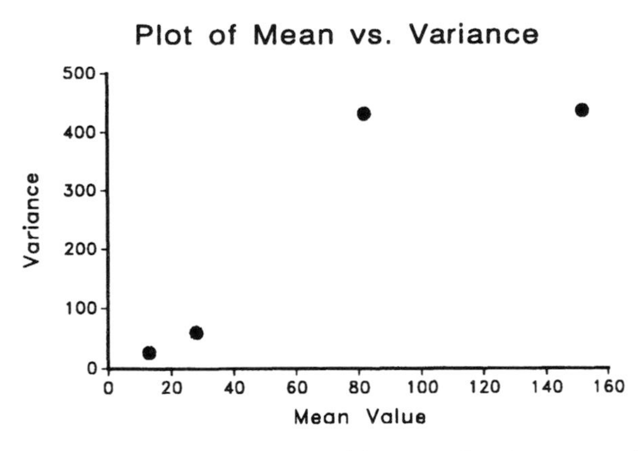

Figure 2.2 Crossplot of data from Table 2.5.1 showing approximate linearity.

For growth-type data, like changes in weight associated with various diets, the most commonly used transformation is ln y. Ln y is also used when the mean is proportional to the standard deviation, *i.e.*, when the coefficient of variation is roughly constant.

TABLE 2.5.5

BARTLETT'S TEST ON $\sqrt{y}$

Type	$s^2_{\sqrt{y}}$	$\ln s^2_{\sqrt{y}}$
A	0.538	-0.6199
B	0.441	-0.8187
C	1.279	0.2461
D	0.734	-0.3092

$M = (9)[4(-0.2904) + 1.5017] = 3.063$
$C = 1.0463$ $\chi^2_3(calc) = 2.93$
$\chi^2_3(.01) = 11.34$ $\chi^2_3(.05) = 7.81$

When dealing with percentage or proportions, p, coming from a binomial (0 or 1) response, the variance is proportional to [mean(1 – mean)] and the correct transformation is arcsin $\sqrt{p}$. Usually one does not employ this transformation if the data falls between $p = .30$ and $p = .70$ as the variance is fairly stable in this region.

If the standard deviation is proportional to the mean squared, the reciprocal $(1/y)$ transformation is often used.

If the mean is negatively correlated with the variance, the transformation $\sqrt{B} - \sqrt{B - y}$, where B is an upper bound for the data, seems to work well.

Simple crossplots of the mean versus the standard deviation or the mean versus the variance, similar to Figure 2.2 above, can help distinguish between the various alternatives. Such simple plots are highly recommended.

A class of transformations to try when the data are heterogeneous and all else fails is the Box-Cox (1964) power family indexed by the parameter λ. The transformation takes the form $(y^\lambda - 1)/\lambda$ for $\lambda \neq 0$ and takes the form $\ln y$ for $\lambda = 0$. Simply vary the parameter λ until the homogeneity assumption is accepted.

2.5.5 Tests for Normality

Generally speaking, the F ratio used in the analysis of variance has been shown to be very robust to departures from normality, Eisenhart (1947). In addition, lack of normality is often associated with heterogeneity and transformations that improve homogeneity also improve normality. Nevertheless, we will present several methods for testing normality.

These tests should be applied to observations within a treatment combination or to the residuals obtained after subtracting off the cell means, but never to the entire set of raw data. Do not apply these tests to the raw data because the raw data may be influenced by the factors of interest and thus will not have a common mean. However, the data within a cell should have a common mean, as should the residual values (having mean 0).

There are numerous tests for normality and we recommend the W test developed by Shapiro and Wilk (1965) for hand calculations and the Anderson-Darling test, Stephens (1974, 1976), for machine calculations. These tests are appropriate for composite tests of normality since one does not have to include the mean and variance as part of the hypothesis. These two tests were compared to others by Shapiro, Wilk, and Chen (1968) and by Pearson, D'Agostino, and Bowman (1977) and generally came out as best. The W test has slightly better power but does not lend itself to computer calculations as easily as does the Anderson-Darling test.

To compute the W statistic, the following steps must be carried out:

1. Order the n observations relabeling as $y_1 \leq y_2 \leq \ldots \leq y_n$.
2. Compute $\sum_{i=1}^{n}(y_i - \bar{y})^2$.
3. If n is even, let $k = n/2$. If n is odd, let $k = (n-1)/2$ and omit the median. Calculate $b = \sum_{i=1}^{k} a_{n-i+1}(y_{n-i+1} - y_i)$ using a_{n-i+1} from Appendix 7.
4. Compute $W = b^2 / \sum_{i=1}^{n}(y_i - \bar{y})^2$.
5. Compare W to the critical values given in Appendix 8. Unlike the other statistics presented, small values of W reject the hypothesis of normality. That is, the data is considered nonnormal if W_{calc} is smaller than W_{crit} found in Appendix 8.

Applying this procedure to the data for tablet B in Table 2.5.1 yields

1. Ordered observations: 8, 9, 10, 10, 10, 12, 12, 16, 19, 24.
2. $\sum_{i=1}^{10}(y_i - \bar{y})^2 = 236$.
3. $b = (0.5739)(24-8) + (0.3291)(19-9) + \ldots + (0.0399)(12-10) = 14.0826$.
4. $W_{calc} = (14.08)^2/236 = 0.840$.
5. $W_{.05} = 0.845$. Thus we would reject the normality assumption.

The Anderson-Darling statistic is computed according to the following rules. It is not quite as powerful as the W statistic but does lend itself to computer calculations. On a computer, it will be necessary to compute cumulative normal probabilities. This is easily accomplished via normal tables or using approximate formula such as those given in Abramowitz and Stegun (1972).

1. Order the n observations relabeling as $y_1 \leq y_2 \leq \ldots \leq y_n$.
2. Compute the standardized variates $w_i = (y_i - \bar{y})/s$, where $\bar{y}$ and s are the sample mean and standard deviation.
3. Compute the cumulative probabilities $Z_i = \Phi(w_i)$, where $\Phi(w_i)$ is the probability a standard normal distribution is less than or equal to w_i.
4. Compute $A^2 = -\{\sum_{i=1}^{n}(2i-1)[\ln(Z_i) + \ln(1 - Z_{n-i+1})]\}/n - n$.
5. Compute $B^2 = A^2(1 + .75/n + 2.25/n^2)$.
6. Compare B^2 to the critical value found in Appendix 9. If the calculated value exceeds the tabulated value, the hypothesis of normality is rejected.

2.6 TYPE II ERRORS AND SAMPLE SIZE CALCULATIONS

A type II or β error is the probability of accepting the null hypothesis of no significant difference when, in fact, there is a significant difference in mean response levels. Naturally the type II error is related to how much difference is present.

Similarly, when an experimenter wants to know how many samples of each factor level to take, he is expressing an interest in being able to detect a certain difference in mean response levels with a given degree of certainty. This again deals with the type II error associated with a given design and a given difference in mean response levels. Finally, the size of effect that the experiment is capable of detecting (with a certain β error usually taken as $\beta = .10$) is often used to compare the effectiveness of several different designs. This comparison will be more clearly spelled out in future chapters.

For the one factor model given by (2.2.1) and under the hypothesis of *different* mean responses, F_{calc} is distributed as a noncentral F statistic with noncentrality parameter related to $J\Phi(A) = J\sum_{i=1}^{I} A_i^2/(I-1)$. Therefore, the probability of a type II error can only be calculated upon complete knowledge of all of the A_i's. It has been our experience that the experimenter is seldom able to specify all of the values of A_i that one wishes to detect. Instead, one can often specify the minimal distance δ ($= \max A_i - \min A_i$) that one wishes to detect.

Knowing the distance of interest, δ, and having an estimate of the variability σ is sufficient to estimate the sample size J required for a given type II error rate β. Likewise, for a given error rate β, a given error variability σ, and a given sample size J, we can compute the minimal detectable difference δ. Both calculations are made with the aid of Appendix 10, defining $\Delta = \delta/\sigma$. The calculations are illustrated below.

In the caplet example given in Table 2.3.3 and analyzed in Table 2.3.4 we rejected the null hypothesis of no significant difference in disintegration times using $J = 6$ repeats of each treatment level. The sample size six was arrived at as follows. Previous studies had indicated that the standard deviation $\sigma \approx \sqrt{\text{MS}(\varepsilon)}$ was approximately 1.6. The experimenter was interested in detecting a difference δ as small as four seconds with a probability $1 - \beta = .90$. Thus we calculate $\Delta = \delta/\sigma = 4/1.6 = 2.50$. Letting $\alpha = .05$, $\beta = .10$, and $I = 4$ treatment combinations of interest, Appendix 10 gives $\Delta = 2.698$ for $J = 5$ and $\Delta = 2.401$ for $J = 6$. Therefore, to be 90% sure of detecting a difference of $\Delta = 2.50$, we chose a total of $J = 6$ repetitions for each treatment level.

To make the above calculations, knowledge of the minimal detectable difference δ and knowledge of the underlying process variation σ had to be known. In many experiments, the magnitude of σ is simply not known. In these cases, a two-staged approach is often used. In the first stage, a small number of repeats are run for each treatment level. This small experiment is used to estimate σ by $\sqrt{\text{MS}(\varepsilon)}$ and then the sample size J can be estimated using Appendix 10. Of course, if the test was already significant, enough evidence has been gathered to reject the null hypothesis and the calculation of sample size is unnecessary.

PROBLEMS

2.1 *An experiment was conducted to determine the effects of a certain air pollutant on the cardiopulmonary functions of mongrel dogs. The measurement used for this test was pulmonary arterial pressure in pounds per square inch, psi, which should increase under the influence of toxins. A total of 15 dogs were randomly chosen, five were injected with 2 mg sodium chloride (table salt) as a control, five were injected with 1 mg sodium chloride, 1 mg sodium sulfate, and five were injected with 2 mg sodium sulfate. Sodium sulfate is a constituent of air pollution. The data obtained from this experiment are as follows.*

	pulmonary arterial pressure (psi)				
2 mg sodium chloride	*12.3*	*11.7*	*12.2*	*12.1*	*12.0*
1 mg of each	*12.8*	*13.2*	*11.9*	*12.9*	*13.1*
2 mg sodium sulfate	*13.3*	*12.7*	*13.2*	*12.8*	*12.0*

(a) Using an α of .05, is there a significant difference in psi for the three treatments? (b) Give a 90% confidence interval on the difference between 2 mg sodium chloride and 2 mg sodium sulfate. (c) Use Duncan's procedure to determine which means differ significantly at the 5% level. (d) For this example, does it make sense to use a two-staged approach to estimate the additional sample size needed to detect a difference in means? Why or why not? (e) Does it make sense to fit orthogonal polynomials to this data?

2.2 A government committee on highway safety was interested in whether roads were getting safer. Ten states were randomly selected out of the 48 contiguous states and, for each of the years 1968 through 1972, the fatality rate (= deaths per 100 million vehicle miles) was calculated.

year (i)	observations (j)	$T_{i\cdot}$	$s_{i\cdot}^2$
1968	5.0, 4.9, 5.9, 5.0, 7.2, 7.0, 4.9, 6.4, 6.3, 7.3	59.9	0.9788
1969	5.4, 3.3, 4.5, 5.5, 5.0, 7.6, 7.6, 6.4, 5.3, 7.1	57.7	1.9557
1970	4.3, 5.1, 4.0, 6.0, 4.4, 4.6, 4.8, 6.4, 5.2, 4.3	49.1	0.6077
1971	5.2, 7.8, 3.0, 5.0, 5.8, 4.0, 4.5, 4.4, 4.4, 3.9	48.0	1.7000
1972	5.0, 5.8, 6.6, 4.1, 3.8, 4.5, 3.8, 4.5, 5.3, 3.8	47.2	0.9040

(a) Use each of the three tests for homogeneity on this data. For consistency in calculating the ln s^2 ANOVA test, use the first five and the last five observations in each set. (b) Test for the normality of the data collected in 1970. (c) Run an ANOVA on this data. (d) Is there a significant difference in fatality rates using $\alpha = .05$?

2.3 (a) For the data listed in problem **2.2**, what is the state to state (also called within year or error) standard deviation? (b) Assuming $\beta = .10$, what is the minimal difference in the annual death rate that can be detected with this data?

2.4 Use Duncan's test on the data in Problem **2.2** to determine which years differ. Interpret the results. Are the results consistent with **2.3**(b)? Why or why not?

2.5 Use orthogonal polynomials to find the order of the best fitting polynomial for the data in Problem **2.2**. Give the ANOVA table in its final form.

2.6 The data in Table 2.5.1 failed both the homogeneity and the normality criteria. Theory indicated that the square root transformation would remove the heterogeneity. (a) Run the modified Levene test on the transformed data to check for homogeneity. (b) Typically a transformation helping homogeneity also helps normality. Run the W test on the transformed data from tablet type B to check for normality. (c) Run an ANOVA and, if appropriate, Duncan's test on the transformed data. Interpret the results.

2.7 In an experiment with $I = 5$ treatment combinations, it is believed that σ will be somewhere around 1.5. If we are interested in being able to detect a *difference of $\delta = 1.6$ units and are willing to let $\alpha = .05$, $\beta = .10$, how many repetitions of the treatment combinations should be run?*

2.8 *Set up the formula for an ANOVA table including source, df, SS, MS, F_{calc}, and F_{crit} for an experiment consisting of four repeats of five different treatment levels of a factor.*

2.9 *Repeat Problem* **2.8** *using notation appropriate for hand calculations.*

2.10 *Use the definitions of $\bar{y}_{i\cdot}$ and $T_{i\cdot}$ to derive equations (2.3.4) and (2.3.5) from the formulas given in Table 2.3.1.*

2.11 *In 1932, P. Emrich designed an experiment to determine if adding a tablespoon of honey to a cup of milk for six straight weeks increased the haemoglobin in children in a children's home. The data for the twelve children, six given honey and six not given honey, is given below. Does honey increase haemoglobin?*

Given Honey	*Not Given Honey*
19, 12, 9, 17, 24, 22	*14, 8, 4, 4, 11, 15*

BIBLIOGRAPHY FOR CHAPTER 2

Abramowitz, A. and Stegun, I. A. (1972). *Handbook of Mathematical Functions with Formulas, Graphs, and Mathematical Tables*, Washington, D.C.: U.S. Government Printing Office. (Chapter 26, particularly 26.2.17)

Anderson, V. L. and McLean, R. A. (1974). *Design of Experiments A Realistic Approach*, New York: Marcel Dekker. (Chapter 1)

Bartlett, M. S. (1937). "Properties of sufficiency and statistical tests," *Proceedings of the Royal Statistical Society A*, **160**, 268–282.

Bartlett, M. S. (1947). "The use of transformations," *Biometrics*, **3**, 39–52.

Bartlett, M. S. and Kendall, D. G. (1946). "The statistical analysis of variance—heterogeneity and the logarithmic transformation," *Journal of the Royal Statistical Society Supplement*, **8**, 128–138.

Box, G. E. P. (1954). "Some theorems on quadratic forms applied in the study of analysis of variance problems, I. Effect of inequality of variance in the one-way classification," *Annals of Mathematical Statistics*, **25**, 290–302.

Box, G. E. P. and Cox, D. R. (1964). "An analysis of transformations," *Journal of the Royal Statistical Society Series B*, **26**, 211–243.

Box, G. E. P., Hunter, W. G. and Hunter, J. S. (1978). *Statistics for Experimenters An Introduction to Design, Data Analysis, and Model Building*, New York: John Wiley & Sons. (Chapter 6)

Bozivich, H., Bancroft, T.A. and Hartley, H. O. (1956). "Power of analysis of variance test procedures for certain incompletely specified models," *Annals of Mathematical Statistics*, **27**, 1017–1043.

Bratcher, T. L., Moran, M. A. and Zimmer, W. J. (1970). "Tables of sample size in the analysis of variance," *Journal of Quality Technology*, **2**, 156–164.

Brown, M. B. and Forsythe, A. B. (1974). "Robust Tests for the Equality of Variances," *Journal of the American Statistical Association*, **69**, 364–367.

Carmer, S. G. and Swanson, M. R. (1973). "Evaluation of ten pairwise multiple comparison procedures by Monte Carlo methods," *Journal of the American Statistical Association*, **68**, 66–74.

Cochran, W. G. (1947). "Some consequencces when the assumptions for the analysis of variance are not satisfied," *Biometrics*, **3**, 22–38.

Conover, W. J., Johnson, M. E., and Johnson, M. M. (1981). "A Comparative Study of Tests for Homogeneity of Variances, With Applications to the Outer Continental Shelf Bidding Data," *Technometrics*, **22**, 351–361.

Das, M. N. and Giri, N. C. (1986). *Design and Analysis of Experiments*, 2nd ed., New York: John Wiley & Sons. (Chapter 3)

Dixon, W. J. and Massey, F. J., Jr. (1983). *Introduction to Statistical Analysis*, 4th ed., New York: McGraw-Hill. (Chapters 8, 10)

Draper, N. R. and Cox, D. R. (1969). "On distributions and their transformation to normality," *Journal of the Royal Statistical Society Series B*, **31**, 472–476.

Draper, N. R. and Smith, H. (1966). *Applied Regression Analysis*, New York: John Wiley & Sons. (Chapter 9)

Duncan, D. B. (1955). "Multiple range and multiple F tests," *Biometrics*, **11**, 1–42.

Duncan, D. B. (1975). "t-tests and intervals for comparisons suggested by the data," *Biometrics*, **31**, 339–359.

Eisenhart, C. (1947). "The assumptions underlying the analysis of variance," *Biometrics*, **3**, 1–21.

Emrich, P. (1932). "Weitere Erfahrungen mit Honigkuren," *Bienenzeitung*, **12**, 1–5.

Gabriel, K. R. (1978) "Comment on Ramsey (1978)," *Journal of the American Statistical Association*, **73**, 485–487.

Ghosh, S., editor. (1990). *Statistical Design and Analysis of Industrial Experiments*, New York: Marcel Dekker.

Hicks, C. R. (1982). *Fundamental Concepts in the Design of Experiments*, New York: Holt, Rinehart, & Winston. (Chapters 2, 7)

Johnson, N. L. and Leone, F. C. (1977). *Statistics and Experimental Design in Engineering and the Physical Sciences*, 2nd ed., New York: John Wiley & Sons. (Chapter 13)

Kastenbaum, M. A., Hoel, D. G. and Bowman, K. O. (1970). "Sample size requirements: one-way analysis of variance," *Biometrika*, **57**, 421–430.

Keuls, M. (1952). "The use of the Studentized range in connection with an analysis of variance," *Euphytica*, **1**, 112-122.

Levene, H. (1960). "Robust Tests for Equality of Variances," *Contributions to Probability and Statistics*, 278–292. I. Olkin, S. G. Ghurye, W. G. Madow, and H. B. Mann (eds.), Stanford: Stanford University Press.

Montgomery, D. C. (1991). *Design and Analysis of Experiments*, 3rd ed., New York, John Wiley & Sons. (Chapters 3, 4)

Newman, D. (1939). "The distribution of the range in samples from a normal population expressed in terms of an independent estimate of the standard deviation," *Biometrika*, **31**, 20–30.

Ostle, B. and Malone, L. C. (1988). *Statistics in Research*, 4th ed., Ames, Iowa: Iowa State University Press.

Owen, D. B. (1965). "A special case of a bivariate non-central t distribution," *Biometrika*, **52**, 437–446.

Pearson, E. S., D'Agostino, R. B. and Bowman, K. O. (1977). "Tests for departure from normality: comparison of power," *Biometrika*, **64**, 231–246.

Petersen, R. G. (1985). *Design and Analysis of Experiments*, New York: Marcel Dekker.

Ramsey, P. H. (1978). "Power differences between pairwise multiple comparisons," *Journal of the American Statistical Association*, **73**, 479–485.

Robson, D. S. (1959). "A simple method for constructing orthogonal polynomials when the independent variable is unequally spaced," *Biometrics*, **15**, 187–191.

Samiuddin, M., Hanif, M. and Asad, H. (1978). "Some comparisons of the Bartlett and cube root tests of homogeneity of variance," *Biometrika*, **65**, 218–221.

Scheffé, H. (1953). "A method for judging all contrasts in the analysis of variance," *Biometrika*, **40**, 87–104.

Scheffé, H. (1959). *The Analysis of Variance*, New York: John Wiley & Sons. (Chapters 3, 10)

Shapiro, S. S. and Wilk, M. B. (1965). "An analysis of variance test for normality (complete samples)," *Biometrika*, **52**, 591–611.

Shapiro, S. S., Wilk, M. B. and Chen, H. J. (1968). "A comparative study of various tests for normality," *Journal of the American Statistical Association*, **63**, 1343–1372.

Snedecor, G. W. and Cochran, W. G. (1991). *Statistical Methods*, 8th ed., Ames, Iowa: Iowa State University Press. (Chapters 2–4, 10)

Steel, R. G. D. and Torrie, J. H. (1980). *Principles and Procedures of Statistics—A Biometrical Approach*, 2nd ed., New York: McGraw-Hill. (Chapter 8)

Stephens, M. A. (1974). "EDF statistics for goodness of fit and some comparisons," *Journal of the American Statistical Association*, **69**, 730–737.

Stephens, M. A. (1976). Asymptotic results for goodness of fit statistics with unknown parameters," *The Annals of Statistics*, **4**, 357–369.

Tadikamalla, P. R. (1974). "Constructing orthogonal polynomials when the independent variable is unequally spaced," *Journal of Quality Technology*, **6**, 113–115.

Tukey, J. W. (1949). "Comparing individual means in the analysis of variance," *Biometrics*, **5**, 99–114.

Tukey, J. W. (1953). *The Problem of Multiple Comparisons*, Princeton, New Jersey: Princeton University.

Waller, R. A. and Duncan, D. B. (1969). "A Bayes rule for the symmetric multiple comparisons problem," *Journal of the American Statistical Association*, **64**, 1484–1499. Corrigenda (1972), *JASA*, **67**, 253–255.

Winer, B. J. (1971). *Statistical Principles in Experimental Design*, 2nd ed., New York: McGraw-Hill. (Chapters 3, 4)

SOLUTIONS TO SELECTED PROBLEMS

2.2 (a) *Bartlett's test:* $M = 4.0593$, $C = 1.0444$, $\chi_4^2 = 3.89$, $\chi_{crit}^2(.01) = 13.28$, $\chi_{crit}^2(.001) = 18.47$.

ANOVA on ln s^2

Source	df	SS	MS	F_{calc}	$F_{4,5}(.01)$	$F_{4,5}(.001)$
Year	4	0.5760	0.1440	0.09	11.4	31.1
Error	5	7.6732	1.5346			
Total	9	8.2492				

ANOVA on $\mid y_{ij} - \tilde{y}_i \mid$

Source	df	SS	MS	F_{calc}	$F_{4,45}(.01)$	$F_{4,45}(.001)$
Year	4	1.2788	0.3197	0.60	3.81	5.60
Error	45	23.8750	0.5306			
Total	49	25.1538				

Therefore, we accept the homogeneity assumption for all three tests.

(b) *W test:* $b = 2.223$, $W_{calc} = 0.904$, $W_{.05} = 0.845$.
Anderson-Darling: $A^2 = 0.440$, $B^2 = 0.483$, $B_{.05}^2 = 0.752$.
The two tests agree. I would not transform the data.

(c) *ANOVA on Fatality Rates*

Source	df	SS	MS	F_{calc}	$F_{4,45}(.05)$
Year	4	14.1628	3.5407	2.88*	2.59
Error	45	55.3150	1.2292		
Total	49	69.4778		*significant at the .05 level	

(d) *Yes.*

2.3 (a) $\sigma = \sqrt{MSE} = 1.1087$. (b) $\delta = \sigma\Delta = 1.1087(1.850) = 2.051$.

2.4 *The results of Duncan's test are given below. They are not necessarily inconsistent with* **2.3** *(b), which states a 90% certainty for detecting a difference as large as 2.051. The observed difference is 1.27. A theoretical difference of 1.27 can be detected with some positive probability (less than 90%).*

Year	Mean
'68	5.99
'69	5.77
'70	4.91
'71	4.80
'72	4.72

2.5

Polnomial Fit for Fatality Rates

Source	df	SS	MS	F_{calc}	$F_{crit}(.05)$
Year	4	14.1628	3.5407	2.88*	2.59
linear	1	12.3201	12.3201	10.02*	4.06
lack of fit	3	1.8427	0.6142	0.50	2.82
Error	45	55.3150	1.2292		
Total	49	69.4778			

*significant at the .05 level

3

Factorial Designs

3.0 INTRODUCTION

This chapter extends the one factor design introduced in the previous chapter to designs including many factors. It is assumed that the reader understands the scientific principles outlined in the first chapter, particularly that the reader has recognized and formulated the problem, has thought about the factors and levels affecting the experiment, has specified the variable to be measured, and has a clear idea about the inference space to which the results will apply.

In factorial designs, all possible combinations of the levels of the factors are applied to the same number of experimental units. When all combinations occur equally often, the result is what is called a *balanced design*. Balanced designs will be the primary focus of this book. Unless specifically noted, the experiment is to be completely randomized. (The consequences of incomplete randomization will be covered in Chapter 5.) Complete randomization includes random selection of the experimental units, random assignment of the treatment combinations to the experimental units, random run order (if appropriate), and random location (if appropriate).

This chapter covers the exact layout of the experiment, the corresponding mathematical model, assumptions, and meanings of the terms, the formulae for degrees of freedom (df), sums of squares (SS), mean squares (MS), and expected mean squares (EMS), derivation of exact, approximate, and conservative tests of the factors and interactions, estimation of variance components, comparisons of means, type II errors, power, and sample size requirements. All procedures will be illustrated with examples.

While they occupy considerable space in this chapter, the reader should not overemphasize the formulae for df, SS, and MS. With ready access to desktop computers, these quantities are rarely computed by hand. Instead, the reader should emphasize the computation of EMS, the resulting tests, comparisons of means, and variance component estimation. These ideas have yet to be routinely included in computer software.

3.1 LAYOUT OF THE EXPERIMENT

The various factors of interest will generally be labeled with capital letters, the capital letter usually being the first letter of the description of the factor of interest. For example, the factor temperature would be labeled T. If several factors have the same first letter, a lower case letter or two can be included to differentiate between the two factors. When talking in general terms, the factors will be labeled alphabetically, A, B, C, *etc.* We label the levels of the first factor of interest with the subscript i, ranging from 1 to I. The levels of the second factor will be denoted with the subscript j, ranging from 1 to J. Repeats of the entire experiment will simply be denoted by the error term ε having a subscript of the next available index. For the error term ε, all other indices should be put in brackets to indicate that different experimental units are used for each combination of the other factors. Brackets indicate *nesting*, used here because the error term is nested within the combinations of levels of the other factors. Nesting will be covered in general in Chapter **4**. For now, only the error term will contain indices that are bracketed. For example, if there are two factors of interest denoted by the indices i and j, the index on the error term ε would be $k(ij)$, read k depending on i and j.

Let $y_{ij\cdots n}$ be the measured response for the nth replicate of treatment level i for factor A, treatment level j for factor B, *etc.* To assure statistical validity of the ensuing tests and estimates, complete randomization should be used. The $IJ \cdots N$ experimental units should be randomly selected from all possible experimental units, the experimental units should be randomly assigned to the treatment combinations, and the experiment should be run in a random fashion. This procedure is known as complete randomization.

While randomization was thoroughly discussed in Chapter **2**, we will repeat the randomization procedure used to assign the units to the treatment combinations. Write down all possible combinations of the factor levels N times (for the N repeats) and label the experimental units 1 through $IJ \cdots N$. Now use Appendix 1 to select a random number between 1 and $IJ \cdots N$ and assign that experimental unit to the first treatment combination in the list. Pick a different random number between 1 and $IJ \cdots N$ and assign that experimental unit to the second treatment combination in the list (even if it is simply one of the N repeats of the same treatment combination). Continue until all units have been assigned to the treatment combinations.

Follow the same randomization procedure for the run order or for the location of the treatment combinations, depending on whether the experiment is run sequentially or simultaneously. While this seems like a lot of needless work, it is up to the statistician to take every step possible to guarantee the validity of the statistical tests and estimates. Complete randomization is one of the best known methods.

3.2 MATHEMATICAL MODEL AND ASSUMPTIONS

For the one factor model of the previous chapter, the model was written as

$$y_{ij} = \mu + A_i + \varepsilon_{j(i)} \tag{3.2.1}$$

where μ is the overall mean, A_i is the differential effect due to the ith treatment level of A having the restriction $\sum_{i=1}^{I} A_i = 0$, and $\varepsilon_{j(i)}$ is the random error term assumed Normal$(0,\sigma^2)$. *Note: We use the same notation for the level of a factor and for the differential effect of that factor.* That is, A_1 stands for the *treatment* associated with the first level of A when referring to the factor and, when used in the mathematical model, A_1 stands for the *effect* of the first level of A. To the practitioner, this seems to be less confusing than using two different symbols for the effect and for the level itself. Tests on the assumptions of homogeneity of variance and normality were given in Chapter **2**.

To be able to generalize model (3.2.1) to many factors, we first need to generalize the concept of a factor. In (3.2.1), specific levels of the factor A were chosen and inferences were made only to those specific levels. In that case, we drew conclusions about and made comparisons between the various mean levels of the factor. Factor A is an example of a *fixed factor*. For a fixed factor, we make the assumption $\sum_{i=1}^{I} A_i = 0$, an assumption that can be made without loss of generality.

Suppose, however, that the experiment consisted of several people measuring an item several times to get an idea of the consistency of the measuring device. The different people can be represented by the factor A_i, A_1 standing for the first person, A_2 the second, *etc.*, and the repeated measurement by the error term $\varepsilon_{j(i)}$. In this case, we do not wish to draw conclusions about the specific individuals running the experiment. We wish, instead, to draw conclusions about all individuals that could use the measuring device. The individual means are not of interest because we expect each person to differ from the next. It is the magnitude of the variation among all individuals using the measuring device that is of interest. We have what is called a *random factor* For a random factor, we make the assumption that A is distributed Normal(0,σ_A^2) where σ_A^2 is unknown and independent of σ^2 defined for the error term.

The difference between a *fixed* and a *random* factor is one of *inference*. The inference for a fixed factor is only to the levels specifically chosen in the experiment. If a fixed factor is significant then we will estimate averages associated with the various levels of that factor and compare all possible average levels to determine which pairs are significantly different. We may also look for trends in the average values. On the other hand, the inference for a random factor is over an entire population or a large collection of items and the chosen values must be representative of that population or collection. If a random factor is significant, we will estimate the magnitude of its variance, *i.e.*, estimate σ_A^2. Individual mean levels of a random factor have no meaning.

If two different factors, say A_i and B_j, are included in the experiment, we will automatically include the *interaction* term, AB_{ij}, in the model. The interaction term consists of both capital letters and all of the subscripts contained in either factor. It is a term in the same fashion as A_i and B_j, not the product of A_i and B_j. The importance of this interaction term will be discussed in detail. For a two factor design, the model becomes

$$y_{ijk} = \mu + A_i + B_j + AB_{ij} + \varepsilon_{k(ij)} \tag{3.2.2}$$

where i ranges from 1 to I indicating I different levels of factor A, j ranges from 1 to J indicating J different levels of factor B, and k ranges from 1 to K indicating that each of the ij combinations of factors A and B occur K times. A total of IJK experimental units will be used in the experiment.

If three different factors, say A_i, B_j, and C_k, are included in the experiment, then all of the *two factor interactions*, AB_{ij}, AC_{ik}, and BC_{jk}, as well as the *three factor interaction*, ABC_{ijk}, are included in the model. The meaning of a three factor interaction will also be discussed below. The model becomes

$$y_{ijk\ell} = \mu + A_i + B_j + AB_{ij} + C_k + AC_{ik} + BC_{jk} + ABC_{ijk} + \varepsilon_{\ell(ijk)} \tag{3.2.3}$$

where $i = 1, \cdots, I$, $j = 1, \cdots, J$, $k = 1, \cdots, K$, $\ell = 1, \cdots, L$, A_i, B_j, and C_k are the main effects, AB_{ij}, AC_{ik}, and BC_{jk} are the two factor interactions, ABC_{ijk} is the three factor interaction, and $\varepsilon_{\ell(ijk)}$ is the random error term. A total of $IJKL$ experimental units will be used in the experiment.

The generalization to more than three factors follows the same pattern as (3.2.1) through (3.2.3), tacking the new factor on after completing the previous factors and interactions and then forming the interactions of this factor with all previous factors and interactions. For example, if a fourth factor were added, we would start with the three factor model given by (3.2.3), add the factor D_ℓ immediately to the right of ABC_{ijk}, and then add the interactions of A_i with D_ℓ (yielding $AD_{i\ell}$), B_j with D_ℓ (yielding $BD_{j\ell}$), $\cdots$, and ABC_{ijk} with D_ℓ (yielding $ABCD_{ijk\ell}$). Of course, y would now have subscript $ijk\ell m$ and ε would have subscript $m(ijk\ell)$.

Two factor interactions are included in the model to account for any *nonadditivities* among the main effects. Three factor interactions account for any nonadditivities between the main effects and two factor interactions. And so on. One of the easiest ways to demonstrate the meaning of interactions for two fixed factors is through two-way plots. The three plots in Figure 3.1 demonstrate, respectively, no interaction and two different types of interactions. The factor A_i is assumed to have 3 levels plotted along the x-axis. The response variable is plotted along the y-axis. The factor B_j acts as a contour variable with constant values connected by straight lines.

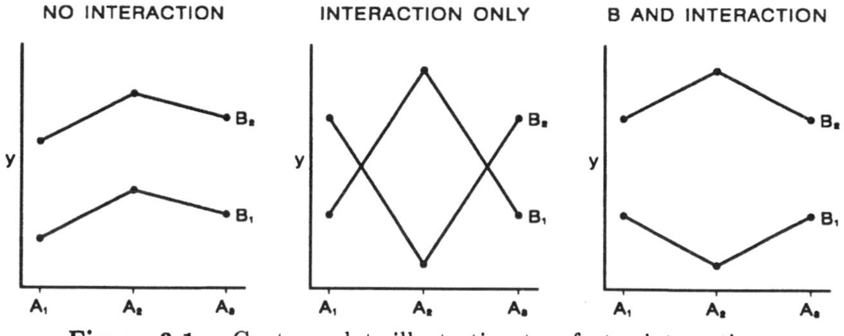

Figure 3.1 Contour plots illustrating two factor interactions.

To further explain, there is no interaction between two fixed factors A_i and B_j if the differential response for every combination of the two factors is simply the sum of the differential responses for each factor alone. If the combination yields anything other than the sum of the individual effects, an interaction is said to exist. Additivity (or lack of interaction), is equivalent to parallel lines as is observed in the first plot. If the response due to B_j depends on which level of A_i is present (or vice versa), as indicated by the nonparallel lines in the second and third plots, an AB_{ij} interaction is said to exist. When an interaction exists, inference is usually to all six AB_{ij} combinations rather than to the three A_i levels alone or the two B_j levels alone.

This extends to higher order interactions as well. For example, an ABC_{ijk} interaction indicates that the response due to C_k depends on which of the AB_{ij} combinations is present. Equivalently, a significant ABC_{ijk} interaction indicates the response due to A_i depends on which BC_{jk} combination is present or the response due to B_j depends on which AC_{ik} combination is present or the response due to C_k depends on AB_{ij}. Generally, one interpretation makes more physical sense than another and this is the interpretation to use.

If factors A_i and B_j are both fixed, we will further assume that

$$\sum_{i=1}^{I} AB_{ij} = \sum_{j=1}^{J} AB_{ij} = 0. \qquad (A_i \text{ and } B_j \text{ both fixed.}) \qquad (3.2.4)$$

These assumptions can be made without loss of generality because, by definition, AB_{ij} is a differential effect and if the sum, say over i, were to equal to some value $S \neq 0$, we could replace B_j with $B_j - S/I$ and the redefined AB_{ij} would sum to 0.

If factors A_i, B_j, and C_k are all considered fixed, we will assume that

$$\sum_{i=1}^{I} ABC_{ijk} = \sum_{j=1}^{J} ABC_{ijk} = \sum_{k=1}^{K} ABC_{ijk} = 0. \qquad (A_i, B_j, C_k \text{ all fixed.}) \qquad (3.2.5)$$

Again these assumptions can be made without loss of generality. The extension to any order interaction, all of whose terms are considered fixed, should be obvious. Such interactions are referred to as *fixed interactions* because each individual factor is fixed.

If factor A_i is considered random, we make the assumption that

$$A_i \text{ are } iid \text{ Normal}(0, \sigma_A^2). \qquad (A_i \text{ random}) \qquad (3.2.6)$$

where σ_A^2 is considered unknown. As in the fixed case, the assumption of zero mean can be made without loss of generality. However, each individual A_i is an observation from a theoretically infinite population, so there is no reason why the summation over the selected A_i will be 0. For a random factor, we will *not* assume $\sum_{i=1}^{I} A_i = 0$. Instead, we have $\mathrm{E}[A_i^2] = \sigma_A^2$ for all i and $\mathrm{E}[A_i A_{i'}] = 0$ for $i \neq i'$. Here $\mathrm{E}[\cdot]$ denotes the expected value. Note, in particular, that (3.2.6) implies the variances are homogeneous and y will be normally distributed.

If both A_i and B_j are considered random, we will further assume

$$AB_{ij} \text{ are } iid \text{ Normal}(0, \sigma^2_{AB}). \qquad (A_i \text{ and } B_j \text{ random.}) \tag{3.2.7}$$

That is, the interaction is simply the added variance in the model not accounted for by the variance of the two main effects. We assume that this variance term is independent of each of the two main effect variance terms, and the two main effects are independent of each other. The generalization to higher order random components follows naturally.

Since AB_{ij} is random in both indices, we will not assume that any summation over an index equals 0. Then, by the independence assumption, we get $\mathrm{E}[AB_{ij}^2] = \sigma^2_{AB}$ and $\mathrm{E}[AB_{ij}AB_{i'j'}] = 0$ for either $i \neq i'$, $j \neq j'$, or both $i \neq i'$ and $j \neq j'$. Such interactions will be called *random interactions* and have all the individual components random.

When some of the factors making up an interaction are fixed and some are random, we have what is called a *mixed interaction*. For such factors, we assume that the interaction is a normal random variable but will impose the summation condition over the indices associated with fixed factors. Such a condition has been the basis of much discussion. Generally, statisticians seem willing to enforce this condition. The reasoning is simple. Since the interaction represents non-additivity of the main effects, the behavior of the interaction should reflect the behavior of the main effects it is representing. In other words, the behavior of the non-additive piece should be the same as the behavior of the additive piece. Searle (1971) takes exception to these conditions. He feels that such conditions cause unrealistic interpretations of the variance terms. In this book, we will enforce the condition that the mixed interaction sum to zero over any fixed index, but recognize that other conditions are possible.

Since, for given values of the random indices, the entire population represented by the fixed indices is known, we will make the assumption that a summation over any fixed index is zero. In essence, when we are summing over a fixed index, all of the other indices are "fixed" at their given levels and, therefore, every index is considered fixed. However, when we sum over a random index, the index of summation is still "random" and therefore there is no reason to assume the expression will sum to zero. As will be seen later, this condition has implications in the derivation of the expected mean squares and the corresponding statistical tests.

For example, if A_i is fixed and B_j random, we will assume

$$\begin{gathered} AB_{ij} \text{ are identically Normal}(0, \sigma^2_{AB}) \text{ with} \\ \sum_{i=1}^{I} AB_{ij} = 0. \qquad (A_i \text{ fixed, } B_j \text{ random}). \end{gathered} \tag{3.2.8}$$

This implies $\mathrm{E}[\sum_{i=1}^{I} AB_{ij}^2] = (I-1)\sigma_{AB}^2$, from which $\mathrm{E}[AB_{ij}^2] = [(I-1)/I]\sigma_{AB}^2$ follows. We also have $\mathrm{E}[AB_{ij}AB_{i'j'}] = 0$ for $j \neq j'$, and $\mathrm{E}[AB_{ij}AB_{i'j}] = -\sigma_{AB}^2/I$. The first equality is a result of the linear constraint on AB_{ij}, Scheffé (1956b). The fourth equality is derived as follows: $(I-1)\mathrm{E}[AB_{ij}AB_{i'j}] = \mathrm{E}[\sum_{i' \neq i} AB_{ij}AB_{i'j}] = \mathrm{E}[AB_{ij}\sum_{i' \neq i} AB_{i'j}] = -\mathrm{E}[AB_{ij}AB_{ij}] = -[(I-1)/I]\sigma_{AB}^2$. As will be seen, these equations have direct bearing on the derivation of expected mean squares and confidence intervals.

If A_i was considered random and B_j fixed, we would assume

$$\begin{aligned} &AB_{ij} \text{ are identically Normal}(0, \sigma_{AB}^2) \text{ with} \\ &\sum_{j=1}^{J} AB_{ij} = 0, \qquad (A_i \text{ random, } B_j \text{ fixed}). \end{aligned} \tag{3.2.9}$$

This implies $\mathrm{E}[\sum_{j=1}^{J} AB_{ij}^2] = (J-1)\sigma_{AB}^2$, from which $\mathrm{E}[AB_{ij}^2] = [(J-1)/J]\sigma_{AB}^2$ follows. Like the last example, $\mathrm{E}[AB_{ij}AB_{i'j'}] = 0$ for $i \neq i'$, and $\mathrm{E}[AB_{ij}AB_{ij'}] = -\sigma_{AB}^2/J$.

As a final example, suppose A_i and C_k were fixed, B_j and D_ℓ were random. Then we would assume

$$\begin{aligned} &ABCD_{ijk\ell} \text{ are identically Normal}(0, \sigma_{ABCD}^2) \text{ with} \\ &\sum_{i=1}^{I} ABCD_{ijk\ell} = \sum_{k=1}^{K} ABCD_{ijk\ell} = 0, \\ &(A_i, C_k \text{ fixed, } B_j, D_\ell \text{ random}). \end{aligned} \tag{3.2.10}$$

The assumptions imply $\mathrm{E}[\sum_{i=1}^{I}\sum_{k=1}^{K} ABCD_{ijk\ell}^2] = (I-1)(K-1)\sigma_{ABCD}^2$, $\mathrm{E}[ABCD_{ijk\ell}^2] = [(I-1)(K-1)/IK]\sigma_{ABCD}^2$, $\mathrm{E}[ABCD_{ijk\ell}ABCD_{i'j'k'\ell'}] = 0$ for $j \neq j'$, $\ell \neq \ell'$, or both $j \neq j'$ & $\ell \neq \ell'$, $\mathrm{E}[ABCD_{ijk\ell}ABCD_{i'jk\ell}] = -[(K-1)/(IK)]\sigma_{ABCD}^2$ for $i \neq i'$, $\mathrm{E}[ABCD_{ijk\ell}ABCD_{ijk'\ell}] = -[(K-1)/(IK)]\sigma_{ABCD}^2$ for $k \neq k'$, and $\mathrm{E}[ABCD_{ijk\ell}ABCD_{i'jk'\ell}] = \sigma_{ABCD}^2/(IK)$ for $i \neq i'$ and $k \neq k'$. These equations all follow from the summation conditions given in (3.2.10).

The consequences of the above assumptions will be fully developed in Sections **3.4** and **3.5** where we derive EMS and determine the appropriate statistical tests. In general, the assumptions of normality and homogeneity of variance are not considered critical since F tests are quite robust to nonnormality and heterogeneity of variance. Tests that detect drastic departures from normality or homogeneity have already been given in Chapter **2** and should be applied whenever the assumptions are in doubt.

3.3 THE ANOVA TABLE FOR FACTORIAL DESIGNS

Table 2.3.1 of Chapter **2** developed the ANOVA table for a one factor design and consisted of the headings Source, df, SS, MS, and F, respectively standing for the source of the variation, degrees of freedom, sum of squares, mean squares, and the calculated F ratio. The ANOVA table can also include a column labeled EMS for the expected mean squares (to be developed in the next section) and a column labeled F_{crit} containing the critical values obtained from Appendix 2. We will depart from formality and, depending on the circumstances, let the ANOVA table contain any or all of these headings. In particular, this section develops formulae for the df, SS, and MS columns. These will be summarized in an ANOVA table.

With the advent of modern computing equipment and canned software packages, the importance of knowing the formulae in this section is greatly diminished. The reader must keep this in mind and not overemphasize the importance of this section. When teaching the use of a specific software package, this section can be skipped completely and replaced with the instructions needed to run the program.

If A is a fixed factor in a one factor experiment, we argued in Section **2.2** of Chapter **2** that A would be considered significant if the parameter $\Phi(A) = \sum_{i=1}^{I} A_i^2/(I-1)$ were greater than zero. Since A_i is the differential effect of the ith treatment level from the mean, we were lead to the natural estimate of $\Phi(A)$: $\sum_{i=1}^{I}(\bar{y}_{i\cdot} - \bar{y}_{\cdot\cdot})^2/(I-1)$. If the A_i were considered random rather than fixed, we would assume that A_i is distributed Normal with mean zero and variance σ_A^2 and the natural estimate of σ_A^2 would be $\sum_{i=1}^{I} \hat{A}_i^2/(I-1) = \sum_{i=1}^{I}(\bar{y}_{i\cdot} - \bar{y}_{\cdot\cdot})^2/(I-1)$, the exact estimate used when A was considered fixed. This is really quite a remarkable result, made possible by the form of Φ and the model assumptions. This result motivates the following general observation: *the df, SS, and MS columns are the same whether the factors are considered random or fixed.*

Since the formulae are the same, we will motivate the calculations assuming every factor is fixed. We believe this makes it is easier to understand the calculations. General rules will also be presented.

Consider a two factor experiment, with factors A_i and B_j, having K repeats of each treatment combination. Recalling that A_i is the differential effect of the ith treatment level of A, the natural estimate of A_i would be $\hat{A}_i = \bar{y}_{i\cdot\cdot} - \bar{y}_{\cdot\cdot\cdot}$, where the averages are taken over j and k for the first mean and over i, j, and k for the second mean. As in Chapter **2**, the sum of squares for A, SS(A), would be $JK\sum_{i=1}^{I} \hat{A}_i^2$, the sum of the squares of the natural estimates of A_i over all of the data. The corresponding df would be $I-1$, I for the estimated averages minus one for subtracting the grand average. In similar fashion, the natural estimate of B_j would be $\hat{B}_j = \bar{y}_{\cdot j \cdot} - \bar{y}_{\cdots}$. The corresponding SS would be $\text{SS}(B) = IK \sum_{j=1}^{J} \hat{B}_j^2$ and the corresponding df would be $\text{df}(B) = J - 1$.

The interaction term AB_{ij} is defined as the differential effect of the (ij)th combination above and beyond the differential effect due to A_i and B_j. The term "differential effect" implies that the grand mean μ is subtracted from the cell mean and the term "above and beyond" implies that A_i and B_j are subtracted from the difference. Carefully following this definition, the natural estimate of AB_{ij} would be $(\bar{y}_{ij\cdot} - \hat{\mu}) - \hat{A}_i - \hat{B}_j = (\bar{y}_{ij\cdot} - \bar{y}_{\cdots}) - (\bar{y}_{i\cdot\cdot} - \bar{y}_{\cdots}) - (\bar{y}_{\cdot j \cdot} - \bar{y}_{\cdots}) = \bar{y}_{ij\cdot} - \bar{y}_{i\cdot\cdot} - \bar{y}_{\cdot j\cdot} + \bar{y}_{\cdots}$. The SS would be $\text{SS}(AB) = K \sum_{i=1}^{I} \sum_{j=1}^{J} (\bar{y}_{ij\cdot} - \bar{y}_{i\cdot\cdot} - \bar{y}_{\cdot j\cdot} + \bar{y}_{\cdots})^2$, the sum of the natural estimate squared over all of the data. The df for this interaction is the number of averages estimated minus one for the grand average minus the df for A and the df for B: $(IJ - 1) - (I - 1) - (J - 1) = IJ - I - J + 1 = (I - 1)(J - 1)$, the product of the df for the two main effects making up the interaction. This last form is particularly easy to remember.

A three factor interaction ABC_{ijk} is the differential effect of the (ijk)th combination minus all three main effects and all three two factor interactions. The result is $(\bar{y}_{ijk\cdot} - \bar{y}_{\cdots\cdot}) - (\bar{y}_{i\cdots} - \bar{y}_{\cdots\cdot}) - (\bar{y}_{\cdot j\cdot\cdot} - \bar{y}_{\cdots\cdot}) - (\bar{y}_{\cdot\cdot k\cdot} - \bar{y}_{\cdots\cdot}) - (\bar{y}_{ij\cdot\cdot} - \bar{y}_{i\cdots} - \bar{y}_{\cdot j\cdot\cdot} + \bar{y}_{\cdots\cdot}) - (\bar{y}_{i\cdot k\cdot} - \bar{y}_{i\cdots} - \bar{y}_{\cdot\cdot k\cdot} + \bar{y}_{\cdots\cdot}) - (\bar{y}_{\cdot jk\cdot} - \bar{y}_{\cdot j\cdot\cdot} - \bar{y}_{\cdot\cdot k\cdot} + \bar{y}_{\cdots\cdot}) = \bar{y}_{ijk\cdot} - \bar{y}_{ij\cdot\cdot} - \bar{y}_{i\cdot k\cdot} - \bar{y}_{\cdot jk\cdot} + \bar{y}_{i\cdots} + \bar{y}_{\cdot j\cdot\cdot} + \bar{y}_{\cdot\cdot k\cdot} - \bar{y}_{\cdots\cdot}$. The SS for ABC, $\text{SS}(ABC)$, is the square of this term summed over all the data and the df for this term, $\text{df}(ABC)$ is $(IJK - 1) - (I - 1) - (J - 1) - (K - 1) - (I - 1)(J - 1) - (I - 1)(K - 1) - (J - 1)(K - 1) = IJK - IJ - IK - JK + I + J + K - 1 = (I - 1)(J - 1)(K - 1)$, the product of the df for the three main effects.

We immediately generalize to the following set of rules, applicable to all the *terms* (main effects and interactions) in balanced designs. That is, the following rules apply to all experiments having all combinations of all levels of the factors repeated the same number of times.

RULE FOR df IN BALANCED DESIGNS

- The df for any term is given by the product of the levels of each bracketed index times the product of (levels minus one) for each index not bracketed.

RULE FOR SS IN BALANCED DESIGNS

- Write down the formula for df as given immediately above except replace capital letters with lower case letters.
- Expand this product.

- Each combination of subscripts in the df expansion corresponds to a mean in the SS. The sign preceding the mean is the same as the sign preceding the subscripts (*i.e.*, + or −). The corresponding mean term will have the same subscripts with all missing subscripts replaced by dots. A value of one in the expansion corresponds to the grand mean, having all subscripts replaced with dots.
- The SS is the square of the linear combination of means summed over all of the data. For simplicity, the sum over a subscript not appearing in the linear combination is replaced by its corresponding capital letter.

RULE FOR MS IN BALANCED DESIGNS

- The MS for a given term in the model is simply the SS divided by the df.

We illustrate the rules with the three factor design given in (3.2.3). The factors are labeled A_i, B_j, C_k, and the repeats are labeled with the index ℓ. The df for A_i, df(A), is simply $I-1$ since the A_i term has subscript i only and i is not bracketed. The corresponding SS(A) is obtained by changing $I-1$ to $i-1$, indicating that the term $(\bar{y}_{i\cdots}-\bar{y}_{\cdots\cdot})^2$ would be summed over all the data. The result is $\sum_{i=1}^{I}\sum_{j=1}^{J}\sum_{k=1}^{K}\sum_{\ell=1}^{L}(\bar{y}_{i\cdots}-\bar{y}_{\cdots\cdot})^2$ which simplifies to $\text{SS}(A) = JKL\sum_{i=1}^{I}(\bar{y}_{i\cdots}-\bar{y}_{\cdots\cdot})^2$. MS($A$) is simply SS($A$)/df($A$), yielding $\text{MS}(A) = JKL\sum_{i=1}^{I}(\bar{y}_{i\cdots}-\bar{y}_{\cdots\cdot})^2/(I-1)$.

The df for the two factor interaction AC_{ik} is $\text{df}(AC) = (I-1)(K-1)$ since the subscripts i and k are associated with AC_{ik}. Replacing the capital letters in the df term with lower case letters and expanding the product yields $ik-i-k+1$. The corresponding SS(AC) would then be $JL\sum_{i=1}^{I}\sum_{k=1}^{K}(\bar{y}_{i\cdot k\cdot}-\bar{y}_{i\cdots}-\bar{y}_{\cdot\cdot k\cdot}+\bar{y}_{\cdots\cdot})^2$ with the mean squares given by $\text{MS}(AC) = JL\sum_{i=1}^{I}\sum_{k=1}^{K}(\bar{y}_{i\cdot k\cdot}-\bar{y}_{i\cdots}-\bar{y}_{\cdot\cdot k\cdot}+\bar{y}_{\cdots\cdot})^2/(I-1)(K-1)$.

The df for the error term $\varepsilon_{\ell(ijk)}$ is $\text{df}(\varepsilon) = IJK(L-1)$ since the i,j, and k indices are bracketed but the ℓ index is not. Replacing the capital letters with lower case letters and expanding yields $ijk\ell - ijk$ so the corresponding $\text{SS}(\varepsilon) = \sum_{i=1}^{I}\sum_{j=1}^{J}\sum_{k=1}^{K}\sum_{\ell=1}^{L}(\bar{y}_{ijk\ell}-\bar{y}_{ijk\cdot})^2$ and MS(ε) is SS(ε) divided by df(ε). Continuing this process for all terms in the model gives Table 3.3.1.

This form of the ANOVA table should be used for accuracy in machine calculations. (Faster machine routines exist but will not be discussed in this book.) However, it is not particularly convenient for hand calculations. Using the T.. notation defined in Section **2.3** (particularly (2.3.1)–(2.3.3)) in place of the $\bar{y}_{\cdot\cdot}$ notation is much easier for hand calculations. To derive the formulae involving the T.. notation, expand the

TABLE 3.3.1

ANOVA FOR THREE FACTOR DESIGNS

Source	df	SS	MS
A_i	$I-1$	$JKL\sum_{i=1}^{I}(\bar{y}_{i\cdots}-\bar{y}_{\cdots\cdot})^2$	$\text{SS}(A)/(I-1)$
B_j	$J-1$	$IKL\sum_{j=1}^{J}(\bar{y}_{\cdot j\cdot\cdot}-\bar{y}_{\cdots\cdot})^2$	$\text{SS}(B)/(J-1)$
AB_{ij}	$(I-1)(J-1)$	$KL\sum_{i=1}^{I}\sum_{j=1}^{J}(\bar{y}_{ij\cdot\cdot}-\bar{y}_{i\cdots}-\bar{y}_{\cdot j\cdot\cdot}+\bar{y}_{\cdots\cdot})^2$	$\text{SS}(AB)/(I-1)(J-1)$
C_k	$K-1$	$IJL\sum_{k=1}^{K}(\bar{y}_{\cdot\cdot k\cdot}-\bar{y}_{\cdots\cdot})^2$	$\text{SS}(C)/(K-1)$
AC_{ik}	$(I-1)(K-1)$	$JL\sum_{i=1}^{I}\sum_{k=1}^{K}(\bar{y}_{i\cdot k\cdot}-\bar{y}_{i\cdots}-\bar{y}_{\cdot\cdot k\cdot}+\bar{y}_{\cdots\cdot})^2$	$\text{SS}(AC)/(I-1)(K-1)$
BC_{jk}	$(J-1)(K-1)$	$IL\sum_{j=1}^{J}\sum_{k=1}^{K}(\bar{y}_{\cdot jk\cdot}-\bar{y}_{\cdot j\cdot\cdot}-\bar{y}_{\cdot\cdot k\cdot}+\bar{y}_{\cdots\cdot})^2$	$\text{SS}(BC)/(J-1)(K-1)$
ABC_{ijk}	$(I-1)(J-1)(K-1)$	$L\sum_{i=1}^{I}\sum_{j=1}^{J}\sum_{k=1}^{K}(\bar{y}_{ijk\cdot}-\bar{y}_{ij\cdot\cdot}-\bar{y}_{i\cdot k\cdot}-\bar{y}_{\cdot jk\cdot}+\bar{y}_{i\cdots}+\bar{y}_{\cdot j\cdot\cdot}+\bar{y}_{\cdot\cdot k\cdot}-\bar{y}_{\cdots\cdot})^2$	$\text{SS}(ABC)/\text{df}(ABC)$
$\varepsilon_{\ell(ijk)}$	$IJK(L-1)$	$\sum_{i=1}^{I}\sum_{j=1}^{J}\sum_{k=1}^{K}\sum_{\ell=1}^{L}(y_{ijk\ell}-\bar{y}_{ijk\cdot})^2$	$\text{SS}(\varepsilon)/IJK(L-1)$
Total	$IJKL-1$	$\sum_{i=1}^{I}\sum_{j=1}^{J}\sum_{k=1}^{K}\sum_{\ell=1}^{L}(y_{ijk\ell}-\bar{y}_{\cdots\cdot})^2$	

squared term in each SS term and algebraically simplify. We are lead to the following general rule, with the previous example illustrated in Table 3.3.2.

RULE FOR HAND CALCULATING SS IN BALANCED DESIGNS

- Write down the formula for df except replace capital letters with lower case letters.
- Expand this product.
- Each combination of subscripts in the df expansion corresponds to a summation in the SS. Each term in the SS is a sum of totals squared (T^2) with the sum over the corresponding indices and dots on the T for all indices not in the set of corresponding indices. The sign in front of the summation (+ or −) is the same as the sign in front of the combination of indices and each T^2 term is divided by the number of observations making up the total (= the missing indices with capital letters replacing the small letters).
- Corresponding to the value of one in the expansion, we have T^2, with dots for all subscripts, divided by the total number of observations. This is commonly denoted CT for the correction term.

3.4 COMPUTATION OF EMS

In this section we expand the ANOVA table of the previous section with the addition of one extra column, the expected mean squares, EMS, column. The EMS column is exactly as indicated, the expected value or average of the mean square column. We will illustrate the calculations with an example dealing with a two factor experiment. Even in this simple case, the algebra becomes extremely tedious. Therefore, we give two general procedures for deriving the EMS. The first is easy to remember and apply. The second requires less paper. In practice, EMS are almost never derived algebraically so even a statistician could jump ahead to the algorithms after reading the next two paragraphs.

The EMS, along with the normality assumption and the orthogonality built into the computations, determines the proper F tests indicated by the F column. Note that the formulae for df, SS, MS, EMS, and the F tests can be determined prior to the collection of any data. Therefore, we know exactly what to expect even before the experiment is run. If the information is insufficient or more information on the effect of some some factor or interaction is desired, we simply redesign the experiment. Following this procedure avoids collecting worthless or near worthless data.

TABLE 3.3.2

ANOVA FOR THREE FACTORS – HAND CALCULATIONS

Source	df	SS	MS
A_i	$I-1$	$\sum_{i=1}^{I} \frac{T_{i\cdots}^2}{JKL} - CT$	$SS(A)/df(A)$
B_j	$J-1$	$\sum_{j=1}^{J} \frac{T_{\cdot j\cdot\cdot}^2}{IKL} - CT$	$SS(B)/df(B)$
AB_{ij}	$(I-1)(J-1)$	$\sum_{i=1}^{I}\sum_{j=1}^{J} \frac{T_{ij\cdot\cdot}^2}{KL} - \sum_{i=1}^{I} \frac{T_{i\cdots}^2}{JKL} - \sum_{j=1}^{J} \frac{T_{\cdot j\cdot\cdot}^2}{IKL} + CT$	$SS(AB)/df(AB)$
C_k	$K-1$	$\sum_{k=1}^{K} \frac{T_{\cdot\cdot k\cdot}^2}{IJL} - CT$	$SS(C)/df(C)$
AC_{ik}	$(I-1)(K-1)$	$\sum_{i=1}^{I}\sum_{k=1}^{K} \frac{T_{i\cdot k\cdot}^2}{JL} - \sum_{i=1}^{I} \frac{T_{i\cdots}^2}{JKL} - \sum_{k=1}^{K} \frac{T_{\cdot\cdot k\cdot}^2}{IJL} + CT$	$SS(BC)/df(BC)$
ABC_{ijk}	$(I-1)(J-1)(K-1)$	$\sum_{i=1}^{I}\sum_{j=1}^{J}\sum_{k=1}^{K} \frac{T_{ijk\cdot}^2}{L} - \sum_{i=1}^{I}\sum_{j=1}^{J} \frac{T_{ij\cdot\cdot}^2}{KL} - \sum_{i=1}^{I}\sum_{k=1}^{K} \frac{T_{i\cdot k\cdot}^2}{JL} - \sum_{j=1}^{J}\sum_{k=1}^{K} \frac{T_{\cdot jk\cdot}^2}{IL} + \sum_{i=1}^{I} \frac{T_{i\cdots}^2}{JKL} + \sum_{j=1}^{J} \frac{T_{\cdot j\cdot\cdot}^2}{IKL} + \sum_{k=1}^{K} \frac{T_{\cdot\cdot k\cdot}^2}{IJL} - CT$	$SS(ABC)/df(ABC)$
$\varepsilon_{\ell(ijk)}$	$IJK(L-1)$	$\sum_{i=1}^{I}\sum_{j=1}^{J}\sum_{k=1}^{K}\sum_{\ell=1}^{L} T_{ijk\ell}^2 - \sum_{i=1}^{I}\sum_{j=1}^{J}\sum_{k=1}^{K} \frac{T_{ijk\cdot}^2}{L}$	$SS(\varepsilon)/df(\varepsilon)$
Total	$IJKL-1$	$\sum_{i=1}^{I}\sum_{j=1}^{J}\sum_{k=1}^{K}\sum_{\ell=1}^{L} T_{ijk\ell}^2 - CT$	$CT = T_{\cdots\cdot}^2/IJKL$

Even if the last section was neglected, this section should *not* be skipped (except possibly the "light bulb" paragraphs). Most software packages do not correctly calculate the F tests. Therefore, it is essential that the user be able to compute the EMS to determine the proper tests. Then, either hand calculate or instruct the computer package to calculate the proper F tests.

To illustrate the tedium of the algebraic derivation of EMS, consider the two factor model given by (3.2.2) with factor A_i fixed and B_j random. We assume $\sum_{i=1}^{I} A_i = 0$, B_j is *iid* Normal(0, σ_B^2), AB_{ij} is Normal(0, σ_{AB}^2) with AB_{ij} independent for different j and independent of B_j but $\sum_{i=1}^{I} AB_{ij} = 0$ for all j, and $\varepsilon_{k(ij)}$ is *iid* Normal(0, σ^2) where $\varepsilon_{k(ij)}$ is independent of B_j and AB_{ij}. Note in particular that $\sum_{i=1}^{I} AB_{ij} = 0$ implies $\mathrm{E}[\sum_{i=1}^{I} AB_{ij}^2] = (I-1)\sigma_{AB}^2$, where E[·] is the expected value function. The expected value of MS(A) is given by

$$\mathrm{EMS}(A) = \mathrm{E}[\mathrm{MS}(A)] = \mathrm{E}\Big[\frac{JK}{I-1}\sum_{i=1}^{I}(\bar{y}_{i\cdot\cdot} - \bar{y}_{\cdot\cdot\cdot})^2\Big]$$

$$= \frac{JK}{I-1}\mathrm{E}\Big[\sum_{i=1}^{I}\Big\{\sum_{j=1}^{J}\sum_{k=1}^{K}\frac{y_{ijk}}{JK} - \sum_{i'=1}^{I}\sum_{j'=1}^{J}\sum_{k'=1}^{K}\frac{y_{i'j'k'}}{IJK}\Big\}^2\Big]$$

$$= \frac{JK}{I-1}\mathrm{E}\Big[\sum_{i=1}^{I}\Big\{\sum_{j=1}^{J}\sum_{k=1}^{K}\frac{\mu + A_i + B_j + AB_{ij} + \varepsilon_{k(ij)}}{JK}$$

$$- \sum_{i'=1}^{I}\sum_{j'=1}^{J}\sum_{k'=1}^{K}\frac{\mu + A_{i'} + B_{j'} + AB_{i'j'} + \varepsilon_{k'(i'j')}}{IJK}\Big\}^2\Big]$$

$$= \frac{JK}{I-1}\mathrm{E}\Big[\sum_{i=1}^{I}\Big\{A_i + \sum_{j=1}^{J}\frac{AB_{ij}}{J} + \sum_{j=1}^{J}\sum_{k=1}^{K}\frac{\varepsilon_{k(ij)}}{JK} - \sum_{i'=1}^{I}\sum_{j'=1}^{J}\sum_{k'=1}^{K}\frac{\varepsilon_{k'(i'j')}}{IJK}\Big\}^2\Big]$$

$$= \frac{JK}{I-1}\Big(\sum_{i=1}^{I}A_i^2 + \mathrm{E}\Big[\sum_{i=1}^{I}\Big\{\sum_{j=1}^{J}\frac{AB_{ij}}{J}\Big\}^2\Big] + \mathrm{E}\Big[\sum_{i=1}^{I}\Big\{\sum_{j=1}^{J}\sum_{k=1}^{K}\frac{\varepsilon_{k(ij)}}{JK}\Big\}^2\Big]$$

$$+ I\mathrm{E}\Big[\sum_{i'=1}^{I}\sum_{j'=1}^{J}\sum_{k'=1}^{K}\frac{\varepsilon_{k'(i'j')}}{IJK}\Big]^2$$

$$- 2\mathrm{E}\Big[\sum_{i=1}^{I}\Big\{\sum_{j=1}^{J}\sum_{k=1}^{K}\frac{\varepsilon_{k(ij)}}{JK}\Big\}\Big\{\sum_{i'=1}^{I}\sum_{j'=1}^{J}\sum_{k'=1}^{K}\frac{\varepsilon_{k'(i'j')}}{IJK}\Big\}\Big]\Big)$$

$$= \frac{JK}{I-1}\Big(\sum_{i=1}^{I}A_i^2 + \mathrm{E}\Big[\sum_{j=1}^{J}\sum_{i=1}^{I}\frac{AB_{ij}^2}{J^2}\Big] + \frac{IJK\sigma^2}{J^2K^2} + \frac{I^2JK\sigma^2}{I^2J^2K^2} - \frac{2IJK\sigma^2}{IJ^2K^2}\Big)$$

$$= JK\Phi(A) + K\sigma_{AB}^2 + \sigma^2 ,$$

with similar derivations for EMS(B), EMS(AB), and EMS(ε). If, instead of assuming B_j random, we assumed B_j fixed, $\sum_{i=1}^{I}\sum_{j=1}^{J} AB_{ij}$ in line 4 above would equal 0 and the resulting EMS would be $JK\Phi(A) + \sigma^2$. If, instead of assuming A_i fixed, we also assumed A_i random, line 4 above would contain additional terms involving $\sum_{i'=1}^{I} A_{i'}$ and $\sum_{i'=1}^{I}\sum_{j'=1}^{J} AB_{i'j'}$ and the final result would be $JK\sigma_A^2 + K\sigma_{AB}^2 + \sigma^2$.

Similar calculations show that all of the expressions for the various MS are orthogonal. In other words, even though the formula for the MS are the same whether the factors are fixed or random, the formulae for the EMS will differ depending on which factors are fixed and which are random.

While the algebraic calculations given above are informative, they are far too cumbersome for everyday use. Instead, we give two algorithms for the derivation of EMS. The first algorithm, developed by Lorenzen (1977), is fast and easy to learn and has the advantage of deriving appropriate F tests prior to the actual derivation of EMS. Its pictorial method may aid in developing approximate and conservative tests. The second algorithm, developed by Bennett and Franklin (1954), is also fast but a little more complicated to learn. It has considerable space advantage over the first algorithm. This algorithm is recommended for those who frequently design experiments.

ALGORITHM 1 FOR DERIVATION OF EMS IN BALANCED DESIGNS

1 Write the model across the top of the page, including every term and its associated subscripts. Write each term having nonzero df down the left side of the page. Assuming bracketed subscripts are random ((i) is random even if i is fixed), label each term in the model F if all subscripts are fixed, R if all subscripts are random, and M if some are random and some are fixed.

2 Go across each row placing a $\star$ in every column where the subscripts at the left are contained in the subscripts at the top. For this step, ignore all brackets.

3 In columns marked F, erase all $\star$'s and place a ■ where the term at the left matches the term at the top.

4 In columns marked M, go down one $\star$ at a time comparing the subscripts at the top with those at the left. If the top contains any fixed subscript not contained at the left, erase the $\star$. Otherwise leave it. Remember that bracketed subscripts are considered random. In other words, start with the subscripts at the top and eliminate the subscripts at the left. If there are any fixed indices remaining, then erase the $\star$.

5 Above each term at the top, place the integer resulting from the division of the total experiment size by all levels of the subscripts contained in that term. These are the column coefficients.

6 To write the EMS for any given term, read the corresponding row from right to left using the column coefficients, replacing any ⋆ with the corresponding σ^2 effect, and replacing any ■ with the corresponding Φ effect.

We illustrate the procedure with the three factor model (3.2.3) written as $y_{ijk\ell} = \mu + A_i + B_j + AB_{ij} + C_k + AC_{ik} + BC_{jk} + ABC_{ijk} + \varepsilon_{\ell(ijk)}$ where we will let $I = 2$, $J = 4$, $K = 3$, and $L = 2$. The experiment size is $2 \times 4 \times 3 \times 2 = 48$. We will take A_i and C_k as fixed and B_j as random. $\varepsilon_{\ell(ijk)}$ will always be random. The steps of the algorithm are demonstrated in Tables 3.4.1 through 3.4.4 on this and the following several pages.

TABLE 3.4.1

EMS FOR THREE FACTORS, A_i, C_k FIXED, B_j RANDOM—STEPS 1–2

	F	R	M	F	F	M	M	R
	A_i	B_j	AB_{ij}	C_k	AC_{ik}	BC_{jk}	ABC_{ijk}	$\varepsilon_{\ell(ijk)}$
A_i	⋆		⋆		⋆		⋆	⋆
B_j		⋆	⋆			⋆	⋆	⋆
AB_{ij}			⋆				⋆	⋆
C_k				⋆	⋆	⋆	⋆	⋆
AC_{ik}					⋆		⋆	⋆
BC_{jk}						⋆	⋆	⋆
ABC_{ijk}							⋆	⋆
$\varepsilon_{\ell(ijk)}$								⋆

At this point, the reader should stop and review Tables 3.4.1 through 3.4.4, making sure the six steps in the algorithm are understood and following the progress from table to table. Of course, there is no need to use three separate tables. Simply use the same table, erasing ⋆'s and changing ⋆'s to ■'s as necessary. As you gain experience, it will be easy to combine steps two and three immediately by not including ⋆'s under any column marked F and placing a ■ directly on the diagonal. The algorithm

TABLE 3.4.2

EMS FOR THREE FACTORS, A_i, C_k FIXED, B_j RANDOM—STEPS 1–3

	F	R	M	F	F	M	M	R
	A_i	B_j	AB_{ij}	C_k	AC_{ik}	BC_{jk}	ABC_{ijk}	$\varepsilon_{\ell(ijk)}$
A_i	■		★				★	★
B_j		★	★			★	★	★
AB_{ij}			★				★	★
C_k				■		★	★	★
AC_{ik}					■		★	★
BC_{jk}						★	★	★
ABC_{ijk}							★	★
$\varepsilon_{\ell(ijk)}$								★

TABLE 3.4.3

EMS FOR THREE FACTORS, A_i, C_k FIXED, B_j RANDOM—STEPS 1–5

	24	12	6	16	8	4	2	1
	F	R	M	F	F	M	M	R
	A_i	B_j	AB_{ij}	C_k	AC_{ik}	BC_{jk}	ABC_{ijk}	$\varepsilon_{\ell(ijk)}$
A_i	■		★					★
B_j		★						★
AB_{ij}			★					★
C_k				■		★		★
AC_{ik}					■		★	★
BC_{jk}						★		★
ABC_{ijk}							★	★
$\varepsilon_{\ell(ijk)}$								★

yields the same EMS as the computationally intense algebraic method since the summation over any fixed index equals zero. This summation to zero is equivalent to erasing a $\star$.

TABLE 3.4.4

ANOVA FOR THREE FACTORS, A_i, C_k FIXED, B_j RANDOM—STEP 6

Source	df	EMS
A_i	1	$\sigma^2 + 6\sigma^2_{AB} + 24\Phi(A)$
B_j	3	$\sigma^2 + 12\sigma^2_B$
AB_{ij}	3	$\sigma^2 + 6\sigma^2_{AB}$
C_k	2	$\sigma^2 + 4\sigma^2_{BC} + 16\Phi(C)$
AC_{ik}	2	$\sigma^2 + 2\sigma^2_{ABC} + 8\Phi(AC)$
BC_{jk}	6	$\sigma^2 + 4\sigma^2_{BC}$
ABC_{ijk}	6	$\sigma^2 + 2\sigma^2_{ABC}$
$\varepsilon_{\ell(ijk)}$	24	σ^2

The second algorithm requires much less paper and utilizes the approach of Bennett and Franklin (1954).

ALGORITHM 2 FOR DERIVATION OF EMS IN BALANCED DESIGNS

1 Write the terms of the model with associated subscripts down the left side of the page. Across the top write the single letter subscripts (i,j,k, *etc.*). Above each subscript place either an F or an R if the factor associated with that subscript is fixed or random. Above that place the number of levels associated with that subscript (I,J,K, *etc.*).

2 Enter a 1 in every slot where the subscript at the top is contained within brackets in the term at the left.

3 Enter a 0 in every slot where the subscript at the top is fixed and also contained in the term at the left. Enter a 1 in every slot where the subscript at the top is random and also contained in the term at the left.

4 Fill in the remaining slots with the number of levels at the top of each column.

5 To compute the EMS for a given term having df > 0, start at the bottom and work up. Only consider terms whose indices include all the indices in the term whose EMS you are deriving. Compute the coefficient of this term by covering the columns corresponding to the indices in the term whose EMS you are deriving and multiplying the values in the remaining columns. If there is a 0 column that is not covered, this term need not be written in the EMS. A factor

is considered fixed and denoted with a Φ only if all of its indices are fixed. Otherwise it is considered random and denoted by the appropriate σ^2 term.

Note that this algorithm can be used to compute EMS for all terms in the model, including those that have zero df. A term that has zero df has no mean squares so, of course, cannot have expected mean squares. For this reason, we will not compute EMS for terms having zero df even though such terms must be left in the algorithm to make the EMS of the other terms come out right.

We illustrate Algorithm 2 with the three factor model (3.2.3) having $I = 2$, $J = 4$, $K = 3$, and $L = 2$ levels for A_i, B_j, C_k, and the number of repeats, respectively. This time we take A_i as fixed, B_j and C_k as random. The results are given in Table 3.4.5.

TABLE 3.4.5

EMS CALCULATIONS, ALGORITHM 2, A_i FIXED, B_j,C_k RANDOM.

		2	4	3	2	
		F	R	R	R	
Source	df	i	j	k	ℓ	EMS
A_i	1	0	4	3	2	$\sigma^2 + 2\sigma^2_{ABC} + 8\sigma^2_{AC} + 6\sigma^2_{AB} + 24\Phi(A)$
B_j	3	2	1	3	2	$\sigma^2 + 4\sigma^2_{BC} + 12\sigma^2_B$
AB_{ij}	3	0	1	3	2	$\sigma^2 + 2\sigma^2_{ABC} + 6\sigma^2_{AB}$
C_k	2	2	4	1	2	$\sigma^2 + 4\sigma^2_{BC} + 16\sigma^2_C$
AC_{ik}	2	0	4	1	2	$\sigma^2 + 2\sigma^2_{ABC} + 8\sigma^2_{AC}$
BC_{jk}	6	2	1	1	2	$\sigma^2 + 4\sigma^2_{BC}$
ABC_{ijk}	6	0	1	1	2	$\sigma^2 + 2\sigma^2_{ABC}$
$\varepsilon_{\ell(ijk)}$	24	1	1	1	1	σ^2

At this point, the reader should stop and make sure the entire procedure is understood by reproducing Table 3.4.5. As an aid in reproducing this table, consider the AC_{ik} row. We cover up columns i and k. Starting from the bottom, we consider only terms with subscripts containing i and k. The first is $\varepsilon_{\ell(ijk)}$ which yields the coefficient 1 and puts the σ^2 term in EMS(AC). Next we consider ABC_{ijk} which has coefficient 2 and puts the $2\sigma^2_{ABC}$ term in EMS(AC). We do not consider BC_{jk} since it does not contain the indices i and k. The AC_{ik} row itself contributes $8\sigma^2_{AC}$ and we have completed the calculations for EMS(AC) since none of the other terms contain both of the indices i and k.

This algorithm takes much less space than the first algorithm and directly yields the EMS column. However, it seems to lead to mistakes more often than the first algorithm. Algorithm 2 has an advantage over Algorithm 1 because it can be extended to factors that are neither fixed nor random. Suppose there are Q levels of interest and we randomly select and test q of them. We then enter the finite population correction term 1−q/Q in every slot where the indices at the left contain the index at the top and the algorithm continues as before. Note that, for a fixed factor, q = Q and the quantity 1−q/Q = 0, while for a random factor, Q = ∞ and the quantity 1−q/Q = 1, agreeing with the rules stated in Algorithm 2.

3.5 TESTS OF SIGNIFICANCE

Consider the row corresponding to the fixed factor A_i in the three factor design given in Table 3.4.4. Under the null hypothesis that A_i has no effect on the outcome of the experiment, $\Phi(A)$ will be 0 and EMS(A) $= \sigma^2 + 6\sigma^2_{AB}$. Scanning the other EMS values, we also find that EMS(AB) $= \sigma^2 + 6\sigma^2_{AB}$. Under the null hypothesis that $\Phi(A) = 0$, both quantities are distributed as χ^2 random variates and can be shown to be independent. The ratio MS(A)/MS(AB) will then follow an F distribution with 1 df for the numerator and 3 df for the denominator and should take a value near 1. If the alternate hypothesis is true and A_i has an effect on the response variable, $24\Phi(A)$ will be positive, and the ratio MS(A)/MS(AB) will be large. Therefore, we reject the null hypothesis of no effect due to A_i and declare that A is significant if MS(A)/MS(AB) exceeds the critical F given in Appendix 2.

If we consider the random factor B_j in Table 3.4.4, the null hypothesis that B_j has no effect is equivalent to $\sigma^2_B = 0$. Under such a null hypothesis, both EMS(B) and EMS(ε) will have expected value σ^2, and the ratio MS(B)/MS(ε) will have an F distribution with 3 and 24 df for numerator and denominator respectively. Under the alternative hypothesis that B_j has an effect, $12\sigma^2_B$ will be positive so we again reject the null hypothesis and conclude that B_j has a significant effect if F_{calc} = MS(B)/MS(ε) exceeds F_{crit} having 3 and 24 df for the numerator and denominator respectively, where F_{crit} can be found in Appendix 2.

3.5.1 Direct Tests

These examples lead to the following general rules for F tests.

RULE FOR DIRECT F TESTS IN BALANCED DESIGNS

- A fixed term Y in the model is said to have an effect if Φ(Y) is positive. A random or mixed term Y has an effect if σ_Y^2 is positive. A direct test for Y exists if there is a second term X in the model whose EMS is identical to EMS(Y) with either σ_Y^2 or Φ(Y) removed. That is, X directly tests Y if EMS(X) = EMS(Y) when either $\sigma_Y^2 = 0$ or $\Phi(Y) = 0$, as the case may be.
- The calculated F value is MS(Y) divided by MS(X) with df(Y) degrees of freedom for the numerator and df(X) degrees of freedom for the the denominator.
- We reject the null hypothesis of no effect and conclude that Y has a significant effect if the calculated F value exceeds the critical F value found in Appendix 2.

When using Algorithm 1 for the computation of EMS, all direct tests can be found upon completion of steps one through four. The denominator X directly testing Y is that term containing all of the off-diagonal stars in the Y row (if such an X row exists). This can be determined prior to the derivation of the EMS, saving some time and effort.

The rule for direct F tests is usually implemented by augmenting the ANOVA table with columns headed F_{calc} and F_{crit}. We illustrate with the three factor design given in (3.2.3) and Table 3.4.4. The result is given in Tables 3.5.1 and 3.5.2.

Again, the reader should stop and make sure the procedure is understood by reproducing Table 3.5.1. In applying the rule for the third row (the AB_{ij} row), Y denotes AB_{ij} and X denotes the error term $\varepsilon_{\ell(ijk)}$.

An alternate method of indicating direct tests is with arrows. The base of the arrow indicates the denominator of the F statistic and the tip of the arrow indicates the numerator. This is particularly useful when using Algorithm 1 for determining EMS as the tests can be determined after completion of step four. The use of arrows also saves space by eliminating a column from the ANOVA table. The results are given in Table 3.5.2, equivalent to Table 3.5.1.

3.5.2 Approximate Tests

Unlike Table 3.5.2, it is not always possible to determine direct tests on all of the terms of interest. In such cases we can often form approximate tests known as F′ tests by combining MS terms to form the proper

TABLE 3.5.1

ANOVA FOR THREE FACTORS, A_i, C_k FIXED, B_j RANDOM

Source	df	EMS	F_{calc}	$F_{crit}(.05)$
A_i	1	$\sigma^2 + 6\sigma^2_{AB} + 24\Phi(A)$	$MS(A)/MS(AB)$	10.13
B_j	3	$\sigma^2 + 12\sigma^2_B$	$MS(B)/MS(\varepsilon)$	3.01
AB_{ij}	3	$\sigma^2 + 6\sigma^2_{AB}$	$MS(AB)/MS(\varepsilon)$	3.01
C_k	2	$\sigma^2 + 4\sigma^2_{BC} + 16\Phi(C)$	$MS(C)/MS(BC)$	5.14
AC_{ik}	2	$\sigma^2 + 2\sigma^2_{ABC} + 8\Phi(AC)$	$MS(AC)/MS(ABC)$	5.14
BC_{jk}	6	$\sigma^2 + 4\sigma^2_{BC}$	$MS(BC)/MS(\varepsilon)$	2.51
ABC_{ijk}	6	$\sigma^2 + 2\sigma^2_{ABC}$	$MS(ABC)/MS(\varepsilon)$	2.51
$\varepsilon_{\ell(ijk)}$	24	σ^2		

TABLE 3.5.2

ANOVA DISPLAY FOR THREE FACTORS, A_i, C_k FIXED, B_j RANDOM

	F	R	M	F	F	M	M	R
	A_i	B_j	AB_{ij}	C_k	AC_{ik}	BC_{jk}	ABC_{ijk}	$\varepsilon_{\ell(ijk)}$
A_i	■		★					★
B_j		★						★
AB_{ij}			★					★
C_k				■		★		★
AC_{ik}					■		★	★
BC_{jk}						★		★
ABC_{ijk}							★	★
$\varepsilon_{\ell(ijk)}$								★

denominator for the test. An approximation developed by Satterthwaite (1946) must be used to obtain the degrees of freedom and critical values. The general rule follows immediately.

RULE FOR APPROXIMATE F′ TESTS

- When there is no direct test on a term Y in the model because no other term X has the proper EMS, a denominator can often be formed using a linear combination of various EMS values, say $\sum_{i=1}^{k} a_i \text{EMS}_i$.

- Using the MS values in place of the EMS values produces an approximate F′ test. The denominator will be denoted $\sum_{i=1}^{k} a_i \text{MS}_i$.
- The df associated with this denominator is approximated by

$$\text{df} = \frac{M^2}{\frac{(a_1 \text{MS}_1)^2}{\text{df}_1} + \frac{(a_2 \text{MS}_2)^2}{\text{df}_2} + \cdots + \frac{(a_k \text{MS}_k)^2}{\text{df}_k}}, \qquad M = \sum_{i=1}^{k} a_i \text{MS}_i$$

- We reject the null hypothesis of no effect and conclude Y has a significant effect whenever F'_{calc} exceeds the F_{crit} value found in or interpolated from Appendix 2.

The approximate df for the denominator is seldom an integer. The most accurate technique is to use interpolation on the values in Appendix 2. If the df approximation comes out less than one, this is an indication that *pooling* should have been performed first. *Pooling* is discussed in Section **3.6**.

As an example, consider the ANOVA in Table 3.4.5, a three factor design with A_i fixed, B_j and C_k random. Using either of the two algorithms yields Table 3.5.3.

As indicated by either the lack of arrows, or the form of the EMS given on the right, there is no direct test for A_i. A direct test would have to contain stars in the AB_{ij}, AC_{ik}, ABC_{ijk}, and $\varepsilon_{\ell(ijk)}$ columns or the equivalent σ^2 terms in the EMS. No such row or EMS exists. However, if we add EMS(AB) and EMS(AC) and then subtract EMS(ABC), the result is the proper denominator for testing A_i. Pictorially, we add the stars in the AB_{ij} and AC_{ik} rows, getting two stars in the ABC_{ijk} and $\varepsilon_{\ell(ijk)}$ columns. After subtracting the ABC_{ijk} row, we get one star in the AB_{ij}, AC_{ik}, ABC_{ijk}, and $\varepsilon_{\ell(ijk)}$ columns, as was desired. Thus, MS(AB) + MS(AC) − MS(ABC) serves as the denominator in the F′ test for significance of A_i. The df for the denominator depends on the data, particularly on the values of MS(AB), MS(AC), and MS(ABC): $\text{df} \approx [\text{MS}(AB) + \text{MS}(AC) - \text{MS}(ABC)]^2/[\text{MS}(AB)^2/3 + \text{MS}(AC)^2/2 + \text{MS}(ABC)^2/6]$. This is typical of the linear combinations used to form the F′ test, the a_i are either +1 or −1.

The resulting ANOVA table, including the F′ test is given in Table 3.5.4.

An alternate approximate test on A_i in Table 3.4.5 is specified by $\tilde{F} = [\text{MS}(A)+\text{MS}(ABC)]/[\text{MS}(AB)+\text{MS}(AC)]$. $\tilde{F}$ has an advantage over

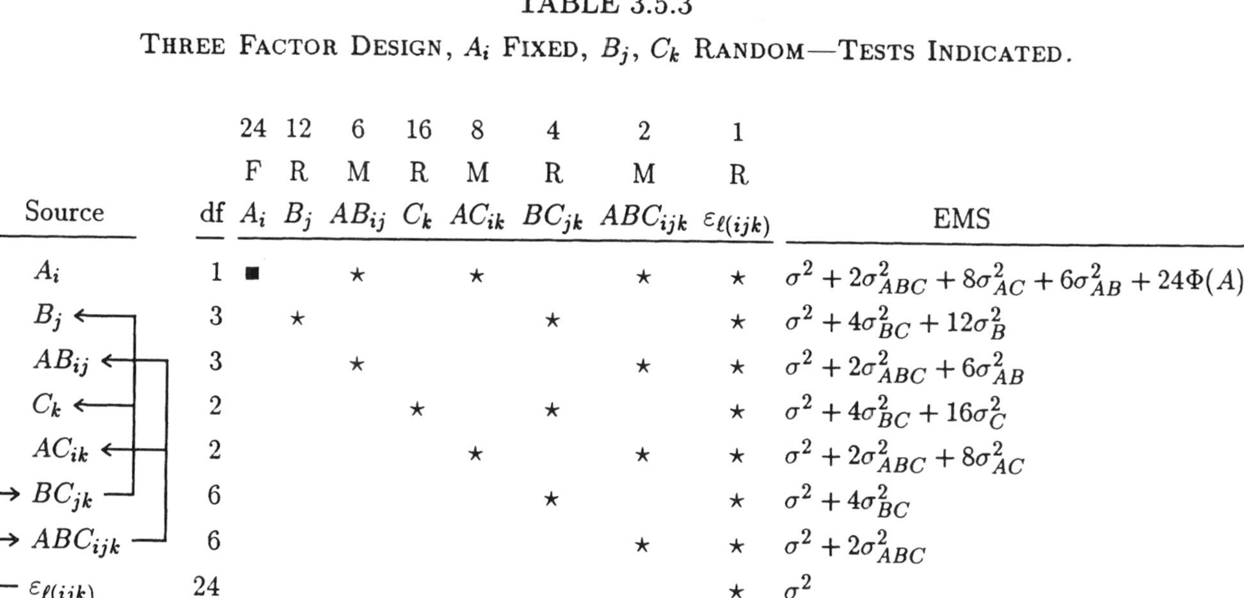

TABLE 3.5.3

THREE FACTOR DESIGN, A_i FIXED, B_j, C_k RANDOM—TESTS INDICATED.

		24	12	6	16	8	4	2	1	
		F	R	M	R	M	R	M	R	
Source	df	A_i	B_j	AB_{ij}	C_k	AC_{ik}	BC_{jk}	ABC_{ijk}	$\varepsilon_{\ell(ijk)}$	EMS
A_i	1	■		⋆		⋆		⋆	⋆	$\sigma^2 + 2\sigma^2_{ABC} + 8\sigma^2_{AC} + 6\sigma^2_{AB} + 24\Phi(A)$
B_j	3		⋆				⋆		⋆	$\sigma^2 + 4\sigma^2_{BC} + 12\sigma^2_B$
AB_{ij}	3			⋆				⋆	⋆	$\sigma^2 + 2\sigma^2_{ABC} + 6\sigma^2_{AB}$
C_k	2				⋆		⋆		⋆	$\sigma^2 + 4\sigma^2_{BC} + 16\sigma^2_C$
AC_{ik}	2					⋆		⋆	⋆	$\sigma^2 + 2\sigma^2_{ABC} + 8\sigma^2_{AC}$
BC_{jk}	6						⋆		⋆	$\sigma^2 + 4\sigma^2_{BC}$
ABC_{ijk}	6							⋆	⋆	$\sigma^2 + 2\sigma^2_{ABC}$
$\varepsilon_{\ell(ijk)}$	24								⋆	σ^2

TABLE 3.5.4

ANOVA FOR THREE FACTOR DESIGN, A_i FIXED, B_j, C_k RANDOM

Source	df	EMS	F_{calc} or F'_{calc}
A_i	1	$\sigma^2 + 2\sigma^2_{ABC} + 8\sigma^2_{AC} + 6\sigma^2_{AB} + 24\Phi(A)$	$\dfrac{\text{MS}(A)}{[\text{MS}(AB) + \text{MS}(AC) - \text{MS}(ABC)]}$
B_j	3	$\sigma^2 + 4\sigma^2_{BC} + 12\sigma^2_B$	$\text{MS}(B)/\text{MS}(BC)$
AB_{ij}	3	$\sigma^2 + 2\sigma^2_{ABC} + 6\sigma^2_{AB}$	$\text{MS}(AB)/\text{MS}(ABC)$
C_k	2	$\sigma^2 + 4\sigma^2_{BC} + 16\sigma^2_C$	$\text{MS}(C)/\text{MS}(BC)$
AC_{ik}	2	$\sigma^2 + 2\sigma^2_{ABC} + 8\sigma^2_{AC}$	$\text{MS}(AC)/\text{MS}(ABC)$
BC_{jk}	6	$\sigma^2 + 4\sigma^2_{BC}$	$\text{MS}(BC)/\text{MS}(\varepsilon)$
ABC_{ijk}	6	$\sigma^2 + 2\sigma^2_{ABC}$	$\text{MS}(ABC)/\text{MS}(\varepsilon)$
$\varepsilon_{\ell(ijk)}$	24	σ^2	

F′ because mean square terms are never subtracted. F′ has an advantage over $\tilde{\text{F}}$ because the numerator degrees of freedom are never approximated. Theoretical papers by Hudson and Krutchkoff (1968), Davenport and Webster (1973), and Lorenzen (1987) show that the two statistics are comparable in a wide variety of situations. Thus, it is not wrong to use either approximation. We will stick with the F′ test, approximating only the denominator, because it makes it easier to present other material later.

3.5.3 Conservative Tests

Conservative tests are often used when there is not a replication of the experiment or for some other reason no direct or approximate test is available. Conservative tests must be used when there is incomplete randomization (covered in Chapter **5**) and when replication does not exist. These tests are called conservative because you can draw conclusions one way but not the other. In some conservative tests, you can draw the conclusion that a term is significant but not that it is insignificant. In other conservative tests, the opposite is true: you can draw the conclusion that a term is insignificant but not that it is significant.

Conservative tests occur because an extra variance term is either in the numerator or the denominator. If this extra variance term is in the denominator, it is possible to conclude the effect of the factor or interaction of interest is significant but impossible to conclude that it is insignificant. If this extra variance term is in the numerator, it is possible to conclude the effect of the factor or interaction of interest is insignificant but not that it is significant.

We illustrate conservative tests with the previous example (Table 3.5.4). Consider first an extra variance term in the denominator. Suppose there was no repetition in the three factor experiment summarized in Table 3.5.4. The df and form of the EMS for terms A_i through ABC_{ijk} would remain the same (although the coefficients would be halved) but there would be 0 df for $\varepsilon_{\ell(ijk)}$ and thus no EMS for $\varepsilon_{\ell(ijk)}$. No direct or approximate test on BC_{jk} would exist. If, however, we thought that ABC_{ijk} might be insignificant ($\sigma^2_{ABC} = 0$), the ratio MS(BC)/MS(ABC) would directly test BC_{jk}. We go ahead and make the test whether or not ABC_{ijk} has an effect and obtain a conservative test. If the test is significant, we conclude $(\sigma^2 + 4\sigma^2_{BC})/(\sigma^2 + 2\sigma^2_{ABC}) > 1$, so most assuredly $(\sigma^2 + 4\sigma^2_{BC})/\sigma^2 > 1$ and we conclude that BC_{jk} is significant. If the test is insignificant, we cannot be sure whether the insignificance is due to an insignificant BC_{jk} term or due to a significant ABC_{ijk} term and therefore cannot conclude that BC_{jk} is insignificant. This sort of conservative test often occurs when there is no replication in the experiment.

Now consider the case of an extra variance term in the numerator. If we form the ratio MS(A)/MS(AC), the EMS column shows that EMS(A) contains EMS(AC) plus $6\sigma^2_{AB} + 24\Phi(A)$. Thus we are really testing A_i and/or AB_{ij}, not just A_i alone. If the test is insignificant, then we can conclude both A_i and AB_{ij} are insignificant since σ^2_{AB} and $\Phi(A)$ are both nonnegative quantities. However, if the test is significant, we do not know whether the significance is due to A_i or due to AB_{ij} or due to both A_i and AB_{ij}. The test is again conservative. (Note that the approximate test given earlier is preferred to the conservative test just given. This conservative test was used for illustrative purposes.) Conservative tests of this nature often occur as a result of incomplete randomization, a topic to be covered in Chapter **5**, not at the present time.

Conservative tests are easy to find using the graphical approach of Algorithm 1. Instead of finding a row containing all off-diagonal stars, find a row that contains either one too many or one too few stars and the resulting ratio is a conservative test of one of the two types. When there is more than one row to select from, always choose the row corresponding to the highest order off-diagonal star as this is the most likely to be insignificant and the smaller the effect of the off-diagonal term, the closer the conservative test approximates a direct test.

3.6 SOMETIMES POOLING

When a term in the model is declared insignificant, valuable information has been gathered. If the term is fixed, then all levels of this factor yield the same average value and it does not matter which level is chosen. If the term is random or mixed, then the fluctuations of this factor do not influence the dependent variable. An insignificant term can be ignored in future experiments involving the same dependent variable, effectively reducing the size of future experiments.

In addition to these conclusions, an insignificant term can be used to modify the existing ANOVA table, sometimes creating direct tests where only approximate or conservative tests previously existed. At a minimum, insignificant terms add df and therefore accuracy to other rows in the ANOVA table. In this book, we will use the sometimes pooling rules given by Bozivich, *et al* (1956) to remove terms from the model and possibly simplify the analysis.

SOMETIMES POOLING RULES

- A term in the model will be declared negligible and a candidate for removal from the model and EMS column if it is insignificant at the $\alpha = .25$ level.
- A term should not be removed from the model if a higher order interaction involving that term is significant at the .25 level.
- When a term is removed from the model, its entire column of stars or boxes is erased and the corresponding σ^2 or Φ term eliminated everywhere in the EMS table. (Erasing stars and boxes applies only if you are using Algorithm 1.)
- Pooling often yields identical rows. These rows should be pooled by summing the corresponding df and SS and forming a new MS by dividing the pooled (summed) SS by the pooled (summed) df.

In the three factor example given in Table 3.5.4, let us suppose that B_j, AB_{ij}, and ABC_{ijk} were all insignificant at the .25 level. Since BC_{jk} was significant at the .25 level, we would not remove B_j from the model. Since both AB_{ij} and ABC_{ijk} are insignificant at the .25 level, we take $\sigma^2_{AB} = \sigma^2_{ABC} = 0$ everywhere in the EMS table. The resulting EMS for AB_{ij}, ABC_{ijk}, and $\varepsilon_{\ell(ijk)}$ are all σ^2, so can be pooled. The resulting df is $3 + 6 + 24 = 33$ and the resulting SS is SS(AB) + SS(ABC) + SS(ε). In addition, $\varepsilon_{\ell(ijk)}$ now tests AC_{ik} and there is a direct test on A_i using AC_{ik} as the denominator. *One should always try pooling before calculating approximate or conservative tests in an attempt to create a direct test where none had previously existed* (see Lorenzen (1987)).

Pooling rules can be applied to an ANOVA table derived as a result of pooling. That is, you can repetitively apply sometimes pooling rules. The table quickly converges.

3.7 ESTIMATION OF VARIANCE COMPONENTS

If a term in the model is significant and either random or mixed, the magnitude of its corresponding variance component is of interest. The method of estimation is natural, use the EMS values to solve explicitly for the variance component of interest and then substitute the MS values for the EMS values to obtain the estimate. Note that the estimate of a variance component is the (numerator of the F or F′ test minus the denominator of the same test) divided by the coefficient of that variance component. Sometimes pooling rules should be used prior to forming the estimates.

RULE FOR ESTIMATION OF VARIANCE COMPONENTS

- To estimate the variance of a random factor, a random interaction, or a mixed interaction Y, use the various EMS rows to algebraically isolate σ_Y^2.
- The estimated variance, $\hat{\sigma}_Y^2$, is found by substituting the calculated MS values for the theoretical EMS values.
- When an F or F′ test on Y exists, $\hat{\sigma}_Y^2$ is simply (the numerator of the test minus the denominator of the test) divided by the coefficient preceding σ_Y^2 in the EMS(Y).
- Pooled variance components are taken as 0.

We illustrate these ideas using Table 3.5.4. To estimate the size of the variance component σ_B^2, we algebraically solve for σ_B^2 using the EMS values. The result is $\sigma_B^2 = [\text{EMS}(B) - \text{EMS}(BC)]/12$. The estimate of σ_B^2 is then given by $\hat{\sigma}_B^2 = [\text{MS}(B) - \text{MS}(BC)]/12$ where MS(B) and MS(BC) are calculated using the collected data. In a similar fashion we find that $\hat{\sigma}_{AB}^2 = [\text{MS}(AB) - \text{MS}(ABC)]/6$, $\hat{\sigma}_C^2 = [\text{MS}(C) - \text{MS}(BC)]/16$, $\hat{\sigma}_{AC}^2 = [\text{MS}(AC) - \text{MS}(ABC)]/8$, $\hat{\sigma}_{BC}^2 = [\text{MS}(BC) - \text{MS}(\varepsilon)]/4$, $\hat{\sigma}_{ABC}^2 = [\text{MS}(ABC) - \text{MS}(\varepsilon)]/2$, and $\hat{\sigma}^2 = \text{MS}(\varepsilon)$.

When there is no direct test, the algebra is not as easy as the above paragraph indicates. More complicated expressions are given in Section **3.9** below. It is also possible that the estimate of variance comes out negative. In such cases, replace the estimate with zero since we know that a variance cannot be negative. Such cases are virtually eliminated if the sometimes pooling rules given in the previous section are used prior to estimation of the variance components. (Variance components removed from the model are taken as zero.)

3.8 COMPARISON OF MEANS

If a term in the model is significant and fixed, comparisons of the means are of interest. We may wish to compute confidence intervals around individual means, around the difference of two preselected means, perform multiple (all possible pairwise) comparisons using Duncan's test, or, for equally spaced quantitative levels, determine polynomial trends.

These procedures are all identical to the procedures presented in Section **2.4** except the appropriate error term must be calculated and used. This error term depends on the exact model describing the experiment and can be derived algebraically by computing the variance of the appropriate mean values. As will be seen, the appropriate error term is the same for comparing two prespecified means and for Duncan's test, but can differ for confidence intervals on single means and polynomial trends. Confidence intervals on single means turn out to be difficult to calculate but the error term used for the other procedures is related to the denominator in the F or F′ test. We illustrate the theoretical calculations with one example, but, since the algebra is again tedious, we quickly resort to general rules.

To illustrate the calculations, consider the three factor design given in Table 3.4.4, with A_i and C_k fixed and B_j random. The model is $y_{ijk\ell} = \mu+A_i+B_j+AB_{ij}+C_k+AC_{ik}+BC_{jk}+ABC_{ijk}+\varepsilon_{\ell(ijk)}$ where $\sum_{i=1}^{I} A_i = 0$, B_j is *iid* N$(0,\sigma_B^2)$, AB_{ij} is N$(0,\sigma_{AB}^2)$ with $\sum_{i=1}^{I} AB_{ij} = 0$, $\sum_{k=1}^{K} C_k = 0$, $\sum_{i=1}^{I} AC_{ik} = \sum_{k=1}^{K} AC_{ik} = 0$, BC_{jk} is N$(0,\sigma_{BC}^2)$ with $\sum_{k=1}^{K} BC_{jk} = 0$, ABC_{ijk} is N$(0,\sigma_{ABC}^2)$ with $\sum_{i=1}^{I} ABC_{ijk} = \sum_{k=1}^{K} ABC_{ijk} = 0$, and $\varepsilon_{\ell(ijk)}$ is *iid* N$(0,\sigma^2)$. All of the random and mixed terms are independent of each other so their cross products will have expected value zero. We recall that Var(AB_{ij}) = E$[AB_{ij}^2] = [(I-1)/I]\sigma_{AB}^2$ and E$[AB_{ij}AB_{i'j'}] = 0$ for $j \neq j'$. Similar expressions hold for BC_{jk} and ABC_{ijk}. Also, E$[\bar{y}_{i\cdots}] = \mu + A_i$ so we compute

$$
\begin{aligned}
\mathrm{Var}(\bar{y}_{i\cdots}) &= \mathrm{E}[\bar{y}_{i\cdots} - \mu - A_i]^2 \\
&= \mathrm{E}\Big[\sum_{j=1}^{J}\sum_{k=1}^{K}\sum_{\ell=1}^{L} \frac{(B_j + AB_{ij} + C_k + AC_{ik} + BC_{jk} + ABC_{ijk} + \varepsilon_{\ell(ijk)})}{JKL}\Big]^2 \\
&= \mathrm{E}\Big[\sum_{j=1}^{J} \frac{B_j}{J} + \sum_{j=1}^{J} \frac{AB_{ij}}{J} + \sum_{j=1}^{J}\sum_{k=1}^{K}\sum_{\ell=1}^{L} \frac{\varepsilon_{\ell(ijk)}}{JKL}\Big]^2 \\
&= \frac{J\sigma_B^2}{J^2} + \frac{(I-1)J\sigma_{AB}^2}{IJ^2} + \frac{JKL\sigma^2}{J^2K^2L^2} \\
&= \frac{\sigma_B^2}{J} + \frac{(I-1)\sigma_{AB}^2}{IJ} + \frac{\sigma^2}{JKL}\,.
\end{aligned}
$$

For the variance of the difference between two means and for multiple comparisons, we use $\mathrm{Var}(\bar{y}_{i\ldots} - \bar{y}_{i'\ldots})$. Here $\mathrm{E}[\bar{y}_{i\ldots} - \bar{y}_{i'\ldots}] = A_i - A_{i'}$ and, recalling $\mathrm{E}[AB_{ij}AB_{i'j}] = -\sigma^2_{AB}/I$, we find that all terms not involving i cancel and get

$$\begin{aligned}
\mathrm{Var}(\bar{y}_{i\ldots} - \bar{y}_{i'\ldots}) &= \mathrm{E}[\bar{y}_{i\ldots} - A_i - \bar{y}_{i'\ldots} + A_{i'}]^2 \\
&= \mathrm{E}\Big[\sum_{j=1}^{J} \frac{AB_{ij}}{J} - \sum_{j'=1}^{J} \frac{AB_{i'j'}}{J} + \sum_{j=1}^{J}\sum_{k=1}^{K}\sum_{\ell=1}^{L} \frac{\varepsilon_{\ell(ijk)}}{JKL} - \sum_{j'=1}^{J}\sum_{k'=1}^{K}\sum_{\ell'=1}^{L} \frac{\varepsilon_{\ell'(i'j'k')}}{JKL}\Big]^2 \\
&= \frac{2(I-1)J\sigma^2_{AB}}{IJ^2} - \frac{2\mathrm{E}\Big[\Big(\sum_{j=1}^{J} AB_{ij}\Big)\Big(\sum_{j'=1}^{J} AB_{i'j'}\Big)\Big]}{J^2} + \frac{2\sigma^2}{JKL} \\
&= \frac{2(I-1)\sigma^2_{AB}}{IJ} + \frac{2J\sigma^2_{AB}}{IJ^2} + \frac{2\sigma^2}{JKL} \\
&= \frac{2(\sigma^2 + KL\sigma^2_{AB})}{JKL}\,.
\end{aligned}$$

The expression for the variance of a single mean is complicated but the expression for the variance of the difference of two means looks familiar. Comparing the above expression to Table 3.5.1, where $I = 2$, $J = 4$, $K = 3$, and $L = 2$, shows that the variance of the difference $\bar{y}_{i\ldots} - \bar{y}_{i'\ldots}$ is twice the expected mean squares testing A_i divided by the number of data points making up each mean. This generalizes to all models and also leads to the proper standard error used for comparing two preselected means and in Duncan's multiple range test.

3.8.1 Comparison of Two or More Means

RULES FOR COMPARISONS OF MEANS IN BALANCED DESIGNS

- When comparing mean levels of a fixed factor or fixed interaction Y, the theoretical standard error, $\sigma_{\bar{y}_1-\bar{y}_2}$, is given by the square root of the EMS testing the effect of that factor or interaction divided by the number of observations in each mean. Note, this is the denominator in the F or F′ test on the effect of the fixed factor or interaction Y.
- If there is a term X in the model having the appropriate EMS, use the square root of MS(X)/(# observations in each mean) to estimate the standard error, $s_{\bar{y}_1-\bar{y}_2}$. The df associated with $s_{\bar{y}_1-\bar{y}_2}$ is df(X). If a linear combination of MS's must be used to estimate the standard error (because only an approximate F′ test on Y exists), Satterthwaite's approximation must be used to estimate the corresponding df.

- A $100(1-\alpha)\%$ confidence interval on two preselected means is given by the difference in the means $\pm\ t_{\alpha/2}\sqrt{2}s_{\bar{y}_1-\bar{y}_2} = \bar{y}_1 - \bar{y}_2 \pm t_{\alpha/2}\sqrt{2MS(X)/n}$, where $t_{\alpha/2}$ can be found in Appendix 3 and both $\bar{y}_1$ and $\bar{y}_2$ are based on n observations.
- For Duncan's test, the least significant range for comparing ordered means spanning k values is $R_k = q_\alpha(k,df)\ s_{\bar{y}_1-\bar{y}_2}$, where df is read from the ANOVA table or estimated from the data and $q_\alpha(k,df)$ can be found in Appendix 4. Duncan's test should only be applied on a significant fixed factor or interaction.

For example, when comparing means associated with factor A_i in the three factor experiment having A_i fixed with 2 levels, C_k fixed with 3 levels, B_j random with 4 levels, and having 2 repeats of each combination, we see from Table 3.4.4 that $s_{\bar{y}_{A1}-\bar{y}_{A2}} = \sqrt{MS(AB)/24}$ with df = 3. When comparing means associated with AC_{ik}, we see that $s_{\bar{y}_{AC1}-\bar{y}_{AC2}} = \sqrt{MS(ABC)/8}$ with df = 6.

When considering the three factor design with A_i fixed, B_j and C_k random (see Table 3.5.4), we find that the standard error for comparing means on A_i is $s_{\bar{y}_{A1}-\bar{y}_{A2}} = \sqrt{[MS(AB) + MS(AC) - MS(ABC)]/24}$, the denominator of the F' test, with the df estimated using Satterthwaite's approximation. This applies for both the comparison of two prespecified means and for multiple comparisons using Duncan's test.

When interactions are significant, we usually apply Duncan's test to all of the possible mean levels in the interaction rather than to the main effects alone. This is because additivity no longer applies and judgements based on main effects alone can be misleading.

Let us now consider a confidence interval on a single mean. As was illustrated by the algebraic calculations, this is a complicated expression that is difficult to generalize. Small wonder most authors have either ignored confidence intervals on a single mean or have given incorrect formulae. The general rule, although somewhat complicated, is given below.

3.8.2 Confidence Intervals on a Single Mean

RULES FOR CONFIDENCE INTERVALS ON A SINGLE MEAN

- When computing a confidence interval on a mean level of a fixed factor or a mean level of a fixed interaction, Y, the theoretical error term, $\sigma^2_{\bar{y}}$, contains the variance of every random term in the model plus every mixed term whose fixed indices are a subset of the indices in Y.

- Each random variance term is divided by the product of all indices contained in that particular term and not in the term Y.
- Each mixed variance term is multiplied by the product of one less than each fixed index in the term and divided by the product of all indices (random and fixed) in the term.
- The estimated error term, $s^2_{\bar{y}}$, will generally be a linear combination of mean squares whose df will have to be calculated using Satterthwaite's approximation.
- The 100(1-α)% confidence interval is given by the mean $\pm$ $t_{\alpha/2}s^2_{\bar{y}}$, where $t_{\alpha/2}$ can be found in Appendix 3.

We illustrate the calculations with our three factor experiment having A_i fixed at 2 levels, B_j random at 4 levels, C_k fixed at 3 levels, and 2 repeats. The random components in the model are B_j and $\varepsilon_{\ell(ijk)}$ while the mixed components are AB_{ij}, BC_{jk}, and ABC_{ijk}. For a confidence interval on a single level of factor A_i, say level 1, the only mixed term selected is AB_{ij} since the other mixed terms contain the index k not found in A_i. The appropriate variance is then $\sigma^2_{\bar{y}_A} = \sigma^2_B/J + (I-1)\sigma^2_{AB}/IJ + \sigma^2/JKL = \sigma^2_B/4 + \sigma^2_{AB}/8 + \sigma^2/24$. Referring back to Table 3.4.4, we estimate the variance by $s^2_{\bar{y}_A} = \{[\text{MS}(B) - \text{MS}(\varepsilon)]/12\}/4 + \{[\text{MS}(AB) - \text{MS}(\varepsilon)]/6\}/8 + \{\text{MS}(\varepsilon)\}/24$ which is further simplified to $s^2_{\bar{y}_A} = \text{MS}(B)/48 + \text{MS}(AB)/48$. Satterthwaite's approximation on the df yields $\text{MS}(B)/48 + \text{MS}(AB)/48]^2 \; / \; \{[\text{MS}(B)/48]^2/3 + [\text{MS}(AB)/48]^2/3\}$, which depends on the data and yields the confidence interval $\bar{y}_{1\cdots} \pm t_{\alpha/2}s_{\bar{y}_A}$.

The variance of a particular AC_{ij} mean is given by $\sigma^2_{\bar{y}_{AC}} = \sigma^2_B/J + (I-1)\sigma^2_{AB}/IJ + (K-1)\sigma^2_{BC}/JK + (I-1)(K-1)\sigma^2_{ABC}/IJK + \sigma^2/JL$. Even this simple example demonstrates the complexities associated with confidence intervals on single means and the reason why most textbooks do not cover the topic. The reader should try to reproduce these examples to be sure the rules are understood.

3.8.3 Polynomial Trends

When a factor of interest is fixed, equally-spaced, quantitative, and significant, polynomial trends can be calculated. We start by extracting the linear trend, then, if appropriate, the quadratic trend, and so on. For one factor only, the calculation of polynomial trends is exactly as given in Section **2.4**. Put the levels of the factor in their natural order and put the means $\bar{y}_i$, based on m observations, into the adjacent column.

Next, look up the orthogonal coefficients, z_i, from Appendix 5 and compute $m(\sum_{i=1}^{I} \bar{y}_i z_i)/(\sum_{i=1}^{I} z_i^2)$. This will not be illustrated. If there is any confusion, refer back to Section **2.4**.

The extension of this procedure to the interaction of two different trends (*i.e.*, linear by linear, linear by quadratic, *etc.*), each on fixed, equally-spaced, and quantitative factors, is straightforward. Let z_i be the orthogonal coefficients for the factor associated with index i, and let z_j be the orthogonal coefficients associated with the factor having index j. The sum of squares is given by $m(\sum_{i=1}^{I} \sum_{j=1}^{J} \bar{y}_{ij} z_i z_j)^2/(\sum_{i=1}^{I} \sum_{j=1}^{J} (z_i z_j)^2)$, where each $\bar{y}_{ij}$ is the mean of m observations. A three factor interaction would contain z_i, z_j, and z_k.

While the extension of polynomial trends to interactions is straightforward, the test for significance may not be straightforward. If the main effect or interaction is tested by a random term (usually the error term), then each polynomial trend or polynomial interaction will be tested by the same random term. This is the easy case.

The more difficult case is when the main effect or interaction is tested by a mixed term. Suppose A_i, fixed, is tested by AB_{ij}, a mixed term. Then A_{linear} would be tested by $A_{linear} \times B_j$, the interaction of the linear component of A by *all* of B. The sum of squares for $A_{linear} \times B_j$ is $\text{SS}(A_{linear} \times B_j) = m \sum_{j=1}^{J} [\sum_{i=1}^{I} z_i(\bar{y}_{ij} - \bar{y}_{i\cdot})]^2/[\sum_{i=1}^{I} z_i^2]$, having $1 \times (J - 1) = J - 1$ degrees of freedom, and where each $\bar{y}_{ij}$ is based on m observations. The extension to larger models should be obvious and is a task best suited to machine calculations. Machine calculations are illustrated in Example 3 in the next section.

3.9 EXAMPLES

Example 1. In an experiment on the firing time of explosive switches, there are three factors of interest, the metal used, the amount of primary initiator, and the packing pressure of the explosive. The variable of interest is the firing time of the switch. The metal used in the switches is composed of recycled material, so this factor has an infinite number of levels and is considered random. Two metal compositions are considered. Three specific amounts of initiator are chosen: 5, 10, and 15 *mg*, and three specific packing pressures are chosen: 12,000, 20,000, and 28,000 *psi*. Both of these factors are considered fixed.

For each of the 18 combinations, two switches are manufactured and tested. Assuming a completely randomized design, the model is given by

$$y_{ijk\ell} = \mu + M_i + I_j + MI_{ij} + P_k + MP_{ik} + IP_{jk} + MIP_{ijk} + \varepsilon_{\ell(ijk)}, \quad (3.9.1)$$

where M_i stands for metals, $i = 1, 2$ assumed Normal($0,\sigma_M^2$), I_j stands for initiators, $j = 1, 2, 3$ meaning 5, 10, and 15 *mg* of initiator respectively, P_k stands for packing pressures, $k = 1, 2, 3$ meaning 12,000, 20,000, and 28,000 *psi* respectively, and $\varepsilon_{\ell(ijk)}$, $\ell = 1, 2$ is the random error term associated with repeats and assumed Normal($0,\sigma^2$). The ANOVA table, associated with (3.9.1) and calculable before any data is collected, is given in Table 3.9.1.

The EMS table indicates that direct tests are available on all main effects and interactions, so this is a good experiment. The order of manufacture and testing of these switches was completely randomized before experimentation began. All 36 combinations were written down using the convention that the rightmost index moved most quickly. The run order was randomized using Appendix 1; starting in column 16 and 17, row 12 and reading down to the bottom of these columns before starting in columns 18 and 19. The results are given in Table 3.9.2.

TABLE 3.9.2

MANUFACTURE AND RUN ORDER FOR EXAMPLE 1.

M I P	Order	M I P	Order	M I P	Order	M I P	Order	M I P	Order	M I P	Order
1 1 1	15	1 2 1	28	1 3 1	4	2 1 1	33	2 2 1	2	2 3 1	6
1 1 1	22	1 2 1	34	1 3 1	12	2 1 1	19	2 2 1	10	2 3 1	31
1 1 2	23	1 2 2	35	1 3 2	11	2 1 2	17	2 2 2	3	2 3 2	20
1 1 2	25	1 2 2	7	1 3 2	8	2 1 2	18	2 2 2	24	2 3 2	26
1 1 3	36	1 2 3	27	1 3 3	9	2 1 3	13	2 2 3	29	2 3 3	16
1 1 3	32	1 2 3	1	1 3 3	21	2 1 3	30	2 2 3	14	2 3 3	5

The first switch manufactured and tested (order 1) uses metal 1, 10 *mg* of initiator, and is packed at 28,000 *psi*. The second switch manufactured and tested (order 2) uses metal 2, 10 *mg* of initiator, and is packed at 12,000 *psi*. The last switch manufactured and tested (order 36) uses metal 1, 5 *mg* of initiator, and 28,000 *psi*. Of course, Table 3.9.2 is not convenient for running the experiment as one would spend too much time searching through the table for the next ordered value. For convenience in running the experiment, Table 3.9.2 should be sorted in the order in which the experiment will be run.

The collected data is summarized in Table 3.9.3. For convenience, all of the totals for the main effects and interactions are included. Hand calculations of the sums of squares immediately follow.

TABLE 3.9.1

ANOVA for Example 1 on Explosive Switches

Source	df	EMS	18 R M_i	12 F I_j	6 M MI_{ij}	12 F P_k	6 M MP_{ik}	4 F IP_{jk}	2 M MIP_{ijk}	1 R $\varepsilon_{\ell(ijk)}$
M_i	1	$\sigma^2 + 18\sigma_M^2$	⋆							⋆
I_j	2	$\sigma^2 + 6\sigma_{MI}^2 + 12\Phi(I)$		■	⋆					⋆
MI_{ij}	2	$\sigma^2 + 6\sigma_{MI}^2$			⋆					⋆
P_k	2	$\sigma^2 + 6\sigma_{MP}^2 + 12\Phi(P)$				■	⋆			⋆
MP_{ik}	2	$\sigma^2 + 6\sigma_{MP}^2$					⋆			⋆
IP_{jk}	4	$\sigma^2 + 2\sigma_{MIP}^2 + 4\Phi(IP)$						■	⋆	⋆
MIP_{ijk}	4	$\sigma^2 + 2\sigma_{MIP}^2$							⋆	⋆
$\varepsilon_{\ell(ijk)}$	18	σ^2								⋆
Total	35									

TABLE 3.9.3

FIRING TIMES AND CALCULATIONS FOR EXAMPLE 1.

		Metals M_i 1			Metals M_i 2		
		Initiators I_j			Initiators I_j		
		1	2	3	1	2	3
Pressures P_k	1	44 39	27 20	35 30	12 7	15 10	22 15
	2	48 40	25 21	29 34	6 11	12 17	27 22
	3	43 41	28 22	31 38	7 12	11 13	21 19

Metals	Initiators	Pressures	Metal-Initiator Cells	
$T_{1\cdots} = 595$	$T_{\cdot 1 \cdot\cdot} = 310$	$T_{\cdot\cdot 1 \cdot} = 276$	$T_{11\cdot\cdot} = 255$	$T_{21\cdot\cdot} = 55$
$T_{2\cdots} = 259$	$T_{\cdot 2 \cdot\cdot} = 221$	$T_{\cdot\cdot 2 \cdot} = 292$	$T_{12\cdot\cdot} = 143$	$T_{22\cdot\cdot} = 78$
	$T_{\cdot 3 \cdot\cdot} = 323$	$T_{\cdot\cdot 3 \cdot} = 286$	$T_{13\cdot\cdot} = 197$	$T_{23\cdot\cdot} = 126$

Metal-Pressure Cells		Initiator-Pressure Cells		
$T_{1\cdot 1\cdot} = 195$	$T_{2\cdot 1\cdot} = 81$	$T_{\cdot 11\cdot} = 102$	$T_{\cdot 21\cdot} = 72$	$T_{\cdot 31\cdot} = 102$
$T_{1\cdot 2\cdot} = 197$	$T_{2\cdot 2\cdot} = 95$	$T_{\cdot 12\cdot} = 105$	$T_{\cdot 22\cdot} = 75$	$T_{\cdot 32\cdot} = 112$
$T_{1\cdot 3\cdot} = 203$	$T_{2\cdot 3\cdot} = 83$	$T_{\cdot 13\cdot} = 103$	$T_{\cdot 23\cdot} = 74$	$T_{\cdot 33\cdot} = 109$

Metal-Initiator-Pressure Cells

$T_{111\cdot} = 83$	$T_{121\cdot} = 47$	$T_{131\cdot} = 65$	$T_{211\cdot} = 19$	$T_{221\cdot} = 25$	$T_{231\cdot} = 37$
$T_{112\cdot} = 88$	$T_{122\cdot} = 46$	$T_{132\cdot} = 63$	$T_{212\cdot} = 17$	$T_{222\cdot} = 29$	$T_{232\cdot} = 49$
$T_{113\cdot} = 84$	$T_{123\cdot} = 50$	$T_{133\cdot} = 69$	$T_{213\cdot} = 19$	$T_{223\cdot} = 24$	$T_{233\cdot} = 40$

Grand Total	Individuals Squared	CT
$T_{\cdots\cdot} = 854$	$\sum_{i=1}^{2} \sum_{j=1}^{3} \sum_{k=1}^{3} \sum_{\ell=1}^{2} y_{ijk\ell}^2 = 25196$	$T_{\cdots\cdot}^2/36 = 20258.78$

$$SS(M) = \frac{595^2 + 259^2}{18} - CT = 3136.0$$

$$\text{SS}(I) = \frac{310^2 + 221^2 + 323^2}{12} - \text{CT} = 513.72$$

$$\text{SS}(MI) = \frac{255^2 + 143^2 + 197^2 + 55^2 + 78^2 + 126^2}{6} - \frac{595^2 + 259^2}{18} - \frac{10^2 + 221^2 + 323^2}{12} + \text{CT} = 969.50$$

$$\text{SS}(P) = \frac{276^2 + 292^2 + 286^2}{12} - \text{CT} = 10.89$$

$$\text{SS}(MP) = \frac{195^2 + 197^2 + 203^2 + 81^2 + 95^2 + 83^2}{6} - \frac{595^2 + 259^2}{18} - \frac{276^2 + 292^2 + 286^2}{12} + \text{CT} = 14.00$$

$$\text{SS}(IP) = \frac{102^2 + 105^2 + 103^2 + \cdots + 109^2}{4} - \frac{310^2 + 221^2 + 323^2}{12} - \frac{276^2 + 292^2 + 286^2}{12} + \text{CT} = 4.61$$

$$\text{SS}(MIP) = \frac{83^2 + 88^2 + \cdots + 40^2}{2} - \frac{255^2 + 143^2 + \cdots + 126^2}{6} - \frac{195^2 + 197^2 + \cdots + 83^2}{6} - \cdots + \cdots - \text{CT} = 38.50$$

$$\text{SS}(\varepsilon) = \sum_{i=1}^{2}\sum_{j=1}^{3}\sum_{k=1}^{3}\sum_{\ell=1}^{2} y_{ijk\ell}^2 - \frac{83^2 + 88^2 + \cdots + 40^2}{2} = 250.00$$

$$\text{SS(Total)} = \sum_{i=1}^{2}\sum_{j=1}^{3}\sum_{k=1}^{3}\sum_{\ell=1}^{2} y_{ijk\ell}^2 - \text{CT} = 25196 - 20258.78 = 4937.22$$

$$\begin{aligned} &\text{SS}(M) + \text{SS}(I) + \text{SS}(MI) + \text{SS}(P) + \text{SS}(MP) + \text{SS}(IP) + \text{SS}(\varepsilon) \\ &= 3136.00 + 513.72 + 969.50 + 10.89 + 14.00 + 4.61 + 38.50 + 250.00 \\ &= 4937.22 = \text{SS(Total)} \end{aligned}$$

The last calculation serves as a check on the arithmetic and is always worth performing. Of course, machine calculations are even better as machines do not make arithmetic errors. The above results are summarized in Table 3.9.4.

According to our sometimes pooling rules, we can pool P_k, MP_{ik}, IP_{jk}, and MIP_{ijk}. We cannot pool I_j since MI_{ij} is significant at the .25 level. (Even stronger than that, MI_{ij} is significant at the .05 level.)

TABLE 3.9.4

ANOVA FOR EXAMPLE 1 ON EXPLOSIVE SWITCHES.

Source	df	SS	MS	F_{calc}	$F_{.05}$	$F_{.25}$
M_i	1	3136.00	3136.00	225.79**	4.41	1.41
I_j	2	513.72	256.86	0.53†	19.00	3.00
MI_{ij}	2	969.50	484.75	34.90**	3.55	1.50
P_k	2	10.89	5.44	0.78†	19.00	3.00
MP_{ik}	2	14.00	7.00	0.50†	3.55	1.50
IP_{jk}	4	4.61	1.15	0.12†	6.39	2.06
MIP_{ijk}	4	38.50	9.62	0.69†	2.93	1.48
$\varepsilon_{\ell(ijk)}$	18	250.00	13.89			
Total	35	4937.22				

**Significant at the .01 level †Candidate for pooling

Therefore, we modify Table 3.9.1 by setting $\Phi(P)$, σ^2_{MP}, $\Phi(IP)$, and σ^2_{MIP} equal to zero, adding the appropriate df and SS terms, and simplifying the EMS column. In Table 3.9.4, all of the terms P_k, MP_{ik}, IP_{jk}, and MIP_{ijk} can be pooled with the error term $\varepsilon_{\ell(ijk)}$ yielding 30 df. The simplified result is given in Table 3.9.5.

TABLE 3.9.5

ANOVA FOR EXAMPLE 1 AFTER POOLING.

Source	df	SS	MS	F_{calc}	$F_{.05}$
M_i	1	3136.00	3136.00	295.85**	4.17
I_j	2	513.72	256.86	0.53	19.00
MI_{ij}	2	969.50	484.75	45.73**	3.32
$\varepsilon_{\ell(ijk)}$	30	318.00	10.60		
Total	35	4937.22			

**Significant at the .01 level

We conclude that, within the range 12,000 to 28,000 *psi*, packing pressure does not affect the firing time of explosive switches. However, the use of metals does affect the firing time as does the interaction between metals and initiator. The significant interaction means that the effect of initiators depends on the (random) composition of the metal. The main effect for initiators was not significant because two metals were chosen that had opposite effects for the first two initiators. This can be seen by comparing the first and second columns of the Metal-Initiator's cell totals or by inspection of the main effect and interaction plots in Figure 3.2. This example illustrates why we do not pool a main effect when an interaction is significant.

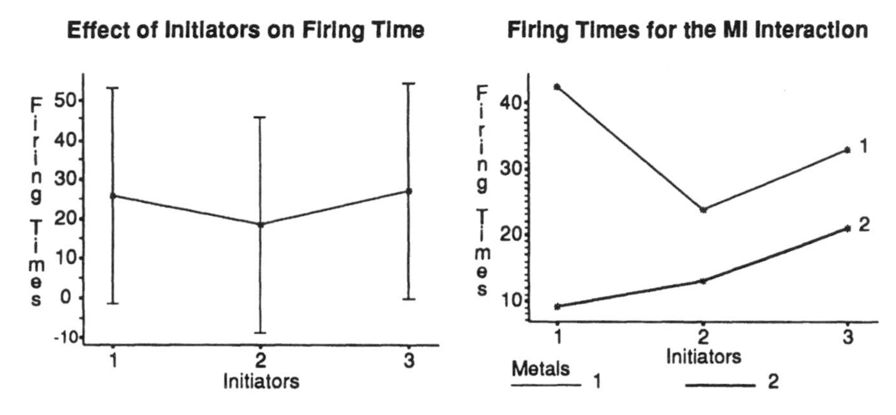

Figure 3.2 Main effect and interaction plots for Example 1.

The estimated variability for metals is $\sigma^2_M = [\text{MS}(M) - \text{MS}(\varepsilon)]/18 = 173.633$, the variability due to the interaction is $\sigma^2_{MI} = [\text{MS}(MI) - \text{MS}(\varepsilon)]/6 = 79.025$, and the variability of the error term is $\sigma^2 = \text{MS}(\varepsilon) = 10.60$. The total variability over all metals, initiators, and repeats is $173.63 + 79.025 + 10.60 = 263.258$ which gives a standard deviation in firing time of $\sqrt{263.258} = 16.225$. Management is now in a position to decide whether the variability in firing time warrants the extra cost of a specific formulation of metal or whether they would continue using scrap metal. Purchasing more expensive machinery to better control packing pressure or initiator amount will not reduce the variability in firing time.

Example 2. An experiment on diesel engines was conducted to determine the causes of variability in brake specific fuel consumption. Four different engines of the same type, four different fuel injectors of the same type, and fuel from two different fuel tanks was used. The 32 combinations were run in a completely randomized fashion so the following model applies:

$$y_{ijk} = \mu + E_i + I_j + EI_{ij} + F_k + EF_{ik} + IF_{jk} + EIF_{ijk} + \varepsilon_{(ijk)}, \quad (3.9.2)$$

where E_i stands for engines, $i = 1, \cdots, 4$, I_j stands for injectors, $j = 1, \cdots, 4$, F_k stands for fuel, $k = 1, 2$, and each combination is run only once. In reality, $\varepsilon_{(ijk)}$ should have a subscript ℓ where ℓ takes the value 1 only, but we generally drop such subscripts from the model. This will not influence any of the rules given in this chapter as long as we remember $\text{df}(\varepsilon) = 0$.

The inference is to all engines of the same type and to all fuel injectors of the same type so E_i and I_j are taken as random. Note that the inference is only the specific type of engine and injector used in the experiment, not to any other type of engine or any other fuel injector design. If the two different tanks of fuel were of the same brand, then the inference is only to that specific brand. If the tanks of fuel were from different (randomly chosen) brands and there were enough brands that we could take the population as infinite, we can also take F_k as random. If two specific brands were tested, we would take F_k as fixed. We will take F_k as random since the same brand of fuel is used and the inference is to all tanks of this particular brand of fuel.

The data and corresponding EMS are given in Tables 3.9.6 and 3.9.7. There are no direct tests on any of the main effects in this design. However, approximate tests on all of the main effects exist. The denominator testing I_j is $\text{MS}(EI)+\text{MS}(IF)-\text{MS}(EIF)$ since $\text{EMS}(EI)+\text{EMS}(IF)-\text{EMS}(EIF) = \sigma^2+\sigma^2_{EIF}+2\sigma^2_{EI}+4\sigma^2_{IF}$, as was desired. The F′ test on I_j is calculated as $(.1905)/(.0675 + .0176 - .1310) = -4.15$ with df for the denominator approximated by $(-.0459)^2/[((.0675)^2/9) + ((.0176)^2/3) + ((.1310)^2/9)] = 0.84$. As indicated by either the negative F′ value or the df less than 1, something is wrong. This illustrates the dangers of applying approximate tests prior to pooling. Always look to pool insignificant terms before computing approximate tests. In this example, we can pool all of the two factor interactions since they are insignificant at the 0.25 level and will be combined with the EIF_{ijk} interaction. The new results are given in Table 3.9.8.

Since all of E_i, I_j and F_k are significant at the .25 level, there are no more candidates for pooling. We conclude that engines contribute significantly to the variability in brake specific fuel consumption but injectors and fuel do not significantly affect the overall variability.

We take the estimates of σ^2_{EI}, σ^2_{EF}, and σ^2_{IF} as zero since they were insignificant at the $\alpha = .25$ level and pooled. The estimates of variability and percent contribution for the other terms in the model are given below Table 3.9.8.

The total variation of the process (not the total variability of the experiment) is .2530, giving a standard deviation of .503. The largest

TABLE 3.9.6

DATA FOR EXAMPLE 2 ON BRAKE SPECIFIC FUEL CONSUMPTION.

		Fuel Tank 1				Fuel Tank 2			
	Injector:	1	2	3	4	1	2	3	4
Engine:	1	43.20	42.80	42.75	42.95	43.45	42.85	42.70	42.95
	2	42.10	42.30	42.55	42.80	43.30	43.00	42.75	42.85
	3	43.85	43.75	43.20	43.70	43.05	43.20	43.55	44.40
	4	43.05	43.20	43.00	43.25	43.30	43.30	43.25	43.70

TABLE 3.9.7

ANOVA FOR EXAMPLE 2.

Source	df	MS	EMS	F_{calc}	$F_{.05}$	$F_{.25}$
E_i	3	1.1599	$\sigma^2 + \sigma^2_{EIF} + 4\sigma^2_{EF} + 2\sigma^2_{EI} + 8\sigma^2_E$	–	–	–
I_j	3	.1905	$\sigma^2 + \sigma^2_{EIF} + 4\sigma^2_{IF} + 2\sigma^2_{EI} + 8\sigma^2_I$	–	–	–
→ EI_{ij}	9	.0675	$\sigma^2 + \sigma^2_{EIF} + 2\sigma^2_{EI}$	0.52†	3.18	1.59
F_k	1	.3101	$\sigma^2 + \sigma^2_{EIF} + 4\sigma^2_{IF} + 4\sigma^2_{EF} + 16\sigma^2_F$	–	–	–
→ EF_{ik}	3	.1415	$\sigma^2 + \sigma^2_{EIF} + 4\sigma^2_{EF}$	1.08†	3.86	1.63
→ IF_{jk}	3	.0176	$\sigma^2 + \sigma^2_{EIF} + 4\sigma^2_{IF}$	0.13†	3.86	1.63
— EIF_{ijk}	9	.1310	$\sigma^2 + \sigma^2_{EIF}$	–	–	–
$\varepsilon_{(ijk)}$	0	–				
Total	31					

†Candidate for pooling

contributer to this variation is engine to engine variability. Injectors and fuel contribute very little to the total variation. The error term and/or the EIF_{ijk} interaction term contribute about 37% of the variation. The error term represents all unaccounted variation and consists of things like the variability of the measuring device, the effect of atmospheric conditions if they differed for the various measurements, the effect of operators if different operators were used, *etc.* Further experimentation is necessary to separate the variability in the error term and in the three factor interaction. One possibility is to repeat a few combinations to form an independent estimate of σ^2, allowing a test on EIF_{ijk}.

TABLE 3.9.8

ANOVA FOR EXAMPLE 2 AFTER POOLING.

Source	df	SS	MS	EMS	F_{calc}	$F_{.05}$	$F_{.25}$
→ E_i	3	3.4796	1.1599	$\sigma^2 + \sigma^2_{EIF} + 8\sigma^2_E$	12.30*	3.01	1.46
→ I_j	3	.5715	.1905	$\sigma^2 + \sigma^2_{EIF} + 8\sigma^2_I$	2.02	3.01	1.46
→ F_k	1	.3101	.3101	$\sigma^2 + \sigma^2_{EIF} + 16\sigma^2_F$	3.29	4.26	1.39
└ EIF_{ijk}	24	2.2637	.0943	$\sigma^2 + \sigma^2_{EIF}$			
$\varepsilon_{(ijk)}$	0	-					
Total	31	6.6249					

*Significant at the .05 level

Variance Term	Estimate	% Contribution
σ^2_E	$(1.1599 - .0943)/8 = .1332$	52.6%
σ^2_I	$(.1905 - .0943)/8 = .0120$	4.8%
σ^2_F	$(.3101 - .0943)/16 = .0135$	5.3%
$\sigma^2 + \sigma^2_{EIF}$	.0943	37.3%
Total	.2530	100.0%

Since all of the two factor interactions were insignificant, it is reasonable to *assume* the three factor interaction is also insignificant. Such an assumption, while quite reasonable, cannot be verified statistically without additional data. The experimenter, based on non-statistical data, may be willing to make such an assumption.

Example 3. In an agronomy experiment, the experimenters were interested in the effect of fertilizer, F or Fe, on the bushel per acre yield of wheat. To be able to infer the results to all fields, $\mathcal{F}$ or Fi, not just the individual field that was used in the experiment, two different 25 acre fields were randomly selected. Within each of the 25 acre fields, five different acres were used for each of the five fertilizer levels: 0, 10, 20, 30, and 40 pounds of fertilizer per acre. The assignment of the 50 individual acres to the fertilizer levels was done in a completely randomized fashion. The model for this experiment is

$$y_{ijk} = \mu + \mathcal{F}_i + F_j + \mathcal{F}F_{ij} + \varepsilon_{k(ij)},$$

$i = 1, 2$ for the two different fields, $j = 1, \cdots, 5$ corresponding to the five fertilizer levels, and $k = 1, \cdots, 5$ for the five acres tested for each field/fertilizer combination. The EMS calculations in Table 3.9.9 showed that the proper inferences could be made. Thus, the experiment was run and the data following Table 3.9.9 was collected.

TABLE 3.9.9

ANOVA AND DATA FOR EXAMPLE 3 ON FERTILIZERS

Source	df	EMS
$\mathcal{F}_i$	1	$\sigma^2 + 25\sigma^2_{\mathcal{F}}$
F_j	4	$\sigma^2 + 5\sigma^2_{\mathcal{F}F} + 10\Phi(F)$
$\mathcal{F}F_{ij}$	4	$\sigma^2 + 5\sigma^2_{\mathcal{F}F}$
$\varepsilon_{k(ij)}$	40	σ^2
Total	49	

	Field 1 ($\mathcal{F}_1$)					Field 2 ($\mathcal{F}_2$)				
Fertilizer:	0	10	20	30	40	0	10	20	30	40
Yield:	20	25	36	35	43	18	35	51	43	41
	25	29	37	39	40	26	34	42	45	38
	23	31	29	31	36	22	31	47	40	34
	27	30	40	42	48	23	30	47	34	45
	19	27	33	44	47	20	32	44	48	45

We analyze the data using PROC ANOVA in the SAS package. The statements and corresponding output appear in Table 3.9.10. Of course, the dots indicate that the same format was followed for entry of the remaining data.

The "PR > F" column gives the probability that an F variate exceeds the calculated value and is often called the p-value. If the p-value is less than .05, the test is significant at the .05 level. If the p-value is greater than .25, the term is insignificant at the 25% level and a candidate for pooling.

Calculations of the SS, F, and p-value are fast and, if the data is entered correctly, very accurate. The disadvantage is that the F values and corresponding p-values printed out by SAS are all formed by dividing the numerator by MS(ERROR). For the current example, this is correct for testing $\mathcal{F}_i$ and $\mathcal{F}F_{ij}$, but not for testing F_j, the factor of most interest.

TABLE 3.9.10

SAS STATEMENTS AND OUTPUT FOR EXAMPLE 3 ON FERTILIZERS

Input Statements

```
DATA;
INPUT Fi Fe Y;
CARDS;
1 1 20
1 1 25
1 1 23
1 1 27
1 1 19
1 2 25
 :
2 5 45
RUN;
PROC ANOVA;
CLASS Fi Fe;
MODEL Y=Fi Fe Fi*Fe;
 TEST H=Fe E=Fi*Fe;
RUN;
```

Output Condensed From SAS

SOURCE	DF	SS	F	PR > F
Fi	1	124.82	7.66	0.0085
Fe	4	2847.08	43.69	0.0001
Fi*Fe	4	279.487	4.29	0.0056
ERROR	40	651.60		
TOTAL	49	3902.98		

SOURCE	DF	SS	F	PR > F
Fe	4	2847.08	10.19	0.0225

When using SAS, the EMS column must be hand computed using the rules given earlier in this chapter, the proper tests must be determined, and the F value must be either calculated by hand or separately calculated by SAS and replaced in the SAS ANOVA table. As indicated by the ANOVA in Table 3.9.9, $\mathcal{F}F_{ij}$ tests F_j. SAS will compute the proper F value and PR > F if you issue the statement TEST H=Fe E=Fi*Fe; between the model statement and the run statement. The H stands for hypothesis (the numerator) and the E stands for error (the denominator). The proper values will appear below the SAS ANOVA table, with the ANOVA table still containing improper values that should be ignored. It is advisable to then cross out the improper F values and write in the proper F values by hand. The correct ANOVA table is given in Table 3.9.11.

Among other mainframe statistical packages, BBN and SPSS are similar to SAS in that they test all terms in the model with the error term and therefore must be hand modified. BMDP uses information on which components are fixed and which are random in order to make the proper tests. On the other hand, BMDP has the least amount of flexibility for

TABLE 3.9.11

ANOVA FOR EXAMPLE 3 ON FERTILIZERS

Source	df	MS	EMS	F_{calc}	p-value
$\mathcal{F}_i$	1	124.82	$\sigma^2 + 25\sigma^2_{\mathcal{F}}$	7.66**	.0085
F_j	4	711.77	$\sigma^2 + 5\sigma^2_{\mathcal{F}F} + 10\Phi(F)$	10.19*	.0225
$\mathcal{F}F_{ij}$	4	69.87	$\sigma^2 + 5\sigma^2_{\mathcal{F}F}$	4.29**	.0056
$\varepsilon_{k(ij)}$	40	16.29	σ^2		
Total	49				

*Significant at the .05 level.
**Significant at the .01 level.

data entry and lacks more advanced ANOVA type analyses. A multitude of PC based software products such as Minitab, Statgraphics, Stat-Ease, StatSoft, NCSS, Systat, Echip, Statistix, and others are currently available in the market. More are being introduced, almost on a monthly basis. With any of these products, it is recommended that Example 3 be run and the output be compared to Table 3.9.11. This will quickly determine whether the output needs to be hand altered or not. All such packages should properly compute the SS and MS values. Each individual product should be assessed before general use.

For Example 3, we find that fields, fertilizer, and the interaction are all significant. The estimate of the variability from field to field is $\hat{\sigma}^2_{\mathcal{F}} = (124.82 - 16.29)/25 = 4.34$ bushels per acre while the variability of the interaction is $\hat{\sigma}^2_{\mathcal{F}F} = (69.87-16.29)/5 = 10.72$. The within field variation, $\hat{\sigma}^2$, is 16.29, making the total variation of the yield $4.34 + 10.72 + 16.29 = 31.35$. We see that 48% of the variation in yield is due to field differences, including field to field variation and the field/fertilizer interaction. The variation within a given field accounts for 52% of the total.

Since fertilizers are significant and fixed, we wish to compare means. A confidence interval on the average yield for 20 pounds of fertilizer per acre is given by the mean $\pm t_{\alpha/2}s_{\bar{y}}$. The mean yield is $\bar{y}_{\cdot 3 \cdot} = 40.6$. The theoretical variation of this mean is $\sigma^2_{\bar{y}} = \sigma^2/IK + \sigma^2_{\mathcal{F}}/I + (J-1)\sigma^2_{\mathcal{F}F}/IJ = .1\sigma^2 + .5\sigma^2_{\mathcal{F}} + .4\sigma^2_{\mathcal{F}F}$. The estimated variance of $\bar{y}_{\cdot 3 \cdot}$ is $s^2_{\bar{y}} = .1\{\text{MS}(\varepsilon)\} + .5\{[\text{MS}(\mathcal{F})-\text{MS}(\varepsilon)]/25\} + .4\{[\text{MS}(\mathcal{F}F)-\text{MS}(\varepsilon)]/5\} = .02\text{MS}(\mathcal{F}) + .08\text{MS}(\mathcal{F}F) = 8.086$. The df for this variance is given by Satterthwaite's formula: df$= [.02(124.82) + .08(69.87)]^2/[\{.02(124.82)\}^2/1 + \{.08(69.87)\}^2/4] = 4.66$. Using interpolation in Appendix 3 gives a 95% confidence interval on the mean yield using 20 pounds of fertilizer per acre of $40.6 \pm 2.641\sqrt{8.086} = 40.6 \pm 7.5 = (33.1, 48.1)$.

For comparing two or more means, the appropriate variance is given by the denominator testing F_j divided by the number of observations in each mean: $\sigma^2_{\bar{y}_1 - \bar{y}_2} = (\sigma^2 + 5\sigma^2_{\mathcal{F}F})/(IK) = (\sigma^2 + 5\sigma^2_{\mathcal{F}F})/10$. The estimate of this variance is $s_{\bar{y}_1 - \bar{y}_2} = \sqrt{\text{MS}(\mathcal{F}F)/10} = \sqrt{6.987} = 2.643$, having 4 df. A 95% confidence interval on the added yield per acre due to the addition of 20 pounds of fertilizer per acre is given by $(\bar{y}_{\cdot 3 \cdot} - \bar{y}_{\cdot 1 \cdot} \pm t_{\alpha/2}\sqrt{2 s^2_{\bar{y}_1 - \bar{y}_2}} = (40.6 - 22.3) \pm 2.776(1.414)2.643 = 18.3 \pm 10.4 = (7.9, 26.2)$.

To compare all possible means for F_j, Duncan's test is used. The means for the five levels are $\bar{y}_{\cdot 1 \cdot} = 22.3$, $\bar{y}_{\cdot 2 \cdot} = 30.4$, $\bar{y}_{\cdot 3 \cdot} = 40.6$, $\bar{y}_{\cdot 5 \cdot} = 40.1$, and $\bar{y}_{\cdot 5 \cdot} = 41.7$. These are placed in descending order below. The least significant ranges are given by

$$
\begin{aligned}
R_5 &= q_{.05}(5,4) s_{\bar{y}_1 - \bar{y}_2} = 4.033(2.643) = 10.66 \\
R_4 &= q_{.05}(4,4) s_{\bar{y}_1 - \bar{y}_2} = 4.033(2.643) = 10.66 \\
R_3 &= q_{.05}(3,4) s_{\bar{y}_1 - \bar{y}_2} = 4.013(2.643) = 10.61 \\
R_2 &= q_{.05}(2,4) s_{\bar{y}_1 - \bar{y}_2} = 3.927(2.643) = 10.38
\end{aligned}
$$

The difference spanning five means is $41.7 - 22.3 = 19.4$ which exceeds 10.66 so is not connected. The two differences spanning four means are $41.7 - 30.4 = 11.3$ and $40.6 - 22.3 = 18.3$ so neither are connected. The differences spanning three means are $41.7 - 40.1 = 1.6$, $40.6 - 30.4 = 10.2$, and $40.1 - 22.3 = 17.8$, which implies 41.7 gets connected to 40.1 and separately 40.6 gets connected to 30.4. When comparing differences spanning two means, we already find that 41.7 is connected to 40.6, 40.6 is connected to 40.1, and 40.1 is connected to 30.4. Therefore we only need compute $30.4 - 22.3 = 8.1$ which is less than R_2 and therefore should be connected. The final result is given below.

Fertilizer Level	Mean			
40	41.7	│		
20	40.6	│	│	
30	40.1	│	│	
10	30.4		│	│
0	22.3			│

We conclude there is no significant difference between 0 and 10 pounds of fertilizer, between 10, 20, and 30 pounds of fertilizer, and between 20, 30, and 40 pounds of fertilizer. There is a significant difference, for example, between no fertilizer and 20 pounds of fertilizer per acre, since no line connects 0 to 20. Duncan's procedure is run in SAS by issuing the command MEANS Fe/DUNCAN E=Fi*Fe; after the MODEL statement. If you do not specify the denominator using E=, SAS will test with MS(ERROR). SAS will not perform Duncan's test on interaction terms even though it will calculate and print the mean values. To get SAS to compute Duncan's test on an interaction, you must "fool" SAS into thinking the interaction is a main effect by creating a new variable that has the same number of levels as the interaction and removing the original main effects from the model. (Duncan's procedure is not currently available on either SPSS or BMDP. Check your local statistical package for availablity of Duncan's or a similar procedure.)

Scanning the means associated with the fertilizer levels (see also Figure 3.3) indicates that the yield increases as the fertilizer level increases. This leads us to fitting polynomial functions to the model. We start with a linear trend. For this, we go to Appendix 5 for n=5 and find the coefficients −2, −1, 0, 1, and 2. Since $\mathcal{F}F_{ij}$ tests F_j, the quantity $\mathcal{F}_i * F_{linear}$, based on $m = 10$ observations, tests F_{linear}. The calculations and resulting ANOVA are given in Tables 3.9.12 and 3.9.13.

The lack of fit is determined by subtraction with the appropriate tests indicated by arrows. Since there is a significant linear term and not a significant lack of fit, we conclude that a linear trend, Figure 3.3, adequately describes the relation between fertilizer and yield. The linear function is given by $\alpha_0 + \alpha_1 X = \bar{y}_{...} + m(\sum \bar{y}_{.j.} z_j)X/(\sum z_j^2) = (175.1)/5 + 10(48.5)X/10 = 35.02 + (48.5)X$, where $X = \lambda(F' - \bar{F'})$, $F' =$ Fe + (the spacing) − (the smallest Fe value)/(the spacing), and $\bar{F'}$ is the average value of F'. The resulting equation, in terms of the fertilizer level Fe, simplifies to 25.32 + (.485)Fe.

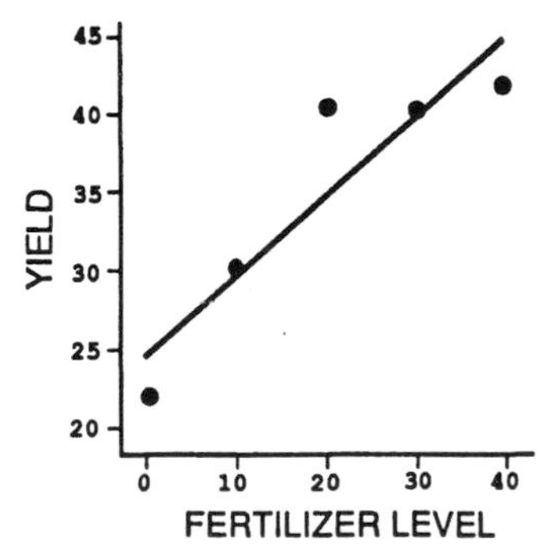

Figure 3.3 Plot of linear trend for fertilizer.

TABLE 3.9.12

CALCULATIONS FOR EXAMPLE 3, LINEAR TREND

Fertilizer Level	$\bar{y}_{\cdot j\cdot}$(m = 10)	$z_{(linear)}$	$\bar{y}$*z
0	22.3	−2	−44.6
10	30.4	−1	−30.4
20	40.6	0	0.0
30	40.1	1	40.1
40	41.7	2	83.4
	175.1	$\sum z^2 = 10$	48.5

$$SS(F_{linear}) = 10(48.5)^2/10 = 2352.25$$

Level	$\bar{y}_{1j\cdot} - \bar{y}_{\cdot j\cdot}$	$\bar{y}_{2j\cdot} - \bar{y}_{\cdot j\cdot}$	z_{linear}	$(\bar{y}_{1j\cdot} - \bar{y}_{\cdot j\cdot}) * z$	$(\bar{y}_{2j\cdot} - \bar{y}_{\cdot j\cdot}) * z$
0	0.5	−0.5	−2	−1.0	1.0
10	−2.0	2.0	−1	2.0	2.0
20	−5.6	5.6	0	0.0	0.0
30	−1.9	1.9	1	−1.9	1.9
40	1.1	−1.1	2	2.2	−2.2
			$\sum z^2 = 10$	1.3	−1.3

$$SS(\mathcal{F} * F_{linear}) = 5[(1.3)^2 + (-1.3)^2]/10 = 1.69$$

TABLE 3.9.13

ANOVA FOR EXAMPLE 3, LINEAR TREND

Source	df	SS	MS	F_{calc}	p-value
$\mathcal{F}_i$	1	124.82	124.82	7.66**	.0085
F_j	4	2847.08	711.77	10.19*	.0225
F_{linear}	1	2352.25	2352.25	1391.86*	.0171
$F_{lack\ of\ fit}$	3	494.83	164.94	1.78	.3235
$\mathcal{F}F_{ij}$	4	279.48	69.87	4.29**	.0056
$\mathcal{F}*F_{linear}$	1	1.69	1.69	0.10	.7491
$\mathcal{F}*F_{lack\ of\ fit}$	3	277.79	92.60	5.68**	.0024
$\varepsilon_{k(ij)}$	40	651.60	16.29		

To obtain Table 3.9.13 using SAS, create a variable, say F1, equal to the value of Fe but not declared as a class variable. Use PROC GLM with type I sums of squares, F1 appearing before Fe, and Fi*F1 appearing before Fi*Fe in the model statement. Fe will now correspond to the lack of fit term for F_i and Fi*Fe will correspond to the lack of fit for $\mathcal{F}F_{ij}$. Again, SAS will use the error term to test the linear trend F1 so the proper test must be either specified or hand-calculated. For a quadratic trend, simply compute another variable, say F2, equal to F1 squared, put this term after F1 in the model, and put Fi*F2 after Fi*F1 in the model. This generalizes to all terms and, for a linear fit, is illustrated in Table 3.9.14 below. An advantage of the SAS procedure is that the linear, *etc.*, terms will be calculated correctly even if the variable is not equally spaced. This is not true for the hand calculation method.

TABLE 3.9.14

SAS Statements and Output, Linear Trend on Fertilizers

Input Statements

```
DATA;
INPUT Fi Fe Y;
F1=Fe;
CARDS;
1 1 20
1 1 25
1 1 23
1 1 27
1 1 19
1 2 25
 ⋮
2 5 45
2 5 45
RUN;
PROC GLM;
CLASS Fi Fe;
 MODEL Y=Fi F1 Fe Fi*F1 Fi*Fe
 /SOLUTION SS1;
 TEST H=F1 E=Fi*F1;
 TEST H=Fe E=Fi*Fe;
RUN;
```

Output Condensed From SAS

SOURCE	DF	TYPE I SS	F	PR > F
Fi	1	124.82	7.66	0.0085
F1	1	2352.25	144.40	0.0001
Fe	3	494.83	10.13	0.0001
Fi*F1	1	1.69	0.10	0.7491
Fi*Fe	3	277.79	5.86	0.0024
ERROR	40	651.60		
TOTAL	49	3902.98		

SOURCE	DF	TYPE I SS	F	PR > F
F1	1	2352.25	1391.86	0.0171
Fe	3	494.83	1.78	0.3235

PARAMETER	ESTIMATE
INTERCEPT	25.32000000
F1	0.485000000

To reproduce Table 3.9.13 from the SAS output given in Table 3.9.14, replace the F VALUE and PR > F for F1 and Fe with the proper values printed below the ANOVA table. Note that F1 is the linear fit and Fe becomes the lack of fit, both having the proper df and SS. If a quadratic term was entered, F2 becomes the quadratic fit and Fe becomes the lack of quadratic fit for fertilizers.

A further refinement on this example is to note that $\mathcal{F}*F_{linear}$ is insignificant at the .25 level so can be pooled with error. Then, F_{linear} is tested by the pooled error term having 41 df. The result is $F_{1,41} = 147.63$ with a p-value of 0.0001. This does not change any of the conclusions of the study.

Example 4. In an experiment on the stress-rupture life of material used in turbine blades of aircraft engines, there were four different materials formed by adding more of a certain alloy and three different temperatures of interest. All levels were equally spaced and taken as fixed. The measured variable is stress-rupture life in hours. Only one observation per combination is taken. The model is $y_{ij} = \mu + M_i + T_j + MT_{ij} + \varepsilon_{(ij)}$. The data and analysis, including conservative tests, are given in Table 3.9.15.

TABLE 3.9.15

DATA AND ANALYSIS FOR EXAMPLE 4.

		Temperature		
		1	2	3
Material:	1	185	182	182
	2	175	183	184
	3	171	184	189
	4	165	191	189

ANOVA FOR EXAMPLE 4

Source	df	SS	MS	EMS	Conservative F_{calc}	$F_{.05}$
M_i	3	8.67	2.89	$\sigma^2 + 3\Phi(M)$	0.06	4.76
T_j	2	354.67	177.34	$\sigma^2 + 4\Phi(T)$	3.65	5.14
MT_{ij}	6	291.33	48.56	$\sigma^2 + \Phi(MT)$		
Total	11	654.67				

Since there are no df for error and error tests the effect of all factors and interactions, there are no direct tests on the effect of any factors. Conservative tests consist of testing M_i and T_j with the interaction MT_{ij}. If the test is significant, then we can conclude that the numerator is significant. However, both tests are insignificant. We are unable to conclude that either M_i or T_j are insignificant since it is possibile that MT_{ij} is significant.

Under such a circumstance, we still wish to get information on the main effects. To get such information, we need to know about the MT_{ij} interaction. Again orthogonal contrasts may provide an answer. The most probable cause of a significant MT_{ij} interaction would be the linear by linear interaction, Tukey (1949b). Therefore, we feel more secure in assuming the higher order components are insignificant and using them to test the main effects and the linear by linear component. To determine the SS for the linear by linear component, we must first derive the orthogonal coefficients. These are found by simply multiplying the coefficients for the linear term on temperature by the coefficients for the linear term on materials. The procedure is illustrated in Table 3.9.16.

We conclude that temperature is significant and the linear by linear piece of the material/temperature interaction is significant. We cannot conclude that material is insignificant since this is still a conservative test. Further analyses on the linear by quadratic terms, quadratic by quadratic term, *etc.*, could be performed. A conservative Duncan's test and conservative confidence bounds on material could be performed using the MT_{remain} term. True Duncan's tests and confidence intervals can only be calculated if an independent estimate of σ^2 were available or a few observations were repeated with the sole purpose of estimating σ^2. One can assume the higher order interactions negligible, but this cannot be verified statistically without further experimentation.

3.10 TYPE II ERRORS AND SAMPLE SIZE CALCULATIONS

In Section **2.6** we illustrated the computation of sample size, or equivalently the power, type II, or β error for a one factor fixed model. We specified the maximum distance δ between the means for which there is thought to be no practical difference. Then, fixing the number of levels, we computed $\Delta = \delta/\sigma$ and used Appendix 10 to find the sample size necessary to obtain a given α and β error. The standard deviation σ must either be known or estimated by $\sqrt{\text{MSE}}$ using a two-staged approach.

For a one-way random model, Appendix 11 can be used in a similar fashion. Let $\delta = \sigma_Y$ be the magnitude of the random component Y we wish to detect and let σ be the magnitude of the error component

TABLE 3.9.16

COEFFICIENTS AND TESTS FOR LINEAR M BY LINEAR T.

		Temperature (linear coefficients)		
		−1	0	1
Material (linear coefficients)	−3	3	0	−3
	−1	1	0	−1
	1	−1	0	1
	3	−3	0	3

$$SS(MT_{lin \times lin}) = \frac{[185(3) + 175(1) + 171(-1) + \cdots + 189(1) + 189(3)]^2}{(3)^2 + (1)^2 + (-1)^2 + \cdots + (1)^2 + (3)^2}$$

$$= \frac{[90]^2}{40} = 202.5 \ .$$

ANOVA FOR EXAMPLE 4, LINEAR BY LINEAR REMOVED.

Source	df	SS	MS	EMS	Cons. F_{calc}	$F_{.05}$
M_i	3	8.67	2.89	$\sigma^2 + 3\Phi(M)$	0.16	5.41
T_j	2	354.67	177.34	$\sigma^2 + 4\Phi(T)$	9.98*	5.79
$MT_{lin \times lin}$	1	202.50	202.50	$\sigma^2 + \Phi(MT_{lin \times lin})$	11.40*	6.61
MT_{remain}	5	88.83	17.77	$\sigma^2 + \Phi(MT_{remain})$		
Total	11	654.67				

*Significant at the .05 level

ε. Compute $\Delta = \delta/\sigma$ and, for a given α and β value, find Δ in the appropriate table, reading the number of repetitions J off the left hand side. Likewise, for given number of repetitions J, we can look up Δ and compute the minimal detectable standard deviation $\delta = \sigma\Delta$. If σ is not known or is impossible to estimate, run a small experiment and estimate σ by $\sqrt{\text{MSE}}$.

For example, in a one-way model $y_{ij} = \mu + A_i + \varepsilon_{j(i)}$ with I = 4, J = 3, A_i random, and $\sigma_\varepsilon = 3.5$, the minimal detectable standard deviation $\delta = \sigma_A$ $(= \sigma_\varepsilon\Delta)$ is $\Delta\sigma_\varepsilon = 2.605(3.5) = 9.12$. To detect a standard deviation of $\delta = 5.0$, we compute $\Delta = 5.0/\sigma_\varepsilon = 1.429$ and find that J = 8 repetitions must be run.

The computation of minimal detectable difference and minimal detectable standard deviation is much more complicated for general designs.

The minimal detectable difference will depend on the number of levels of the factor or interaction of interest, the df for the numerator, the appropriate denominator (not necessarily the error term) and its degrees of freedom, the coefficient of the Φ or σ^2 term in the numerator, and the α and β errors. For the one-way model, I and J determine all of the coefficients and degrees of freedom so a simple tabulation was possible.

For general balanced designs, let M be the total number of observations in the experiment. If we are interested in a fixed factor or interaction Y having L levels and D df and Y is directly tested by another factor X, the minimal detectable difference $\delta = \sqrt{\Phi(Y)}$ is given by

$$\delta = \phi\sqrt{(D+1)\text{EMS}(X)/(D \times C)} \qquad (3.10.1)$$

where, for a given α and β, ϕ must be read from charts supplied by Odeh and Fox (1991), depending on df(Y), df(X), and $C = M/L$, the number of times each level of Y is repeated. Note that C is also the coefficient preceding the $\Phi(Y)$ term in the EMS. Wheeler (1974) suggests a crude approximation to ϕ for $\alpha = .05$ and $\beta = .10$ which gives a simple to use formula. For df(X) ≤ 5, take $\phi = 2.8$, while for df(X) > 5, take $\phi = 2.2$. This leads to a simple approximation of the minimal detectable difference. Bowman and Kastenbaum (1974) point out that this approximation becomes more and more conservative as both D and C get large and the actual tables are not that difficult to use.

$$\delta \approx 2.8\sqrt{(D+1)\text{EMS}(X)/(D \times C)},\ \alpha = .05,\ \beta = .10,\ \text{df}(X) \leq 5, \qquad (3.10.2)$$

$$\delta \approx 2.2\sqrt{(D+1)\text{EMS}(X)/(D \times C)},\ \alpha = .05,\ \beta = .10,\ \text{df}(X) > 5. \qquad (3.10.3)$$

To compute the total size of the experiment, M, merely solve for C in equation (3.10.1) and note that $M = L \times C$ where L is the number of levels of the term Y. The resulting equation is

$$M = L \times C = \phi^2 L(D+1)\text{EMS}(X)/(\delta^2 D). \qquad (3.10.4)$$

Note that this must be performed for every fixed factor or interaction Y of interest, taking the largest experiment size calculated. Also note that ϕ will depend on the coefficient C which depends on the size of the experiment, so that several iterations are necessary before M is obtained.

Using the Wheeler approximation gives

$$M \approx 7.8L(D+1)\text{MS}(X)/(\delta^2 D),\ \alpha = .05,\ \beta = .10,\ \text{df}(X) \leq 5, \qquad (3.10.5)$$

$$M \approx 4.8L(D+1)\text{MS}(X)/(\delta^2 D),\ \alpha = .05,\ \beta = .10,\ \text{df}(X) > 5. \qquad (3.10.6)$$

Here MS(X) must be estimated and, to further complicate matters, may depend on the design and the number of levels of each factor. The best bet is to write down an initial design, compute M for each fixed factor or interaction of interest using (3.10.4), (3.10.5), or (3.10.6) and see if the number of observations in the initial design exceeds each calculated M. If not, increase the size of the design, recalculate the EMS table and start again. If it is too large, decrease the size of the design. This takes considerable effort.

Sample size calculations for the effect of random factors and interactions are generally ignored in the literature. Yet, these calculations are actually easier. Suppose we are interested in a random term Y, directly tested by X with ν_1 and ν_2 df, and with coefficient C immediately preceding σ_Y^2. Let $\Delta = \sigma_Y/\sigma_X = \sigma_Y/[\text{EMS}(X)]^{1/2}$. Under the alternate hypothesis, $\text{F}_{calc} = \text{MS}(Y)/\text{MS}(X) = (1 + C\Delta^2)F_{\nu_1,\nu_2}$. Setting $\Pr(\text{F}_{calc} > \text{F}_{.05}) = .90$ leads to the basic equation

$$\text{F}_{.90} = \text{F}_{.05}/\big(1 + C\Delta^2\big). \tag{3.10.7}$$

For a given experimental set up, the minimal detectable standard deviation δ ($=\sigma_Y$) is approximated by

$$\delta \approx \sqrt{\text{MS}(X)[\text{F}_{.05} - \text{F}_{.90}]/[C\,\text{F}_{.90}]}. \tag{3.10.8}$$

In a similar fashion, one can use the minimal detectable standard deviation δ to iteratively solve for the required sample size. This iterative technique is again complicated by the fact that the degrees of freedom ν_1 and ν_2, as well as the coefficient C, will change as the size of the experiment is changed. We solve for the coefficient C and get

$$M = L \times C = L \times \frac{\text{F}_{.05} - \text{F}_{.90}}{\Delta^2 \text{F}_{.90}}. \tag{3.10.9}$$

where L is the number of levels of Y and $\Delta = \delta/\sqrt{(\text{EMS}(X))}$. Rounding up on M, we get new df for ν_2 according to the rules for df given in Section **3.3**. This gives new values for $\text{F}_{.05}$, $\text{F}_{.90}$, and C which must be put back into equation (3.10.9). Generally, two or three iterations does the trick.

When approximate tests are used in a particular experiment, df are approximated using Satterthwaite's formula only after data has been collected. This makes it difficult to work with the type II errors, which depend on the df. However, one can *approximate* the approximate df using the harmonic mean of the df making up the approximate test. This will give some idea about the power of the approximate test. Suppose k terms make up the approximate test. The harmonic mean would be

$$\mathrm{df}_{harmonic} = \frac{k}{\sum_{i=1}^{k} 1/\mathrm{df}_i}. \quad (3.10.8)$$

3.11 REDESIGNING THE EXPERIMENT

Because changing the levels of any factor in an experiment changes the power for every single term in the experiment, it is extremely difficult to design the best possible experiment from theoretical terms. In addition, the variabilities of all random or mixed terms in the model are seldom known. However, use of minimal detectable sizes is extremely valuable when comparing several different designs. The experimenter can then logically make the trade-off between the size of the experiment and the size of an effect that the experiment is capable of detecting. The experimenter can also see how different experiments emphasize different factors or interactions and choose the design giving the proper emphasis.

Appendices 12 and 13 have been generated to easily facilitate computation of the minimal detectable difference, $\Delta = \delta/\sqrt{(\mathrm{EMS}(X))}$, for $\alpha = .05$ and $\beta = .10$. The minimal detectable difference Δ refers to the effect of the term, either $\delta = \sqrt{\Phi(Y)}$ or $\delta = \sigma_Y$, and is stated in units of the standard deviation of the denominator X in the appropriate F test. Use Appendix 12 for fixed effects ($\sqrt{\Phi(Y)}$), and Appendix 13 for random effects, (σ_Y).

Appendices 12 and 13 are used the same way. To find the minimal detectable size Δ for a given term Y, read the df for the numerator and the df for the denominator from the ANOVA table. Look up the appropriate value in the table and divide this value by the square root of the coefficient C preceding the term in the EMS for that term. Note also that C is the number of observations that go into the means for each value of Y. This Δ is the relative magnitude of the effect that can be detected with this particular experiment. It is to be used for comparative purposes only.

To further simplify comparisons, we recommend classifying the Δ values into easy to interpret categories illustrated in Figure 3.4. When Δ is less than 0.5, we say the experiment is capable of detecting *extremely small* differences among the factor levels. When Δ falls between 0.5 and 1.5, we say the experiment is capable of detecting *small* differences among the factor levels. When Δ falls between 1.5 and 3.0, we say the experiment is capable of detecting *medium* differences among the factor levels.

When Δ falls between 3.0 and 5.0, we say the experiment is capable of detecting *large* differences. When Δ exceeds 5.0, the experiment is only capable of detecting *extremely large* differences. Generally speaking, statistical procedures are not needed to detect *extremely large* differences. These differences will be detected by even casual observation of the process, making the experiment a waste of time and effort.

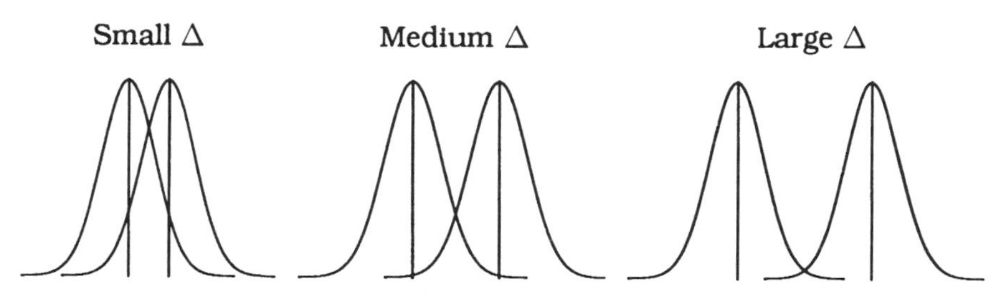

Figure 3.4. Detectability of 1.5σ, 3σ, and 5σ respectively.

Consider Example 1 dealing with metals, initiators, and pressures with the ANOVA given in Table 3.9.1. We can compute the minimal detectable differences for this experiment *before* any data is collected. Metals, M_i, is a random term having numerator and denominator df of 1 and 18 respectively and a coefficient $C = 18$. We look up the tabulated value in Appendix 13 and compute the minimal detectable difference $\Delta = 16.46/\sqrt{18} = 3.88$. This falls into the *large* detectable category. Initiators, I_j, is a fixed term having numerator and denominator df of 2 and 2 respectively and a coefficient $C = 12$. We look up the tabulated value in Appendix 12 and compute $\Delta = 6.711/\sqrt{12} = 1.94$. Similar calculations for the remaining terms give the ANOVA in Table 3.11.1.

Based on the information given in Table 3.11.1, the user can quickly evaluate the design. One immediately notes that there is a *large* difference for metals and *medium* differences for all other factors and interactions. If it is felt that metals is likely to have a large effect on the firing times of the switches, or there is not much interest in the effect of metals, this would be a fine design. However, one strategy for improving the switches is to replace the recycled material with material of a specific composition. Thus, good information about metals is needed. One possible way to improve the detectability of metals is to test three metal compositions instead of two. The results of this redesign are given in Table 3.11.2.

Comparing Tables 3.11.1 and 3.11.2 gives the trade-offs between two designs. Design 1 requires 36 runs while design 2 requires 54 runs and one extra batch of metal. However, for the extra effort, design 2 is capable of detecting small effects where design 1 can only detect medium or large

TABLE 3.11.1

ANOVA FOR EXAMPLE 1 ON EXPLOSIVE SWITCHES

Source	df	Δ	Size	EMS
M_i	1	3.88	LARGE	$\sigma^2 + 18\sigma_M^2$
I_j	2	1.94	MEDIUM	$\sigma^2 + 6\sigma_{MI}^2 + 12\Phi(I)$
MI_{ij}	2	2.33	MEDIUM	$\sigma^2 + 6\sigma_{MI}^2$
P_k	2	1.94	MEDIUM	$\sigma^2 + 6\sigma_{MP}^2 + 12\Phi(P)$
MP_{ik}	2	2.33	MEDIUM	$\sigma^2 + 6\sigma_{MP}^2$
IP_{jk}	4	1.79	MEDIUM	$\sigma^2 + 2\sigma_{MIP}^2 + 4\Phi(IP)$
MIP_{ijk}	4	2.27	MEDIUM	$\sigma^2 + 2\sigma_{MIP}^2$
$\varepsilon_{\ell(ijk)}$	18			σ^2
Total	35			

TABLE 3.11.2

ANOVA FOR EXPLOSIVE SWITCHES WITH THREE METALS

Source	df	Δ	Size	EMS
M_i	2	1.31	SMALL	$\sigma^2 + 18\sigma_M^2$
I_j	2	0.92	SMALL	$\sigma^2 + 6\sigma_{MI}^2 + 18\Phi(I)$
MI_{ij}	4	1.25	SMALL	$\sigma^2 + 6\sigma_{MI}^2$
P_k	2	0.92	SMALL	$\sigma^2 + 6\sigma_{MP}^2 + 18\Phi(P)$
MP_{ik}	4	1.25	SMALL	$\sigma^2 + 6\sigma_{MP}^2$
IP_{jk}	4	1.08	SMALL	$\sigma^2 + 2\sigma_{MIP}^2 + 6\Phi(IP)$
MIP_{ijk}	8	1.50	SMALL	$\sigma^2 + 2\sigma_{MIP}^2$
$\varepsilon_{\ell(ijk)}$	27			σ^2
Total	53			

effects. The statistician makes all of the calculations and points out the advantages and disadvantages of each design but cannot select the design for the user. Only the user knows the added effort in running 54 units rather than 36 units and only the user can have an idea of the magnitude of effects likely to occur using the levels selected for the experiment. Thus, the statistician and the user must work hand in hand to discover the best possible design for the user's particular needs.

Of course, neither of these two designs is necessarily the best for the user's needs. The only way to discover if a better design exists is to try different levels of the factors and compare the new design to existing designs. Again, all of the trade-offs must be made with the user's help. At first, the only way to be sure to get a good design is to compare many different designs. With experience, fewer and fewer designs will be tried before deciding on a best-possible trade-off.

A word of caution is appropriate before turning the reader loose on the real world. There will be circumstances where redesigning the experiment changes the denominator of the various tests. This can only be detected by inspection of the EMS terms. When the denominator changes, further knowledge about the variances in the model is necessary to compare designs. For example, one design may have small detectability with a large denominator while the other has a large detectability with a small denominator. Only knowledge about the magnitudes of the denominators will help make the decision. Upon careful comparison of the EMS values in Tables 3.11.1 and 3.11.2, we see that the denominators of all tests (the bases of the arrows) are the same. Thus, direct comparison of the sizes is meaningful. If some denominators differed, a more detailed comparison is necessary.

PROBLEMS

3.1 *Part of the reason randomization is slow is the vast number of unused random numbers caused by the 1 to 1 association. A faster way to randomize is to associate more than one random number with each of the experimental units. (a) Starting in columns 6 and 7, row 11 and working down, randomize the numbers 1 through 12 using the 1 to 1 method of Example 1. How many random numbers were searched? (b) Now associate 01, 13, 25, 37, 49, 61, 73, and 85 with 1; 02, 14, 26, 38, 50, 62, 74, and 86 with 2; etc. Starting with columns 41 and 42, row 27 and working down, again randomize the numbers 1 through 12. How many random numbers were searched this time? (c) Is the extra time spent associating numbers worth the decrease in search time in the table?*

3.2 *Another way to randomize N numbers is to select the first N unique random numbers and let the lowest number correspond to 1, the next lowest correspond to 2, etc. Starting in columns 23 and 24, row 22, and working down, randomize the numbers 1 through 12 using this method. Compare the time using this method to the times obtained in Problem* **3.1**.

3.3 *Give experimental circumstances under which complete randomization refers to the run order, the location, and both run order and location.*

3.4 *A factor often used in designs is Weeks. Give a set of circumstances under which Weeks could be considered a fixed factor and under which Weeks could be considered a random factor.*

3.5 *Write the general model for a four factor completely randomized design having three repeats. Use the letters A_i having two levels, B_j having three levels, C_k having four levels and D_ℓ having two levels. How many experimental units are required for this experiment?*

3.6 *Using A_i fixed at three levels and B_j fixed at two levels, sketch a picture having A_i significant, B_j and AB_{ij} insignificant. Sketch a picture having A_i and B_j insignificant but AB_{ij} significant. Sketch a picture having B_j insignificant, A_i and AB_{ij} significant.*

3.7 *Explain how color can be used to graphically display three factor interactions. Demonstrate by constructing a graph having a three factor interaction and another graph having a two factor interaction but not a three factor interaction.*

3.8 *For a three factor interaction ABC_{ijk} where A_i and B_j are fixed and C_k is random, list all the conditions made and compute $E[ABC_{ijk}ABC_{i'j'k'}]$ for all i, i', j, j', k, and k'.*

3.9 *Expand the formula for SS(AB) given in Table 3.3.1 and substitute the T notation for the $\bar{y}$ notation to get SS(AB) as given in Table 3.3.2.*

3.10 *Use the formulae in Table 3.3.2 to show $SS(AB) = \sum_{i=1}^{I}\sum_{j=1}^{J} T_{ij\cdot\cdot}^2/KL - CT - SS(A) - SS(B)$, paralleling the definition of a two factor interaction.*

3.11 *Again use Table 3.3.2 to show $SS(ABC) = \sum_{i=1}^{I}\sum_{j=1}^{J}\sum_{k=1}^{K} T_{ijk\cdot}^2/L - CT - SS(A) - SS(B) - SS(C) - SS(AB) - SS(AC) - SS(BC)$, paralleling the definition of a three factor interaction.*

3.12 *Use the rules given in Section* **3.3** *to derive the formulae for df, SS, and MS for the two factor completely randomized design given by (3.2.2).*

3.13 *Using the formula for MS(AB) derived in Problem* **3.12**, *assuming A_i fixed and B_j random, give all of the conditions of model (3.2.2) and algebraically derive EMS(AB).*

3.14 *Using the formula for MS(A) derived in Problem* **3.12**, *assuming A_i fixed and B_j random, give all of the conditions of model (3.2.2) and algebraically derive EMS(A).*

3.15 *Assuming A_i fixed and B_j random, use the rules in Section* **3.4** *to derive the EMS values for model (3.2.2).*

3.16 *Using the formulae derived in Problem* **3.12**, *show the product of MS(A) times MS(AB) equals zero for all y values. By the normality assumption, this shows the statistics MS(A) and MS(AB) are orthogonal. The EMS* *table shows that, under the null hypothesis of no effect due to A_i ($\Phi(A) = 0$), EMS(A) =EMS(AB). These two show that, under the null hypothesis, MS(A)/MS(AB) has an F distribution.*

3.17 *Starting with the definition of the four factor interaction $ABCD_{ijk\ell}$ as the differential effect of the $(ijk\ell)^{th}$ mean minus all main effects, minus all two factor interactions, and minus all three factor interactions, algebraically show the estimate is equivalent to that found by expanding the df formula $(I-1)(J-1)(K-1)(L-1)$.*

3.18 *Use algorithm 2 to derive the EMS in Table 3.4.4 and use algorithm 1 to derive the EMS in Table 3.4.5.*

3.19 *For a two factor model having A_i random at two levels, B_j fixed at four levels, and three repeats, give the model and ANOVA table including source, df, SS, MS, EMS, F_{calc}, $F_{.05}$, and $F_{.25}$ columns. Indicate direct tests by arrows.*

3.20 *For a three factor experiment having A_i random at three levels, B_j random at four levels, C_k random at two levels, and no repeats of the 24 treatment combinations, derive the EMS values and indicate the direct and approximate tests. Even though the experiment is not replicated, tests on all factor effects except ABC_{ijk} are available!*

3.21 *For a three factor experiment having A_i fixed at three levels, B_j fixed at four levels, C_k fixed at two levels, and only one experimental unit of each of the 24 treatment combinations, derive the EMS values and indicate the conservative tests. In this case, no direct or approximate tests are available without replication of the experiment!*

3.22 *For a three factor experiment with no repeats, derive the exact, approximate, and conservative tests under the following circumstances. When the test is conservative, indicate which conclusion can be made and which conclusion cannot be made. (a) All three factors are fixed. (b) Only the first factor is random. (c) Only the first factor is fixed. (d) All three factors are random.*

3.23 *Repeat Problem* **3.22** *assuming the three factor interaction is zero. This shows how important a simple assumption can be in some experimental set ups. Unfortunately, the statistician cannot make such an assumption without data and a specific test. However, if the experimenter fully understands what a three factor interaction is, this assumption may be made on the basis of non-statistical evidence. This shows the importance of communication between the statistician and the client.*

3.24 *In a three factor completely randomized design, indicate which terms can be assumed negligible and pooled under the following circumstances. Use sometimes pooling rules. (a) AB, AC, and ABC insignificant at the .25 level. (b) C, BC, and ABC insignificant at the .25 level. (c) AC and BC* *insignificant at the .25 level. (d) B, AB, BC, and ABC insignificant at the .25 level. (e) A and ABC insignificant at the .25 level.*

3.25 *Give formulae for each of the variance terms estimable in the design summarized in Table 3.4.4.*

3.26 *Use the design summarized in Table 3.4.4 to find* $\sigma^2_{\bar{y}_{AC1}-\bar{y}_{AC2}}$ *used to compare two or more means in the* AC_{ik} *interaction and to find* $\sigma^2_{\bar{y}_{AC}}$ *used for confidence intervals on a single* AC_{ik} *mean. Give the appropriate estimates and degrees of freedom.*

3.27 *Use the design summarized in Table 3.4.4 to find the coefficients appropriate for the* $A_{linear} \times C_{quadratic}$ *interaction. What term is used to test* $A_{linear} \times C_{quadratic}$*?*

3.28 *Using the three factor design in Table 3.4.4 with* A_i *and* C_k *fixed and* B_j *random, algebraically derive* $\mathrm{Var}(\bar{y}_{i\cdot k\cdot} - \bar{y}_{i'\cdot k'\cdot})$ *and* $\mathrm{Var}(\bar{y}_{i\cdot k\cdot})$ *and compare to Problem* **3.26**.

3.29 *Gage R & R. An important quality control concept concerning measuring devices is known as gage repeatability and reproducibility. Different operators measure different parts a number of times using the same gage. Repeatability is defined as the standard deviation due to error (repeats). Reproducibility is defined as the standard deviation due to operators and the operator-part interaction* ($= [\sigma^2_O + \sigma^2_{OP}]^{1/2}$)*. Gage R & R is defined as the standard deviation due to error, operators, and the part-operator interaction. A good gage has gage R & R less than 30% of the standard deviation due to parts. An experiment on 0 to 50 mph acceleration times of automobiles was run to determine gage R & R, with the results given below. Assuming the experiment was completely randomized, estimate repeatability, reproducibility, the standard deviation due to automobiles, and gage R & R. Is this a good gage?*

	Operator 1		Operator 2		Operator 3		Operator 4		Operator 5	
Auto-	Repeat		Repeat		Repeat		Repeat		Repeat	
mobile	1	2	1	2	1	2	1	2	1	2
1	3.77	3.76	4.13	4.34	3.81	3.96	4.20	4.56	4.28	4.20
2	3.90	4.05	3.81	4.05	4.04	3.98	4.24	4.24	4.48	4.42
3	4.03	4.19	4.20	3.74	4.54	4.20	4.86	4.45	4.44	4.35
4	4.09	4.22	4.02	4.47	4.27	4.28	4.45	4.12	4.49	4.44
5	4.38	4.26	4.09	4.21	4.13	4.06	4.28	4.22	4.31	4.66
6	4.21	4.14	4.13	4.38	4.26	4.38	4.45	4.39	4.46	4.40
7	4.00	3.96	4.06	3.99	4.20	4.25	4.69	4.35	4.70	4.66
8	3.94	4.52	3.81	3.94	4.34	4.25	4.85	5.08	4.38	4.66
9	4.16	4.34	4.53	4.41	4.37	4.83	4.73	4.34	4.45	4.70
10	4.03	3.91	3.84	4.09	3.91	4.41	4.34	5.06	4.49	4.69

3.30 *Use equation (3.10.8) to approximate the minimal detectable standard deviation for fuel tanks, injectors, engines, and the injector/engine interaction in Example 2. Note that you have to use the approximation to Satterthwaite's approximation for degrees of freedom on the main effects.*

3.31 *In Example 4, calculate the linear temperature by quadratic material SS and the quadratic temperature by linear material SS and give the new ANOVA table. In general, it is not desirable to use such higher order terms to form new conservative tests. Give two reasons why not.*

3.32 *In Example 2, power (measured by friction horsepower) was measured along with brake specific fuel consumption. Use ANOVA to estimate the contribution of each variance component to the overall variation in friction horsepower.*

		Fuel Tank 1				Fuel Tank 2			
	Injector:	1	2	3	4	1	2	3	4
Engine:	1	67.4	64.9	67.4	67.0	69.5	67.0	67.8	67.0
	2	65.3	66.6	66.6	66.1	69.5	68.3	67.8	68.3
	3	67.4	67.4	67.4	67.4	64.0	66.6	68.7	67.4
	4	67.4	67.4	66.1	65.7	64.5	67.4	68.7	67.0

3.33 *If there were four different materials in Example 4 rather than equally spaced intervals, it no longer makes sense to refer to the linear, quadratic, etc., effect of M_i. Yet we would still like to make a Tukey-like test by removing the linear temperature by (all of) material piece of the MT_{ij} interaction, having $1 \times 3 = 3$ df. This is accomplished by applying the linear coefficients for temperature to the means (which equals the data itself since there are no other factors or repeats) of each material and computing the sums of squares of the resulting numbers. The resulting formula for $SS(M \times T_{linear})$ is $\sum_{j=1}^{3}[\sum_{i=1}^{4} z_i(\bar{y}_{ij} - \bar{y}_{i\cdot})]^2/[\sum_{i=1}^{4} z_i^2]$, having J-1 degrees of freedom. (If there were a third subscript k, premultiply the expression by K and average over this subscript too. The general expression extends immediately.) Perform this Tukey-like test on the data in Example 4, give the new ANOVA table, and draw whatever conclusions can be made.*

3.34 *An experiment was run by a bacteriology organization to measure the effect of enzymes on growth of bacteria. Two different enzymes, 3975 and 2712, were applied to five different plates in concentrations of .125%, .25%, .5%, and 1% and the diameter of the colony of bacteria was measured. Assuming the experiment was completely randomized, use the following data to address the following. (a) What model describes this experiment and what assumptions are made? (b) Give the ANOVA table including source, df, SS, MS, EMS, and F values. (c) Give the ANOVA table after using sometimes pooling rules. (d) What conclusions can be made about enzymes and concentrations? (e) Give a 95% confidence interval on the* *diameter of a colony of bacteria if a .5% concentration of enzyme 3975 is used. (f) Give a 95% confidence interval on the difference between the bacteria diameter using a .5% concentration of 3975 and a .5% concentration of 2712. (g) Run Duncan's test on enzymes and concentrations. Was it necessary to run Duncan's test on enzymes? (h) Since each concentration is the same multiple of the previous concentration, concentrations are equally spaced on a log scale. Determine the linear SS and lack of fit using log concentration. How do you interpret this result?*

		Enzyme 3975				Enzyme 2712			
Concentration (%):		.125	.25	.5	1	.125	.25	.5	1
	1	24.0	25.2	26.4	27.6	25.0	26.0	26.8	27.8
	2	24.2	25.4	26.2	27.4	24.8	25.6	26.6	27.8
Plate:	3	24.6	25.4	26.4	27.6	25.0	25.6	26.8	28.2
	4	24.0	25.4	26.4	27.4	25.2	25.6	27.4	28.2
	5	24.8	25.0	26.4	26.8	27.0	27.6	28.4	29.8

3.35 *It visually appears as if there is a quadratic trend in Figure 3.3. Repeat the plot using 95% confidence intervals. It is now easy to see why the quadratic trend was not significant. This points out the danger of plotting just the mean values and not the confidence intervals.*

3.36 *An experiment was conducted on burr heights after a trimming operation for a certain sheet metal process. Two different coils of sheet metal, three different trimming angles, two different trimming speeds, and four repeats were used for the experiment. The data (in thousands of an inch) is given below. (a) Assuming a completely randomized design, write the model for this experiment. (b) If the inference is to be over all coils, decide which factors are random and which are fixed and use an EMS algorithm to determine all direct tests. (c) Perform the ANOVA, pooling if possible, and draw conclusions. (d) Run a Duncan's test wherever appropriate. (e) Find the order of the best fitting polynomial describing the relationship between angle of cut and burr height and give the polynomial.*

	Coil 1			Coil 2		
Angle:	15°	30°	45°	15°	30°	45°
low speed	31 26 34 27	31 31 34 39	35 36 34 28	29 30 29 23	32 38 30 29	37 42 37 36
high speed	29 23 26 26	28 32 29 29	34 37 32 35	22 23 29 25	29 30 28 26	33 35 33 32

3.37 *An experiment on the roughness of the finish on painted panels had two factors: sanding technique and supplier. Twelve panels of each combination were run and the roughness was measured in μ inches. Run an analysis of variance and use Duncan's test to help interpret the results. Do you see anything disturbing about the data?*

	SUPPLIER 1	SUPPLIER 2
WET SANDED	24 23 27 22 10 13 11 7 11 12 11 14	16 19 19 22 21 25 13 20 22 20 21 20
ORBITAL SANDED	23 28 26 23 24 24 22 25 29 11 22 19	27 29 26 26 24 21 21 29 26 32 28 25
NOT SANDED	7 7 7 7 7 8 7 7 7 6 7 8	8 9 8 9 10 9 7 10 7 7 8 12

3.38 *Test for homogeneity of variance among the ten field/fertilizer combinations in Example 3 of this chapter.*

3.39 *An experiment dealing with driving simulators had three factors of interest, the delay associated with two different computational methods, whether or not motion was simulated, and two different size vehicles. The accuracy of an avoidance maneuver was recorded. (a) Suppose two subjects were selected to try the maneuver for all eight combinations of delay, motion, and size. Write out the ANOVA table including source, df, EMS, and indicating F tests. (b) Under the same set up, the experimenters were worried about any learning effect biasing the results. Thus, they selected 16 different subjects and had two different subjects test each of the eight combinations. Write out the ANOVA table for this experiment. (c) Compare experiments (a) and (b). What would you recommend to the experimenters?*

3.40 *Comparison with a known standard. For fixed effects, the ANOVA test of the effect of a factor or interaction tests the equality of the parameters in the model. The parameters refer to differential mean effects. Experiments are sometimes run to test, not whether the means are equal, but whether the means are equal to a known standard. The analyses of such experiments are virtually the same as the usual ANOVA. The difference is that the overall mean $\bar{y}_{...}$ gets replaced with the known standard μ_0. All of the tests remain the same. Rewrite Table 3.3.1 if we are interested in comparing the effects to the known value 1.0. Note that SS(Total) is no longer the sum of the other SS terms. Why not? These tests are easy in SAS. Subtract μ_0 from each observation in the data step and specify /NOINT as an option in PROC GLM.*

3.41 *Taguchi suggestion. Finding the set of controllable conditions under which the process is least sensitive to uncontrollable conditions is a desirable quality goal. In sheet metal forming, the springback of formed parts was of interest. The average springback is easily adjusted by changing the bend angle, so consistency of springback is of utmost importance. Three different controllable factors, Gage - 12 and 15 mill, coded as 0 and 1 respectively; Type of steel - AK and standard, coded as 0 and 1; and Supplier - also coded as 0 and 1, were tested in a factorial fashion. The uncontrollable conditions of interest included amount of mill oil, position in the coil, factory temperature, and composition of the steel. While not controllable* *in production, the latter factors can be controlled in a laboratory setting. For each of the three controllable factors of interest, the same five extremal conditions were chosen for the uncontrollable factors. These five extremal conditions were used to compute one data point, $y = \ln s^2$. The goal of the experiment was to find the combination of G_i, T_j, and S_k that gave the most consistent springback, i.e., that minimized $\ln s^2$. The results are given below. Calculate the ANOVA table for this problem and indicate all direct, approximate, and conservative tests. Using sometimes pooling rules where appropriate, make recommendations to minimize variation.*

Treatment				Treatment				Treatment			
G_i	T_j	S_k	$\ln s^2$	G_i	T_j	S_k	$\ln s^2$	G_i	T_j	S_k	$\ln s^2$
0	0	0	1.76	1	0	0	1.13	0	1	0	1.46
0	0	1	1.38	0	0	0	1.11	1	1	0	0.40
1	0	1	0.59	1	0	0	1.14	0	1	1	0.90
0	1	0	1.39	0	0	1	0.90	1	1	1	0.14
1	1	0	0.38	1	0	1	0.62	0	1	1	1.00
1	1	1	0.06								

3.42 *Component switching. In an attempt to understand why one part is good and another part bad, engineers often take components from the good part and put them in the bad part (and vice-versa) to see if the bad part becomes good. (Or the good part becomes bad.) Assume there are two parts, P_1, good, and P_2, bad, and three components labeled A, B, and C. Design a three factor completely randomized experiment. Include a layout sheet that gives the engineer explicit instructions how to carry out the experiment and an ANOVA table with test indicated by arrows. State any assumptions that you make. Explain what it means for the A component to be significant. What would it mean if there was an AB interaction?*

3.43 *Compute the minimal detectable standard deviation for fields and the minimal detectable difference for fertilizers in Example 3.*

3.44 *Use the definitions of α (=.05) and β (=.10) error and the definition of an F variate to derive equations (3.10.7) and (3.10.8) from first principles.*

3.45 *In an experiment dealing with the amount of shrink in molten metal used for casting complicated parts, three factors were considered, the amount of aluminum in the metal, 40 and 44 ppm, the pouring spout type, p and m, and the temperature of the molten metal, 750 and 790 degrees. Two different castings for each combination were poured and the percentage shrink recorded. Write the mathematical model appropriate for this experiment, decide which factors are fixed and which are random, and analyze this data using ANOVA techniques. Where appropriate, plot the means and draw confidence limits. Estimate all random and mixed components. Summarize the findings in a report appropriate for metal casting experts who have no statistical background.*

Amount of Shrink in Molten Metal

aluminum:		40		44	
spout type:		p	m	p	m
temperature:	750	0.98 0.90	0.93 0.85	0.62 0.82	0.82 1.17
	790	0.52 0.62	0.58 0.60	0.68 0.57	0.77 0.68

3.46 *A persistent problem around our household is weight loss. There seems to be no difference in weight on the days we exercise and avoid snacks and the days we "pig out." We recognize that variation in the scale exists and wondered if there was also variation due to the spot on which the scale rested. Readings were taken at two different spots on the floor for a four week dieting period. An unknown constant has been subtracted from the data to protect our marriage. (a) Model and analyze the data from this experiment. Pull out the linear and quadratic trends to estimate the weight lost per day. Now plot the means and look for interesting patterns. This has been described as the increasing plateau phenomenon and shows how frustrating dieting can be. (b) If you wish to know your weight to within 1 pound, how many times should you step on the scale? Explain your answer.*

spot	readings on days 1 through 14
1	20.5, 18.0, 18.0, 18.0, 19.0, 19.0, 16.5, 15.5, 16.0, 15.5, 13.5, 14.5, 14.5, 12.5
1	20.0, 18.0, 17.5, 17.5, 18.0, 18.0, 16.0, 15.5, 15.0, 13.5, 13.5, 13.0, 13.0, 13.0
2	18.5, 18.5, 18.0, 17.5, 17.5, 18.0, 16.0, 15.0, 15.0, 13.5, 14.0, 13.0, 14.5, 13.5
2	19.0, 17.5, 17.0, 17.0, 17.5, 16.5, 15.5, 15.0, 14.0, 13.0, 13.0, 13.0, 12.5, 15.5

spot	readings on days 15 through 28
1	14.5, 15.5, 12.0, 13.0, 13.5, 14.5, 14.5, 11.5, 12.5, 12.0, 12.0, 10.5, 10.5, 11.0
1	15.0, 15.0, 12.5, 13.5, 11.0, 12.5, 12.5, 12.0, 11.0, 11.0, 13.0, 11.5, 10.5, 10.5
2	14.0, 14.0, 12.5, 13.0, 13.0, 12.0, 12.5, 12.0, 12.5, 11.0, 11.5, 11.0, 11.5, 11.0
2	13.5, 13.5, 12.0, 12.0, 12.0, 12.5, 11.5, 11.5, 11.0, 11.0, 11.5, 11.5, 10.0, 10.0

3.47 *In Example 4, assuming $\sigma^2 = 17.77$, find the number of repeats of the entire experiment necessary to determine a temperature effect as small as five hours. Use Wheeler's approximation as well as the exact computation.*

3.48 *For Problem* **2.7**, *compare the exact sample size obtained using Appendix 10 to the approximate sample size using each of Wheeler's approximations, (3.10.5) and (3.10.6). Why is there a difference?*

3.49 *In a one factor (random) design with 5 levels for the factor, use Appendix 11 to determine the sample size needed to detect a standard deviation of $\delta = 2.5$ if we know that $\sigma = 2$. Now use Appendix 2 and iterate on the* *number of repeats to calculate the sample size. You should get the same answer both ways.*

3.50 *For Problem* **2.7**, *compare the exact sample size obtained using Appendix 10 and using Appendix 12. If the two answers differ, explain why.*

3.51 *Compute* Δ *for each of the terms in Table 3.9.7 using the approximation to the approximate df given in (3.10.8) at the end of Section 3.10 where appropriate. (Do not use the pooled model or the actual MS's since the Deltas are supposed to be calculated before any data is collected.*

3.52 *Compute* Δ *for each of the terms in Table 3.9.8 after pooling. Compare to the* Δ *in Problem* **3.51**. *If one had sufficient knowledge about the process to know none of the interactions were important before data was collected, this increased detectability would have existed with the same experimental design. This shows some of the power associated with additional (non-statistical) knowledge of the process. This also shows the importance of pooling.*

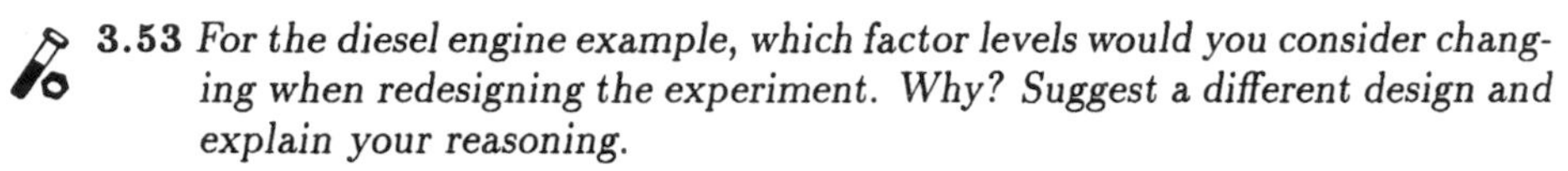

3.53 *For the diesel engine example, which factor levels would you consider changing when redesigning the experiment. Why? Suggest a different design and explain your reasoning.*

3.54 *For Example 3, would you be better off with 2 fields and 5 acres per field/fertilizer combination or with 5 fields and 2 acres per field/fertilizer combination? Keep in mind the experimenters were interested in the effect of fertilizers and the ease of running the experiment. Justify your answer.*

BIBLIOGRAPHY FOR CHAPTER 3

Anderson, V. L. and McLean, R. A. (1974). *Design of Experiments A Realistic Approach*, New York: Marcel Dekker. (Chapters 2, 4)

Bennett, C. A. and Franklin, N. L. (1954). *Statistical Analysis in Chemistry and the Chemical Industry*, New York: John Wiley & Sons. (Chapters 7, 8)

Bowman, K. O. and Kastenbaum, M. A. (1974). "Potential pitfalls of portable power," *Technometrics*, **16**, 349–352.

Box, G. E. P., Hunter, W. G. and Hunter, J. S. (1978). *Statistics for Experimenters An Introduction to Design, Data Analysis, and Model Building*, New York: John Wiley & Sons. (Chapters 6, 7, 10, 11)

Bozivich, H., Bancroft, T. A. and Hartley, H. O. (1956). "Power of analysis of variance test procedures for certain incompletely specified models," *Annals of Mathematical Statistics*, **27**, 1017–1043.

Cohen, A. (1974). "To pool or not to pool in hypothesis testing," *Journal of the American Statistical Association*, **69**, 721–725.

Cornfield, J. and Tukey, J. W. (1956). "Average values of mean squares in factorials," *Annals of Mathematical Statistics*, **27**, 907–949.

Das, M. N. and Giri, N. C. (1986). *Design and Analysis of Experiments*, 2nd ed., New York: John Wiley & Sons. (Chapter 3)

Davenport, J. M. and Webster, J. T. (1973). "A comparison of some approximate F-tests," *Technometrics*, **15**, 779–789.

Dixon, W. J. and Massey, F. J., Jr. (1983). *Introduction to Statistical Analysis*, 4th ed., New York: McGraw-Hill. (Chapter 10)

Draper, N. R. and Smith, H. (1966). *Applied Regression Analysis*, New York: John Wiley & Sons. (Chapter 9)

Fox, M. (1956). "Charts of the power of the F-test," *Annals of Mathematical Statistics*, **27**, 484–497.

Ghosh, S., editor. (1990). *Statistical Design and Analysis of Industrial Experiments*, New York: Marcel Dekker.

Guenther, W. C. (1964). *Analysis of Variance*, Englewood Cliffs, NJ: Prentice-Hall. (Chapter 5)

Hicks, C. R. (1982). *Fundamental Concepts in the Design of Experiments*, New York: John Wiley & Sons. (Chapters 5–10)

Hudson, J. D., Jr. and Krutchkoff, R. G. (1968). "A Monte Carlo investigation of the size and power of tests employing Satterthwaite's synthetic mean squares," *Biometrika*, **55**, 431–433.

John, P. M. W. (1971). *Statistical Design and Analysis of Experiments*, New York: The Macmillan Co. (Chapters 3–5)

Johnson, N. L. and Leone, F. C. (1977). *Statistics and Experimental Design in Engineering and the Physical Sciences*, 2nd ed., New York: John Wiley & Sons. (Chapter 13)

Kempthorne, O. (1975). "Fixed and mixed models in the analysis of variance," *Biometrics*, **31**, 473–486.

Lentner, M. M. (1965). "Listing expected mean square components," *Biometrics*, **21**, 459–466.

Lorenzen, T. J. (1977). "Derivation of expected mean squares and F-tests in statistical experimental design," *Research Publication GMR-2442*: Mathematics Department, General Motors Research Laboratories, Warren, MI 48090-9055.

Lorenzen, T. J. (1987). "A comparison of approximate F′ tests under pooling rules," *Research Publication GMR-5928*: Mathematics Department, General Motors Research Laboratories, Warren, MI 48090-9055.

Montgomery, D. C. (1991). *Design and Analysis of Experiments*, 3rd ed., New York, John Wiley & Sons. (Chapters 7, 8)

Odeh, R. E. and Fox, M. (1991). *Sample Size Choice - Charts for Experiments with Linear Models*, New York: Marcel Dekker.

Ostle, B. and Malone, L. C. (1988). *Statistics in Research*, 4th ed., Ames, Iowa: Iowa State University Press.

Peng, K. C. (1967). *The Design and Analysis of Scientific Experiments*, Reading, MA: Addison-Wesley. (Chapters 1–3, 5)

Petersen, R. G. (1985). *Design and Analysis of Experiments*, New York: Marcel Dekker.

Satterthwaite, F. E. (1946). "An approximate distribution of estimates of variance components," *Biometrics Bulletin*, **2**, 110–114.

Scheffé, H. (1956a). "A 'mixed model' for the analysis of variance," *Annals of Mathematical Statistics*, **27**, 23–36.

Scheffé, H. (1956b). "Alternative models for the analysis of variance," *Annals of Mathematical Statistics*, **27**, 251–271.

Scheffé, H. (1959). *The Analysis of Variance*, New York: John Wiley & Sons. (Chapters 4, 7–9)

Searle, S. R. (1971). *Linear Models*, New York: John Wiley & Sons. (Chapters 4, 6, 7, 9)

Steel, R. G. D. and Torrie, J. H. (1980). *Principles and Procedures of Statistics—A Biometrical Approach*, 2nd ed., New York: McGraw-Hill. (Chapter 8)

Taguchi, G. (1986). *Introduction to Quality Engineering*, White Plains, NY: Kraus International Publications. (Chapters 4–6)

Taguchi, G. (1988). *System of Experimental Designs*, White Plains, NY: Kraus International Publications. (Chapters 1, 5, 16, 21)

Tukey, J. W. (1949a). "Comparing individual means in the analysis of variance," *Biometrics*, **5**, 99–114.

Tukey, J. W. (1949b). "One degree of freedom for non-additivity," *Biometrics*, **5**, 232–242.

Welch, B. L. (1956). "On linear combinations of several variances," *Journal of the American Statistical Association*, **51**, 132–148.

Wheeler, R. E. (1974). "Portable power," *Technometrics*, **16**, 193–201.

Wheeler, R. E. (1975). "The validity of portable power," *Technometrics*, **17**, 177–179.

Wilk, M. B. and Kempthorne, O. (1955). "Fixed, mixed, and random models," *Journal of the American Statistical Association*, **50**, 1144–1167.

Winer, B. J. (1971). *Statistical Principles in Experimental Design*, 2nd ed., New York: McGraw-Hill. (Chapters 5–7)

Zyskind, G. (1962). "On structure, relation, Σ, and the expectation of mean squares," *Sankyhā A*, **24**, 115–148.

SOLUTIONS TO SELECTED PROBLEMS

3.4 *When considering the growth of animals, weeks would be considered fixed. When repeating an entire experiment a week later to measure the variability over time, weeks would be considered random. Of course, many other examples can be given. Two people may even interpret the same example differently, depending on their implied inference space.*

3.5 $y_{ijk\ell m} = \mu + A_i + B_j + AB_{ij} + C_k + AC_{ik} + BC_{jk} + ABC_{ijk} + D_\ell + AD_{i\ell} + BD_{j\ell} + ABD_{ij\ell} + CD_{k\ell} + ACD_{ik\ell} + BCD_{jk\ell} + ABCD_{ijk\ell} + \varepsilon_{m(ijk\ell)}$; $i = 1,2$; $j = 1,2,3$; $k = 1,2,3,4$; $\ell = 1,2$; $m = 1,2,3$. *There are 144 experimental units.*

3.6

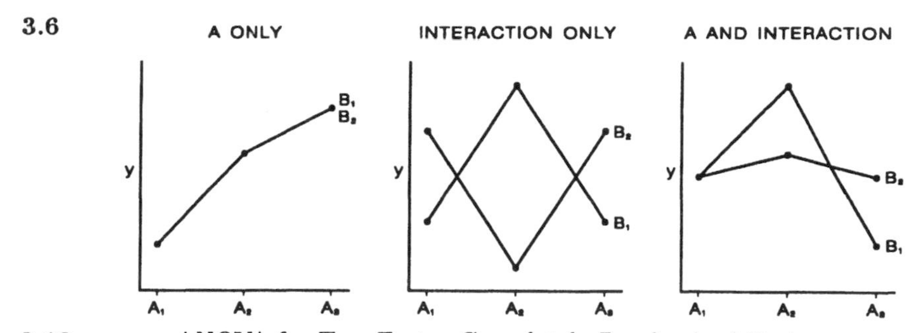

3.12 *ANOVA for Two Factor Completely Randomized Design*

Source	*df*	*SS*	*MS*
A_i	I-1	$JK\sum_{i=1}^{I}(\bar{y}_{i..} - \bar{y}_{...})^2$	SS(A)/(I-1)
B_j	J-1	$IK\sum_{j=1}^{J}(\bar{y}_{.j.} - \bar{y}_{...})^2$	SS(B)/(J-1)
AB_{ij}	(I-1)(J-1)	$K\sum_{i=1}^{I}\sum_{j=1}^{J}(\bar{y}_{ij.} - \bar{y}_{i..} - \bar{y}_{.j.} + \bar{y}_{...})^2$	SS(AB)/(I-1)(J-1)
$\varepsilon_{k(ij)}$	IJ(K-1)	$\sum_{i=1}^{I}\sum_{j=1}^{J}\sum_{k=1}^{K}(\bar{y}_{ijk} - \bar{y}_{...})^2$	SS(ε)/IJ(K-1)

3.15 *ANOVA for Two Factor Completely Randomized Design*

Source	*EMS*
A_i	$\sigma^2 + K\sigma^2_{AB} + JK\Phi(A)$
B_j	$\sigma^2 + IK\sigma^2_B$
AB_{ij}	$\sigma^2 + K\sigma^2_{AB}$
$\varepsilon_{k(ij)}$	σ^2

3.22 (a) All main effects and two factor interactions are conservatively tested by ABC. ABC is not testable. (b) AB tests B, AC tests C, and ABC tests BC. ABC conservatively tests A, AB, and AC. ABC is not testable. (c) B and C are tested by BC, and AB and AC are tested by ABC. A is approximately tested by $AB + AC - ABC$. BC is conservatively tested by ABC. ABC is not testable. (d) AB, AC, and BC are tested by ABC. A is approximately tested by $AB + AC - ABC$, B is approximately tested by $AB+BC-ABC$, and C is approximately tested by $AC+BC-ABC$. ABC is not testable. For all conservative tests, you may conclude the numerator is significant but cannot conclude the numerator is insignificant.

3.23 All conservative tests become direct tests. All approximate tests are still approximate, you simply subtract the error instead of the ABC term.

3.24 (a) Pool AB, AC, and ABC with the error term. (b) Pool BC and ABC with the error term. You cannot pool C since AC is not insignificant at the .25 level. (c) Cannot pool anything. (d) Pool B, AB, BC, and ABC with the error term. (e) Pool ABC with the error term.

3.25 $\hat{\sigma}^2_B = [MS(B) - MS(\varepsilon)]/12$. $\hat{\sigma}^2_{AB} = [MS(AB) - MS(\varepsilon)]/6$. $\hat{\sigma}^2_{BC} = [MS(BC) - MS(\varepsilon)]/4$. $\hat{\sigma}^2_{ABC} = [MS(ABC) - MS(\varepsilon)]/2$. $\hat{\sigma}^2 = MS(\varepsilon)$.

3.26 For 2 or more means, the theoretical variance is $\sigma^2_{\bar{y}_{AC1}-\bar{y}_{AC2}} = (\sigma^2 + 2\sigma^2_{ABC})/8$. The estimate, $s^2_{\bar{y}_{AC1}-\bar{y}_{AC2}} = MS(ABC)/8$ having 6 df. For a single mean, the estimate $\sigma^2_{\bar{y}_{AC}} = \sigma^2_B/J+(I-1)\sigma^2_{AB}/IJ+(K-1)\sigma^2_{BC}/JK+(I-1)(K-1)\sigma^2_{ABC}/IJK + \sigma^2/JL$. The estimate, $s^2_{\bar{y}_{AC}} = MS(B)/48 + MS(AB)/48 + MS(BC)/24 + MS(ABC)/24$ with df approximately $[MS(B)/48 + MS(AB)/48 + MS(BC)/24 + MS(ABC)/24]^2$ / $\{[MS(B)/48]^2/3 + [MS(AB)/48]^2/3 + [MS(BC)/24]^2/6 + [MS(ABC)/24]^2/6\}$.

3.27 The coefficients for AC_{11}, AC_{12}, AC_{13}, AC_{21}, AC_{22}, and AC_{23} are -1, 2, -1, 1, -2, and 1. $A_{linear} \times C_{quadratic}$ is tested by $A_{linear} \times B \times C_{quadratic}$ with 1 and 3 df.

3.35

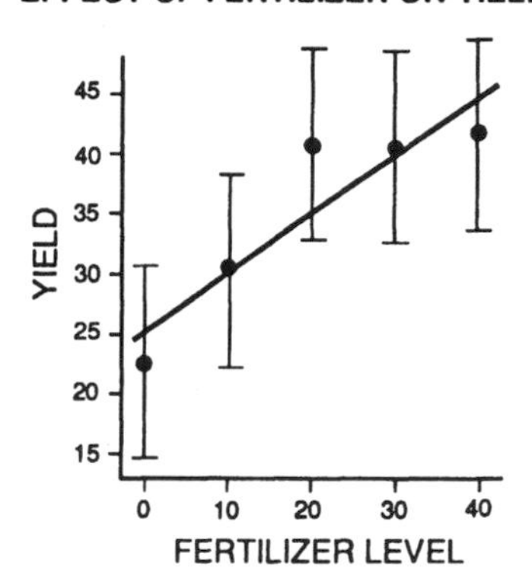

3.51 *The Δ are 1.86, 1.86, 1.84, 5.82, 2.19, and 2.19 repectively for E, I, EI, F, EF, and IF. The main effects are based on the approximation to the df, rounding down to the next integer. With the exception of F, all terms have medium detectability.*

3.53 *A design that I consider superior has 3 levels for each of E, I, and F. In only 27 runs, all factors and interactions have medium detectability (2.65 for main effects and 2.17 for interactions). A good 36 run design has one factor at 4 levels, the other two at 3 levels. The factor at 4 levels has* $\Delta = 1.56$ *with the other factors having* $\Delta = 1.98$. *The interactions have* $\Delta = 1.60$ *or* $\Delta = 1.71$. *Let the 4 level factor be the one requiring the most knowledge or, in the absence of such information, the cheapest to repeat.*

4

Nested Designs

4.0 INTRODUCTION

The previous chapter covered factorial designs, characterized by every level of every factor appearing equally often with every level of every other factor. In this chapter we consider the *nesting* concept. A factor, B, is considered nested in another factor, A, if the levels of B *differ* depending on which level of A is present. Under such a circumstance, we say that B is nested within A. To distinguish between the factorial concept and the nesting concept, we say that a factor is *crossed* if every level of that factor appears equally often with every level of every other factor and we say that a factor is *nested* if its levels *depend on* one or more factors.

Nesting often occurs naturally in a process and must merely be recognized. Such is the case when several parts are made from different materials. A part is made from one type of material or the other, it cannot be switched back and forth. Thus, it must be nested.

Other times nesting is built into a design for specific reasons. For example, different students may be used in an experiment on teaching methods because there is concern about learning effects. In this case, the students *could* be taught using all of the teaching methods so a factorial design is possible. However, the experimenter does not *wish* to use a factorial design since the results could be confounded by learning effects.

There are many synonyms that may indicate nesting in a design. Some of these are *depends on, different for, within, in,* and *each.* For example, when an experiment is described as having four *different* subjects for *each* treatment, subjects will be nested in treatments. If the experiment is described as having four subjects for *each* treatment, the language is ambivalent but it is likely that subjects are nested within treatments. You have to ask whether the experiment uses the *same* subjects for each treatment or *different* subjects, corresponding to crossing and nesting respectively.

As we will see, both the nested concept and the factorial (crossed) concept can exist in the same design. Nesting need not be associated with one factor only; a factor can be nested within combinations of two or more factors and even nested within a factor that is nested itself. All combinations are possible.

Nesting will be noted in the subscript of the factor. We again use the next available index but add parentheses around the subscripts associated with the nesting factor(s). Parentheses can be read *depending on* or *nested within*. Use of the subscript notation will be thoroughly discussed.

In this chapter, we will first demonstrate how to recognize a nested factor. Then we will show how a nested factor is denoted and its effect on the mathematical model and associated tests. The general rules that were derived in Chapter **3** will apply to nested designs so will not be restated. Examples of all concepts are given.

4.1 LAYOUT OF THE EXPERIMENT

Consider the following experiment. To determine if there was an inherent sexual difference in IQ tests, a psychologist randomly selected ten males and ten females and gave them two different IQ tests. Of course, the order of the tests was completely randomized. It was assumed that no learning took place from test to test.

The factors in this experiment are sex, S, people, P, and test, T. The subscripts have been purposely left off of the factors since these are to be determined. Clearly there are two sexes and two different IQ tests so we have, respectively, S_i, $i = 1, 2$ and T_k, $k = 1, 2$. However, there are a total of twenty people, ten of which are male and a different ten of which are female. Since the males are different people than the females, people are nested within sex and we have $P_{j(i)}$, $j = 1, \cdots, 10$. Note that the capital letter representing the factor, P, gives no indication of the nesting. Rather, the subscript $j(i)$ (read *j within i* or *j depending on i*) denotes the nesting, indicating that persons are nested within sex. Note also that j ranges from one to ten, indicating that there are ten *different* people for *each* sex (for *each* i) resulting in a total of twenty people used in the experiment.

Had we let j run from one to twenty, we would have indicated that twenty different people for each sex, or a total of forty people, were used in the experiment. This would be incorrect.

Had we not indicated the nesting, each person would have to take the test both as a male and as a female. While this is not an impossibility, it requires an expensive and dangerous operation that is not recommended for scientific experiments. (In case you missed it, that was a joke!)

The key to recognizing the nesting in this example is the recognition that the ten males had to be *different* people than the ten females. A given person takes the experiment as a male or as a female, not as both. For example, the male labeled one is a *different* person than the female labeled one. Therefore, the factor people depends on the factor sex, we have people nested within sex, and j depends on i.

Once you know what to look for, nesting is fairly easy to detect when the factors appear in a nice sequential order (as in the last example). However, when the factors are presented out of order, nesting is harder to detect. This is illustrated by the next example, a slight modification of the first example.

Suppose the psychologist feared that learning would take place from test to test and influence the results of the experiment. In that case, each person should only take one test. To keep the total size of the experiment the same, twenty males and twenty females would have to be tested. Of the twenty males, ten would be randomly selected to take test 1 and the other ten would take test 2. Of the twenty females, ten would take test 1 and the other ten would take test 2.

This is a different experiment even though the three factors S, P, and T are the same. It is clear there is nesting since a person cannot be both male and female. It is tempting to nest people within sex with a subscript j ranging from one to twenty, since there are twenty males and twenty females in the experiment. This would be incorrect since it would imply each person takes both test 1 and test 2, for a total of eighty test scores.

In fact, each person only takes one test and this is the key to recognizing people as nested within *both* S and T. The factors in this experiment are S_i, $i = 1, 2$; T_k, $k = 1, 2$; and $P_{j(ik)}$, $j = 1, \cdots, 10$. Note that there are ten *different* people for each ik combination for a total of $10 \times 2 \times 2 = 40$ different people in the experiment. There are twenty males and twenty females. Twenty people take test 1 and twenty take test 2. Checking the total number of levels of nested factors is a good way to prevent errors. Nesting can occur within one or many factors.

Nesting can also occur within nested factors. A good example comes from the polling industry. They wish to make inferences about different states. To do so, they sample different counties from each state. Within each county they sample towns and within each town they sample individuals. The nesting is hierarchical in nature, individuals nested in towns, nested in counties, nested in states. The factors for such an experiment would be denoted S_i, $C_{j(i)}$, $T_{k(ij)}$, and $I_{\ell(ijk)}$. Note that towns are nested within counties so that k depends on j. But since j depends on i, k automatically depends on i. It is important to include both i and j in the subscript for towns to make the rules in Chapter **3** come out correctly.

Some confusion always seems to arise about the error term ε. As we have been using it in Chapters **2** and **3**, ε is nested within all of the factors of the experiment. This is correct since we are measuring a *different* experimental unit for each repeat of a particular combination of factors. Factorial experiments do not seem to cause difficulties.

Experiments involving nesting can cause difficulties. Take the polling example for instance. If we define the experimental procedure as interviewing an individual, then individuals is not a factor but simply a repeat of the experimental process. In this case, the factors are S_i, $C_{j(i)}$, $T_{k(ij)}$, and $\varepsilon_{\ell(ijk)}$ and there is no factor I. However, if we include the factor $I_{\ell(ijk)}$ in the experiment, the error term becomes $\varepsilon_{(ijk\ell)}$. (There is really a subscript m having one level but we generally omit subscripts having only one level.) The difficulty arises in deciding which representation is correct.

Actually, both representations are correct and yield the same experiment, analysis, and interpretation. Our recommendation is to write down a factor like individuals unless you are 100% certain that this factor merely represents repeats of the entire experiment. Adding one too many factors will never misrepresent an experiment. Forgetting a factor may misrepresent the experiment and could cause incorrect conclusions.

These ideas are further reinforced in the examples in Section **4.4**.

4.2 GENERATING THE MATHEMATICAL MODEL

For factorial designs, we generate the mathematical model by forming all possible interactions and combining the subscripts for each interaction. The same procedure is used for nested factors except a nested factor cannot interact with any factor(s) it is nested within. This is because the nested factor has different levels for each level of the factor it is nested within so there is no basis for computing an interaction.

The first example involving IQ scores illustrates the procedure quite well. The factors in the experiment are S_i, $i = 1, 2$; $P_{j(i)}$, $j = 1, \cdots, 10$; and T_k, $k = 1, 2$. Since P is nested within S (indicated by the subscript $j(i)$), there can be no PS interaction and the model becomes

$$y_{ijk} = \mu + S_i + P_{j(i)} + T_k + ST_{ik} + PT_{jk(i)} + \varepsilon_{(ijk)}, \tag{4.2.1}$$

with $i = 1, 2$; $j = 1, \cdots, 10$; and $k = 1, 2$. When forming an interaction with a factorial and a nested factor, like PT, all of the subscripts get combined, including the subscripts within the parentheses. By convention we put all of the subscripts outside of the parentheses first and all of the subscripts within parentheses last. For the PT interaction, the subscripts become $jk(i)$ as indicated in equation (4.2.1).

A subscript can never appear inside and outside the parentheses for the same term in the model. Should such a phenomenon occur, you can be sure you have formed an interaction that cannot exist. For example, should one try to form the SP interaction, the subscript would become $ij(i)$. Since the subscript i cannot depend on itself, the SP interaction cannot exist.

For complicated nesting patterns, interactions can occur that have different indices nested in different terms. In this case, simply combine all of the indices into one set of parentheses. An example illustrates the procedure. Suppose B is nested in A and D is nested within C. The factors are represented as A_i, $B_{j(i)}$, C_k, and $D_{\ell(k)}$. B cannot interact with A and D cannot interact with C. The BD interaction is represented as $BD_{j\ell(ik)}$, with the nesting subscripts i and k combined into one set of parentheses.

The second example involving IQ scores, where each person takes only one test (*i.e.*, where people are nested in both sex and test), has the following mathematical model:

$$y_{ijk} = \mu + S_i + P_{j(ik)} + T_k + ST_{ik} + \varepsilon_{(ijk)}, \tag{4.2.2}$$

with $i = 1, 2$; $j = 1, \cdots, 10$; and $k = 1, 2$. $P_{j(ik)}$ does not interact with either S_i, T_k, or any interaction involving S_i or T_k since $P_{j(ik)}$ is nested within both S_i and T_k. While this model is technically correct and will lead to the correct ANOVA, tests, and estimates, it is bothersome to have k appear as a subscript before it appears in the main effect T_k. It is less bothersome to have the nested factor appear after S_i , T_k, and ST_{ik}:

$$y_{ijk} = \mu + S_i + T_k + ST_{ik} + P_{j(ik)} + \varepsilon_{(ijk)}, \tag{4.2.3}$$

$i = 1, 2$; $k = 1, 2$; and $j = 1, \cdots, 10$. It is even better if we recognize the proper nesting up front and label the indices in lexicographic order—S_i, T_j, and $P_{k(ij)}$—giving the preferred model:

$$y_{ijk} = \mu + S_i + T_j + ST_{ij} + P_{k(ij)} + \varepsilon_{(ijk)}, \tag{4.2.4}$$

$i = 1, 2$; $j = 1, 2$; and $k = 1, \cdots, 10$.

As a final example, we write the mathematical model for the polling example with the factor individuals included in the design. The factors are S_i, $C_{j(i)}$, $T_{k(ij)}$, and $I_{\ell(ijk)}$ which generates the model

$$y_{ijk\ell} = \mu + S_i + C_{j(i)} + T_{k(ij)} + I_{\ell(ijk)} + \varepsilon_{(ijk\ell)}. \tag{4.2.5}$$

Since each factor is nested within the prior factor, there are no interactions present at all. Such experiments are called hierarchical designs.

4.3 THE ANOVA TABLE, TESTS, AND ESTIMATES

As in Chapter **3**, the ANOVA table consisting of df, SS, and MS can be generated directly from the mathematical model. All of the rules of that chapter apply to this chapter as well. To compute the EMS, which leads to tests, power, and estimates of effects, we have to know whether the factors are considered random or fixed. In most textbooks, and virtually every example, nested factors are considered random. The rules for EMS given in Chapter **3** were derived *assuming* nested factors will be random. When a nested factor is not random, special care must be taken in the representation of the factors. This will be discussed in Sections **4.3.2** and **4.3.3**.

Section **4.3.1** covers the common case when all nested factors are random. Derivation of EMS, tests, and estimates of parameters follow the rules derived in Chapter **3**. Sections **4.3.2** and **4.3.3** are appropriate when some nested factor is considered fixed. Section **4.3.2** shows what to do when the levels of the nested factor are different but the *intention* is to have the levels the same. Section **4.3.3** shows what to do when the intention is to have different levels. Examples are given to help clarify the differences between the sections.

4.3.1 Random Nested Factors

Generally speaking, when a factor is nested it is representative of an entire population and considered random. For example, the nested factor may be repeats of a certain level of a factor. Certainly, the inference is to all possible repeats, not the specific repeats chosen, and is therefore random. Another common example of nesting occurs with sub-units or sub-processes. The inference again is to all possible sub-units so the nested factor is random.

The rules for deriving EMS, tests, power, and effects are given in Chapter **3**. A reader not yet familiar with the rules for EMS should turn back at this time and reread the rules until they are fully understood. In particular, pay close attention to the rules involving parentheses.

The ANOVA for the first example involving IQ scores is given in Table 4.3.1. The mathematical equation is given by equation (4.2.1). The factor sex is fixed since there are only two possibilities and the factor people is random since the inference is to all males and all females. The factor test could be considered either random or fixed depending on the implied inference. The inference here is supposed to be to all IQ tests of the same form, so test (T) will be considered a random factor.

As can be seen in Table 4.3.1, there is an approximate test on S_i using $P_{j(i)} + ST_{ik} - PT_{jk(i)}$ and direct tests on the remaining terms using the $PT_{jk(i)}$ term. As long as you are not interested in the PT interaction, this is a good design even though there are no repeats of the entire experiment.

The ANOVA table for the polling example is given in Table 4.3.2. We take states, S_i, as fixed since we know which states are to be polled. All of the rest of the factors are considered random since they are representative of the entire population from which they were randomly selected. We assume there are enough counties, towns, and individuals to treat these as theoretically infinite populations.

TABLE 4.3.2

ANOVA FOR THE POLLING EXAMPLE GIVEN BY (4.2.5)

Source	df	SS	EMS
S_i	$I-1$	$JKL \sum_{i=1}^{I} (\bar{y}_{i\cdots} - \bar{y}_{\cdots\cdot})^2$	$\sigma^2+\sigma_I^2 + L\sigma_T^2 + KL\sigma_C^2 + JKL\Phi(S)$
$C_{j(i)}$	$I(J-1)$	$KL \sum_{i=1}^{I} \sum_{j=1}^{J} (\bar{y}_{ij\cdot\cdot} - \bar{y}_{i\cdots})^2$	$\sigma^2 + \sigma_I^2 + L\sigma_T^2 + KL\sigma_C^2$
$T_{k(ij)}$	$IJ(K-1)$	$L \sum_{i=1}^{I} \sum_{j=1}^{J} \sum_{k=1}^{K} (\bar{y}_{ijk\cdot} - \bar{y}_{ij\cdot\cdot})^2$	$\sigma^2 + \sigma_I^2 + L\sigma_T^2$
$I_{\ell(ijk)}$	$IJK(L-1)$	$\sum_{i=1}^{I} \sum_{j=1}^{J} \sum_{k=1}^{K} \sum_{\ell=1}^{L} (y_{ijk\ell} - \bar{y}_{ijk\cdot})^2$	$\sigma^2 + \sigma_I^2$
$\varepsilon_{(ijk\ell)}$	0	—	—
Total	$IJKL-1$	$\sum_{i=1}^{I} \sum_{j=1}^{J} \sum_{k=1}^{K} \sum_{\ell=1}^{L} (y_{ijk\ell} - \bar{y}_{\cdots\cdot})^2$	

In hierarchical designs, each factor tests the preceding factor. In addition, the degrees of freedom for each factor progressively increase. This is very helpful in the design stages of the experiment, implying that fewer and fewer repeats are necessary for the latter factors. For example, if five states were chosen but only two counties, two towns, and two individuals were chosen, the df for states would be 4, the df for counties would be 5, the df for towns would be 10, and the df for individuals would be 20. By increasing the number of individuals sampled from two to three, the

TABLE 4.3.1

ANOVA for (4.2.1)—S_i Fixed, $P_{j(i)}$ and T_k Random

Source	df	Δ	SS	EMS
S_i	1	1.52	$20\sum_{i=1}^{2}(\bar{y}_{i..}-\bar{y}_{...})^2$	$\sigma^2+\sigma_{PT}^2+10\sigma_{ST}^2+2\sigma_P^2+20\Phi(S)$
$P_{j(i)}$	18	1.24	$2\sum_{i=1}^{2}\sum_{j=1}^{10}(\bar{y}_{ij.}-\bar{y}_{i..})^2$	$\sigma^2+\sigma_{PT}^2+2\sigma_P^2$
T_k	1	3.68	$20\sum_{k=1}^{2}(\bar{y}_{..k}-\bar{y}_{...})^2$	$\sigma^2+\sigma_{PT}^2+20\sigma_T^2$
ST_{ik}	1	5.20	$10\sum_{i=1}^{2}\sum_{k=1}^{2}(\bar{y}_{i.k}-\bar{y}_{i..}-\bar{y}_{..k}+\bar{y}_{...})^2$	$\sigma^2+\sigma_{PT}^2+10\sigma_{ST}^2$
$PT_{jk(i)}$	18	—	$\sum_{i=1}^{2}\sum_{j=1}^{10}\sum_{k=1}^{2}(y_{ijk}-\bar{y}_{ij.}-\bar{y}_{i.k}+\bar{y}_{i..})^2$	$\sigma^2+\sigma_{PT}^2$
$\varepsilon_{(ijk)}$	0	—	—	—
Total	39		$\sum_{i=1}^{2}\sum_{j=1}^{10}\sum_{k=1}^{2}(y_{ijk}-\bar{y}_{...})^2$	

df for individuals would jump from 20 to 40. By increasing the number of individuals sampled from three to four, the df for individuals would increase to 60. And so on. Sampling more individuals requires minimal additional effort.

Most major statistical packages calculate df and MS values for nested factors. The user must learn the notational requirements of the system being used. For example, the SAS system does not require the use of subscripts. Rather, nesting is noted with parentheses around the factor names. In particular, the ANOVA table for the first example concerning IQ tests, equation (4.2.1), would be obtained using the following commands: PROC ANOVA; CLASS S P T; MODEL Y = S P(S) T S*T P*T(S); RUN;. The notation P(S) indicates that persons are nested within the sex factor and P*T(S) represents the interaction between persons and tests. As in the previous chapter, the F-values must be modified since SAS treats all factors as fixed and tests them with the error term. For this particular example, there are no degrees of freedom for error so SAS will not compute any F-values. SAS will produce the proper F-values indicated in Table 3.4.1 with the addition of the commands TEST H = P(S) T S*T E = P*T(S);. There is currently no way to make SAS compute the approximate test.

If you only have access to a computer package that computes ANOVA for factorial designs, you can hand modify the output to obtain the ANOVA for designs containing nesting. The trick is to realize that the df and SS for a nested factor consists of the df and SS for the main effect plus all of its interactions with the factors it is nested within. The computation of the MS for a nested factor is the sum of the appropriate SS divided by the sum of the appropriate df.

Equation (4.2.1) illustrates the point quite well. Treating this design as factorial in nature yields df and SS for S, P, $S*P$, T, $S*T$, $P*T$, and $S*P*T$. We then compute SS($P(S)$) = SS(P) + SS($S*P$), df($P(S)$) = df(P) + df($S*P$), SS($P*T(S)$) = SS($P*T$) + SS($S*P*T$), and df($P*T(S)$) = df($P*T$) + df($S*P*T$). Calculation of MS and F values is exactly as before. The generalization to any nested design should be quite straightforward.

4.3.2 Sliding Factors

Sometimes factors will appear to be nested when they are not meant to be. This can happen when, for example, the low and high values of one factor depend on the level of another factor, either from some known scientific principle or because the second factor implies different parts must be used. Yet we are still interested in knowing the overall effect of the factor that appears to be nested.

Such factors will be called *sliding factors*, Taguchi (1988). They will be treated as if they are factorial in nature. That is, despite the fact that sliding factors appear to be nested, they will be treated as if they were crossed with the other factors. Specifically, the indices for sliding factors will not contain parentheses and sliding factors will interact with all other factors.

Sliding factors occur most frequently due to labeling practices. An example illustrates the point. Suppose an experiment deals with the formability of body panels. (Formability is the ability to bend a flat panel into an arbitrary shape.) Two different factors, material and thickness, will be considered. Specifically, the materials considered are sheet metal and sheet molded compound, SMC. A thin piece of sheet metal is .7 mm thick. A thick piece of sheet metal is 1.2 mm thick. Because SMC is not as strong as sheet metal, a thin piece of SMC is 1.5 mm thick while a thick piece is 5 mm thick. Because the levels of thickness are .7 and 1.2 for sheet metal and 1.5 and 5 for SMC, it would appear that thickness is nested in material.

Yet, if one labels the factor thickness as thin and thick, thickness would appear to be crossed with material. (Both thin and thick appear with each material). This is a dilemma because thickness cannot both be nested and crossed at the same time.

The key to solving this dilemma is recognizing the intent of the experimenter. In this case, the experimenter was interested in discovering if the thickness of the panel affected formability. The actual labels were of no consequence. As usual, should the material/thickness interaction be significant, the experimenter would then have to consider the four material/thickness combinations to determine which is most formable.

Along the same lines as the formability example, sliding factors may simply be a surrogate for a factor that cannot be easily measured. An example is given by the ideal gas law, $PV \sim RT$, with the volume V fixed. The experiment concerns two different rates, R, and two different temperatures, T. However, it is difficult to measure R so the surrogate factor pressure, P, is used. P will appear as a sliding factor since, for a given value of R, P will depend on T.

Recognition of sliding factors does not involve statistical reasoning. Whenever a fixed factor appears to be nested, the statistician should ask the experimenter about the intent to determine if that factor is actually a sliding factor that should be treated as if it were crossed.

4.3.2 Fixed Nested Factors

Sometimes a fixed factor is nested and the intent is to have the levels different depending on another factor. An example is two different suppliers, each of which produces parts from different materials. The two factors are suppliers and materials, with materials depending on the supplier. There are, in fact, four different materials. The two materials from one supplier are not intended to correspond to the two materials from the other supplier.

Unless there is strong reason to do otherwise, fixed nested factors should be treated as one factor having many levels. In this particular case, incorporate the suppliers so there is only one factor, supplier/material, having four different levels. Again, there should be no parentheses in the subscripts for supplier/material.

If there is a strong desire to consider two different factors, one can only proceed if the nested factor is considered a sliding factor. Again, this assumption cannot be made on any statistical basis and cannot be verified with any statistical test. Should the assumption be false, the interpretation of the data will be incorrect and this could steer the experimenter in the wrong direction.

In the supplier/material example given above, it has sometimes been suggested that the analysis for a one-half fraction of a two by four level design be used. (See Chapters **7** and **8** for the fractionation concept.) This completely confounds one of the degrees of freedom for material with supplier. The assumption that the one degree of freedom for material is negligible is equivalent to the sliding factor assumption. Again, the assumption cannot be made on a statistical basis and cannot be verified statistically.

4.4 EXAMPLES

Example 1. An engineer (Beeson (1965)) was interested in comparing the effect of four different cardiac valve types, T, and six different pulse rates, P, on the maximum flow gradient (mmHg) in a simulator of the human circulatory system. Two different valves, V, of each type were chosen and run at all six pulse rates. The experiment was carried out in a completely randomized fashion (meaning the valves had to be put in the machine and taken out again quite often).

A picture of the experimental layout often helps visualize the design and identify nested factors. In this case, the valves were numbered from one to eight, the first two being valve type 1, the next two being valve type 2, *etc.* Both the layout and the data are given in Table 4.4.1.

TABLE 4.4.1

LAYOUT AND DATA FOR EXAMPLE 1 ON CARDIAC VALVES

Valve Type:	1		2		3		4	
Valve #:	1	2	3	4	5	6	7	8
1	y=2	y=3	y=4	y=2	y=6	y= 5	y= 7	y= 5
2	4	4	4	4	5	5	5	4
Pulse 3	5	7	4	3	5	6	6	5
Rate: 4	3	5	5	3	8	10	9	10
5	7	7	8	5	9	9	10	11
6	6	6	6	7	7	8	8	9

Note that the different numbers associated with valves immediately indicates that valves are nested in valve types. However, had the experimenter numbered each valve 1 and 2 under each valve type, it would have visually appeared as if valves were crossed with valve type. This is an important lesson to learn: *a visual layout of the experiment may or may not indicate nesting among factors.* The only sure way is to ask about each factor: Are these the same valves for each type? Are these the same pulse rates for each valve type? Are these the same pulse rates for each valve?

The factors for this example are T_i, $i = 1, \cdots, 4$, fixed; $V_{j(i)}$, $j = 1, 2$, random; and P_k, $k = 1, \cdots, 6$, fixed. The subscript $j(i)$ indicates that valves are nested in valve types. Since j runs from one to two for each of the four valve types, there are a total of eight different valves. An accounting such as this is another excellent way to check if the nesting has been properly noted.

The analysis for the data in Table 4.4.1 is given in Table 4.4.2. It is important to stop at this point and verify that you can produce all of the entries in this table, whether you calculate these quantities by hand or by computer. All of the necessary rules have already been given in Chapter **3**. However, no examples in that chapter dealt with nested factors. If you have difficulty, stop, go back to Chapter **3**, and rereview the appropriate sections.

TABLE 4.4.2

ANOVA for Example 1 on Cardiac Valves

Source	df	MS	EMS	F_{calc}	p-Value
$\to T_i$	3	24.0833	$\sigma^2 + 6\sigma_V^2 + 12\Phi(T)$	13.76*	0.0142
$V_{j(i)}$	4	1.7500	$\sigma^2 + 6\sigma_V^2$	—	—
$\to P_k$	5	21.0833	$\sigma^2 + \sigma_{VP}^2 + 8\Phi(P)$	28.11**	0.0001
$\to TP_{ik}$	15	2.5500	$\sigma^2 + \sigma_{VP}^2 + 2\Phi(TP)$	3.40**	0.0059
$VP_{jk(i)}$	20	0.7500	$\sigma^2 + \sigma_{VP}^2$	—	—
$\varepsilon_{(ijk)}$	0	—	σ^2		
Total	47				

* significant at the .05 level

** significant at the .01 level

As can be seen in Table 4.4.2, there are no direct tests on valves or the valve/pulse interaction. This was known prior to the collection of the data and was not a concern to the engineer, who figured that valves would be different anyhow. Also in Table 4.4.2, the EMS for types contains valve to valve, σ_W^2, as well as measurement to measurement, σ^2, variation. This makes the valve the inferential experimental unit, see Section **1.6**. Notice that valve contains both σ_W^2 and σ^2 and forms the denominator of the test. Many experiments have nested factors which are not very important to the experimenter. The nested factors are the inferential experimental unit used to test the factors that are of primary interest.

The engineer was most interested in selecting the valve type that had the highest overall flow gradient. This meant selecting the valve type with the highest overall average. Since valve type was significant, this can be accomplished using Duncan's test. The standard error for valve type is given by $\sqrt{\text{MS}(V)/12} = 0.38188$ having four df. The least significant ranges for comparing two, three, and four means are 1.4996, 1.5325, and 1.5401 respectively.

Valve Type	Mean	
4	7.4	|
3	6.9	|
1	4.9	|
2	4.6	|

Duncan's test shows valve types 3 and 4 are indistinguishable and overall better than valve types 1 and 2. The results are illustrated in Figure 4.1.

Since there is a significant interaction between valve types and pulse rates, it is a good idea to look at a picture of the interaction to make sure one pulse rate is not disastrous for either of valve types three or four. The non-parallel lines in Figure 4.2 correspond to the interaction between valve types and pulse rates. Since types three and four generally exceed types one and two, no major problems are foreseen. You may wish to further explore the behavior of the valves around pulse rate 3. Type four would probably be chosen as long as its cost is comparable to that of type three.

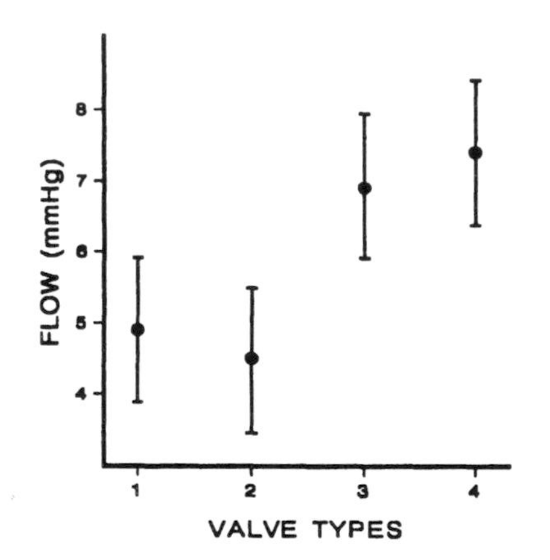

Figure 4.1 Mean flow gradients.

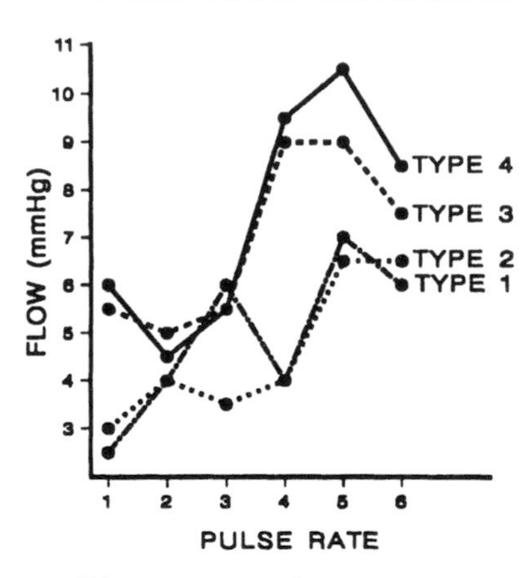

Figure 4.2 Interaction plot.

Example 2. In order to punch a non-flat surface from a flat piece of sheet metal, the sheet metal must stretch. Too much stretch causes the metal to tear and become worthless. To measure the stretch, dots are stenciled on a flat piece of metal, the part is punched, and the eccentricities of the resulting ellipses are measured using a hand-held camera. This experiment deals with the accuracy of the stenciling and measuring process prior to punching the metal piece.

The stencils of dots come from large rolls of clear plastic. A piece is cut off the large roll and applied to the flat metal. Then, several dots are measured for eccentricity. Ideally, the eccentricity should be zero, *i.e.*, the dots should be perfectly round. Of course, they are not. The goal of the experiment is to find out which factors contribute to the overall variability in the stenciling and measurement process.

Many thoughts were offered as to the cause of the variability. Perhaps the person applying the stencil makes a difference. Perhaps the location of the stenciled dots within the roll matters. Results may also differ from camera to camera, operator to operator, or even the angle at which the camera is held.

After much discussion, the following experiment was devised. Two different people would be selected to apply the stencils to the metal. Each would apply two stencils from the start of the roll and two stencils from the middle of the roll. From each stenciled metal piece, two dots would be randomly selected. Two different operators would then measure each dot twice at 60° and 90° using two different cameras. (Each operator uses the same two cameras.) What are the factors, the number of levels, are they random or fixed, and is there any nesting? A layout of the experiment is given in Table 4.4.3.

Prior to giving the solution to this example, the reader should stop reading and try to identify the nesting(s). Reread the experimental layout and look for the key words indicating nesting. This is a fairly hard example. If you can understand this one, you have probably grasped the nesting concept. Understanding this example after you have seen the explanation is easy and may not indicate complete understanding. Try to model the experiment first.

Now that you have tried to model this experiment, we will indicate the reasoning process. The first factor encountered is people, P_i, $i = 1, 2$, random since the inference is to all people who apply the stencil, not just the people used in the experiment.

The next factor encountered in the description is stencils. The stencil taken from the beginning of the roll must be different than the stencil taken from the middle so stencils are nested within location on the roll. In addition, the stencil cannot be reused by each person (if you are not sure about relations such as this, *ask the experimenter*), so stencils are also nested within people. We get the next two factors, location, L_j, $j = 1, 2$, fixed since the locations were specifically selected, and stencils, $S_{k(ij)}$, $k = 1, 2$, random. This implies that eight different stencils are to be used, four taken from the start of the roll and four from the middle. Each person is to apply two of these four stencils, one from each location. Again, this should be verified with the experimenter.

The next factor encountered is dots. Since the key phrase "from each stencil···" was used, dots are nested within stencils. That is, there are two different dots from each stencil. Even if the key phrase "from each"

TABLE 4.4.3
LAYOUT FOR EXAMPLE 2

Apply Stencil to Metal

People:	1	2
Location = Start:	Stencil 1 Stencil 2	Stencil 1 Stencil 2
Location = Middle:	Stencil 1 Stencil 2	Stencil 1 Stencil 2

Measure the Ellipse on each Stencil

Dot:	1				2			
Operator:	1		2		1		2	
Camera:	1	2	1	2	1	2	1	2
Angle = 60°	M 1 M 2	M 1 M 2	M 1 M 2	M 1 M 2	M 1 M 2	M 1 M 2	M 1 M 2	M 1 M 2
Angle = 90°	M 1 M 2	M 1 M 2	M 1 M 2	M 1 M 2	M 1 M 2	M 1 M 2	M 1 M 2	M 1 M 2

was not used, you should be able to figure out that the dots cannot be the same so must be nested within stencils. Since stencils are already nested within people and location, dots are nested within stencils, people, and location. The inference is to all possible dots so dots is a random factor. The result is $D_{\ell(ijk)}$, $\ell = 1, 2$, random.

Since each operator measures each dot with each camera at each angle, operator, camera, and angle are all crossed factors. Their representations are respectively given by O_m, $m = 1, 2$, random; C_n, $n = 1, 2$, random; and A_o, $o = 1, 2$, fixed.

Finally, each measurement is taken twice yielding $\varepsilon_{p(ijk\ell mno)}$, $p = 1, 2$, random. Alternatively, measurement could be taken as a factor and denoted $M_{p(ijk\ell mno)}$, $p = 1, 2$, random. If we use M, the error term is given by $\varepsilon_{(ijk\ell mnop)}$ and the evaluation of the design is the same.

To reconcile this experiment as a check on the nesting, there should be a total of eight stenciled metal pieces, 16 dots, and a total of 256 measurements made. These came directly from the nested terms: stencils, dots, and measurements, respectively. Two people should make stencils, there should be two locations within the roll of stencils, two operators, two cameras, and two angles should be used. These come from the crossed terms.

Example 3. The purpose of this experiment is to discover the effect of an anesthetic on tissue injury in a laboratory rat. The rat has to be sacrificed to measure the extent of the injury. The factors under consideration are days after injection, D, 1 day, 3 days, and 14 days; type of rat, T, type 1 and type 2; dose amount, A, low and high; and injection material, M, anesthetic and a saline solution. To save expense due to laboratory rats, a dose amount was selected, the anesthetic was injected in the left leg, and the saline was injected in the right leg. The experimenter was willing to assume there was no inherent difference between the left and the right leg and the injection in one leg did not influence the tissue damage in the other leg. Five rats were used for each day/type/amount combination for a total of 60 rats used in the experiment. What is the appropriate model? Is this a good experiment?

When considering the amount of anesthetic injected into the tissue of the rat, the low and high amount of the saline control are the same—zero. Potentially, this makes the amount of anesthetic nested within injection material. However, the *intent* of the experimenter was to discover if the amount of liquid also influenced the tissue injury, not just the amount of anesthetic. (As a statistician you cannot make assumptions about the intent of the experiment, you must specifically *ask* the intent.) Therefore, dose amount is considered a sliding factor and treated as crossed in the experiment. (Note: if *dose amount* had simply been labeled *amount*, the factor levels low and high would have had the same meaning for anesthetic and control and we never would have suspected it could be nested within injection material. Always proceed cautiously.)

Since a different rat is used for each day/type/amount combination but the same rat receives both injection materials, rats, R, are nested within days, types, and amounts but crossed with injection material. The correct mathematical model becomes

$$\begin{aligned} y_{ijk\ell m} = \mu &+ D_i + T_j + DT_{ij} + A_k + DA_{ik} + TA_{jk} + DTA_{ijk} + R_{\ell(ijk)} \\ &+ M_m + DM_{im} + TM_{jm} + DTM_{ijm} + AM_{km} + DAM_{ikm} \\ &+ TAM_{jkm} + DTAM_{ijkm} + RM_{\ell m(ijk)} + \varepsilon_{(ijk\ell m)}, \end{aligned} \tag{4.4.1}$$

where D_i, $i = 1, 2, 3$, is fixed; T_j, $j = 1, 2$, is fixed; A_k, $k = 1, 2$, is fixed; $R_{\ell(ijk)}$, $\ell = 1, \cdots, 5$, is random; and M_m, $m = 1, 2$, is fixed. The ANOVA for Example 3 is given in Table 4.4.4.

TABLE 4.4.4

ANOVA FOR EXAMPLE 3 ON ANESTHETICS.

Source	df	Δ	Detectable Size	EMS
→ D_i	2	0.41	EXT. SMALL	$\sigma^2 + 2\sigma_R^2 + 40\Phi(D)$
→ T_j	1	0.43	EXT. SMALL	$\sigma^2 + 2\sigma_R^2 + 60\Phi(T)$
→ DT_{ij}	2	0.58	SMALL	$\sigma^2 + 2\sigma_R^2 + 20\Phi(DT)$
→ A_k	1	0.43	EXT. SMALL	$\sigma^2 + 2\sigma_R^2 + 60\Phi(A)$
→ DA_{ik}	2	0.58	SMALL	$\sigma^2 + 2\sigma_R^2 + 20\Phi(DA)$
→ TA_{jk}	1	0.60	SMALL	$\sigma^2 + 2\sigma_R^2 + 30\Phi(TA)$
→ DTA_{ijk}	2	0.82	SMALL	$\sigma^2 + 2\sigma_R^2 + 10\Phi(DTA)$
$R_{\ell(ijk)}$	48	—	—	$\sigma^2 + 2\sigma_R^2$
→ M_m	1	0.43	EXT. SMALL	$\sigma^2 + \sigma_{RM}^2 + 60\Phi(M)$
→ DM_{im}	2	0.58	SMALL	$\sigma^2 + \sigma_{RM}^2 + 20\Phi(DM)$
→ TM_{jm}	1	0.60	SMALL	$\sigma^2 + \sigma_{RM}^2 + 30\Phi(TM)$
→ DTM_{ijm}	2	0.82	SMALL	$\sigma^2 + \sigma_{RM}^2 + 10\Phi(DTM)$
→ AM_{km}	1	0.60	SMALL	$\sigma^2 + \sigma_{RM}^2 + 30\Phi(AM)$
→ DAM_{ikm}	2	0.82	SMALL	$\sigma^2 + \sigma_{RM}^2 + 10\Phi(DAM)$
→ TAM_{jkm}	1	0.85	SMALL	$\sigma^2 + \sigma_{RM}^2 + 15\Phi(TAM)$
→ $DTAM_{ijkm}$	2	1.16	SMALL	$\sigma^2 + \sigma_{RM}^2 + 5\Phi(DTAM)$
$RM_{\ell m(ijk)}$	48	—	—	$\sigma^2 + \sigma_{RM}^2$
$\varepsilon_{(ijk\ell m)}$	0	—	—	—
Total	119			

As can be seen in Table 4.4.4, there is a direct test having 48 df on all of the main effects and interactions not involving rats. This corresponds to either a *small* or an *extremely small* detectable difference and is generally plenty of information. If the variation due to ε, R, or RM is large, then the detectable sizes need to be this small. If the variations are moderate or small, then the experiment size can be reduced by testing fewer rats. The decision must be made based on the knowledge of the experimenter and the cost of additional rats. Redesigns are highly suggested. For example, even with 2 rats, the detectable sizes are still in the small range, Δ falling between .72 and 1.44, for everything except the $DTAM$ interaction.

Hand calculation of sums of squares, mean squares, and F-tests for experiments of this size should not even be attempted. An arithmetic error will occur with near certainty. Use your local statistical computing package for calculations and double check the entry of the data for errors.

Confidence limits for a given level of T or A are given by the rules developed in Chapter **3**: the mean response $\pm\, t_{\alpha/2}\sqrt{\sigma^2/60 + \sigma_R^2/30}$ which is approximated by the mean $\pm\, t_{\alpha/2}\sqrt{MS(R)/60}$, with 48 df for the t statistic. The confidence limit for a given level of D is the same except $MS(R)/60$ gets replaced by $MS(R)/40$.

The confidence limit for a given level of M contains the theoretical variance terms $\sigma^2/60 + \sigma_{RM}^2/120 + \sigma_R^2/60$. This is approximated by $[MS(R) + MS(RM)]/120$, having df that must be estimated using Satterthwaite's approximation. Satterthwaite's approximation must also be used on all interactions involving M.

Since there are direct tests on all main effects and interactions, confidence intervals on the differences of two means and Duncan's test on the means can be performed for all main effects and interactions. This is a good design for all possible comparisons.

Had the experimenter feared an influence from one leg to the other, rats, R, would also have to be nested in M and there would be 120 rats in the experiment rather than 60. Rats would then test everything with 96 df. The Δ values fall between 0.4 and 1.14 with all main effects falling in the *extremely small* range and all interactions falling in the *small* range. Three rats per combination gives the same df for the denominator as the original experiment. Two rats per combination is probably sufficient. Either of these new designs will yield confidence intervals on all means and differences of means and are thus good designs.

PROBLEMS

4.1 *Suppose an experiment deals with people, P, training method, M, and time after completion of training, T. Identify the nesting under each of the following circumstances. For this problem, ignore any difficulties with inferences and with experimental layout. (a) Each person completes each training method and is tested at all times. (b) Each person completes only one training method and is tested at all times. (c) Each person completes only one training method and is tested at only one time.*

4.2 *An experiment consists of fuel injectors, I, engines, E, and atmospheric conditions represented by weeks, W. One fuel injector is put in one engine and the vehicle is tested. Assuming complete randomization, identify the nesting in each of the following experimental layouts. (a) Four engines and twelve injectors are selected. Three different injectors are tested in each engine. A week later, the exact procedure is repeated with the same engines and injectors. (b) Four engines and eight injectors are selected. Four injectors are selected and tested in all four engines. A week later, the four remaining injectors are tested in the same four engines. (c) Four engines and eight injectors are selected. Two engines and four injectors are selected, each engine being tested with two different injectors. A week later, the experiment is repeated with two new engines and four new injectors. (d) Four engines and four injectors are selected. All four injectors are tested in two of the engines the first week. The second week, the same injectors are put in each of the other two engines. (e) Four engines and four injectors are selected. All four injectors are tested in each engine. A week later, all four injectors are tested in each engine again. (f) Two engines and eight injectors are selected. Four injectors are chosen, two to be tested in each engine. A week later, the other four injectors are tested in the same engines, again two injectors per engine. (g) Four engines and four injectors are chosen. Two injectors are used in each of two engines the first week. The next week, the other two injectors are used in each of the other two engines.*

4.3 *An experiment consists of operators, O, riveting two different parts, A and B, together. Identify the nesting, subscripts, and appropriate mathematical model for each of the following experimental layouts. (a) Ten A parts and ten B parts are selected. Operator 1 selects five of each and rivets and takes apart the pieces until all 25 combinations have been made. Operator 2 does the same with the remaining parts. (b) Five A parts and five B parts are selected. Both operator 1 and operator 2 rivet all 25 possible combinations. (c) Twenty A and twenty B parts are selected. Operator 1 rivets ten A parts to ten B parts without taking any apart. Operator 2 does likewise on the remaining ten parts. A total of twenty parts are riveted. (d) Operator 1 rivets ten A parts to ten B parts. These are taken apart and operator 2 rivets the same ten combinations back together again. (e) Ten A parts and ten B parts are selected. Operator 1 rivets five A parts to five B parts. Operator 2 rivets the remaining five A and B parts together. Nothing is* *taken apart. (Hint on (c), (d), and (e): There are three ways to look at this problem, all yielding the same result. You could have B, with one level only, nested in A. You could have A, with one level only, nested in B. Finally, you could make the experimental procedure consist of selecting one A and one B part and call this factor $[AB]$.)*

4.4 *Explain why a nested factor cannot interact with a factor it is nested within.*

4.5 *Write out the mathematical model for an experiment having B nested in A, C nested in B, and E nested in D.*

4.6 *Write out the mathematical model for an experiment having both B and C nested in A, and D crossed with everything.*

4.7 *Write out the mathematical model for an experiment having B and D nested in A, and C nested in B.*

4.8 *Write out the mathematical model for an experiment having C nested in both A and B and D crossed with everything.*

4.9 *Write out the mathematical model for each of the experimental layouts in Problem **4.2**. Include ranges on the subscripts.*

4.10 *Write out the ANOVA table, including source, df, SS, Δ, EMS, and arrows indicating the tests, for equation (4.2.4). Let T_k be a random factor.*

4.11 *Write out the ANOVA table, including source, df, SS, Δ, EMS, and arrows indicating tests, for Problem **4.5**. Assume each factor has two levels, A is fixed, B, C, D, and E are random, and there are no repeats.*

4.12 *Write out the ANOVA table, including source, df, SS, EMS, Δ, and arrows indicating the tests, for Problem **4.6**. Assume each factor has two levels, A is fixed, B, C, and D are random, and there are no repeats.*

4.13 *Write out the ANOVA table, including source, df, SS, Δ, EMS, and arrows indicating the tests, for Problem **4.7**. Assume each factor has two levels, A is fixed, B, C, and D are random, and there are no repeats.*

4.14 *Write out the ANOVA table, including source, df, SS, Δ, EMS, and arrows indicating the tests, for Problem **4.8**. Assume each factor has two levels, A is fixed, B, C, and D are random, and there are no repeats.*

4.15 *Can a random factor also be a sliding factor? Explain or give an example.*

4.16 *What should you do if there is a factor called supplier having two levels, supplier 1 supplies two materials, and supplier 2 supplies three different materials?*

4.17 *Use the rules of Chapter **3** to show the SS and df for $B_{j(i)}$ in the model $y_{ijk} = \mu + A_i + B_{j(i)} + \varepsilon_{k(ij)}$ is the same as the SS and df for B_j plus AB_{ij} in the model $y_{ijk} = \mu + A_i + B_j + AB_{ij} + \varepsilon_{k(ij)}$.*

4.18 *Suppose Example 3 had been run with rats, R, also nested in injection material, M, and there were three rats for each type/amount/material* *combination. Give the theoretical and estimated variation for a given mean level of material M.*

4.19 *Assuming engines, injectors, and weeks are all random, determine the tests for each of the experimental layouts given in Problem* **4.2**.

4.20 *A genetics study on fruit flies was conducted to find the cause of variation in body weight. For each of four genome samples, G, four males, M, and twelve females, F, were selected. Three females were mated with each male. The eggs from each female were divided into two bottles, B. After the eggs became adults, five individuals, I, were selected and weighed. A total of 480 fruit flies were weighed. Write the mathematical model describing this experiment. Considering all factors random, write out the variance components estimates for G, M, F, B, and I.*

4.21 *An experiment on the strength of aluminum was performed on four different (fixed) alloys. From each alloy, one heat (batch of molten metal) was mixed. From each heat, many ingots were made. Three ingots were selected at random and four samples from each ingot were selected and tested for strength. Write the mathematical model and ANOVA table (source, df, Δ, and EMS) for this four factor experiment. What is wrong with the experiment? Keeping in mind that heats are most expensive and samples are least expensive, redesign this experiment as you see fit.*

4.22 *An experiment had three different groups of ten subjects, S. Each group of ten was trained using a different training method, M. After training, each group was asked to do four tasks, T. All students did the same four tasks. Write the mathematical model and determine the tests for this experiment. You must decide which factors are random and which are fixed and explain why you made your decision.*

4.23 *Suppose the students in Problem* **4.22** *did the tasks before training as well as after training. If you define the response variable as the difference in time to perform the task, the evaluation is exactly the same as in* **4.22**. *Model the experiment using time as a new factor. If training method 1 is "no training," then the time difference for training method 1 represents a task learning effect. Write the formula for a confidence bound on the task learning effect. Write the formula for a confidence bound on the task learning effect for task 2.*

4.24 *Regarding Problem* **4.23**, *which method of analysis, i.e., the difference in time or considering time as a factor, is likely to give better results. Explain your answer.*

4.25 *A nutrition experiment considered the effect of three different rations on the weight gain of hens. Eight different hens were assigned to each ration and the difference in weight gain was measured for five weeks. Model this experiment and determine the tests. If each hen was weighed twice at each time period, model this new experiment.*

4.26 *In an ammunition depot, an igniter was operating ineffectively and an experiment was devised to help locate the problem. After thorough discussion, the following factors were decided upon: the thickness of the encloser disk, thick and thin; powder lots from the manufacturer, lot 1 and lot 2; containers from within lots, 1 and 2; and the moisture content of the ingredients, 0.2, 0.4, and 0.8%. Five igniters were made from each combination. Model this experiment and determine the detectable differences.*

4.27 *An advertising firm wanted to test the effectiveness of three sizes, S, and four color combinations, C, on the attractiveness of billboards in five random locations, L, of the country. Five judges scored each of the 20 location/color combinations. That is, a total of 100 judges were used, each judge scoring three different sizes for a total of 300 scores. (a) Model the experiment and indicate the tests. (b) What is the appropriate standard error for size, color, and the size/color combination? (c) What is the standard error for Duncan's test on color? (d) Can you redesign this experiment for better tests with minimal added cost?*

4.28 *A biologist was interested in the effect of a man-made pollutant, hydrocarbon, and a natural phenomenon, barometric pressure, on the weight gain of six week old rats. Three levels of hydrocarbon—none, low, and high—and two levels of barometric pressure—low and high—were used. The change in weight was measured for five consecutive weeks. Model this experiment and indicate the tests.*

4.29 *A design often used in drug testing is known as a* crossover *design. The following is an example. Twelve people were used to test the effectiveness of two drugs. Six were randomly chosen to receive drug one first. The other six received drug two first. After a sufficient washout period, the six who received drug one were given drug two and the six who received drug two were given drug one. (This is known as the crossover.) The factors here are order, drugs, and people. Find the nesting in this experiment. Explain the implications if order was significant, if there was an order/drug interaction, and if the mean for order one exceeded the mean for order two. (Notice that there is something strange about randomization. This will be covered in the next chapter.)*

4.30 *Suppose there were three drugs in Problem* **4.29**. *Six orders would have to be considered. How many subjects would be needed to have a reasonable design? Now suppose there were four drugs. How many orders are possible? You can obviously see the problem with more and more crossovers. It should make life easier if there is no carryover for more than one period, i.e., no carryover from period one to period three, etc., and this is often a reasonable assumption. Redesign the three drug problem with the no carryover for more than one period assumption.*

4.31 *Component switching. A common method for problem solving in complicated parts is known as component switching. Start with a failed part and* *a good part. Disassemble each and reassemble using some bad components and some good components to try to figure out which component causes the part to fail. Each component becomes a factor in the experiment having two levels, good and failed, and the experiment was modeled in Problem* **3.42**. *When there are several good and failed parts, each component corresponds to two factors: the component itself, having two levels, and repeats for that component, nested within the component. Write the factors and model for an experiment having three good and three failed parts, each part consisting of components A, B, and C. How many combinations are possible? Are there any assumptions made in your model? If so, state them.*

4.32 *An experiment on the effect of carbon black dust (which causes coal miner's disease) and its washout effect was conducted on rats. The exposure times to carbon black dust are 0 weeks, 1 week, 3 weeks, and 6 weeks. After exposure, washout periods of 0, 1, 7, 21, 50, 80, 141, 210 and 365 days are considered. The rat has to be sacrificed to measure the amount of dust remaining in the lung. Design and model the experiment, including the total number of rats used. In addition to the usual tests on exposures and washout, indicate what specific tests would be made and why. Be as comprehensive as possible in your answer.*

4.33 *An experiment was conducted to determine the effect of substrates and primers on the appearance of a painted surface. Nine plaques were made from each of four different substrates. There were three primers, each applied to three of the nine plaques. The distinctness of image (a measure of the mirror-like appearance of the surface) was measured in four different directions on three pre-selected locations on each plaque. Model this experiment, indicating the tests and detectable sizes.*

4.34 *To determine the roughness of plastics, three different substrates were tested before and after painting. Two different panels were made from each substrate. Each panel was measured for roughness at five randomly selected locations, different locations before and after painting. Model this data and determine the tests. What standard error would be used for a given substrate? What standard error would be used for Duncan's test on substrates? What standard error would be used to test the difference in roughness before and after painting?*

4.35 *An experiment was conducted to determine the effect of lead amount, A, trace material amount, T, and location, L, of "buttons" used in maintenance free batteries. Physically, the lead and tracc materials are added to a base and poured into a mold, forming many rods. Two buttons are cut off the top (location 1) and bottom (location 2) of three randomly selected rods. Each button is tested for cranking power, higher being better. There are two levels of lead amount, two levels of trace material amount, and the experiment is completely randomized. Give the mathematical model and the ANOVA table including source, df, Δ, EMS, and tests.*

4.36 *The following information was collected in the experiment described in Problem* **4.35**: $SS(A) = 1348.00$, $SS(T) = 175.83$, $SS(AT) = 335.58$, $SS(R) = 164.42$, $SS(L) = 4.19$, $SS(AL) = 0.25$, $SS(TL) = 0.34$, $SS(RL) = 11.01$, *and* $SS(ATL) = 3.00$. *Using our sometimes pooling rules, determine which factors and interactions significantly affect cranking power. Use the following information on means to decide what levels of the factors to use and give a 95% confidence interval on the average cranking power.* $\overline{A}_1 = 158.52$, $\overline{A}_2 = 147.92$, $\overline{T}_1 = 151.31$, $\overline{T}_2 = 155.14$, $\overline{AT}_{11} = 159.25$, $\overline{AT}_{12} = 157.79$, $\overline{AT}_{21} = 143.37$, $\overline{AT}_{22} = 152.48$, $\overline{L}_1 = 152.93$, *and* $\overline{L}_2 = 153.52$.

4.37 *An experiment was conducted to study the effect of an irritant on the formation of collagen in the lungs of rats. Four different levels of the irritant were considered with three different exposure times for each irritant. Three different rats were to be used for each irritant/time combination as the rats had to be sacrificed to measure each lung. For a given rat, many sections of lungs, called fields, could be measured. The key question is, how many fields should be measured? (a) Compute the detectability,* Δ, *for irritant, time, and their interaction for 2, 3, 5, and 10 fields measured. (b) Note that these* Δ*'s are not comparable since the denominator testing them changes as the number of fields is changed. To get meaningful numbers, the variances of the random factors must be estimated using a pre-test on one irritant/time combination. Two rats and five fields for each rat were used. Each field was measured twice and the response variables lung area and non-collagen area were recorded. The results are given below. Use this pre-test to estimate the three variabilities for each response variable. (c) Was the experimenter correct in only measuring each field once in the original experiment? Why or why not? (d) Give detectabilities (in actual units, not standard deviation units) for lung area and non-collagen area for 2, 3, 5, and 10 fields measured. (e) Why do extra fields decrease non-collagen area detectability considerably but do not decrease lung area detectability much? (f) What would you recommend to decrease lung area detectability? Be specific.*

		measurement 1		measurement 2	
rat id	field	lung area	non-coll. area	lung area	non-coll. area
870	1	1816	252	1813	255
870	2	1754	346	1753	345
870	3	1344	160	1342	158
870	4	1064	134	1064	131
870	5	1644	98	1639	100
871	1	1782	221	1782	219
871	2	2429	189	2428	203
871	3	1935	148	1930	146
871	4	1916	199	1915	199
871	5	1907	76	1907	75

4.38 *An experiment by a bacteriology organization was run on two enzymes, 3975 and 2712, and four concentrations, .12, .25, .5, and 1%. The experiment requires four days, considered a factor in the study. On the first day, an enzyme was picked and five different plates for each of the four concentrations were made. On the second, third, and fourth day, the procedure was repeated so each enzyme gets run on two different days. Each plate is measured for the diameter of the largest colony of bacteria. Model and analyze this experiment. Do as complete a job as you see fit, including designing another experiment if necessary. State your assumptions about the needs of the experimenter.*

Day 1 (Enzyme 3975)						Day 2 (Enzyme 2712)					
Plate:	1	2	3	4	5	Plate:	1	2	3	4	5
1.0%	27.6	27.4	27.6	27.4	26.8	1.0%	29.4	28.4	27.4	27.0	27.2
0.5%	26.4	26.2	26.4	26.4	26.4	0.5%	27.6	28.4	27.4	27.0	27.2
.25%	25.2	25.4	25.4	25.4	25.0	.25%	28.0	25.8	25.4	25.0	25.4
.12%	24.0	24.2	24.6	24.0	24.8	.12%	26.0	24.8	25.0	24.0	24.4

Day 3 (Enzyme 2712)						Day 4 (Enzyme 3975)					
Plate:	1	2	3	4	5	Plate:	1	2	3	4	5
1.0%	27.8	27.8	28.2	28.2	29.8	1.0%	27.4	27.6	27.4	28.4	29.0
0.5%	26.8	26.6	26.8	27.4	28.4	0.5%	26.0	26.8	26.6	27.0	28.6
.25%	26.0	25.6	25.6	25.6	27.6	.25%	25.4	25.6	25.4	26.2	26.8
.12%	25.0	24.8	25.0	25.2	27.0	.12%	24.4	24.4	24.2	25.4	25.6

4.39 *An experiment concerning the effect of ozone on the percent change in forced expiratory volume for asthmatics and normal health humans was run in the fashion indicated below. Model and analyze this data. The letters under the Subject column are the initials of the patient.*

			OZONE EXPOSURE							
			0 ppm TIME (Minutes)				0.4 ppm TIME (Minutes)			
HLTH	SEX	SUB	0	30	60	90	0	30	60	90
Asthmatic	Female	AK	-5.6	1.0	-33.9	-26.6	1.6	5.8	-21.1	-30.8
		JS	-12.9	-14.0	-20.3	-25.1	-5.1	-10.9	-33.0	-35.9
		LA	-1.4	3.8	3.5	2.8	-1.8	-0.4	-0.4	1.1
	Male	BT	6.1	5.5	8.3	16.2	2.4	8.2	-18.2	-35.8
		DW	2.7	-5.1	-31.4	-21.7	-15.5	-13.2	-50.0	-39.7
		JR	4.6	6.2	4.3	8.9	-4.4	-2.9	-9.3	-11.1
Normal	Female	GS	-3.8	-4.1	-1.4	0.2	-2.2	-1.4	-1.0	-1.9
		KG	-3.9	-1.8	-0.9	3.0	-6.6	-6.0	-18.4	-17.0
		RF	-6.7	-8.0	-1.6	-1.3	-8.8	-5.4	0.0	-3.0
	Male	DT	-7.6	-2.5	-2.2	-1.0	-6.0	-4.8	-7.6	-5.6
		JK	-3.4	-1.2	0.6	1.4	-3.9	-1.6	-1.8	-3.3
		MB	-4.3	-5.6	-5.8	-2.5	-0.9	2.6	-1.3	-6.2

BIBLIOGRAPHY FOR CHAPTER 4

Anderson, V. L. and McLean, R. A. (1974). *Design of Experiments A Realistic Approach*, New York: Marcel Dekker. (Chapters 6, 7)

Beeson, J. (1965). "A simulator for evaluating prosthetic cardiac valves," M.S. thesis, W. Lafayette, IN 47906: Purdue University Library.

Bennett, C. A. and Franklin, N. L. (1954). *Statistical Analysis in Chemistry and the Chemical Industry*, New York: John Wiley & Sons. (Chapter 7)

Daniel, C. (1976). *Applications of Statistics to Industrial Experimentation*, New York: John Wiley & Sons. (Chapter 16)

Guenther, W. C. (1964). *Analysis of Variance*, Englewood Cliffs, NJ: Prentice-Hall. (Chapter 5)

Hicks, C. R. (1982). *Fundamental Concepts in the Design of Experiments*, New York: Holt, Rinehart, & Winston. (Chapters 11, 12)

John, P. W. M. (1971). *Statistical Design and Analysis of Experiments*, New York: The Macmillan Co. (Chapters 4, 5)

Johnson, N. L. and Leone, F. C. (1977). *Statistics and Experimental Design in Engineering and the Physical Sciences*, 2nd ed., New York: John Wiley & Sons. (Chapter 13)

Montgomery, D. C. (1991). *Design and Analysis of Experiments*, 3rd ed., New York, John Wiley & Sons. (Chapter 13)

Ostle, B. and Malone, L. C. (1988). *Statistics in Research*, 4th ed., Ames, Iowa: Iowa State University Press.

Peng, K. C. (1967). *The Design and Analysis of Scientific Experiments*, Reading, MA: Addison-Wesley. (Chapter 4)

Scheffé, H. (1959). *The Analysis of Variance*, New York: John Wiley & Sons. (Chapter 5)

Searle, S. R. (1971). *Linear Models*, New York: John Wiley & Sons. (Chapters 4, 6, 9)

Steel, R. G. D. and Torrie, J. H. (1980). *Principles and Procedures of Statistics—A Biometrical Approach*, 2nd ed., New York: McGraw-Hill. (Chapter 9)

Taguchi, G. (1988). *System of Experimental Designs*, White Plains, NY: Kraus International Publications. (Chapters 5, 9)

Winer, B. J. (1971). *Statistical Principles in Experimental Design*, 2nd ed., New York: McGraw-Hill. (Chapter 7)

SOLUTIONS TO SELECTED PROBLEMS

4.2 *(a) Injectors nested in engines. (b) Injectors nested in weeks. (c) Engines nested in weeks. Injectors nested in engines and weeks. (d) Engines nested in weeks. (e) No nesting. (f) Injectors nested in engines and weeks. (g) Injectors nested in weeks. Engines nested in weeks.*

4.5 *The mathematical model describing the experiment is* $y_{ijk\ell m} = \mu + A_i + B_{j(i)} + C_{k(ij)} + D_\ell + AD_{i\ell} + BD_{j\ell(i)} + CD_{k\ell(ij)} + E_{m(\ell)} + AE_{im(\ell)} + BE_{jm(i\ell)} + CE_{km(ij\ell)} + \varepsilon_{(ijk\ell m)}$.

4.11 *Approximate tests are:* $CD + BE - CE$ *tests* BD, $BD + AE - BE$ *tests* AD, $BD + E - BE$ *tests* D, $C + BD - CD$ *tests* B, *and* $B + AD - BD$ *tests* A. *The detectable differences are computed using the harmonic average of the df making up the approximate tests. The direct tests are indicated by arrows in the ANOVA table below and continued on the next page. None of the detectable sizes for this experiment are that good. The detectable sizes for the terms involving the factor D are very poor with the tests on* D_ℓ *and* $AD_{i\ell}$ *being extremely large. If information about D is required in this experiment, the experiment is unacceptable and must be redesigned.*

ANOVA for Problem **4.5**

Source	df	Δ	EMS
A_i	1	2.59	$\sigma^2_{CE} + 2\sigma^2_{BE} + 4\sigma^2_{AE} + 2\sigma^2_{CD} + 4\sigma^2_{BD} + 8\sigma^2_{AD} + 4\sigma^2_C + 8\sigma^2_B + 16\Phi(A)$
$B_{j(i)}$	2	3.29	$\sigma^2_{CE} + 2\sigma^2_{BE} + 2\sigma^2_{CD} + 4\sigma^2_{BD} + 4\sigma^2_C + 8\sigma^2_B$
$C_{k(ij)} \leftarrow$	4	2.51	$\sigma^2_{CE} + 2\sigma^2_{CD} + 4\sigma^2_C$
D_ℓ	1	7.56	$\sigma^2_{CE} + 2\sigma^2_{BE} + 8\sigma^2_E + 2\sigma^2_{CD} + 4\sigma^2_{BD} + 16\sigma^2_D$
$AD_{i\ell}$	1	10.7	$\sigma^2_{CE} + 2\sigma^2_{BE} + 4\sigma^2_{AE} + 2\sigma^2_{CD} + 4\sigma^2_{BD} + 8\sigma^2_{AD}$
$BD_{j\ell(i)}$	2	3.97	$\sigma^2_{CE} + 2\sigma^2_{BE} + 2\sigma^2_{CD} + 4\sigma^2_{BD}$
$\rightarrow CD_{k\ell(ij)}$	4	2.66	$\sigma^2_{CE} + 2\sigma^2_{CD}$
$E_{m(\ell)} \leftarrow$	2	2.81	$\sigma^2_{CE} + 2\sigma^2_{BE} + 8\sigma^2_E$
$AE_{im(\ell)} \leftarrow$	2	3.97	$\sigma^2_{CE} + 2\sigma^2_{BE} + 4\sigma^2_{AE}$
$\rightarrow BE_{jm(i\ell)}$	4	2.66	$\sigma^2_{CE} + 2\sigma^2_{BE}$
$CE_{km(ij\ell)}$	8	—	σ^2_{CE}

ANOVA for Problem **4.5** *Continued*

Source	SS
A_i	$16\sum_{i=1}^{2}(\bar{y}_{i....}-\bar{y}_{.....})^2$
$B_{j(i)}$	$8\sum_{i=1}^{2}\sum_{j=1}^{2}(\bar{y}_{ij...}-\bar{y}_{i....})^2$
$C_{k(ij)}$	$4\sum_{i=1}^{2}\sum_{j=1}^{2}\sum_{k=1}^{2}(\bar{y}_{ijk..}-\bar{y}_{ij...})^2$
D_ℓ	$16\sum_{\ell=1}^{2}(\bar{y}_{...\ell.}-\bar{y}_{.....})^2$
$AD_{i\ell}$	$8\sum_{i=1}^{2}\sum_{\ell=1}^{2}(\bar{y}_{i..\ell.}-\bar{y}_{i....}-\bar{y}_{...\ell.}+\bar{y}_{.....})^2$
$BD_{j\ell(i)}$	$4\sum_{i=1}^{2}\sum_{j=1}^{2}\sum_{\ell=1}^{2}(\bar{y}_{ij.\ell.}-\bar{y}_{ij...}-\bar{y}_{i..\ell.}+\bar{y}_{i....})^2$
$CD_{k\ell(ij)}$	$2\sum_{i=1}^{2}\sum_{j=1}^{2}\sum_{k=1}^{2}\sum_{\ell=1}^{2}(\bar{y}_{ijk\ell.}-\bar{y}_{ijk..}-\bar{y}_{ij.\ell.}+\bar{y}_{ij...})^2$
$E_{m(\ell)}$	$8\sum_{\ell=1}^{2}\sum_{m=1}^{2}(\bar{y}_{...\ell m}-\bar{y}_{...\ell.})^2$
$AE_{im(\ell)}$	$4\sum_{i=1}^{2}\sum_{\ell=1}^{2}\sum_{m=1}^{2}(\bar{y}_{i..\ell m}-\bar{y}_{i..\ell.}-\bar{y}_{...\ell m}+\bar{y}_{...\ell.})^2$
$BE_{jm(i\ell)}$	$2\sum_{i=1}^{2}\sum_{j=1}^{2}\sum_{\ell=1}^{2}\sum_{m=1}^{2}(\bar{y}_{ij.\ell m}-\bar{y}_{i..\ell m}-\bar{y}_{ij.\ell.}+\bar{y}_{i..\ell.})^2$
$CE_{km(ij\ell)}$	$\sum_{i=1}^{2}\sum_{j=1}^{2}\sum_{k=1}^{2}\sum_{\ell=1}^{2}\sum_{m=1}^{2}(y_{ijk\ell m}-\bar{y}_{ij.\ell m}-\bar{y}_{ijk\ell.}+\bar{y}_{ij.\ell.})^2$

4.16 *Drop the factor called supplier and treat materials as a fixed factor having five levels. If it makes sense to do so, use a contrast to test supplier by comparing the average of the first two levels with the average of the last three levels.*

4.19 *Use subscripts i, j, and k for W, E, and I respectively. (a) WE_{ij} tests W_i. $WE_{ij}+I_{k(j)}-WI_{ik(j)}$ tests E_j. $WI_{ik(j)}$ tests WE_{ij} and $I_{k(j)}$. (b) $WE_{ij}+I_{k(i)}-EI_{jk(i)}$ tests W_i. WE_{ij} tests E_j. $EI_{jk(i)}$ tests WE_{ij} and $I_{k(i)}$. (c) $E_{j(i)}$ tests W_i. $I_{k(ij)}$ tests $E_{j(i)}$. (d) $E_{j(i)}+WI_{ik}-EI_{jk(i)}$ tests W_i. $EI_{jk(i)}$ tests $E_{j(i)}$ and WI_{ik}. WI_{ik} tests I_k. (e) $WE_{ij}+WI_{ik}-WEI_{ijk}$ tests W_i. $WE_{ij}+EI_{jk}-WEI_{ijk}$ tests E_j. $WI_{ik}+EI_{jk}-WEI_{ijk}$ tests I_k. WEI_{ijk} tests WE_{ij}, WI_{ik}, and EI_{jk}. (f) WE_{ij} tests W_i and E_j. $I_{k(ij)}$ tests WE_{ij}. (g) $E_{j(i)}+I_{k(i)}-EI_{jk(i)}$ tests W_i. $EI_{jk(i)}$ tests $E_{j(i)}$ and $I_{k(i)}$.*

4.35 $y_{ijk\ell m}=\mu+A_i+T_j+AT_{ij}+R_{k(ij)}+L_\ell+AL_{i\ell}+AT_{j\ell}+ATL_{ij\ell}+RL_{k\ell(ij)}+\varepsilon_{m(ijk\ell)}$; $i=1,2$; $j=1,2$; $k=1,2,3$; $\ell=1,2$; $m=1,2$; A_i, T_j, L_ℓ *fixed; and* $R_{k(ij)}$, $\varepsilon_{m(ijk\ell)}$ *random.* $R_{k(ij)}$ *tests* A_i, T_j, *and* AT_{ij} *with* $\Delta=0.76$, *0.76, and 1.07.* $RL_{k\ell(ij)}$ *tests* L_ℓ, $AL_{i\ell}$, $AT_{j\ell}$, *and* $ATL_{ij\ell}$ *with* $\Delta=0.76$, *1.07, 1.07, and 1.52.* $\varepsilon_{m(ijk\ell)}$ *tests* $R_{k(ij)}$ *and* $RL_{k\ell(ij)}$ *with* $\Delta=1.08$ *and 1.53.*

4.36 *Factors significant at the .01 level are A_i, T_j, AT_{ij}, and $R_{k(ij)}$. All other factors and interactions are insignificant. Both locations are the same.*

Although there is no significant difference between A at level 1, T at level 1 and A at level 1, T at level 2, I would select the highest mean—A at level 1, T at level 1. A 95% confidence interval is $159.25 \pm 2.302\sqrt{20.5525/12} = 159.25 \pm 3.01$.

5

Restrictions on Randomization

5.0 INTRODUCTION

So far we have been extremely careful to completely randomize every experiment considered in this text. This includes randomly assigning the experimental units to the treatment combinations, running the experiment in random order, or randomly assigning the location of treatment combinations. We now consider the effect of a failure to completely randomize the experiment.

Failure to completely randomize an experiment does not mean the randomization concept is ignored. Rather, complete randomization is replaced with separate randomizations within and between certain treatment combinations. Such separate randomizations sometimes occur naturally, *e.g.*, with factors like days or plots that represent a time or location component, and sometimes are used to make the experiment quicker, easier, or less expensive to run. Of course, there is a price to be paid for such restrictions on randomization and it is up to the experimenter to decide if the convenience is worth the price.

In this chapter we will first learn how to identify restrictions on randomization. Then we will learn how to modify the mathematical model to account for restrictions on randomization and how the restrictions affect the analysis of the data. A mathematical proof of the effect of restrictions on randomization is included for those having a strong statistical background. The proof can be skipped without loss of understanding of the principle. Do not skip the last two paragraphs of Section **5.3** because they summarize the conditions under which randomization is and is not important. All procedures will be illustrated with examples and there are plenty of realistic problems at the end of the chapter.

5.1 IDENTIFYING RESTRICTIONS ON RANDOMIZATION

An experiment was run to determine the effect of oven temperature and color of paint on the distinctness of image of painted panels. Three oven temperatures and four colors were selected. Two panels of each combination of temperature and color were tested. The ovens were large and it required several hours to change the temperature from one level to another. Complete randomization requires many changes of the oven temperature so the experiment would take a long time to run. Therefore, it was decided that one temperature should be randomly chosen and all eight panels (two repeats of four colors) painted and baked before changing the oven to another temperature.

It is clear that this experiment was not completely randomized and yet randomization was used to select the temperature and to determine the baking and painting order of the two repeats of the four colors. In other words, the experiment was not completely randomized but separately randomized among the levels of oven temperature and within each level of oven temperature.

For this painted panel experiment, we have restricted the randomization, the restriction being associated with the factor temperature. We know that the restriction error is associated with temperature since we selected and fixed one temperature and then randomly ran all of the color/repeat combinations. We say there is a restriction on randomization *associated with the factor temperature.* Alternatively, we say that the experiment is run in *blocks* where blocks are associated with temperature, or we say the experiment is *blocked on temperature*, or we simply say that temperature is a *blocking factor*.

Let us consider the painted panel experiment again under a slightly different set of conditions. Suppose there is only one spray gun and it takes considerable effort to clean the spray gun and mix a new color. It would then make sense to clean the spray gun and mix a new color while the oven temperature is changing. One could then choose a temperature/color combination and paint and bake both repeats before going to the next temperature/color combination. There is no need to individually fix a temperature or fix a color. Rather, both can be changed at the same time.

For these new conditions, a combination of temperature and color is randomly selected and fixed while all repeats are made. Therefore, there is a restriction on randomization associated with the temperature/paint combination. Alternately, temperature/paint is a blocking term.

Generalizing from this simple example, a restriction on randomization can be associated with any *term* in the experiment. Such a restriction implies that one level of that term (possibly a combination of factors if that term represents an interaction) is fixed and all combinations containing that level are randomly run prior to the selection of another level. There is a one-to-one association between the randomization procedure used in an experiment and the blocking term. It is extremely important to be able to detect the restrictions on randomization and associate them with the proper terms in the model.

The same statement applies to all types of randomization, not just randomization of run order. For example, restrictions on randomization can apply to location in agricultural experiments carried out in fields. There is likely to be differences in, say, soil conditions in different plots within a field. Therefore, plots should be considered as a factor in the experiment. However, plots in the same field are likely to be more similar than plots in different fields. There is a restriction on randomization associated with fields since plots are restricted to come from a given field but the location of the plots within a given field can be completely randomized. Again, we can separately randomize fields and separately randomize within each field.

For some reason location related restrictions on randomization seem more difficult to understand than time related restrictions, even though the concept is identical. Therefore, it is crucial that the above example be understood even if you never intend to do an agricultural example. So stop now and rereview the above paragraphs if there is the slightest doubt about the concept of restrictions on randomization.

More than one restriction on randomization can occur in an experiment. The paint panel experiment can be altered to illustrate the possibility. Suppose it is easier to change color than to change temperature. The experimenter may wish to fix a temperature first. Having done so, the experimenter next selects a color and paints and bakes both repeats before selecting another color. After all of the colors are painted and baked, the oven temperature is changed and the process starts all over again. In this case, there is a restriction associated with oven temperature and another restriction associated with the temperature/color interaction term.

The restriction associated with temperature is easy to see. However, it is harder to see that the second restriction on randomization is associated with the temperature/color term and not with the color factor alone. The reason the second restriction is associated with the interaction term is that, when the color was fixed, the temperature was *already* fixed, so that *both* temperature and color were actually fixed. Thus, the restriction on randomization is actually associated with the temperature/color term.

Generally speaking, multiple restrictions on the randomization will be hierarchical in nature, associated first with a factor, then with an interaction term involving that factor, and so on. This is not an iron clad rule, but examples disobeying this rule tend to be artificial in nature. (See Problem **5.26** for a realistic example.)

5.2 EFFECT ON THE MATHEMATICAL MODEL AND TESTS

Once restrictions on randomization have been noted and associated with the appropriate term(s) in the model, determining the effect is actually quite easy. One simply adds a Greek term in the mathematical model immediately after the term associated with the restriction on randomization, Anderson (1970) and Anderson and McLean(1974b). This term is considered random, is Greek to distinguish it from a normal factor, and is called a *restriction error.* The restriction error contains the next available subscript and is nested in all of the factors in the term it follows. Its subscript takes the value one only. All of the usual rules for df, SS, MS, EMS, and F tests apply to restriction errors.

The rationale for nesting the restriction error within the term it follows comes from the process of randomizing *within* each level of the term. As we learned in the previous chapter, *within* is generally associated with nesting. Alternatively, the randomization *depends* on the level of the term the restriction error follows.

Note that the restriction error has only one level. This implies that it has zero df, no SS, MS, or EMS. The magnitude of the restriction error is not estimable so we can never know its true effect. We often drop the subscript associated with the restriction error since it plays no role. If it is dropped, you must remember that the restriction error has zero df and not try to calculate MS or EMS values.

Formally, let us consider the three factor experiment (3.2.3) with factors A_i, $i = 1, 2$, fixed; B_j, $j = 1, 2, 3, 4$, random; C_k, $k = 1, 2, 3$, fixed; and each combination repeated twice. If complete randomization is used, the EMS are as given in Table 3.4.4, reproduced in the third column of Table 5.2.1 for convenience.

Suppose the experiment was not completely randomized but blocked on A_i. That is, one level of A was randomly selected and all of the combinations of B_j, C_k, and repeats were tested in a random fashion prior to selection of the other level of A. The new mathematical model becomes

$$y_{ijk\ell m} = \mu + A_i + \gamma_{\ell(i)} + B_j + AB_{ij} + C_k + AC_{ik} + BC_{jk} + ABC_{ijk} + \varepsilon_{m(ijk\ell)}, \quad (5.2.1)$$

where $\ell = 1$ only. The term $\gamma_{\ell(i)}$ is the restriction error immediately following A_i and nested within i. You will also notice that the restriction error does not interact with any other term in the model. This is by assumption only. It certainly *could* interact with other factors. The restriction error term merely gets added into the model after A_i. Since $\gamma_{\ell(i)}$ is random, the EMS are as given in the fourth column of Table 5.2.1. The reader should reproduce this table to be sure the process is fully understood.

TABLE 5.2.1

ANOVA FOR THREE FACTORS, A_i, C_k FIXED, B_j RANDOM

Source	df	EMS—Randomized	EMS—Blocked on A
A_i	1	$\sigma^2 + 6\sigma^2_{AB} + 24\Phi(A)$	$\sigma^2 + 6\sigma^2_{AB} + 24\sigma^2_{\gamma} + 24\Phi(A)$
$\gamma_{\ell(i)}$	0	N/A	—
B_j	3	$\sigma^2 + 12\sigma^2_B$	$\sigma^2 + 12\sigma^2_B$
AB_{ij}	3	$\sigma^2 + 6\sigma^2_{AB}$	$\sigma^2 + 6\sigma^2_{AB}$
C_k	2	$\sigma^2 + 4\sigma^2_{BC} + 16\Phi(C)$	$\sigma^2 + 4\sigma^2_{BC} + 16\Phi(C)$
AC_{ik}	2	$\sigma^2 + 2\sigma^2_{ABC} + 8\Phi(AC)$	$\sigma^2 + 2\sigma^2_{ABC} + 8\Phi(AC)$
BC_{jk}	6	$\sigma^2 + 4\sigma^2_{BC}$	$\sigma^2 + 4\sigma^2_{BC}$
ABC_{ijk}	6	$\sigma^2 + 2\sigma^2_{ABC}$	$\sigma^2 + 2\sigma^2_{ABC}$
$\varepsilon_{m(ijk\ell)}$	24	σ^2	σ^2

Comparing the last two columns in Table 5.2.1 shows the effect of the restriction error: the direct test on A_i disappears. The best that can be done now is to test A_i with AB_{ij}, a conservative test. This conservative test simultaneously tests $\Phi(A)$ and σ^2_{γ}, showing that A_i is *completely confounded* with the restriction error $\gamma_{\ell(i)}$. If the test is insignificant, then both the restriction error and A_i are insignificant. If the test is significant, we cannot tell if A_i, $\gamma_{\ell(i)}$, or both are significant.

Without belaboring the point by more examples, the effect of blocking on a term or restricting the randomization on a term is to turn the direct test on that term into a conservative test on the same term. If the test on that term is insignificant, then that term is insignificant. However, if the test is significant, we cannot tell if the the significance is attributable to that term or to the restriction on the randomization, or to both.

Since information about a blocking term is always confounded with the restriction error, it is always wise to *design* the experiment by blocking on a factor or combination of factors of lesser importance. This is why experiments are usually blocked on factors like time or plot. We know that time or plot may have an influence on the outcome of the experiment, but we do not care to estimate its effect. Thus, when we block on time or plot, we are losing information about factors that are of secondary interest. We get the convenience of incomplete randomization without the loss of valuable information.

Sometimes lack of randomization is forced upon you by the nature of the experiment. An example is the factor "time after exposure." Naturally, one hour comes before two hours which comes before three hours, so an ordering is imposed. In these situations, you can do little more than simply note the restriction error.

Restriction errors are not always bad. Suppose there is a day to day effect on the response variable. If complete randomization is used in such an experiment, the day to day effect will become part of the error term, inflating the error term and making it more difficult to find true effects. On the other hand, if Day is identified as a factor in the experiment, its effect is removed from the error term and more precise estimates will be possible. It is of little concern that Days are associated with the restriction error eliminating the direct test since we had little interest in the Day effect anyhow.

Further discussion is given in Examples 1 and 2 in the last section of this chapter. These examples help nail down the importance of randomization and the importance of specifically designing for any lack of randomization.

5.3 PROOF OF THE EXISTENCE OF RESTRICTION ERRORS

This section gives a proof of the existence of restriction errors and is primarily taken from Lorenzen (1984). It should only be read by those with a strong mathematical background who are interested in knowing why a restriction on randomization is equivalent to adding a random term in the mathematical model and why this term does not interact with any other term. All other readers are advised to skip to the last two paragraphs of this section (past the light bulbs) which summarize the findings.

For simplicity we will consider the One-Way model given by

$$y_{ij} = \mu + A_i + \varepsilon_{j(i)}, \qquad (5.3.1)$$

with A_i fixed, $i = 1, \cdots, I$ and $j = 1, \cdots, J$. If all model assumptions are correct, *i.e.*, $\varepsilon_{j(i)}$ are iid Normal(0,σ^2), we know that EMS(A) = $\sigma^2 + J\Phi(A)$, EMS(ε) = σ^2, and the error term directly tests A_i.

Suppose the assumptions of the One-Way model are not met. In particular, assume there is an unknown (therefore uncontrolled) factor U influencing y. If U is a random variable, *i.e.*, U is iid Normal(0,σ_U^2), then we can lump U and ε together as a new error term having variation $\sigma_{new}^2 = \sigma^2 + \sigma_U^2$. The result is an inflation of the error term, making it harder to detect the effect of A, but still resulting in a legitimate F test. Note that this is true whether or not we randomize.

If U is not strictly a random variable, then randomization plays a role in the analysis and interpretation. Suppose U is a serially correlated random variable. That is, suppose U is a random variable whose current value is correlated with its previous value. Many factors behave in this fashion: weather conditions like temperature, humidity, or barometric pressure, warm-up or wear out in a machine, operator fatigue, learning in human cognizance studies, soil conditions in agricultural studies, and so on. We will examine the effect of a serially correlated U when complete randomization is used and when the experiment is blocked on A_i.

Let the subscript ℓ represent the order of the experiment. Then ℓ ranges from 1 to IJ and the serial correlation can be represented as $\text{corr}(U_\ell U_{\ell'}) = \rho^{|\ell-\ell'|}$, where ρ is the serial correlation. Let $z_{i\ell}$ be the response if the ith treatment level of A is applied to the ℓth experimental unit. The model describing $z_{i\ell}$ is

$$z_{i\ell} = \mu + A_i + U_\ell + \varepsilon_{i\ell}, \tag{5.3.2}$$

where $\ell = 1, \cdots, IJ$, A_i is fixed, and U_ℓ is the uncontrolled factor effect.

The model (5.3.2) is strictly a conceptual model and its corresponding experiment can never be run. To be able to run the experiment (5.3.2), the first unit run must have treatment A_1, and the first unit run must have treatment A_2, and so on. This is impossible. Rather, model (5.3.2) is related to model (5.3.1) according to the randomization procedure used to decide which experimental unit gets which treatment. To that end, define an indicator function D by

$$D_{i\ell} = \begin{cases} 1 & \text{if the } \ell\text{th unit receives treatment } i, \\ 0 & \text{otherwise.} \end{cases} \tag{5.3.3}$$

Then models (5.3.1) and (5.3.2) are related according to the following two algebraic identities:

$$\sum_{j=1}^{J} y_{ij} = \sum_{\ell=1}^{IJ} D_{i\ell} z_{i\ell} \tag{5.3.4}$$

$$\sum_{j=1}^{J} y_{ij}^2 = \sum_{\ell=1}^{IJ} D_{i\ell} z_{i\ell}^2 \tag{5.3.5}$$

The general approach is to use the randomization procedure to assign probabilities to $D_{i\ell}$ and compute the expected values of the mean squares defined for y_{ij} using equations (5.3.2) through (5.3.5). The only thing that differs in the completely randomized case and the blocking on A case is the expectation of $D_{i\ell}$.

Suppose the experiment is completely randomized. Simple combinatorics lead to the following expectations:

$$\mathrm{E}(D_{i\ell} D_{i'\ell'}) = \begin{cases} 1/I & \text{for } i = i', \ell = \ell', \\ 0 & \text{for } i \neq i', \ell = \ell', \\ (J-1)/[I(IJ-1)] & \text{for } i = i', \ell \neq \ell', \\ J/[I(IJ-1)] & \text{for } i \neq i', \ell \neq \ell'. \end{cases} \tag{5.3.6}$$

Computation of EMS follows from tedious algebra, given in Lorenzen (1984). The resulting EMS, along with the EMS for the usual One-Way model, are given in Table 5.3.1. As can be seen by comparing the last two columns, under complete randomization a serially correlated variable inflates the error term but does not invalidate the F test. This shows the protection afforded by the randomization procedure.

TABLE 5.3.1

ANOVA FOR ONE-WAY MODEL WITH ASSUMPTION VIOLATIONS

Source	EMS (Usual Assumptions)	EMS (Complete Randomization)
A_i	$\sigma^2 + J\Phi(A)$	$\sigma^2 + \sigma_{U,\rho}^2 + J\Phi(A)$
$\varepsilon_{j(i)}$	σ^2	$\sigma^2 + \sigma_{U,\rho}^2$

where $\sigma_{U,\rho}^2 = \left[[(IJ)(IJ-1)]^{-1} \sum_{\ell=1}^{IJ} \sum_{\ell'=1}^{IJ} (1 - \rho^{|\ell-\ell'|})\right] \sigma_U^2$

Now suppose the experiment is blocked on A_i. This means we randomly select a level of A and run all of the repeats before selecting another level of A. The only thing different about this experiment is the probability associated with $D_{i\ell}$. To easily account for the blocking, let m be a subscript denoting the block and n be a subscript denoting the unit within a block. The subscript ℓ is related to m and n by the formula

$\ell = (m-1)J + n$. Again, simple combinatorics lead to the following expectations:

$$\mathrm{E}(D_{i\ell}D_{i'\ell'}) = \begin{cases} 1/I & \text{for } i = i',\ m = m', \\ 0 & \text{for } i \neq i',\ m = m' \text{ or } i = i',\ m \neq m', \\ 1/[I(I-1)] & \text{for } i \neq i',\ m \neq m'. \end{cases} \tag{5.3.7}$$

Again, computation of the EMS follows tedious algebra which leads to Table 5.3.2. As can be seen by comparing the last two columns, under blocking on A a serially correlated variable both inflates the error term *and adds an extra term to the numerator.* This extra term exactly corresponds to the restriction error introduced in the previous section and invalidates the usual F test. The usual F tests now tests the combined effects $\Phi(A)$ and $\breve{\sigma}^2_{U,\rho}$. Calling $\breve{\sigma}^2_{U,\rho}$ the restriction error, A is completely confounded with the restriction error. Blocking does not afford protection against unknown serially correlated variates.

TABLE 5.3.2

ANOVA FOR ONE-WAY MODEL WITH ASSUMPTION VIOLATIONS

Source	EMS (Usual Assumptions)	EMS (Blocked on A_i)
A_i	$\sigma^2 + J\Phi(A)$	$\sigma^2 + \tilde{\sigma}^2_{U,\rho} + J\breve{\sigma}^2_{U,\rho} + J\Phi(A)$
$\varepsilon_{j(i)}$	σ^2	$\sigma^2 + \tilde{\sigma}^2_{U,\rho}$

where $\tilde{\sigma}^2_{U,\rho} = \left[[(IJ)(J-1)]^{-1} \sum_{m=m'=1}^{I} \sum_{n=1}^{J} \sum_{n'=1}^{J} (1 - \rho^{|\ell-\ell'|})\right]\sigma^2_U$, $\breve{\sigma}^2_{U,\rho} = \left[[J(1-J)]^{-1} \sum_{n=1}^{J} \sum_{\substack{n'=1 \\ n'\neq n}}^{J} \rho^{|n-n'|} - [IJ^2(I-1)]^{-1} \sum_{m=1}^{I} \sum_{\substack{m'=1 \\ m'\neq m}}^{I} \sum_{n=1}^{J} \sum_{n'=1}^{J} \rho^{|\ell-\ell'|}\right]\sigma^2_U$, and $\ell = (m-1)J + n$,

In a similar fashion, complete randomization can be shown to protect against any other type of correlation, any fixed (over run order) effects of U_ℓ, and any heterogeneity (again over run order) in U_ℓ. On the other hand, blocking on A_i leads to the presence of a restriction error in any of those circumstances.

All of the results derived for the simple One-Way model can also be derived for the two-way model, and therefore for all balanced designs. Thus, we have shown that restrictions on randomizations are equivalent to the addition of a random term immediately after the term associated with the restriction in all balanced designs.

In summary, restriction errors are mathematically derivable quantities that represent the effect of any uncontrolled factors that influence the response variable and are correlated over the time it takes to run the experiment or the physical location of the experimental units. If all uncontrolled factors are truly random, randomization plays no role in the experiment. If there are uncontrolled factors that change slowly over time or location, then failure to randomize can cause terms to be declared significant when they are truly insignificant.

Whether or not there are uncontrolled factors that change slowly over time or location is not a statistical question. Only the experimenter familiar with the area of application can have an idea about the effect of factors not specifically controlled in the experiment. If the experimenter has no idea about those uncontrolled factors, then such factors are assumed to exist and one must completely randomize the experiment or place restrictions only on terms of secondary importance.

5.4 EXAMPLES

Example 1. Randomized Complete Block Designs. In many experiments there is only one treatment, T_i, of interest. However, it is feared that the experimental units run in close proximity to each other, either in a time or location sense, will have less variation than the experimental units run further apart. To properly account for the proximity problem, the treatments are applied in several "blocks", where each set of treatments is run in close proximity within each "block", and "blocks" are not run in close proximity. The inferential experimental unit would then be "blocks" because the variation within "blocks" is unrealistically small which would cause the inference space to be unrealistically small.

The treatments are randomized within each "block". This is illustrated in Figure 5.1. The observations in each enclosed box must be completed before going to the next box.

B_1	B_2		B_J
T_1	T_1		T_1
T_2	T_2	$\cdots$	T_2
$\vdots$	$\vdots$		$\vdots$
T_I	T_I		T_I

Figure 5.1 Layout for randomized complete block designs. Complete each box before proceeding to the next.

The key to this type of experiment is to realize that "blocks" is simply a random factor. When one fears that experimental units run in close proximity will have less variation than those run further apart, one is indicating that "blocks" will have an influence on the response variable.

To model this experiment, treat "blocks" as a factor, B_j, in the experiment. "Blocks" is not a factor in the usual sense because it is not applied to the experimental units. Yet "blocks" is thought to influence the response variable so must be considered in the experiment. "Blocks", as used in this sense, can have many other names; a few of which include "time", "days", "weeks", "locations", "fields", "plots", "chambers", "pens", and so on.

Since randomization is performed separately within each "block", there is a restriction on randomization associated with "blocks". This makes "blocks" a blocking factor, the origin of our terminology. We apply this terminology to *any* term which restricts the randomization process.

Taking treatments as fixed and blocks as random, the ANOVA for a randomized complete block design is given in Table 5.4.1. There is no test for "blocks". If we test "blocks" with the TB_{ij} interaction, we get a doubly conservative test. If the test is insignificant, either "blocks" and the restriction error are negligible or the TB_{ij} interaction is non-negligible. If the test is significant, either "blocks", the restriction error, or both are significant. We cannot tell which is the case.

TABLE 5.4.1

ANOVA FOR A RANDOMIZED COMPLETE BLOCK DESIGN

Source	df	EMS
→ Treatment (T_i)	$I-1$	$\sigma^2 + \sigma^2_{TB} + J\Phi(T)$
"Blocks" (B_j)	$J-1$	$\sigma^2 + I\sigma^2_\delta + I\sigma^2_B$
Restriction error ($\delta_{k(j)}$)	0	—
— Interaction (TB_{ij})	$(I-1)(J-1)$	$\sigma^2 + \sigma^2_{TB}$
Error ($\varepsilon_{\ell(ijk)}$)	0	—

If we failed to recognize "blocks" as a factor, we would have completely randomized the design and written it as a One-Way model with T_i having $I-1$ df and an error term having $I(J-1)$ df, as in Table 5.4.2.

Comparing Table 5.4.2 to Table 5.4.1, we see that the One-Way model has more df for testing treatments than the Randomized Complete Block

TABLE 5.4.2

ANOVA FOR A ONE-WAY DESIGN

Source	df	EMS
Treatment (T_i)	$I-1$	$\sigma_1^2 + J\Phi(T)$
Error_1 ($\varepsilon_{j(i)}$)	$I(J-1)$	σ_1^2

Design. This would appear to make the One-Way model superior. However, the error term in the One-Way model is not the same as the error in the Randomized Complete Block Design. In the One-Way model the error term σ_1^2 includes σ^2, σ_{TB}^2, and σ_B^2 from the Randomized Complete Block Design. This is because factors not considered in and not fixed in an experiment become part of the error term under complete randomization. The denominator for testing treatments in the Randomized Complete Block Design only contains σ^2 and σ_{TB}^2, so must be smaller. Obviously some sort of trade-off between the df, detectable size, and the size of the appropriate denominator exists. This is simply another example of the design and redesign process.

Example 2. Split-Plot Designs Looking at Example 1, we note that there are no tests on "blocks". If "blocks" are of interest, say because each block represents a treatment applied to the entire block, we have a problem due to confounding of the treatment effect with the restriction error. Basically, if an effect is significant, we will not know whether the significance is due to the treatment or due to the restriction on randomization.

To counteract this problem, replicate the experiment several times using different randomizations each time. The layout is illustrated in Figure 5.2 and is commonly called a split-plot design. Historically, replicates become the "fields", what was called "blocks" in Example 1 become "plots", and the treatments within each "plot" become the "split-plot". The terminology came from agricultural experiments. This type of experiment was created to handle fields in different regions, soil trends within each field, and one additional factor. However, it is easily adaptable to two factors by replacing soil trends with the second factor.

While the procedure and evaluation is often confusing to students who have to try to link their experiment to agricultural terms, the design and evaluation is simple with our restriction error approach. The name split-plot adds nothing to the experiment and was included in this book merely for historic reasons. We recognize three factors: replications, R_k; blocks, B_j; and treatment, T_i. The highest order restriction on randomization is

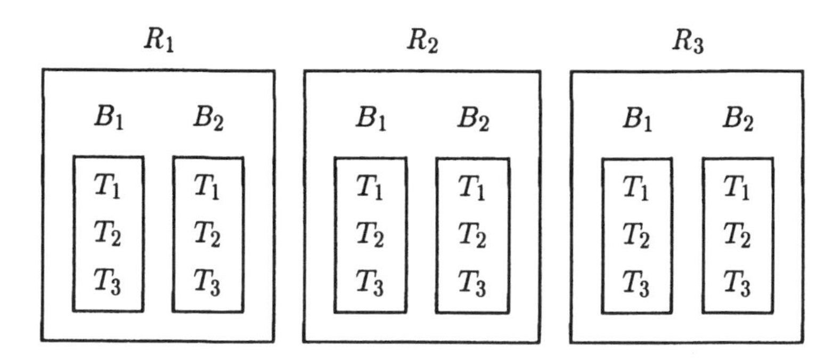

Figure 5.2 Layout for a split-plot design. Complete a box before going to a box of the same size.

on R_k since one randomized complete block design is completed before the next one starts. Within each replicate, there is a second restriction error on blocks. The mathematical model is

$$y_{ijk} = \mu + R_i + \delta_{j(i)} + B_k + RB_{ik} + \gamma_{\ell(ik)} + T_m + RT_{im} + BT_{km} + RBT_{ikm} + \varepsilon_{(ijk\ell m)},$$

with evaluation as indicated in Table 5.4.3. This table assumes treatments and blocks are fixed while replicates are random. Of course, the EMS and tests would change if treatments or blocks were random or there was any nesting.

Rather than memorize the layout of the split-plot design, we prefer that students understand the factors and restrictions in the experiment. In this experiment, there are two factors of interest, blocks and treatments. There is a restriction error on blocks. This eliminates the test on blocks. In order to get some information on blocks, the entire experiment is replicated. Since something might have changed by the time the experiment is replicated, replicate is considered as a third factor. It is random. There is a restriction error on replicates.

Not only is this approach easier to comprehend (see Problems 5.37 through 5.39 for an illustration), it leads one into considering alternative designs that may be better than the original split-plot design. For example, if possible, would one be better off running several repeats of the treatments *within* a block rather than in separate replicates? Should one consider complete randomization within a replicate instead of the restricted randomization? Etc.

TABLE 5.4.3

ANOVA FOR EXAMPLE 2, A SPLIT-PLOT DESIGN

Source	df	Δ	EMS
R_i	2		$\sigma^2 + 3\sigma_\gamma^2 + 6\sigma_\delta^2 + 6\sigma_R^2$
$\delta_{j(i)}$	0		—
→ B_k	1	2.26	$\sigma^2 + 3\sigma_\gamma^2 + 3\sigma_{RB}^2 + 9\Phi(B)$
└ RB_{ik}	2		$\sigma^2 + 3\sigma_\gamma^2 + 3\sigma_{RB}^2$
$\gamma_{\ell(ik)}$	0		—
→ T_m	2	1.59	$\sigma^2 + 2\sigma_{RT}^2 + 6\Phi(T)$
└ RT_{im}	4		$\sigma^2 + 2\sigma_{RT}^2$
→ BT_{km}	2	2.25	$\sigma^2 + \sigma_{RBT}^2 + 3\Phi(BT)$
└ RBT_{ikm}	4		$\sigma^2 + \sigma_{RBT}^2$
$\varepsilon_{(ijk\ell m)}$	0		—

In addition, slight pertubations of the problem will not cause any difficulties when one uses the No-Name approach. For example, suppose blocks were actually a random factor and block 1 in repeat 1 is not the same as block 1 in repeat 2. Then, block is really nested in repeat and we have often seen this called a split-plot design even though the layout and analysis differ from the previous example. In the No-Name approach, merely indicate that blocks are nested within repeats and continue on as always.

While you should be aware that certain names are associated with certain designs, reliance on the use of named designs will decrease your effectiveness, limit your ability to recognize factors and key concepts, and greatly reduce the number of experiments at your command.

Example 3. An experiment was designed to measure the stretch of a piece of metal punched in a die. Four factors were considered: lubricant, L_i, none, mill oil, and an added lubricant; thickness of the steel, T_j, 8, 10, and 12 mill; steel type, S_k, standard and AK steel; and crimp type, C_ℓ, crimped and uncrimped. Ten pieces were punched for each combination of factors. Since it takes a long time to wipe down a die, one lubricant was selected and all combinations of thickness, steel type, and crimp type

were run before another lubricant was used. The factors thickness, steel type, and crimp type were completely randomized but all ten pieces were punched before going to the next combination.

As a confirmatory measure, the entire experiment was repeated a week later, using different randomizations and different metal. What model describes this experiment and what information is lost by the lack of randomization?

While not specifically mentioned by the experimenter, there is a fifth factor in this experiment, weeks, W_m, $m = 1, 2$, random. The first step in understanding a complicated but realistic example such as this one is to draw a picture illustrating the way this experiment is to be run, Figure 5.3. We will use the convention that each box must be completed before moving on to the next box of the same size.

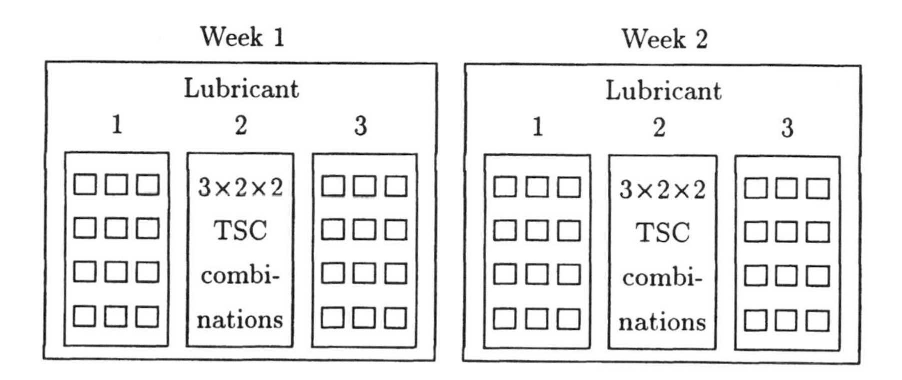

Figure 5.3 Layout for Example 3, metal punch stretch. Boxed material is to be completed before starting the next box of the same size. Small boxes represents 10 consecutive punches.

There are large boxes around the levels of weeks because week 1 must be completed before week 2 can begin. This indicates the first restriction error, associated with weeks.

The next largest box is around lubricant levels because changing lubricant is a time consuming task. Notice that the lubricant box is completely within a week box so the second restriction error is associated with week/lubricant combinations. The twelve TSC combinations appear within each lubricant box, as indicated under lubricant levels 1 and 3 and labeled under lubricant level 2.

Finally the ten repeats take place within each TSC combination (the smallest box) within each L and W boxes. Therefore, the final restriction error is associated with the highest order interaction, i.e., within each LTSCW combination.

The mathematical model describing this experiment is

$$\begin{aligned} y_{ijk\ell mnopq} = \mu &+ L_i + T_j + S_k + C_\ell + \text{interactions} \\ &+ W_m + \delta_{n(m)} + LW_{im} + \gamma_{o(im)} + \text{interactions} \\ &+ LTSCW_{ijk\ell m} + \eta_{p(ijk\ell m)} + \varepsilon_{q(ijk\ell mnop)}, \end{aligned}$$

where n, o, and p take the value 1 only. The ANOVA is given in Table 5.4.4.

TABLE 5.4.4

ANOVA FOR EXAMPLE 3 ON METAL PUNCH STRETCH

Source	df	Δ	EMS
L_i	2	0.43	$\sigma^2 + 10\sigma_\eta^2 + 120\sigma_\gamma^2 + 120\sigma_{LW}^2 + 240\Phi(L)$
T_j	2	0.43	$\sigma^2 + 10\sigma_\eta^2 + 120\sigma_{TW}^2 + 240\Phi(T)$
S_k	1	0.55	$\sigma^2 + 10\sigma_\eta^2 + 180\sigma_{SW}^2 + 360\Phi(S)$
C_ℓ	1	0.55	$\sigma^2 + 10\sigma_\eta^2 + 180\sigma_{CW}^2 + 360\Phi(C)$
interactions	29	$\vdots$	$\vdots$
W_m	1	—	$\sigma^2 + 10\sigma_\eta^2 + 120\sigma_\gamma^2 + 360\sigma_\delta^2 + 360\sigma_W^2$
$\delta_{n(m)}$	0	—	—
LW_{im}	2	—	$\sigma^2 + 10\sigma_\eta^2 + 120\sigma_\gamma^2 + 120\sigma_{LW}^2$
$\gamma_{o(im)}$	0	—	—
TW_{jm}	2	—	$\sigma^2 + 10\sigma_\eta^2 + 120\sigma_{TW}^2$
SW_{km}	1	—	$\sigma^2 + 10\sigma_\eta^2 + 180\sigma_{SW}^2$
$CW_{\ell m}$	1	—	$\sigma^2 + 10\sigma_\eta^2 + 180\sigma_{CW}^2$
interactions	25	$\vdots$	$\vdots$
$LTSCW_{ijk\ell m}$	4	—	$\sigma^2 + 10\sigma_\eta^2 + 10\sigma_{LTSCW}^2$
$\eta_{p(ijk\ell m)}$	0	—	—
$\varepsilon_{q(ijk\ell mnop)}$	648	—	σ^2

The restriction error $\delta_{n(m)}$ indicates that all combinations of the other factors in this experiment are run in the same week. The restriction error corresponds to every other factor not explicitly controlled in this experiment that changes over the course of a week. Note that W_m also corresponds to every other factor that changes over the course of a week. It would be perfectly acceptable to ignore this restriction error since its effects are exactly the same as the effects for W_m. We do not recommend ignoring the restriction error for two reasons. One, the restriction error term in the mathematical model indicates the experiment is not completely randomized. Two, we have found that students sometimes ignore restriction errors that should not be ignored. It is never a mistake to include a restriction error that could be ignored but it is often a mistake to ignore a restriction error that is important.

The restriction error $\gamma_{o(im)}$ indicates that, in a given week, all combinations of factors having the same lubricant level are run together. This restriction error corresponds to any break-in or wear out in the die.

Finally, the restriction error $\eta_{p(ijk\ell m)}$ indicates that the repeats are run back to back. You will notice that the restriction error effect σ_η^2 appears in the EMS for every term except ε. This restriction error corresponds to the set up variation, not present in the error term since the punch was not torn down and set up between each measurement. As we can see by the tests in Table 5.4.4, the inferential experimental unit for factors is weeks, not repeats.

Mathematically, the experiment is the same regardless of the order in which the factors appear in the mathematical model. However, the ordering of the factors can have a large influence on the intuition associated with a given design. This is useful when redesigning the experiment. We illustrate with this example.

Based on Figure 5.3, we let weeks be the first factor in the model because the largest box is around the levels of weeks. Likewise, lubricant should be the second factor in the model and the thickness, steel, and crimp factors are interchangeable. Without changing the indices (change them or not at your own discretion, it does not matter), rewrite the model as

$$\begin{aligned} y_{ijk\ell mnopq} = \mu &+ W_m + \delta_{n(m)} + L_i + WL_{mi} + \gamma_{o(mi)} + T_j + WT_{mj} \\ &+ S_k + WS_{mk} + C_\ell + WC_{m\ell} + \text{interactions} \\ &+ WLTSC_{mijk\ell} + \eta_{p(mijk\ell)} + \varepsilon_{q(mijk\ell nop)}, \end{aligned}$$

yielding Table 5.4.5.

Tables 5.4.4 and 5.4.5 are identical but it is much easier to see the effect of the restrictions on randomization in Table 5.4.5. We see that all of the terms involving L_i, T_j, S_k, and C_ℓ, as well as their interactions,

TABLE 5.4.5

ANOVA for Example 3 on Metal Punch Stretch

Source	df	Δ	EMS
W_m	1	0.48	$\sigma^2 + 10\sigma_\eta^2 + 120\sigma_\gamma^2 + 360\sigma_\delta^2 + 360\sigma_W^2$
$\delta_{n(m)}$	0	—	—
→ L_i	2	0.43	$\sigma^2 + 10\sigma_\eta^2 + 120\sigma_\gamma^2 + 120\sigma_{LW}^2 + 240\Phi(L)$
└ WL_{mi}	1	0.48	$\sigma^2 + 10\sigma_\eta^2 + 120\sigma_\gamma^2 + 120\sigma_{LW}^2$
$\gamma_{o(mi)}$	0	—	—
→ T_j	2	0.43	$\sigma^2 + 10\sigma_\eta^2 + 120\sigma_{TW}^2 + 240\Phi(T)$
└ WT_{mj}	2	0.48	$\sigma^2 + 10\sigma_\eta^2 + 120\sigma_{TW}^2$
→ S_k	1	0.55	$\sigma^2 + 10\sigma_\eta^2 + 180\sigma_{SW}^2 + 360\Phi(S)$
└ WS_{mk}	1	1.16	$\sigma^2 + 10\sigma_\eta^2 + 180\sigma_{SW}^2$
→ C_ℓ	1	0.55	$\sigma^2 + 10\sigma_\eta^2 + 180\sigma_{CW}^2 + 360\Phi(C)$
└ $WC_{m\ell}$	1	1.16	$\sigma^2 + 10\sigma_\eta^2 + 180\sigma_{CW}^2$
interactions	51	⋮	⋮
→ $LTSC_{ijk\ell}$	4	1.50	$\sigma^2 + 10\sigma_\eta^2 + 10\sigma_{LTSCW}^2 + 20\Phi LTSC$
└ $WLTSC_{mijk\ell}$	4	0.89	$\sigma^2 + 10\sigma_\eta^2 + 10\sigma_{LTSCW}^2$
$\eta_{p(mijk\ell)}$	0	—	—
$\varepsilon_{q(mniojk\ell p)}$	648	—	σ^2

are directly tested by their interaction with W_m. This makes weeks the inferential experimental unit. Because the restriction errors are present, there are no direct tests on W_m or any interaction involving W_m. All terms involving W_m are conservatively tested by the error term.

This is not a very good experiment as it currently stands. There are only two df for testing L_i and T_j, and only one df for testing S_k and C_ℓ. This is a result of using only two inferential experimental units. Yet there are 648 df for conservatively testing terms involving W_m, terms of lesser importance. This is because the inferential experimental unit for all terms involving weeks is repeats and there are ten repeats.

Obviously, the emphasis in this experiment is in the wrong place. A much better design from a statistical point of view is to have only two repeats. This reduces the df for error to 72, still plenty of information. However, when one considers the detectable size Δ, one discovers that Δ increases from 0.43 to 0.97 for L and T, and from 0.55 to 1.22 for S and

C. On the surface, this may be considered a problem if small differences are of interest, especially since it is cheap to punch metal once the die is set up.

But one has to be careful with strict numeric comparisons of Δ. With machine set up experiments, it is likely that the variation due to set up is considerably larger than the run-to-run variation. That is, σ_η^2 is likely to be considerably larger than σ^2. Now, a factor like T is tested by WT so the detectability in terms of actual units is $\Delta\sqrt{EMS(WT)}$. For 10 repeats this is $0.43\sqrt{\sigma^2 + 10\sigma_\eta^2 + 120\sigma_{WT}^2} \approx \sqrt{.18\sigma^2 + 1.85\sigma_\eta^2 + 22\sigma_{WT}^2}$. For 2 repeats, $0.97\sqrt{\sigma^2 + 2\sigma_\eta^2 + 24\sigma_{WT}^2} \approx \sqrt{.94\sigma^2 + 1.88\sigma_\eta^2 + 23\sigma_{WT}^2}$. The difference between the coefficients for σ_η^2 and σ_{WT}^2 for ten and two repeats is strictly rounding. Increasing the number of repeats merely decreases the effect of σ^2. If σ^2 is already small, as is probably the case with back to back parts, the benefit of more repeats will be miniscual.

On the other hand, if one repeated only two parts but carried out the experiment for five weeks, the actual unit detectability would be $0.28\sqrt{\sigma^2 + 2\sigma_\eta^2 + 24\sigma_{WT}^2} \approx \sqrt{.08\sigma^2 + .16\sigma_\eta^2 + 1.88\sigma_{WT}^2}$. This is considerably smaller than either of the previous two designs. Intuitively, since WT tests T, increase the levels of W to improve the test.

But, as usual, it is more expensive to run the experiment for five weeks than for two weeks. Informed trade-offs must again be made.

Example 4. A chemist wishes to make inferences about five methods over all laboratories. A random sample of twenty laboratories is selected and each method is run ten times. Each laboratory uses a completely randomized procedure for all fifty set ups of the equipment.

Since each laboratory uses its own randomization, it seems pretty clear that there is a restriction on the randomization. The issue in this example is whether or not it is possible to run the experiment in a completely randomized fashion. The restriction here is due to location of the laboratories, not the run order of the treatment combinations. Since the location of the laboratories cannot be physically changed, the experiment cannot be run in a completely randomized fashion.

There are many examples in which the experiment cannot be run in a completely randomized fashion. Whenever location is a factor (no matter what it is called) and location cannot be changed, there must be a restriction error. Whenever run order is a factor (again, no matter what it is called) and run order cannot be changed, there is a restriction error. A common example occurs in the medical profession when a subject is measured at different *time* intervals after exposure to a treatment. There will be a restriction error associated with the factor *time* since the same subject is measured at consecutive time intervals. A good statistician is always on the lookout for surrogates to location or run order as these factors almost always cause restriction errors. These factors are important as they increase the inference space for the experiment.

The inference for laboratories in Example 4 includes location effects, environmental effects, personnel effect, *etc.* In fact, the inference for laboratories includes everything that would go into the restriction error. Therefore, one could justifiable say that there is no restriction error for the example; it is already included in the laboratories factor.

While this reasoning is correct, it quickly leads to confusion whether to include the restriction error or not. Our recommendation is to *always* include the restriction error. Then it is crystal clear that the inference is to laboratories *and* the restriction error consisting of all those factors that change from laboratory to laboratory. The restriction error serves as a flag on the factor called laboratories, making sure the inference is to laboratories *and* all uncontrolled factors that vary along with laboratories. That is, the inference is to laboratories and the restriction error.

PROBLEMS

5.1 *Reconsider the cardiac valve experiment of Chapter 4 with four types of valves, two valves of each type, six different pulse rates, and two repeats of each combination. Identify the restriction errors for each of the following layouts. (a) A pulse rate is fixed, the eight valves (two each of four different types) are randomly ordered, and the repeat measurements are taken back to back. (b) A pulse rate is fixed and the sixteen measurements (two each for the two different valves of each type) are randomly ordered and run. (c) One of the eight valves is randomly selected and the twelve runs (two repeats of each of the six pulse rates) are randomly ordered and run. (d) A valve type and one of the two valves is randomly selected. The twelve runs (two repeats of the six pulse rates) are randomly run before the other valve of the same type is selected. (e) One of the eight valves and a pulse rate is selected and measured twice. The other pulse rates are tested before changing the valve. (f) A valve and a pulse rate is randomly selected with the repeat measurements back to back. (g) The 96 combinations are written down on a piece of paper with the repeats adjacent to each other and the pulse rates ordered one through six. These are labeled 1 to 96. A random number table is used to determine which label is run first, which label is run second, and so on. (h) One of the twelve type/pulse rate combinations is randomly selected. One of the two valves is then selected and both repeats run before testing the other valve.*

5.2 *An experiment involving injectors, engines, and fuel type was being considered. Identify the nesting and restrictions on randomization for each of the* *following descriptions. Each combination is tested only once. (a) Four engines and twelve injectors are selected. Three injectors go with each engine. An engine is selected along with one of its injectors. Both fuels are tested before the next injector is loaded into the same engine. (b) Four engines and four injectors are selected. All sixteen combinations are tested using fuel 2 before using fuel 1. (c) Four engines and four injectors are selected. Two of the engines use fuel 1 and the other two use fuel 2. All four injectors are put in each engine, testing one engine at a time. (d) Four engines and eight injectors are selected. Injectors one through four are sequentially tested in each of engines one through four (in that order) using fuel 2. Then, injectors five through eight are sequentially tested in the same engines (this time in a different order) using fuel 2. (e) Two different injectors are used in each of two different engines for each of the two fuels. That is, eight injectors and four engines are used. Pick an engine and test both injectors before switching to another engine.*

5.3 *Describe how the experiment should be run for each of the following math models.*

(a) $y_{ijk\ell} = \mu + A_i + B_j + \delta_{k(j)} + AB_{ij} + \varepsilon_{\ell(ijk)}$.
(b) $y_{ijk\ell} = \mu + A_i + \delta_{j(i)} + B_k + AB_{ik} + \varepsilon_{\ell(ijk)}$.
(c) $y_{ijk\ell} = \mu + A_i + B_j + AB_{ij} + \delta_{k(ij)} + \varepsilon_{\ell(ijk)}$.
(d) $y_{ijk\ell m} = \mu + A_i + \delta_{j(i)} + B_k + AB_{ik} + \gamma_{\ell(ik)} + \varepsilon_{m(ijk\ell)}$.
(e) $y_{ijk\ell mno} = \mu + A_i + B_{j(i)} + \delta_{k(ij)} + C_{\ell(ij)} + D_{m(ij\ell)} + \gamma_{n(ij\ell m)} + \varepsilon_{o(ijk\ell m)}$.

5.4 *Describe how the experiment should be run for each of the following math models.*

(a) $y_{ijk\ell m} = \mu + A_i + B_j + \delta_{k(j)} + AB_{ij} + \gamma_{\ell(ij)} + \varepsilon_{m(ijk\ell)}$.
(b) $y_{ijk\ell m} = \mu + A_i + B_j + AB_{ij} + \delta_{k(ij)} + C_\ell + AC_{i\ell} + BC_{j\ell} + ABC_{ij\ell} + \varepsilon_{m(ijk\ell)}$.
(c) $y_{ijk\ell mn} = \mu + A_i + \delta_{j(i)} + B_k + AB_{ik} + C_{\ell(i)} + BC_{k\ell(i)} + \gamma_{m(ik\ell)} + \varepsilon_{n(ijk\ell m)}$.
(d) $y_{ijk\ell m} = \mu + A_i + B_{j(i)} + C_k + AC_{ik} + \delta_{\ell(ik)} + BC_{jk(i)} + \varepsilon_{m(ijk\ell)}$.
(e) $y_{ijk\ell mn} = \mu + A_i + \delta_{j(i)} + B_{k(i)} + C_\ell + AC_{i\ell} + \gamma_{m(i\ell)} + BC_{k\ell(i)} + \varepsilon_{n(ijk\ell m)}$.

5.5 *Justify the use of the parentheses in restriction errors using the depending on definition of parentheses.*

5.6 *By convention, an index whose value is one is commonly left out of the model. Explain how this would affect the rules developed for the ANOVA table.*

5.7 *Does it matter if the indices associated with restriction errors only are left out of the error term? Why or why not?*

5.8 *Suppose an experiment has factors called subjects and time after exposure. Explain how the experiment could be completely randomized. Why is this not generally a good idea?*

5.9 *Use combinatorical arguments to derive equation (5.3.6) when complete randomization is used and equation (5.3.7) when the experiment is blocked on A_i.*

5.10 *Reconsider the model in Section* **5.3**. *If the factor U were known and controlled in the experiment, what would happen to the test on A? This is why a good consultant tries to control as many factors as possible in the experiment.*

5.11 *Suppose there was time related heterogeneity in the experiment, i.e., U was distributed Normal($0,\sigma_\ell^2$) with σ_ℓ^2 different for different times ℓ, how would the usual tests be influenced under complete randomization and under blocking on A?*

5.12 *Repeat the derivation of Section* **5.3** *for a two-way fixed model blocked on A to demonstrate that the restriction error does not interact with any other term in the model.*

5.13 *For several different values of I and J, graph the magnitude of the coefficients of $\sigma_{U,\rho}^2$, $\tilde{\sigma}_{U,\rho}^2$, and $\check{\sigma}_{U,\rho}^2$ against the serial correlation ρ. What happens when the unknown factor is random ($\rho = 0$) and when the unknown factor is a constant ($\rho = 1$)?*

5.14 *Suppose the analysis of data for Example 2 showed that the conservative tests on weeks and all interactions involving weeks were insignificant at the .25 level. How many df would be available for testing the terms of interest?*

5.15 *What would Table 5.4.2 look like if there were ten weeks and two repeats of each treatment combination rather than the reverse?*

5.16 *Suppose Example 2 were run with ten weeks and two repeats and the conservative tests on weeks and all the interactions involving weeks were insignificant at the .25 level. How many df would be available for testing the terms of interest? Compare this answer to the answer for Problem* **5.14**. *Now you see that you should not simply compare ease of running the experiment with amount of information available for tests of interest. You should also consider the likelihood that a factor like weeks will be insignificant and what will happen if it is insignificant and gets pooled.*

5.17 *Compute the ANOVA table for a completely randomized experiment having I treatments and J repeats and compare it to Table 5.4.1. What does this say about the process known as blocking? This is an important concept to learn.*

5.18 *Use the rules in Chapter 3 to find the standard error for a single treatment mean in Example 1. Also find the standard error for the difference of two means and for multiple comparisons. This is a somewhat surprising result that seems counterintuitive.*

5.19 *Consider the randomized complete block design (Example 1) where each treatment is repeated (randomly) within each block. What sort of information is available about blocks? Give the estimates of the standard error for treatments, and give the formula for estimating df.*

5.20 *Based on Example 2, what kind of a layout would be associated with a split-split-plot design. Identify the restriction errors.*

5.21 *An experiment considered the effect of hydrocarbons and barometric pressure on the weight gain of laboratory rats. Three levels of hydrocarbon were used in conjunction with two levels of barometric pressure. Six controlled chambers, one for each hydrocarbon/barometric pressure combination, were used. Five rats were placed in each chamber and the change in weight measured for five consecutive weeks. Write down the model describing this experiment and indicate all tests.*

5.22 *Suppose the experiment given in Problem* **5.21** *was replicated using different rats in the same chambers. Write down the new model and indicate all tests. Compare this to the answer for Problem* **5.21**.

5.23 *An agronomist has three fields to plant with three different types of corn, one type per field. Each field consists of eight plots, four across and two down, each randomly labeled one through eight. The first four labels get fertilizer type 1 and the last four get fertilizer type 2. Labels 1, 2, 5, and 6 get herbicide 1 while the remainder get herbicide 2. Give the appropriate model, indicating all restriction errors and tests. How do you interpret the restriction errors you have used? Give the inference space.*

5.24 *Consider Problem* **5.23** *with one difference, the plots are labeled left to right one through four in the first row and five through eight in the second row. Give the appropriate model, indicating all restriction errors and tests. How do you interpret the restriction errors you have used?*

5.25 *There is really a restriction error in the crossover design given in Problem* **4.29**. *Think carefully about the design. With which term is it associated?*

5.26 *Seeger (1986). A sowing machine has four bills which simultaneously plant four rows of seed. The sowing machine drives up four plots, each the width of the four rows, turns around and drives down four adjacent plots. Two different types of seeds are to be tested. You must plant the same type of seed in all four rows of a plot. Draw a picture of the plots, identify the factors, and design this experiment indicating all nesting and restriction errors.*

5.27 *Find the restriction errors that probably should have been noted in Problems* **4.31** *through* **4.35** *and* **4.37**.

5.28 *A painted panel is to be measured for gloss. There are three basecoat formulations and two topcoat formulations to be studied. The panel is divided into six areas, three across and two down. The three basecoat formulations are sprayed from the top to the bottom, across two areas each. The two topcoat formulations are sprayed from left to right, across three areas each. Within each area, two locations are measured for gloss. These are measured in a random fashion. Draw a picture of the layout of this experiment and write down the appropriate model, including restriction errors. The restriction errors in this experiment actually refer to horizontal and vertical effects. Redesign this experiment dividing this panel into as* *many areas as you see fit. Obtain as much information as possible and indicate the layout of the new experiment.*

5.29 *An experiment in metal casting considered the time to fill the mold with sand, 28 and 35 seconds; the vibration direction for packing the sand, N/S and E/W; and the time to pour the molten metal, 18 and 30 seconds. Four castings are simultaneously poured. Since the vibration direction is accomplished by moving a motor, it was decided by a flip of a coin that all N/S vibrations would be done before any E/W vibrations. Assuming one measurement of each casting, determine the appropriate mathematical model.*

5.30 *A company wishes to increase the light intensity of its photoflash cartridge. The four treatments considered are 1) 1/8 inch thick wall, ignition point at the center, 2) 1/16 inch thick wall, ignition point at the center, 3) 1/8 inch thick wall, ignition point at the edge, and 4) 1/16 inch thick wall, ignition point at the edge. Five batches of the basic formulation were made up and three cartridges for each treatment combination were manufactured from each batch. That is, twelve cartridges were made from each of five different batches. Find all nesting and restrictions on randomization, indicating all tests.*

5.31 *An agronomist was interested in the effect of four different sprays on the control of weeds in corn. Five locations on the research farm were selected. Within each location, three randomly chosen plots were used for each spray. Write the mathematical model describing this experiment and carefully describe the inference space. Can you redesign this experiment to get more information on sprays?*

5.32 *An experiment dealing with the pressure of an oil pump at idle speed considered four factors, the type of metal used in the pump, the end play, and the clearances of two different valves. Once a pump is mounted in an engine, it is relatively easy to change end play and clearances. Thus, the experimenter wished to block on type of material yet needed a direct test on type of material. Make a compromise suggestion for restricting randomization that makes the experiment easier to run yet still obtains a direct test on type of material.*

5.33 *In an experiment dealing with three fixed factors, the treatment combinations were randomly selected and then the repeats were run back to back. Use the EMS with and without the restriction error to decide how badly this restriction affected the experiment.*

5.34 *In the previous experiment, there are two different conservative tests that can be made: using the error term as the denominator and using the highest order interaction as the denominator. What can be concluded using each conservative test? What do you think about making both tests? In what order should these be made?*

5.35 *If in Problem* **5.33**, *the conservative test on the highest order interaction turns out insignificant at the 0.25 level, what will happen to the conservative test on all of the other terms? How does this fit in with Problem* **5.34**?

5.36 *In many textbooks, authors talk about "repeating an experiment" and "replicating an experiment." Physically, "repeating an experiment" means that a factor combination is set up and all repeats are run back to back. "Replicating an experiment" refers to running all combinations again, generally at a later time period. Indicate where the restriction error occurs in both models and give a meaning for each restriction error.*

5.37 *A Split-Plot Design An agronomist was investigating the effect of three legumes on the weight gain in cattle. Fields were chosen in five different counties, often called whole-plots. Each whole-plot was divided into four fields, called split-plots. Each split-plot had three fenced in areas, one for each legume. A total of 180 steers of comparable age, weight, and breed were randomly assigned to the fields, three steers per field, to represent repeat measurements. Write the appropriate model and indicate the tests.*

5.38 *A heart valve experiment had five types of valves. There were four valves of each valve type. Three different pulse rates were considered for each valve, which was measured for flow rate three consecutive times. Due to the inconvenience of loading and unloading valves, all valves of the same type were run together and the pulse rate/repeat combinations were run before switching to another valve of the same type. Write the appropriate model carefully considering the restrictions on randomization.*

5.39 *Did you recognize that Problem* **5.37** *and Problem* **5.38** *were exactly the same design? It is very difficult to recognize them as the same design even though the models take the same form. This is an example of the power of recognizing concepts rather than propagating names. Restate Problem* **5.38** *to parallel Problem* **5.37**.

5.40 *Split-Split-Plot Design If a split-plot design (see Problem* **5.37**) *is further split, one gets a split-split-plot design. Again, our method of identifying restriction errors handles the layout and analysis perfectly as long as the restriction errors are properly identified. One need not identify this as a split-split-plot design. In fact, we discourage the use of names as crutches because they avoid understanding the key concepts. As an example, consider a metal forming experiment in which 3 chemistries were tested. From each chemistry, two heats were run and a metal ingot made. From the top, middle, and bottom of the ingot, call this factor location, a wheel was cut. Each wheel was cut into six specimens, three put into one environmental condition and the other three put into a different environmental condition. There is a restriction error on heats because the three wheels came from the same ingot. There is a second restriction error on the heat/wheel combination because, within the same heat, all specimens come from the* *same wheel. Write the math model and all tests for this example. Take chemistry, location, and environmental condition as fixed. It is your option whether you wish to call specimens error or treat them as another factor with error having only one level. It makes no difference as long as you identify the nesting properly.*

5.41 *Restriction Errors and Nesting can be the same thing! Sometimes we get confused between restriction errors and additional nested factors having only one level. Do not worry which you choose, both will give the same answer! We illustrate with Problem* **5.40**. *Treat ingots as a random factor having one level, nested within heats, and wheel as a random factor having one level, nested within each ingot/location combination. Again, identification of specimens is optional. Write the math model and all tests for this representation and compare to your model in Problem* **5.40**. *They are identical except for labeling differences. This provides strong intuitive justification for the use of restriction errors. As my teenage son would say: "Cool."*

5.42 *Restriction Errors can be tested. Problem* **5.40** *contains several restriction errors and it is not obvious how they could ever be tested. However, Problem* **5.41** *gives an equivalent representation and it is obvious that one simply needs to test two ingots from each heat and two wheels from each ingot/location combination to get tests on heats and wheels. Using this example as a guide, explain how one would test a restriction error.*

5.42 *Crossover Design In Problem* **4.29**, *assume that all 12 people are given their appropriate drug on the same day and the washout period is the same for all people. Indicate the restriction on randomization. Does the restriction error hurt anything in this design or is it really just another name for the crossover effect?*

BIBLIOGRAPHY FOR CHAPTER 5

Anderson, V. L. (1970). "Restriction errors for linear models (an aid to develop models for designed experiments)," *Biometrics*, **26**, 255–268.

Anderson, V. L. and McLean, R. A. (1974). *Design of Experiments A Realistic Approach*, New York: Marcel Dekker. (Chapters 5, 6, 7)

Anderson, V. L. and McLean, R. A. (1974). "Restriction errors: another dimension in teaching experimental statistics," *The American Statistician*, **28**, 145–152.

Bennett, C. A. and Franklin, N. L. (1954). *Statistical Analysis in Chemistry and the Chemical Industry*, New York: John Wiley & Sons. (Chapter 7)

Daniel, C. (1976). *Applications of Statistics to Industrial Experimentation*, New York: John Wiley & Sons. (Chapter 16)

Guenther, W. C. (1964). *Analysis of Variance*, Englewood Cliffs, NJ: Prentice-Hall. (Chapter 3)

Hicks, C. R. (1982). *Fundamental Concepts in the Design of Experiments*, New York: Holt, Rinehart, & Winston. (Chapters 12–14)

John, P. W. M. (1971). *Statistical Design and Analysis of Experiments*, New York: The Macmillan Co. (Chapter 3)

Lorenzen, T. J. (1984). "Randomization and blocking in the design of experiments," *Communications in Statistics—Theory and Methods*, **13**, 2601–2623.

Montgomery, D. C. (1991). *Design and Analysis of Experiments*, 3rd ed., New York, John Wiley & Sons. (Chapter 14)

Peng, K. C. (1967). *The Design and Analysis of Scientific Experiments*, Reading, MA: Addison-Wesley. (Chapter 6)

Scheffé, H. (1959). *The Analysis of Variance*, New York: John Wiley & Sons. (Chapter 9)

Seeger, P. (1986). "Design and Analysis of Experiments with Sugar-Beet Seeds," *Applied Statistics*, **35**, 262–268.

SOLUTIONS TO SELECTED PROBLEMS

5.1 *(a) Restrictions on pulse and the valve/pulse combination. (b) Restriction on the pulse rate. (c) Restriction on valve. (d) Restrictions on type and on valve. (e) Restrictions on valve and on the valve/pulse combination. (f) Restriction on the valve/pulse combination. (g) No restrictions. Labeling does not affect the randomization process. (h) Restrictions on the type/pulse combination and on the valve/pulse combination.*

5.3 *(a) Randomly select a level of B. Do all combinations of A and repeats before switching levels of B. (b) Randomly select a level of A. Do all combinations of B and repeats before switching levels of A. (c) Randomly select an AB combination. Do all repeats before selecting another AB combination. (d) Randomly select a level of A and a level of B. Do all repeats. Now pick another level of B and do all repeats. Do all levels of B before switching levels of A. (e) Randomly select a (nested) level of B. Next, randomly select a level of D associated with the selected level of B and do all repeats. Select another associated level of D and run all repeats. Run all associated levels of D before switching levels of B.*

5.8 *Each subject has to perform the experiment for each individual time after exposure, ignoring the measurement at all other times after exposure. This is not generally a good idea because it requires a lot of extra work, useful information is ignored, and there may be a learning or fatigue factor associated with performing the task that many times.*

5.17 ANOVA FOR A COMPLETELY RANDOMIZED DESIGN

Source	*df*	*EMS*
Treatment (T_i)	$I-1$	$\sigma^2 + J\Phi(T)$
Error ($\varepsilon_{j(i)}$)	$I(J-1)$	σ^2

On first glance, the completely randomized design seems superior as it has more df for testing treatments. However, the σ^2 for the completely randomized design includes σ^2 for the randomized complete block design as well as σ^2_{TB} and σ^2_B. This inflates the denominator and makes it harder to detect a significant difference in T_i. The trade-off depends on the ratios of the various random terms. However, if σ^2_{TB} and σ^2_B are negligible, they can be pooled with error to get the same df as the completely randomized design. Therefore, if one suspects any kind of a blocking factor exists, it should generally be included in the experiment, even though a restriction error is required.

5.18 *For a single mean, $\sigma_{\bar{y}_T} = \sqrt{\sigma^2/IJ + (I-1)\sigma^2_{TB}/IJ + \sigma^2_\delta/J + \sigma^2_B/J}$, which cannot be estimated without information on the error term $\varepsilon_{\ell(ijk)}$. For two means, $\sigma_{\bar{y}_{T1}-\bar{y}_{T2}} = \sqrt{\sigma^2/I + \sigma^2_{TB}/I} \approx \sqrt{MS(TB)/I}$, which does not require information on the error term.*

5.21 *This problem was designed to kick the brain into high gear. First, you must recognize weeks as a fixed factor since weeks are ordered and the ordering is important. We have seen several solutions which are roughly equivalent. Our intention was to recognize chambers as a restriction error. In that case, the model is* $y_{ijklm} = \mu + H_i + B_j + HB_{ij} + \delta_{k(ij)} + R_{\ell(ij)} + W_m + \gamma_{n(m)} + HW_{im} + BW_{jm} + HBW_{ijm} + RW_{\ell m(ij)} + \varepsilon_{(ijk\ell mn)}$. *There are direct tests on HW, BW, and HBW. There are conclude insignificant only tests on H, B, HB, R, and W. In addition, there are conclude significant only tests on H and B. As an alternative solution, recognize that weeks and the restriction error* γ *are really the same thing and ignore* γ. *Then, the conclude insignificant only test on W becomes a direct test. A third solution is to explicitly recognize chambers as a factor in the experiment, nested within the HB interaction and having one level. Rats are nested within hydrocarbon, barometric pressure, and chamber. All terms involving chamber have 0 df. The difference is that all direct tests become conclude insignificant only tests and there are conclude significant only tests on HW and BW. The factor chamber behaves exactly like a restriction error with the exception that it interacts with other terms. If we assume chambers do not interact with weeks, the results are identical!*

5.36 *The procedure known as "repeating an experiment" has a restriction error associated with the highest order combination. This restriction error appears with every term except the error and represents reassembly variation, not present in the error term since parts are run back to back. The procedure known as "replicating an experiment" has an extra factor called time with a restriction error on time. Time and the restriction error have the same meaning.*

6

Play It Again, Sam

You must remember this,
a kiss is just a kiss,
a sigh is just a sigh,
the fundamental things apply,
as time goes by.

HERMAN HUPFELD

6.0 INTRODUCTION

In this chapter we revisit Chapter **1** having the added benefit of the technical material in Chapters **2** through **5**. We will revisit Chapter **1** using a single example, the Engine/Injector/Fuels example from Chapter **3**. Chapter **6** will serve as a guideline to the fundamental thought process, design, redesign, analysis, and interpretation of experiments.

Experience has indicated that the material in Chapter **1** is understood at the time of reading and then promptly forgotten as the students get more involved with the technical details in Chapters **2** through **5**. Yet, without the proper thought process, mastery of the technical details can still lead to a poor experiment. This single example shows how the fundamental things apply.

6.1 RECOGNITION THAT A PROBLEM EXISTS

The purpose of the Engine/Injector/Fuels problem is to determine the cause of variability in the brake specific fuel economy of a certain engine. Variability is an underlying cause of many problems. A customer with an engine that gets poor fuel economy is an unhappy customer. And an unhappy customer talks to other potential customers, possibly resulting in the loss of many sales. In addition, fuel economy standards and regulations are based on a sample of vehicles and it is desired to have high fuel economy with a high degree of certainty. Variability decreases the degree of certainty, forcing high cost alternative strategies.

In addition to specific problems associated with variability, most companies have a goal of never-ending improvement. This translates to constantly reducing variability because variability corresponds to poor quality. Thus, on general grounds alone, study of causes of variability are desirable.

6.2 FORMULATION OF THE PROBLEM

A cross-functional team consisting of the manager, several design engineers, representatives from the suppliers, manufacturing engineers, fluid flow experts, and the statistician was assembled. Within the first half hour of discussing the problem, at least fifty potential causes of variability were identified. It became obvious that several hours could be spent on this problem and hundreds of potential causes could be identified. Trying to design an experiment to get information on hundreds of factors and interactions requires a monumental amount of effort, both statistically and experimentally. Thus, the creation of the list of causes was terminated by the statistician as being unmanageable.

Upon further discussion, it was discovered that even the contribution of the major components to the variability in fuel economy was not known. By discussing only the major components, the list of potential causes was reduced to two. Upon discovering which of the two major components was the largest contributor, further study of the system subcomponents would be necessary. Thus, the team decided that a multi-step experimentation strategy would be appropriate.

A rough rule of thumb, mentioned earlier in this book, is to expend no more than 25% of the total experimental effort on a first experiment. For this particular example, there are *so* many possible causes that considerably less than 25% of the total experimental effort should be spent on the first experiment.

6.3 SPECIFYING THE VARIABLE TO BE MEASURED

In thinking about a measurement process, one quickly realizes that weather conditions, road conditions, and the driving schedule all affect fuel economy. All of these conditions vary from customer to customer and even day to day for the same customer. Fortunately, this problem had existed for many years and a prescribed dynamometer test had already been developed. This test determines fuel economy and is felt to be indicative of normal customer use. The dynamometer results are continuous, fairly repeatable, and accepted as a standard by both government and industry.

The time required to specify a dependent variable for this example was minimal. In many other cases, hours of tough discussion among the experts is required before a continuous, repeatable, and accurate dependent variable is determined. Often times, different requirements force the use of different response variables and compromises will have to be made.

6.4 AGREEING ON FACTORS AND LEVELS

As mentioned earlier, the cross-functional team agreed to study a few major components rather than hundreds of individual components. This general agreement needs to be translated into specific factors and specific levels at which to study these factors.

The two major components to be studied are the engine system and the fuel system. Most of the individual components are quite difficult to change—requiring a complete teardown and rebuilding of the engine. It was decided not to tear down engines and swap individual components until knowledge of the effect of the major components on variability was known. Thus, a complete assembled Engine was selected as one factor in the experiment. If Engines turned out to be a major contributor to the variability in fuel economy, then the added expense of tearing down and assembling Engines would be considered. If Engines were not a major contributor, then much experimental effort would be saved.

One individual component of the Engine that is easy to change and is thought to influence fuel economy is Injectors. This was selected as a second factor to be studied in the experiment.

Fuels was selected as a third factor as this was fairly easy to change and representative of the fuel system in the vehicle.

The next discussion centered around the selection of the levels to be used in the experiment. First, the team had to choose between random selection of the levels or purposely trying to select best and worst levels. Of course, these two methods of selecting factor levels serve similar purposes. Actions that reduce variability should also reduce the difference between the best and worst case, and vice-versa. If we select the best and worst case and there is no significant difference, then we can be sure there is no significant differences between any of the levels. On the other hand, if there was an effect, random selection would correspond to the variability observed in practice while it would be extremely difficult to relate best and worst cases to actual variability. Since we are interested in finding effects and we wish the inference to apply to customers in their normal use, factor levels must be randomly selected. Random selection reflects the purpose of the experiment, to determine the variability in fuel economy caused by the major components. Selecting the best and worst case implies a different purpose: trying to demonstrate that the factor effect is not significant, possibly to be able to substitute cheaper material without loss of quality.

The number of levels of each factor is to be determined in the redesign phase (Section **6.10**) and is an integral part of the compromise between experimental effort and the accuracy of the results. As a first pass, an experiment consisting of two levels of each factor and two repeats will be considered.

6.5 DEFINITION OF THE INFERENCE SPACE

The experiment is to be run on a specific type of engine with a specific supplier of injectors and fuels. (The exact model of engine and the suppliers are privileged information and not crucial to the understanding the scientific approach used to design experiments.) Brake specific fuel consumption is measured by a specific gage on a specific dynamometer run under a specific driving schedule. Thus, the results of the experiment *only* apply to this engine, these suppliers, this gage, and this particular driving schedule. The statistician must clearly indicated these limitations in the inference space.

But the experimenters have no desire to draw conclusions *only* about such a limited inference space. They would like to apply the conclusions to normal driving conditions, not just to a specific driving schedule as measured by a specific gage on a specific dynamometer. This is natural and to be expected. Seldom will the actual experimental setup match the exact conditions under which conclusions are desired. The statistician is obliged to make the differences in inference known to the experimenter. Then the experimenter has two choices: to use application specific knowledge (knowledge about engines, dynamometers, and driving conditions) to generalize the results or to work with the statistician to design an experiment to verify that the results generalize. If the results are not thought to generalize, or do not generalize in an additional experiment, one has to question why the original experiment is being run at all. Go back and start with Section **6.3**, Specifying the Variable to be Measured.

For our example, the experimenters were willing to live with the inference space being applied to a specific supplier of Injectors and of Fuels. If they were not willing to live with this inference space, then suppliers of Injectors and of Fuels would become additional factors in the experiment and Injectors and Fuels would be nested factors.

6.6 RANDOM SELECTION OF THE EXPERIMENTAL UNITS

In addition to carefully considering the inference space for this experiment, one must carefully consider the selection of all random factors. Let us use the factor Engines as an example. Since Engines is a random factor, we are told to randomly select the Engines for experimental purposes. While this is a trivial sentence to state, random selection is not always a trivial procedure. In the case of Engines, experimental units (Engines) are sequentially assembled in another plant and shipped to the assembly point. Ideally, one should build a year's worth of Engines, stack them in a warehouse somewhere, number the Engines, and use a random number table to select the two Engines for experimentation purposes.

Obviously this ideal procedure is a practical impossibility. On the other hand, one has a very uneasy feeling about selecting back to back Engines off the assembly line. Would back to back Engines really represent the true variability one would observe over a week's production period? How about a month's production, or a year's production? It is very little extra effort to randomly select Engines from an hour's production period. It is more work to randomly select Engines from a day's production period. And so on.

As in the previous section, the inference space is directly related to the way in which the Engines were randomly selected. It is the statistician's responsibility to see that a large enough production period is used to comfortably represent the desired Engine variability and that, within this production period, random selection procedures are used.

For this particular example, a week's production was selected as the best compromise between true variation and the required effort. It was felt that all variation likely to occur during a year's production was likely to occur during a week's production and that randomly selecting Engines from a week's production was a manageable task.

The selection of Injectors closely parallels the selection of Engines so no further discussion is required.

The selection of Fuels follows a slightly different line. Fuel within a specific tank is thought to be homogeneous, but Fuels from different tanks are thought to differ. Thus, the experimental unit is really tanks and the thought process must be shifted from Fuels to tanks. There are two underground tanks that hold Fuel for use in this experiment. These are thought to be representative of all possible tanks, so will be taken as random. Again, this was a judgement call motivated by the team of expert's desire to draw conclusions about all possible tanks of Fuel. Without this judgement, the statistician must treat Fuel as fixed with two levels and the formal inference space is limited to these two particular tanks of Fuel.

6.7 LAYOUT OF THE DESIGN

As we discovered in Chapter **5**, complete randomization is the ideal way to run an experiment from the standpoint of knowledge gained. Unfortunately, it is not generally the easiest way to run an experiment from an effort point of view. Chapter **5** gave us the tools to understand different run orders and their effect on the conclusions that can be drawn in the experiment.

As a first pass, complete randomization will be assumed. The effect of various restrictions on randomization on the ease of running the experiment and the information to be gained from the experiment will be covered in the redesign section. Note that this is not always the case. If a factor like Weeks was used in the experiment, there is a natural restriction error and complete randomization is impossible. These kinds of factors must be identified immediately and accounted for in the mathematical model. The redesign phase can then consider additional types of restrictions on randomization.

As an additional reminder, it is the statistician's responsibility to see that the randomization (restricted or not) is properly carried out.

6.8 DEVELOPMENT OF THE MATHEMATICAL MODEL

By this point, the reader is totally familiar with the process of writing out the mathematical model describing the process. Assuming complete randomization, the appropriate model is

$$y_{ijk\ell} = \mu + E_i + I_j + EI_{ij} + F_k + EF_{ik} + IF_{jk} + EIF_{ijk} + \varepsilon_{\ell(ijk)} \quad (6.8.1)$$

where i, j, k, and $\ell = 1, 2$, E_i, I_j, and F_k are all random.

6.9 PRELIMINARY EVALUATION OF THE DESIGN

The preliminary (prior to the collection of any data) evaluation of equation (6.8.1) is given in Table 6.9.1. Creation of this table should be almost second nature by now. A total of 16 runs is required.

6.10 REDESIGNING THE EXPERIMENT

Now the fun and the real work begins. There are no hard and fast rules that lead to the selection of the best possible design for a given

TABLE 6.9.1

ANOVA FOR DESIGN 1, $I = J = K = L = 2$.

Source	df	Δ	Size	EMS
E_i †	1	28.36	EXT. LGE.	$\sigma^2 + 2\sigma^2_{EIF} + 4\sigma^2_{EF} + 4\sigma^2_{EI} + 8\sigma^2_E$
I_j ††	1	28.36	EXT. LGE.	$\sigma^2 + 2\sigma^2_{EIF} + 4\sigma^2_{IF} + 4\sigma^2_{EI} + 8\sigma^2_I$
→ EI_{ij}	1	40.11	EXT. LGE.	$\sigma^2 + 2\sigma^2_{EIF} + 4\sigma^2_{EI}$
F_k †††	1	28.36	EXT. LGE.	$\sigma^2 + 2\sigma^2_{EIF} + 4\sigma^2_{IF} + 4\sigma^2_{EF} + 8\sigma^2_F$
→ EF_{ik}	1	40.11	EXT. LGE.	$\sigma^2 + 2\sigma^2_{EIF} + 4\sigma^2_{EF}$
→ IF_{jk}	1	40.11	EXT. LGE.	$\sigma^2 + 2\sigma^2_{EIF} + 4\sigma^2_{IF}$
EIF_{ijk} ←	1	12.55	EXT. LGE.	$\sigma^2 + 2\sigma^2_{EIF}$
$\varepsilon_{(ijk)}$	8	—	—	σ^2
Total	15			

† $EI_{ij} + EF_{ik} - EIF_{ijk}$ approximately tests E_i

†† $EI_{ij} + IF_{jk} - EIF_{ijk}$ approximately tests I_j

††† $EF_{ik} + IF_{jk} - EIF_{ijk}$ approximately tests F_k

set of circumstances. The more one creates and evaluates designs, the better the chances of finding the best possible design. With experience, readers will develop their own particular strategy. We have observed that beginners require something in the vicinity of eight to ten redesigns before their comfort level is reached. Experts require four to five designs. This particular example required seven designs. The particular reasoning employed in this example will not generalize to any other example but may help readers develop a few strategies of their own.

Upon examining Table 6.9.1, one notices that $\varepsilon_{\ell(ijk)}$ is only used to test the highest order interaction, EIF_{ijk}. Since we are planning to perform additional experiments as soon as we find out which major component contributes most to the variability, there is little interest in the third order interaction. Yet, we had to double the size of the experiment just to get df for $\varepsilon_{\ell(ijk)}$ to be able to test the highest order interaction. Thus, our first redesign is to reduce the number of repetitions to one, resulting in the eight run experiment in Table 6.10.1.

As was expected, tests are available on all terms in the model except the third order interaction. Design 2, Table 6.10.1, is a wonderful design

TABLE 6.10.1

ANOVA FOR DESIGN 2, $I = J = K = 2$, $L = 1$.

Source	df	Δ	Size	EMS
E_i †	1	40.11	EXT. LGE.	$\sigma^2 + \sigma^2_{EIF} + 2\sigma^2_{EF} + 2\sigma^2_{EI} + 4\sigma^2_E$
I_j ††	1	40.11	EXT. LGE.	$\sigma^2 + \sigma^2_{EIF} + 2\sigma^2_{IF} + 2\sigma^2_{EI} + 4\sigma^2_I$
→ EI_{ij}	1	56.72	EXT. LGE.	$\sigma^2 + \sigma^2_{EIF} + 2\sigma^2_{EI}$
F_k †††	1	40.11	EXT. LGE.	$\sigma^2 + \sigma^2_{EIF} + 2\sigma^2_{IF} + 2\sigma^2_{EF} + 4\sigma^2_F$
→ EF_{ik}	1	56.72	EXT. LGE.	$\sigma^2 + \sigma^2_{EIF} + 2\sigma^2_{EF}$
→ IF_{jk}	1	56.72	EXT. LGE.	$\sigma^2 + \sigma^2_{EIF} + 2\sigma^2_{IF}$
EIF_{ijk}	1	—	—	$\sigma^2 + \sigma^2_{EIF}$
$\varepsilon_{\ell(ijk)}$	0	—	—	—
Total	7			

† $EI_{ij} + EF_{ik} - EIF_{ijk}$ approximately tests E_i

†† $EI_{ij} + IF_{jk} - EIF_{ijk}$ approximately tests I_j

††† $EF_{ik} + IF_{jk} - EIF_{ijk}$ approximately tests F_k

from the standpoint of the effort required to run the experiment. Only eight runs are required to complete the entire experiment.

However, Design 2 is literally a worthless experiment from the standpoint of the detectability Δ. While all terms except the third order interaction have tests available, all detectable sizes are *extremely large*. In fact, this experiment can only detect differences as large as 40 standard deviations. Any difference of the magnitude of 40 standard deviations is easy to detect without the help of any statistical training.

It should be obvious that the size of the experiment must be increased in some fashion. Always keep in mind this general rule: It is wise to increase the size of an experiment by increasing the number of repeats or the number of levels of the random factors. We discovered in Design 1 that repeats were not really necessary (although repeats did help Δ), so conclude that we should increase any or all of the random factors, Engines, Injectors, or Fuels. However, from a physical point of view, there are only two underground tanks of Fuel. This makes it very difficult to increase the levels of Fuel. Therefore, we are led to the conclusion that Engines, Injectors, or both should be increased.

Since we are trying to get information about the variability in fuel economy caused by the various factors, the same amount of information should be available for each factor. This leads us to increase simultaneously the number of levels of Engines, Injectors, and Fuels. However, we are constrained because fuels must be fixed at two levels. Thus, we will consider Engines and Injectors at 3, 4, and 5 levels each. The results are summarized in Table 6.10.2, with tests as previously indicated in Table 6.10.1.

TABLE 6.10.2

ANOVA FOR DESIGNS 3, 4, AND 5. $K = 2$, $L = 1$

	$I=J=3$, 18 runs			$I=J=4$, 32 runs			$I=J=5$, 50 runs		
Source	df	Δ	Size	df	Δ	Size	df	Δ	Size
E_i	2	3.80	LARGE	3	1.86	MEDIUM	4	1.19	SMALL
I_j	2	3.80	LARGE	3	1.86	MEDIUM	4	1.19	SMALL
EI_{ij}	4	3.55	LARGE	9	1.84	MEDIUM	16	1.32	SMALL
F_k	1	10.09	EXT. LGE.	1	5.82	EXT. LGE.	1	3.88	LARGE
EF_{ik}	2	4.59	LARGE	3	2.19	MEDIUM	4	1.46	SMALL
IF_{jk}	2	4.59	LARGE	3	2.19	MEDIUM	4	1.46	SMALL
EIF_{ijk}	4	—	—	9	—	—	16	—	—
$\varepsilon_{\ell(ijk)}$	0	—	—	0	—	—	0	—	—
Total	17			31			49		

Glancing at the Size columns in Table 6.10.2, there is obviously a trade-off to be made between the size of the experiment and the detectability of the experiment, with Δ decreasing as the experiment size increases. The team will generally want to select the smallest design to save time and effort so it is the statistician's responsibility to make the proper computations and constantly remind them of the importance of detectability.

The team questioned why there was an unbalanced emphasis in this design, with less detectability on Fuels than on the other factors. It was pointed out that this resulted from the physical limitation of having two tanks of Fuel. One possible way to avoid this disadvantage is to run the experiment in blocks having the same Fuel in each block. The first block would have to be completed and a new tank of Fuel received before the next block could be started. In this fashion, there need not be a limit of two on the number of levels for Fuels.

This question led to a discussion on the concept of restrictions on randomization and the principles set forth in Chapter **5**. For this particular example, a restriction error on Fuels only affects the test on Fuels. As we saw in Chapter **5**, this is not true in general. The effect of Fuels becomes completely confounded with the restriction error effect. Thus, unless the restriction error effect can be assumed negligible, no information will be available for testing the effect of Fuels.

To see if such a restriction error can be assumed negligible, the statistician asked the team if they thought the fuel consumptions could have any sort of time trend, *e.g.*, either a break in effect or a wear down effect. The team expressed concern over a possible break in effect. Over a short period of time, very little break in is expected in an engine. However, over the several month period it would take to use up several tanks of Fuel, break in is expected to occur. Thus, restricting the run order to get more tanks and more levels for Fuels is not a viable option.

While discussing randomization with the team, one member indicated that Injectors are easier to change than Engines and Fuels. If one simply swapped Injectors, the problem of constantly setting up Engines and draining and refilling tanks of Fuel would be avoided. The statistician immediately recognized this as a restriction error associated with the EF_{ik} term in the model. The new ANOVA, taking Design 4 as the best design so far, is given in Table 6.10.3. This is the seventh design considered so far. (Design 6 was Design 4 with a restriction error on Fuels, rejected because all information on Fuels would be lost.)

As usual, the restriction error is completely confounded with the EF_{ik} interaction, effectively removing all information about that term. In addition, the restriction error shows up in the EMS for E_i and F_k, potentially removing all information about these terms. In this case, we are extremely fortunate because the same approximate tests for E_i and F_k occurring in Design 4 also occur in Design 7, despite the restriction error.

To help understand the differences between Design 4 and Design 7, the ANOVA tables were laid side by side and discussed. Those differences were generally understood but some members had difficulty understanding the different randomizations involved in the designs. Figure 6.1, along with the layout sheet given in Table 6.11.1, helped dispell all confusion. In Design 4, all 32 Engine/Injector/Fuel combinations are formed and these 32 combinations randomized. In Design 7, 8 Engine/Fuel combinations are formed and randomized. Then, for each Engine/Fuel combination, the 4 Injectors are combined and separately randomized. In both designs all 32 combinations are formed and tested, but the run order will be different.

TABLE 6.10.3

ANOVA FOR DESIGN 7. I=J=4, K=2, L=1, RESTRICTION ON EF

Source	df	Δ	Size	EMS
E_i †	3	1.86	MEDIUM	$\sigma^2 + \sigma^2_{EIF} + 4\sigma^2_{\delta} + 4\sigma^2_{EF} + 2\sigma^2_{EI} + 8\sigma^2_E$
I_j ††	3	1.86	MEDIUM	$\sigma^2 + \sigma^2_{EIF} + 4\sigma^2_{IF} + 2\sigma^2_{EI} + 8\sigma^2_I$
→ EI_{ij}	9	1.84	MEDIUM	$\sigma^2 + \sigma^2_{EIF} + 2\sigma^2_{EI}$
F_k †††	1	5.82	EXT. LGE.	$\sigma^2 + \sigma^2_{EIF} + 4\sigma^2_{IF} + 4\sigma^2_{\delta} + 4\sigma^2_{EF} + 16\sigma^2_F$
EF_{ik}	3	—	—	$\sigma^2 + \sigma^2_{EIF} + 4\sigma^2_{\delta} + 4\sigma^2_{EF}$
$\delta_{m(ik)}$	0	—	—	—
→ IF_{jk}	3	2.19	MEDIUM	$\sigma^2 + \sigma^2_{EIF} + 4\sigma^2_{IF}$
EIF_{ijk}	9	—	—	$\sigma^2 + \sigma^2_{EIF}$
$\varepsilon_{\ell(ijk)}$	0	—	—	—
Total	31			

† $EI_{ij} + EF_{ik} - EIF_{ijk}$ approximately tests E_i

†† $EI_{ij} + IF_{jk} - EIF_{ijk}$ approximately tests I_j

††† $EF_{ik} + IF_{jk} - EIF_{ijk}$ approximately tests F_k

For this example, the restriction error corresponds to the variability in fuel consumption caused by draining and refilling the tank with Fuel and with loading and unloading the Engine on the test stand. This variability will be present in the effects of E_i, F_k, and EF_{ik}, but not present in any of the other terms in the model. The team felt that draining and refilling the tank with the same Fuel would have little effect on the test results. However, they felt that loading and unloading the Engine on the test stand could introduce variability. Therefore, the restriction error could not be assumed negligible. Nevertheless, it was felt that the time and cost savings associated with restricting the randomization was worth the loss of information about the Engine/Fuel interaction. If the combined effect came out insignificant, then both the EF_{ik} effect and the restriction error effect can be declared insignificant and nothing was lost by restricting the randomization. If this combined term turned out to be significant, the cause would be unknown. If so, a small experiment will be run in a completely randomized fashion to separate the EF_{ik} effect from the $\delta_{m(ik)}$ effect.

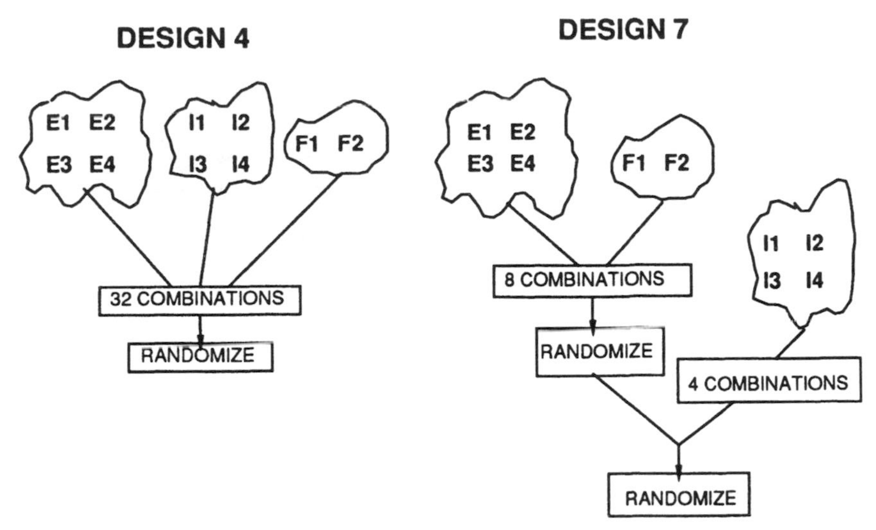

Figure 6.1 Representation of the randomization in Designs 4 and 7.

Thus, Design 7 was accepted as the best possible design to run given the unique conditions of this experiment.

6.11 COLLECTING THE DATA

To ensure the experiment was properly run, the layout sheet in Table 6.11.1 was generated. A computer was used to randomize the experiment. The randomization procedure had to reflect the restriction error on the Engine/Fuel interaction. The first step is to randomly select an Engine/Fuel combination, and then to randomize all four Injectors within the selected Engine/Fuel combination. To randomize the Engine/Fuel combinations, we labeled the combinations 1 through 8 and let the computer randomize the order. Then, we labeled the Injectors 1 through 4 and randomized the order eight times, corresponding to each Engine/Fuel combinations. The result is given in the layout sheet in Table 6.11.1.

It was clearly pointed out to the test personnel that the testing had to be run in the exact order specified in the layout sheet. To be sure the randomized order was followed and the Engines and Injectors were properly selected, the statistician oversaw the selection procedure and part of the run procedure.

Since only 32 data points were collected, the results were hand written on the layout sheet to be directly typed into the computer for analysis. The results were double checked for accuracy, both on the layout sheet and in the computer.

TABLE 6.11.1

LAYOUT SHEET FOR $I = J = 4,\ K = 2,\ L = 1$, RESTRICTION ON EF

Observation	Engine	Injector	Fuel	Brake Specific Fuel Economy
1	E2	I4	F2	
2	E2	I3	F2	
3	E2	I1	F2	
4	E2	I2	F2	
5	E1	I2	F2	
6	E1	I1	F2	
7	E1	I4	F2	
8	E1	I3	F2	
9	E4	I2	F1	
10	E4	I1	F1	
11	E4	I4	F1	
12	E4	I3	F1	
13	E3	I4	F2	
14	E3	I2	F2	
15	E3	I1	F2	
16	E3	I3	F2	
17	E4	I4	F2	
18	E4	I3	F2	
19	E4	I2	F2	
20	E4	I1	F2	
21	E2	I4	F1	
22	E2	I1	F1	
23	E2	I3	F1	
24	E2	I2	F1	
25	E1	I4	F1	
26	E1	I1	F1	
27	E1	I2	F1	
28	E1	I3	F1	
29	E3	I2	F1	
20	E3	I1	F1	
31	E3	I4	F1	
32	E3	I3	F1	

6.12 ANALYZING THE DATA

The collected data, sorted for viewing convenience, is given in Table 6.12.1. A simple plot of the data by run order is given in Figure 6.2. This simple plot helps find any typographical errors as extreme points immediately jump out. It is difficult to use this plot for any other purposes because every term in the model influences the raw data. No gross outliers are evident in this plot.

TABLE 6.12.1

DATA FOR BRAKE SPECIFIC FUEL ECONOMY.

	Fuel Tank 1				Fuel Tank 2			
Injector:	1	2	3	4	1	2	3	4
Engine: 1	43.20	42.80	42.75	42.95	43.45	42.85	42.70	42.95
2	42.10	42.30	42.55	42.80	43.30	43.00	42.75	42.85
3	43.85	43.75	43.20	43.70	43.05	43.20	43.55	44.40
4	43.05	43.20	43.00	43.25	43.30	43.30	43.25	43.70

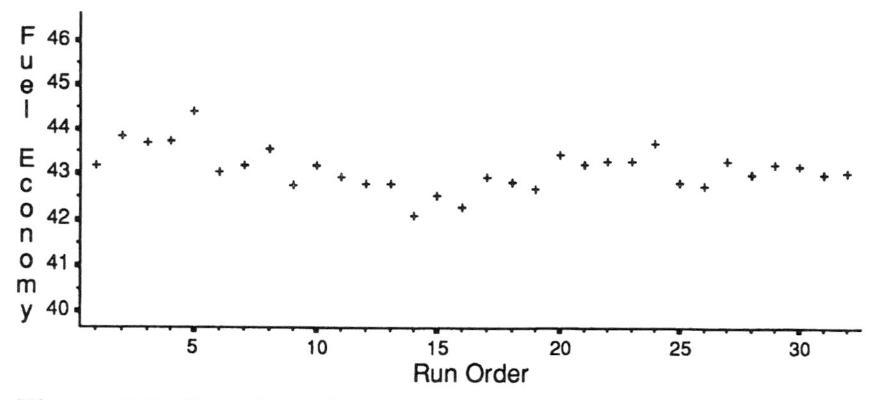

Figure 6.2 Raw data plot used to check for gross outliers in the data

A formal analysis is given in Table 6.12.2. Here we have calculated the approximate tests even though it is wise to look to pool interactions first. In the calculations, all of the denominator degrees of freedom come out less than 1.0 and there is a negative F-value for I_j, all indicating that pooling is necessary. Notice that all of the p-values for the interactions exceed 0.25, indicating they are insignificant and can be pooled. In particular, the combined effect of EF_{ik} and $\delta_{m(ik)}$ is negligible, indicating both terms are negligible. Thus, the decision to put in the restriction error and save considerable time and expense in running the experiment was a wise one. In addition, since $\delta_{m(ik)}$ is insignificant, we know there is a negligible variation added by the process of filling the Fuel tank and loading the Engine on the test stand.

TABLE 6.12.2

ANOVA FOR ENGINE/FUEL/INJECTOR DESIGN 7.

Source	df	SS	MS	F_{calc}	p-value
E_i	0.67	3.480	1.160	14.858	0.2894
I_j	0.84	0.571	0.190	-4.151	—
EI_{ij}	9	0.608	0.068	0.515	0.8311
F_k	0.09	0.310	0.310	11.022	0.7568
EF_{ik} and/or $\delta_{m(ik)}$	3	0.425	0.142	1.081	0.4054
IF_{jk}	3	0.053	0.018	0.134	0.9372
EIF_{ijk}	9	1.179	0.131	—	—
$\varepsilon_{(ijk)}$	0	—	—	—	—

After pooling the two and three factor interactions in accordance with our sometimes pooling rules, the results are given in Table 6.12.3 with estimates of the variances given in the same table. The contribution column shows the percent of the total variation in the system caused by each of the factors under study. The error term, $\varepsilon_{(ijk)}$ accounts for all other factors not directly controlled in the experiment. To summarize the results for management, the three graphs given in Figure 6.3 were generated. The variability in fuel economy caused by each of the main effects is reflected in the spread of the means on the graph. The variability associated with the error term is reflected in the error bars around each mean. When laid next to each other, as in Figure 6.3, a quick visual impression of the effect of each factor and the significance is formed. Clearly, more variability is caused by Engine differences than by Injector or Fuel differences.

6.13 CONCLUSIONS

For the specific diesel Engine tested under a specific dynamometer schedule, there is a highly significant effect due to Engines. No other factor or interaction is significant. Engine to engine differences account

TABLE 6.12.3

POOLED ANOVA FOR ENGINE/FUEL/INJECTOR DESIGN 7.

Source	df	MS	F_{calc}	p-value	Variance	% Contribution
E_i	3	1.160	12.297	0.0001**	0.13	53%
I_j	3	0.190	2.020	0.1380	0.01	5%
F_k	1	0.310	3.287	0.0823	0.01	5%
$\varepsilon_{(ijk)}$	24	0.094	—	—	0.09	37%

** Significant at the .01 level.

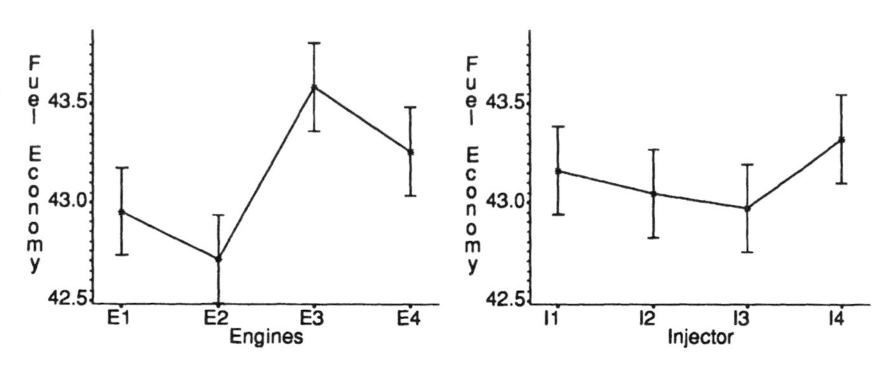

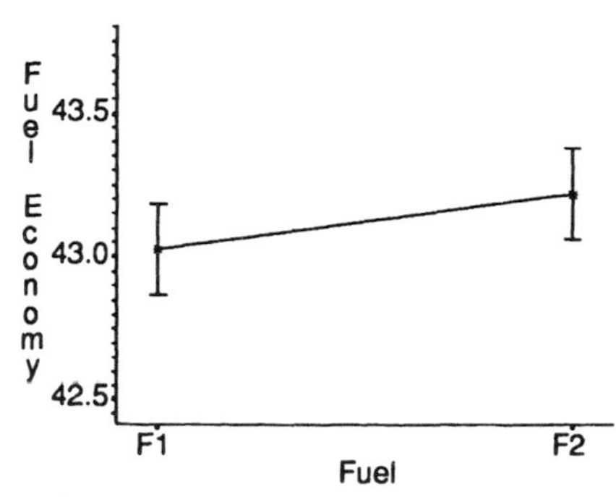

Figure 6.3 Main effects plots showing relative contributions of factors.

for more than 50% of the variability in brake specific fuel economy. Neither Injectors nor Fuel nor any interaction contribute significantly to the variability. Variability in the testing protocol (given by $\varepsilon_{(ijk)}$) accounts for nearly 40% of the observed variability.

Based on this experiment, the team recommends a follow-up study of the major components in the Engine in order to further understand the causes of variability in fuel consumption. Again, a team of experts on the functioning of Engines should be gathered for the purpose of identifying subcomponents that could act as factors. Of course, a statistician will play a key role in the process of designing the next experiment.

6.14 IMPLEMENTATION

The experiment indicated that neither Injectors nor Fuels significantly contributed to the variation in fuel economy. Therefore, no future action, such as sorting of parts or working with suppliers to improve their process, was deemed necessary. In follow-up experiments, there is no reason to worry about using different Injectors or different fuel.

Engines did show a significant effect and needs to be further studied. In designing the next experiment, we now have a good estimate of the variability and can determine the Δ corresponding to any absolute differences. In particular, one of the team's goals is to find the components that contribute 5% or more to the overall variability. Since error contributes about 50% of the variation, we find that $\Delta^2 = 5\%/50\% = 0.1$. Thus, future experiments will have to have a detectability of $\sqrt{0.1} = 0.3$, falling in the *extremely small* category. This corresponds to being able to detect any difference with a standard deviations as small as $\Delta\sigma_\varepsilon \approx 0.1$ mpg.

As an alternative to designing the next experiment with this small a Δ value, procedures should be considered to reduce the testing variability. Improving the test procedure lets one design for larger Δ, requiring smaller experiment sizes. For example, if error could be cut from 50% to 10%, then we only need a Δ of $\sqrt{5\%/10\%} = 0.7$ to detect a 0.1 mpg contributor to the overall variation. Again, design of experiments techniques should be considered as a helpful tool in improving the test procedure, if improvements are thought possible.

PROBLEMS

6.1 *Draw a cause and effect diagram (See Chapter* **1***) for the fuel economy example used in this chapter. Use your knowledge of engines to add at least five more items to make this chart interesting.*

6.2 *Use the cause and effect diagram from* **6.1** *to show how factors for a follow-up study would be selected.*

6.3 *For the fuel economy example in this chapter, a test engineer specified the variable to be measured. In general terms, can a statistician help specify the variable to be measured? If so, how?*

6.4 *How can one be assured that a response variable is accurate and repeatable?*

6.5 *For fixed factors, explain the trade-offs between selection of levels and the expected detectability.*

6.6 *Explain how many levels should be selected for factors that are random, factors that are fixed and quantitative, and factors that are fixed and qualitative.*

6.7 *Suppose two different suppliers supplied Injectors for this engine. Explain how you would change the experiment to include the suppliers in the inference space.*

6.8 *Dynamometers are usually indoors and have a limited temperature range of operation. Explain how one would change the experiment if one was not certain what effect temperature has on the variability in fuel consumption.*

6.9 *How would you change the experiment if you believed that different dynamometers may give different readings? Suppose you thought you knew the best and the worst dynamometer. Would you randomly select dynamometers or select the best and worst case? Explain.*

6.10 *The trick of putting the restriction on the EF interaction worked very well in this experiment because tests on E and F still existed. Would this trick work if one of these factors was fixed? Would it work if both are fixed?*

6.11 *Create the ANOVA table for Design 6, having $I = J = 4$, $K = 2$, $L = 1$, and a restriction on F. (You knew this would be left as an exercise for the reader.)*

6.12 *Compute the ANOVA table for four tanks of fuel, still having the restriction that only two underground tanks exist. That is, two tanks are available at the start of the testing but we have to wait until these tanks are empty before we can test the remaining two tanks. Let $I = J = 4$ and $L = 1$.*

6.13 *Compute the ANOVA table if there are two different Suppliers of Injectors, there are two Injectors for each Supplier, there are four Engines, two Fuels, and there is a restriction error on EF. Compare this design to Design 7 for experimental effort and detectability. We can get information on Suppliers virtually for free!*

6.14 *The data collection (layout) sheet in Table 6.11.1 fits nicely on a single sheet of paper. Some people put each individual run on a separate piece of paper. Can you think of some reasons for putting the layout sheet on one piece of paper and other reasons for putting it on 32 separate pieces of paper?*

6.15 *Figure 6.4 below plots the raw data for each Injector. These plots are commonly used, but not recommended. Why not?*

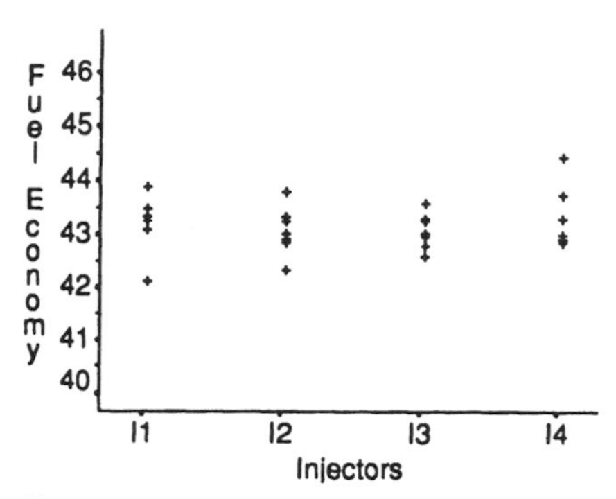

Figure 6.4 Raw data plot for Injectors.

6.16 *Suppose we implemented a change based on a designed experiment and the results are not born out in practice. What would be the probable explanation? What would you do about it?*

6.17 *Suppose we wished to find out how much variability is caused by loading and unloading the engine from the test stand. How would this be handled in an experiment?*

6.18 *Suppose the analysis given in Section* **6.12** *had shown EF_{ik} and/or $\delta_{m(ik)}$ to be significant. How would you design a small experiment to separate the effects?*

6.19 *Variability in the testing protocol resulted in an estimated variability of nearly 40%. Suppose this is unacceptably large. What would you, as a statistician, recommend?*

6.20 *One possible solution to Problem* **6.19** *is to repeat the test several times, removing and reloading the engine each time, and averaging the results. How many tests would be necessary to reduce the variability to 10% of the total?*

6.21 *It has been stated that selection of a high and low value of a variable is worth a random selection of 100 values. To show this is total bunk, compute the detectable size for I_j in model (6.8.1) assuming I_j is fixed at two levels and assuming I_j is random at 100 levels. For this example,* *the ratio of these two Δ's indicate how many standard deviations apart the selected high and low values must be to make this a valid statement. (The chances of selecting two normal values this far apart by chance is approximately* 10^{-30}. *And we haven't even considered the effect on the detectable sizes of any of the other factors or interactions.)*

6.22 *To try and get the symmetry back into the Engine/Injector/Fuel example, consider running Design 4 with two blocks of two different fuels each. Compare the power of this design to the ANOVA in Table 6.10.2. (Hint: there are now 4 factors, nesting, and a restriction error.)*

BIBLIOGRAPHY FOR CHAPTER 6

Anderson, V. L. and McLean, R. A. (1974). *Design of Experiments A Realistic Approach*, New York: Marcel Dekker. (Chapter 3)

Box, G. E. P., Hunter, W. G. and Hunter, J. S. (1978). *Statistics for Experimenters An Introduction to Design, Data Analysis, and Model Building*, New York: John Wiley & Sons. (Chapter 1)

Cox, D. R. (1958). *Planning of Experiments*, New York: John Wiley & Sons. (Chapters 1, 4, 5, 8)

Deming, W. E. (1960). *Sample Design in Business Research*, New York: John Wiley & Sons. (Chapter 1)

Finney, D. J. (1975). *An Introduction to the Theory of Experimental Design*, Chicago, Illinois: University of Chicago Press. (Chapters 1, 8, 9)

Fisher, R. A. (1973). *The Design of Experiments*, 14th ed., New York: Hafner. (Chapters I–IV)

Ghosh, S., editor. (1990). *Statistical Design and Analysis of Industrial Experiments*, New York: Marcel Dekker.

Hicks, C. R. (1982). *Fundamental Concepts in the Design of Experiments*, 3rd ed., New York: Holt, Rinehart and Winston. (Chapter 1)

John, P. W. M. (1971). *Statistical Design and Analysis of Experiments*, New York: The Macmillan Co. (Chapter 1)

Montgomery, D. C. (1991). *Design and Analysis of Experiments*, 3rd ed., New York, John Wiley & Sons. (Chapter 1)

Ostle, B. and Malone, L. C. (1988). *Statistics in Research*, 4th ed., Ames, Iowa: Iowa State University Press.

Steel, R. G. D. and Torrie, J. H. (1980). *Principles and Procedures of Statistics: A Biometrical Approach*, 2nd ed.,New York: McGraw–Hill. (Chapter 6)

SOLUTIONS TO SELECTED PROBLEMS

6.5 *As fixed levels are chosen further and further apart, the expected effect should increase, requiring a less severe Δ and generally a smaller experiment. Spreading the levels too far apart may result in a process that doesn't even run or is not practical from some other aspect of the problem. Spreading levels too far apart may also give poor information about what happens between the levels. In general, balancing all the trade-offs involved in selecting good factor levels is a tricky business.*

6.6 *The number of levels of random factors will be determined in the redesign stage. A good starting point is usually two levels. The number of levels of a fixed quantitative factor depends on the knowledge of the likely effect of the factor. If it is thought the factor affects the response variable in a linear fashion, two levels suffice. If curvature is thought to exist, at least three levels should be used. And so on. Sometimes the number of levels of fixed quantitative factors are changed in the redesign stage of experimentation. The number of levels for fixed qualitative factors is fixed at the number of categories associated with this factor. This is rarely changed in the redesign stage of the experiment.*

6.7 *You would need to add a factor called Supplier to the experiment. Supplier would be fixed, qualitative, and have two levels. Injectors would be nested within Supplier.*

6.10 *For this example, if one of the factors in the restricted interaction was fixed, information about the other factor would be completely confounded with the restriction error. If both factors in the restricted interaction were fixed, information about both factors would be completely confounded with the restriction error. So the answer is no and no.*

6.15 *Contrast Figure* **6.4** *with the corresponding plot in Figure* **6.3**. *The spread in Figure* **6.4** *implied by the data is much greater than the actual spread as indicated by the confidence intervals in Figure* **6.3**. *When plotting raw data, the effects of all other factors and interactions are not removed. This serves to inflate the apparent spread in the data and masks true effects. If you see a difference in raw data plots, the statistics will surely detect it. The reverse is simply not true.*

6.22 *If you got this one, you have really understood the concepts in the previous chapter because they are all in this problem. The key concepts are recognizing that Blocks is another random factor with 2 levels that we will label B_m, that there is a restriction error on Blocks, and that Fuels is a random factor with 2 levels, nested in Blocks. Rearranging the table so terms common to both models are adjacent to one another gives the following comparison. Keep in mind the blocked design requires 64 runs rather than 32 runs and you will have to wait to finish the experiment until the second block of fuel arrives.*

ANOVA Comparing Design 4 to Problem 6.22

Source	Design 4			Problem **6.22**		
	df	Δ	Size	df	Δ	Size
E_i	3	1.86	MEDIUM	3	1.32	SMALL
I_j	3	1.86	MEDIUM	3	1.32	SMALL
EI_{ij}	9	1.84	MEDIUM	9	1.30	SMALL
F_k or $F_{k(m)}$	1	5.82	EXT. LGE.	2	1.64	MEDIUM
EF_{ik} or $EF_{ik(m)}$	3	2.19	MEDIUM	6	1.28	SMALL
IF_{jk} or $IF_{jk(m)}$	3	2.19	MEDIUM	6	1.28	SMALL
EIF_{ijk} or $EIF_{ijk(m)}$	9	—	—	18	—	—
B_m				1	—	
EB_{im}				3	1.60	MEDIUM
IB_{jm}				3	1.60	MEDIUM
EIB_{ijm}				9	1.53	MEDIUM
$\varepsilon_{\ell(ijk)}$ or $\varepsilon_{\ell(ijkm)}$	0	—	—	0	—	—
Total	31			63		

7

Two Level Fractional Designs

7.0 INTRODUCTION

Up to this point we have considered only full factorial and nested designs. These are designs in which every level of every factor appears with every allowable level of every other factor. A complete, model based approach has been developed to lead the experimenter through the design, the evaluation of the design, the redesign, the layout, and the analysis of the collected data.

While this approach is extremely useful and totally general, it does have several disadvantages. For one, when there are a large number of factors, too many runs are required. For example, if there are ten factors thought to influence the response variable, then, even if all factor levels were reduced to just two levels and only one repeat was used, 1024 runs would be required. Few experimenters can afford that many runs. Even if that number of runs is possible, the task of handling that many data points is ominous.

In addition to the size problem, the full factorial approach assumes no knowledge on the part of the experimenter. Often times, subject matter expertise exists and can be substituted for sheer size in an experiment. In essence, knowledge can often be used to reduce the required experimental effort.

In this chapter (and the next), we will show exactly how knowledge can be substituted for experimental effort. This will be done through the process of assuming certain terms in the experiment are negligible. Under such assumptions, we are only interested in certain terms in the model and can design an experiment to just estimate the terms of interest.

In using these fractional designs, caution must be exerted. All of the terms (main effects and interactions) in the full model are always present and their effects will influence the response variable. Just because we have assumed a term's effect negligible does not mean it will actually be negligible. We shudder every time we reflect on the number of times subject matter experts have told us that terms are negligible and, after collecting data, have come to realize that the same terms were indeed important. *Perceived* knowledge can be a very dangerous thing and, if at all possible, all assumptions should be checked.

Even when subject matter knowledge is not available, fractional designs are often run with the intention of running follow-up designs. For example, when very little is known about a process, a team of experts can easily come up with twenty or thirty factors thought to influence the process. Since very little is known about the process, it is impossible to pare down the list to a manageable number. Thus, a first experiment might be run to identify the most important factors. These factors will be more carefully studied in a separate, full factorial experiment. It is likely that a few of the factors have a large influence on the response variable and many of the factors have, at best, a small influence. If strong interactions do exist, too many factors will be recognized as important and we will need a full factorial design to comprehend the full complexities of the situation. Bad assumptions translate to larger follow-up experiments.

There are many approaches taken to designing and analyzing fractional experiments. We will use the modular arithmetic approach in this text. This approach is slightly more difficult for this chapter but considerably easier for the next chapter. In addition, most computer packages use the modular arithmetic approach. Readers interested in more standard hand approaches should consult the textbooks listed in the Bibliography of this chapter.

The layout sheet for fractional designs is identical to the layout sheet for full factorial designs except only a fraction of all possible combinations appear. This is the explanation for the phrase fractional experiment.

We will be assuming that complete randomization is used in the first six sections of this chapter. Section **7.7** considers the effect of restriction errors on the design and analysis. As in Chapter **5**, the often confused concept of blocking is shown to be nothing more than a restriction on randomization.

Section **7.1** introduces the term *confounding* using the mathematical model of the experiment. Section **7.2** introduces modular arithmetic and shows how to generate the confounding pattern. Section **7.3** shows how to compute EMS for fractional designs. The EMS values lead to the tests and Δ values for the tests. Although, to our knowledge, these do not appear in any other book, the EMS values are actually quite easy to compute. Section **7.4** shows how to generate the fraction of all possible combinations that will actually be run. Section **7.5** shows how to analyze the data, specifically the SS, MS, and F values. Several techniques will be given. Sections **7.3** through **7.5** parallel the structure developed in Chapter **3**. Section **7.6** deviates from the normal stream of things to show several methods for selecting a fractional design. With the general availability of computer software, this section is becoming less important. Section **7.7** gets back into the normal flow of things by considering restriction errors. Nesting, as well as factors with more than two levels, will be considered in Chapter **8**. Section **7.8** gives a few examples to nail down the concepts.

7.1 CONFOUNDING IN THE MATHEMATICAL MODEL

We will illustrate the ideas behind the fractionation concept using a three factor model. Labeling the factors A_i, B_j, and C_k, the complete mathematical model is given by

$$y_{ijk} = \mu + A_i + B_j + AB_{ij} + C_k + AC_{ik} + BC_{jk} + ABC_{ijk} + \varepsilon_{(ijk)} \, . \quad (7.1.1)$$

Since each factor has two levels, each term has one degree of freedom and a total of 8 data points must be run to estimate the eight terms μ through ABC_{ijk}.

Suppose, in the above model, the interactions were thought to be negligible compared to the main effects. (We saw such a case in Example 2 of Chapter **3**.) Then one could regroup the mathematical model, say, in the following fashion:

$$y_{ijk} = [\mu + ABC_{ijk}] + [A_i + BC_{jk}] + [B_j + AC_{ik}] + [C_k + AB_{ij}] + \varepsilon_{(ijk)} \, . \quad (7.1.2)$$

Note that, instead of eight individual terms requiring one data point each, there are four groups requiring only four data points. This results in a smaller experiment, *i.e.*, a fractional experiment.

However, this smaller experiment did not come without a price. Notice that we can no longer estimate the effect of the mean μ but must estimate the combined effect of μ and ABC_{ijk}. Likewise, we must estimate the combined effect of A_i and BC_{jk}. We cannot estimate the individual effects. When such a grouping exists we say that the mean is *confounded* or *aliased* with ABC, that A is confounded or aliased with BC, that B is confounded with AC, and so on. There is no way statistically to separate the effect of A from the effect of BC. However, if expert knowledge is used to assume the effect of BC is negligible (*i.e.*,

$\Phi(BC) = 0$), then the effect of the confounded group is due to A alone. Note, however, that such assumptions are non-statistical in nature and, in this particular case, cannot be verified without further experimentation.

If we took the above example one step further, we could have even more confounding and even a smaller experiment. For example, we could have

$$y_{ijk} = [\mu + ABC_{ijk} + B_j + AC_{ik}] + [A_i + BC_{jk} + C_k + AB_{ij}] + \varepsilon_{(ijk)} \ . \quad (7.1.3)$$

Note that there are now two groups so only two data points are required. However, the confounding is so severe that we cannot even estimate the main effects without confounding them with other main effects. While requiring a very small experiment size, this fraction yields almost no useful information due to the severe confounding.

In general, smaller designs will require more fractionation and therefore entail more confounding and stronger assumptions. There is a natural trade-off between knowledge and size of the design, with more knowledge about terms that can be assumed negligible leading to smaller designs. The exact size required depends on the knowledge of the experimenter and the complexity of the problem. If interactions between the factors exist, experiments will have to be larger. If there is no knowledge about interactions, they cannot be assumed negligible and larger designs will have to be run.

A strong warning needs to be issued at this point. It is crucial to recognize the difference between actual knowledge and perceived knowledge. We constantly hear subject matter experts claim that there are a few two factor interactions but there are definitely no three factor interactions. Yet we have observed at least one significant three factor interaction almost 50% of the time three factor interactions could be tested. The reason is twofold. One, the expert did not actually *know* there were no three factor interactions, and two, the increased sample size required to test three factor interactions made the tests more sensitive. It is the statistician's responsibility to temper the desire to run a small experiment and to provide enough leeway in an experiment to at least test a few of the assumptions.

As processes get more complex, more and higher order interactions are going to exist. With the current push to constantly improve processes, simple processes are already going to be understood and optimized. Therefore, we predict that more and more processes will contain higher order interactions that need to be properly modeled to be fully understood and optimized.

7.2 GENERATING THE CONFOUNDING PATTERN

In the previous section, we illustrated how confounding effects can lead to a smaller design. However, that section indicated more flexibility in the confounding than actually exists. In order to maintain the orthogonality that was present in factorial and nested designs, only certain confounding patterns will be allowed. These confounding patterns will be determined on the basis of the terms confounded with the mean and *modular arithmetic*. There are other equivalent techniques that can be used to generate the confounding pattern. Some are easier to use for two level designs. However, the modular arithmetic approach used in this text will naturally extend to the more complicated cases considered in the next chapter. We have found that a little extra work in this chapter saves a lot of confusion in the next chapter.

Modular arithmetic is also called remainder arithmetic because its acceptable values are the remainders after dividing by the module, generally abbreviated "mod". For example, mod 2 must result in either the value 0 or 1 as these are the only two possible remainders after dividing by 2. Mod 3 arithmetic must result in the values 0, 1, or 2 as these are the only possible remainders after dividing by 3. Mod 23 arithmetic must result in one of the values 0 through 22 as these are the only possible remainders after dividing by 23.

A few examples illustrate the use of modular arithmetic. If you have never seen modular arithmetic before, you should find it an easy concept to pick up. Modular arithmetic is the value of the remainder after dividing the original number by the modulus. For example, the value of 5 mod 2 is 1 since 5 divided by 2 is 2 with a remainder of 1. Similarly, 18 mod 2 = 0 since there is no remainder after dividing 18 by 2. Trivially, 0 mod 2 = 0 and 1 mod 2 = 1. The value of 5 mod 3 is 2 because 2 is the remainder after dividing 5 by 3, and the value of 18 mod 3 is 0 since there is no remainder after dividing 18 by 3. As a final example, 654 mod 23 = 10 since 654 divided by 23 is 28 with a remainder of 10.

Modular arithmetic can be found on some hand calculators and on all computers. With a standard calculator, divide the original number by the modulus, subtract the integer portion of the result, and multiply the fraction by the modulus. For example, 654/23 = 28.4348 and (.4348)23 = 10 so 654 mod 23 = 10. None of the examples in this text are anywhere near this complicated and can be easily calculated in your head. For this chapter, we use mod 2 arithmetic so you only need to know if the number is even or odd. If the number is even, the result is 0. If the number is odd, the result is 1. You will not even need a calculator to work any modular arithmetic in this chapter.

The terms that are confounded with the mean will determine the entire confounding pattern using multiplication and mod 2 arithmetic in the exponent. You will find the procedure much easier than the explanation.

The mean, μ is represented as I, called the sl identity, the *generator*, the *identity generator*, the *confounding rule*, the *defining relation*, or the *identity relation*. To determine the confounding pattern, we will use an = sign, but we will represent the confounded term effect with an &/or to indicate the effect is due to any or all of the confounded terms. In equation (7.1.2), μ is grouped with ABC_{ijk} so we would say $\text{I} = ABC$ and the combined effect is due to μ &/or ABC.

The identity I represents the value 1. All other confoundings are found by multiplying the term by the identity I mod 2 in the exponent. For example, A_i is confounded with BC_{jk} because $A = A * \text{I} = A * ABC = A^2BC = A^0BC = BC$. The key step is to realize that $2 = 0$ mod 2 and A to the 0th power is 1. We denote the confounding $A = BC$ and derive the confounding by multiplying A by ABC. The modular arithmetic merely indicates that any factor that is squared (A in this case) gets eliminated from the term. Formally, A is removed because 2 mod 2 = 0 and anything raised to the 0th power equals 1. Informally, all squared factors get removed from the term.

There is a symmetry in the confounding pattern because $BC = BC * ABC = A$. Therefore, once the confounding pattern for A is generated, none of the confounding patterns for the terms confounded with A need be considered.

The complete confounding pattern for identity generator $\text{I} = ABC$ is given below.

$$\begin{aligned} &\text{Identity Generator:} && \text{I} = ABC \\ &\text{Confoundings:} && A = A * ABC = BC \\ & && B = B * ABC = AC \\ & && C = C * ABC = AB \end{aligned} \qquad (7.2.1)$$

This confounding pattern indicates the effects of the three factor interaction will be completely confounded with the effect of the mean, the effect of A will be completely confounded with the effect of BC, the effect of B will be completely confounded with AC, and the effect of C will be completely confounded with AB.

We generally write the confounding pattern with the lowest order term appearing first ($A = BC$ rather than $BC = A$). The *resolution* is the smallest number of letters appearing in the identity generator, III in this case. A resolution of III indicates that at least one main effect will be confounded with a two factor interaction. For this example, all main effects are confounded with two factor interactions.

Since each term is confounded with one other term, only half of the full factorial design needs to be run and this is called a 1/2 fraction. Alternately, since there are three factors all at two levels each, this is called a 2^{3-1} design. The resolution is sometimes added indicating this is a 2^{3-1}_{III} (read 2 to the 3 − 1 resolution three) design. This notation indicates there are three factors having two levels each (2^3), a 1/2 fraction is run (2^{-1}), the design has four factor level combinations ($2^{3-1} = 4$), and at least one main effect is confounded with a two factor interaction (resolution III).

As we just saw, one generator gives a 1/2 fraction and each term consists of two confoundings. Two generators will give a 1/4 fraction and each term will consist of four confoundings. Three generators will give a 1/8 fraction and each term will consist of eight confoundings. The procedure for finding all confoundings is illustrated below. Again, the resolution is defined as the smallest number of letters in any generator.

To illustrate a 1/4 replicate let us consider a 2^{6-2} design. There will be 16 runs ($2^{6-2} = 16$) and two generators are necessary. Label the factors A through F and take $ABCD$ and $CDEF$ as generators. (For now, assume we are given generators. We will learn how to select generators in Section **7.6**.) Each generator equals I, but since I∗I = I, there is an implied third generator: $ABCD * CDEF = ABEF$. This becomes the complete identity generator.

A clever trick for finding the complete identity generator is to expand the product of (I + each individual generator) into its basic components. In this case, the product is given by (I+$ABCD$)(I+$CDEF$) since the individual generators are $ABCD$ and $CDEF$. The expansion yields I+$ABCD+CDEF+ABEF$, indicating that the complete identity generator is $ABCD$, $CDEF$, and $ABEF$. The complete identity generator is usually the first line of the confounding pattern. Using multiplication and mod 2 arithmetic on the exponents as before, we can compute the complete confounding pattern, given in two columns below and starting with the complete identity generator.

$$
\begin{array}{ll}
\mathrm{I} = ABCD = CDEF = ABEF & AC = BD = ADEF = BCEF \\
A = BCD = ACDEF = BEF & AD = BC = ACEF = BDEF \\
B = ACD = BCDEF = AEF & AE = BCDE = ACDF = BF \\
C = ABD = DEF = ABCEF & AF = BCDF = ACDE = BE \\
D = ABC = CEF = ABDEF & CE = ABDE = DF = ABCF \\
E = ABCDE = CDF = ABF & CF = ABDF = DE = ABCE \\
F = ABCDF = CDE = ABE & ACE = BDE = ADF = BCF \\
AB = CD = ABCDEF = EF & ACF = BDF = ADE = BCE
\end{array}
\qquad (7.2.2)
$$

Each individual confounding pattern is written in the order it is generated from the identity. The first term in each confounded string is the lowest order term, the terms are in standard order, but each individual string is not in standard order. Within each string, we usually write the terms alphabetically by order. For instance, we would write $A = BCD = BEF = ACDEF$ rather than the above order. Then, with a quick glance, we can get an idea how low an order term is confounded with each primary term. The properly sorted confounding pattern is repeated below.

$$\begin{array}{ll} \mathrm{I} = ABCD = ABEF = CDEF & AC = BD = ADEF = BCEF \\ A = BCD = BEF = ACDEF & AD = BC = ACEF = BDEF \\ B = ACD = AEF = BCDEF & AE = BF = ACDF = BCDE \\ C = ABD = DEF = ABCEF & AF = BE = ACDE = BCDF \\ D = ABC = CEF = ABDEF & CE = DF = ABCF = ABDE \\ E = ABF = CDF = ABCDE & CF = DE = ABCE = ABDF \\ F = ABE = CDE = ABCDF & ACE = ADF = BCF = BDE \\ AB = CD = EF = ABCDEF & ACF = ADE = BCE = BDF \end{array} \qquad (7.2.3)$$

The above design is resolution IV (specifically, 2_{IV}^{6-2}) since the smallest number of factors in the identity generator is four. This means at least one main effect is confounded with a three factor interaction and at least one two factor interaction is confounded with another two factor interaction. It happens to be the case that *all* main effects are confounded with three factor interactions but this need not be true in general. Resolution IV means at least one main effect is confounded with a three factor interaction, not necessarily all main effects.

The resolution is not an intrinsic property of the size of the design but rather of the selected generators. If the selected generators had been $ABEF$ and $BCDEF$, the implied third generator would be ACD and the resulting 2^{6-2} design would be resolution III. One confounding would be $A = CD$, not desirable unless it is known that the CD interaction is small.

To further illustrate the confounding principle, consider a 2^{6-3} design having three generators and 8 runs. Take as the generators ABC, CDE, and BEF. The first two generators imply a fourth generator $ABC * CDE = ABDE$. The first and third generators imply a fifth generator $ABC * BEF = ACEF$, the second and the third generators imply a sixth generator $CDE * BEF = BCDF$, and all three generators imply a seventh generator $ABC * CDE * BEF = ADF$. A general procedure to follow is generator 1, generator 2, generator 1 times generator 2, generator 3, generator 1 times generator 3, and so on. This gives the identity generator $\mathrm{I} = ABC = CDE = ABDE = BEF = ACEF = BCDF = ADF$.

We would get the same identity generator by expanding (I+ABC)(I+CDE)(I + BEF) and replacing the + signs with = signs. The result is a 2^{6-3}_{III} design with at least one main effect confounded with a two factor interaction. The complete confounding pattern will be left as an exercise for the reader.

In general, a 2^{q-p} design will have q generators and each term has 2^q confoundings with a resolution that depends on the q selected generators. A general rule for computing the entire confounding pattern is given below.

RULE FOR COMPUTING CONFOUNDING PATTERNS
IN TWO LEVEL FRACTIONAL DESIGNS

- For a 2^{p-q} fractional factorial design, start by selecting q generators.
- Find the complete identity generator by expanding the product of (I + each individual generator). Use mod 2 arithmetic in the numerator of each term. There should be a total of 2^q terms, one of the terms being I.
- For each term not already confounded with another term, compute the confounded string by multiplying that term by each element in the identity generator. Again, use mod 2 arithmetic in the numerators.
- For easy readability, sort each confounded string alphabetically by order of the terms.

7.3 EMS AND TESTS IN FRACTIONAL DESIGNS

In all of the references we have seen, fractional designs are only used when all factors are fixed. In this case, higher order interactions are pooled and used to test lower order effects. Yet nothing in the methodology prevents one from using random factors except the lack of a simple algorithm for computing the EMS. As it turns out, the rule for computing EMS is amazingly simple.

RULE FOR EMS IN 2 LEVEL FRACTIONAL FACTORIAL DESIGNS

- Use Algorithm 1 in Chapter 3, steps 1 through 6, to compute the EMS for every term in the full factorial that is not in the identity generator.
- To write the EMS for a confounded group of terms, add each individual EMS computed in the step above.

When the coefficient on a fixed term is less than 1, the actual quadratic being tested will not be defined exactly as represented by this rule. However, for all practical purposes, the representation given by our rule is equivalent to the actual quadratic. (See Problems **7.7** and **7.21**.) Thus we will keep the rules simple by using the representative notation.

ALTERNATIVE RULE FOR EMS IN 2 LEVEL FRACTIONAL DESIGNS

- Use Algorithm 2 in Chapter **3** to compute the EMS for every term in the full factorial that is not in the identity generator. For the slot associated with the error subscript, enter the fraction of the design, 2^{-q} in a 2^{p-q} design.
- To write the EMS for a confounded group of terms, add each individual EMS computed in the step above.

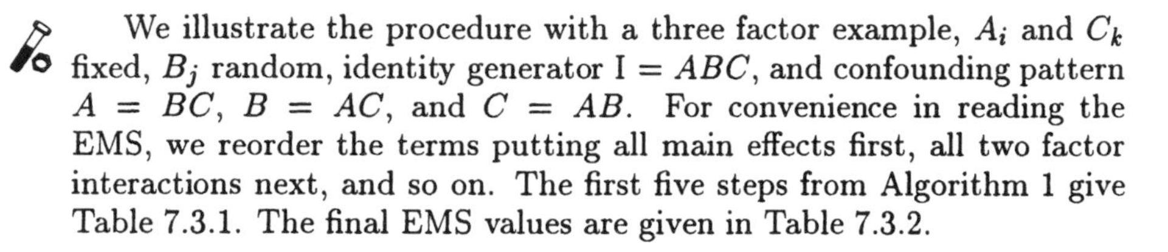

We illustrate the procedure with a three factor example, A_i and C_k fixed, B_j random, identity generator $I = ABC$, and confounding pattern $A = BC$, $B = AC$, and $C = AB$. For convenience in reading the EMS, we reorder the terms putting all main effects first, all two factor interactions next, and so on. The first five steps from Algorithm 1 give Table 7.3.1. The final EMS values are given in Table 7.3.2.

TABLE 7.3.1

EMS FOR 2^{3-1}, A_i, C_k FIXED, B_j RANDOM; AFTER STEPS 1-5

	2	2	2	1	1	1	1/2	1/2
	F	R	F	M	F	M	M	R
	A_i	B_j	C_k	AB_{ij}	AC_{ik}	BC_{jk}	ABC_{ijk}	$\varepsilon_{(ijk)}$
A_i	■			★				★
B_j		★						★
C_k			■			★		★
AB_{ij}				★				★
AC_{ik}					■		★	★
BC_{jk}						★		★

As usual, the reader should go back and reproduce these tables to make sure the process is fully understood. If there is any difficulty, reread the algorithm.

TABLE 7.3.2

ANOVA FOR 2^{3-1}, I = ABC, A_i, C_k FIXED, B_j RANDOM

Source	df	EMS
A_i &/or BC_{jk}	1	$\sigma^2 + \sigma^2_{BC} + \sigma^2_{AB} + 2\Phi(A)$
B_j &/or AC_{ik}	1	$\sigma^2 + (1/2)\sigma^2_{ABC} + \Phi(AC) + 2\sigma^2_B$
C_k &/or AB_{ij}	1	$\sigma^2 + \sigma^2_{BC} + \sigma^2_{AB} + 2\Phi(C)$
$\varepsilon_{(ijk)}$	0	—

Once the EMS are derived, determination of the tests and the detectability, Δ, are exactly as in Chapter **3**. These steps should cause no difficulty.

We are also in a position to look for tests. Our first concern is with the primary terms. For this example, the primary terms are the main effects A_i, B_j, and C_k. To test for the primary term A_i, we look for a row containing $\sigma^2 + \sigma^2_{BC} + \sigma^2_{AB}$. None is found. Likewise, no test is available for either of the primary terms B_j or C_k. It is extremely rare to find tests on the primary terms, but always worth the few seconds it takes to check.

It is rare to find tests on the primary terms due to confounding in the design because terms confounded with the primary term will always appear in the EMS for the confounded primary term. Therefore, we will next look for a test on the confounded string. The first confounded string is $A = BC$ so we look for a test on the combined effect of A_i &/or BC_{jk}. We set both $\Phi(A)$ and σ^2_{BC} to 0 to search for the proper denominator—$\sigma^2 + \sigma^2_{AB}$. Again, no tests are available. This is true for all of the confounded strings.

Generally speaking, there will not be tests on either the primary terms or the confounded strings since only a fraction of the possible combinations are run and there are usually no df for error. There are several options: repeat the experiment, run the experiment and collect the data hoping to be able to pool some insignificant tests, or use non-statistical expertise to assume certain terms are negligible. We will explore each alternative.

Let us consider repeating the entire experiment. The results are given in Table 7.3.3. Notice that, even with two repeats, no direct tests on any of the primary terms or any of the confounded strings exist. Testing each term by the error term results in indirect, conclude insignificant only, tests on each of the confounded strings. This is the opposite type of test than desired. Thus, repeating the entire experiment was not a useful exercise for this example.

TABLE 7.3.3

ANOVA FOR 2^{3-1}, I = ABC, A_i, C_k FIXED, B_j RANDOM, 2 REPEATS

Source	df	EMS
A_i &/or BC_{jk}	1	$\sigma^2 + 2\sigma^2_{BC} + 2\sigma^2_{AB} + 4\Phi(A)$
B_j &/or AC_{ik}	1	$\sigma^2 + \sigma^2_{ABC} + 2\Phi(AC) + 4\sigma^2_B$
C_k &/or AB_{ij}	1	$\sigma^2 + 2\sigma^2_{BC} + 2\sigma^2_{AB} + 4\Phi(C)$
$\varepsilon_{\ell(ijk)}$	4	σ^2

If circumstances had been changed and all of these factors had been fixed, the EMS would be as in Table 7.3.4. Notice that each of the confounded strings is now directly tested by error. This will be the case when all of the factors are fixed and probably explains why most textbooks only consider the simpler all factors fixed case.

TABLE 7.3.4

ANOVA FOR 2^{3-1}, I = ABC, A_i, B_j, C_k FIXED, 2 REPEATS

Source	df	EMS
→ A_i &/or BC_{jk}	1	$\sigma^2 + 2\Phi(BC) + 4\Phi(A)$
→ B_j &/or AC_{ik}	1	$\sigma^2 + 2\Phi(AC) + 4\Phi(B)$
→ C_k &/or AB_{ij}	1	$\sigma^2 + 2\Phi(AB) + 4\Phi(C)$
$\varepsilon_{\ell(ijk)}$	4	σ^2

Even in the all factors fixed case, a total of eight runs must be made. If eight runs can be made, you are almost always better off running the full factorial rather than repeating a 1/2 fraction. For the example having B_j random, the full factorial, Table 7.3.5, results in direct tests on A_i, C_k, and AC_{ik}. There are conclude significant only tests on B_j, AB_{ij}, and BC_{jk}. These are the preferred type of indirect tests for most experiments. In addition to obtaining tests, there is no confounding to muddy up the conclusions and no assumptions are necessary.

For the case when all three factors are fixed, the highest order interaction is used to indirectly test each of the terms in the model. Again, there is no confounding and no assumptions are necessary. (Although one could assume the highest order interaction negligible and the indirect tests would turn into direct tests.)

TABLE 7.3.5

ANOVA FOR 2^3, A_i, C_k FIXED, B_j RANDOM, 1 REPEAT

Source	df	EMS
A_i	1	$\sigma^2 + 2\sigma^2_{AB} + 4\Phi(A)$
B_j	1	$\sigma^2 + 4\sigma^2_B$
AB_{ij}	1	$\sigma^2 + 2\sigma^2_{AB}$
C_k	1	$\sigma^2 + 2\sigma^2_{BC} + 4\Phi(C)$
AC_{ik}	1	$\sigma^2 + \sigma^2_{ABC} + 2\Phi(AC)$
BC_{jk}	1	$\sigma^2 + 2\sigma^2_{BC}$
ABC_{ijk}	1	$\sigma^2 + \sigma^2_{ABC}$
$\varepsilon_{\ell(ijk)}$	0	—

Note: tests on BC_{jk}, AB_{ij}, and B_j are indirect.

An alternate strategy often suggested when there are no tests is to run the experiment and try to pool terms after the data is collected in order to obtain tests. Rules for pooling differ drastically and should be performed with great care. The worst suggestion we have seen is by Taguchi (1988, p. 341), who recommends pooling one-half of the effects. We see no reason why one-half of the effects should be negligible and find it inconsistent to pool main effects when interactions involving those main effects need not be pooled. The best suggestions we have seen involve half-normal plots, Daniel (1959), of mean squares values. These should only be used when all factors are fixed. An example of a half-normal plot is given in Figure 7.1. Start by sorting the mean squares in ascending order. The lower axis contains the square root of the calculated mean squares for each effect. The side axis contains the half-normal scores, $\mathrm{F}^{-1}((n+i-.5)/2n)$, where i is the ith ordered effect, n is the number of effects estimated, and $\mathrm{F}(\cdot)$ is the cumulative normal distribution. The line is usually fitted by eye to the bulk of the data near 0, but formal techniques do exist.

The idea behind half-normal plots is fairly simple. When all factors are fixed, the expected mean squares takes the form σ^2 plus the fixed effect. Therefore, all terms that are candidates for pooling will have expectation σ^2 and the square root will fall roughly along a straight line on half-normal paper. Factors that have an effect will fall to the right of this straight line. Of course, we must maintain the "sometimes pooling" rules and never pool a term if an interaction involving that term cannot also be pooled.

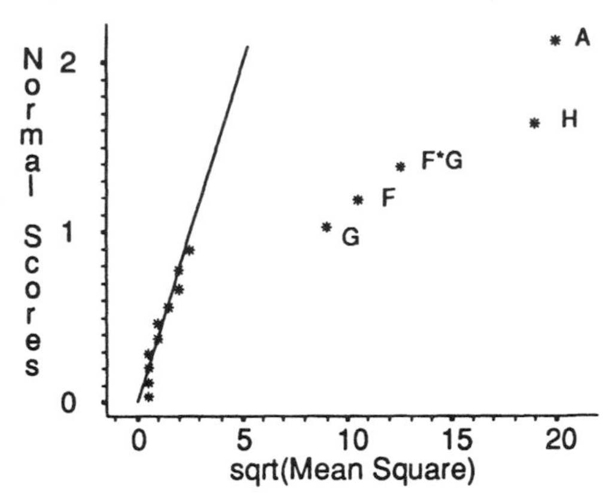

Figure 7.1 Example of half-normal plot of effects.

In Figure 7.1, all terms except A, F, G, H, and $F*G$ are candidates for pooling because the effects of these terms fall along a straight line through the origin. Of course, this is an idealized case because the remainder of the effects fall in such a nice straight line. It is rare to see such a clear-cut example and considerable judgement is generally necessary when using these plots.

But, by far, the most popular technique for obtaining tests is through the use of assumptions. For example, if we can *assume* all interactions in model 7.1.2 are negligible, then the effects are strictly due to the main effects A, B, and C. The mathematical model simplifies to

$$y_{ijk} = \mu + A_i + B_j + C_k + \varepsilon_{(ijk)} \tag{7.3.1}$$

and the EMS for A and C fixed, B random simplifies to

Source	df	EMS
A_i	1	$\sigma^2 + 2\Phi(A)$
B_j	1	$\sigma^2 + 2\sigma_B^2$
C_k	1	$\sigma^2 + 2\Phi(C)$
$\varepsilon_{(ijk)}$	0	—

These EMS are obtained from Table 7.3.2 by simply setting the effects of all interactions to 0. Assuming a term negligible is equivalent to removing it from the mathematical model and removing its effect everywhere in the EMS.

If you recall the rules for pooling given in Chapter **3**, you will remember that pooling is equivalent to removing the term and all of its effects. However, pooling and assumptions are *not* the same thing. Pooling uses the evidence of data to remove terms and effects from the model. Assumptions are based on subject matter expertise or experience and have typically not been supported by data. Assumptions may or may not be true and are often made based on misunderstanding of the meaning of interactions or on hearsay evidence. If at all possible, all assumptions should be checked using the data collected in the experiment. Therefore, we make the following suggestions for implementation of assumptions.

RULE FOR USE OF ASSUMPTIONS IN EXPERIMENTAL DESIGNS

- Compute the confounding pattern and all EMS values as if no assumptions were made.
- Remove all effects and terms assumed negligible to obtain all tests valid under the set of assumptions.
- If possible, use the remainder of the assumptions to test each individual assumption.

We use our previous simple examples to illustrate the rule even though this may not be the best strategy. We use these examples for illustration purposes only. The examples in Section **7.8** show more common uses of the rule. If we assume all interactions are negligible in Table 7.3.2, only the main effects are left in the model but there are still no tests available. If we assume all interactions are negligible in Table 7.3.3, only the main effects are left in the model and these are tested by error having 4 df. There will be no way to test the assumptions that the interactions are negligible since we have no tests on any interactions. If we assume all interactions are negligible in Table 7.3.5, we can pool all of the interactions into the error term and have 4 df for testing the main effects. But, in addition, we can pool all of the other interactions to test each individual interaction with 3 df. For example, we can pool AC, BC, and ABC to get 3 df for testing the assumption that AB is negligible. The other tests on the assumptions are similar. (One can justifiably argue that there is already a test on AC without assumptions and that assuming ABC negligible results in a direct test on AB and BC so that the rule does not make sense. One can also argue that higher order interactions are less likely to be significant than lower order interactions so it does not make sense to test ABC with the pooled two factor interactions. We do not disagree with either of these arguments but simply wish to stress the importance of testing assumptions whenever possible. If the assumptions truly hold, the tests given by the above rule are valid. We leave it to the experience of each individual to develop alternatives to the above rule provided assumptions are tested whenever possible.)

Assumptions should only be made on the basis of knowledge of the system under study or on the basis of experience. Although there are many general statements made about assumptions, each system under study is going to be unique. Therefore, no general recommendations can uniformly hold. The only thing that can be said in general is that subject matter knowledge must be used and perceived knowledge is not the same as actual knowledge. It is critical that all checkable assumptions actually be checked.

General assumptions made on the basis of experience are going to depend on the individual background of each person and will be a source of debate forever. Some people claim that a proper response can be chosen so that no interactions ever exist. Some claim that two factor interactions are common but three factor interactions are rare. Others claim that three factor interactions are common. One thing that can be asserted is that higher order interactions are more likely to occur as processes become more complicated. And, as we attempt to control processes more carefully, processes are going to become more complicated. Thus, we are moving toward the need for designs estimating more interactions. This is consistent with our experience.

A statistician plays an extremely important role in the selection of terms to assume negligible. Given a choice, the experimenter will always make more assumptions in order to obtain a smaller experiment. It is only human nature. The statistician must be prepared to carefully define interactions and point out the dangers of assuming terms negligible when they in fact are important. Specifically, the statistician must generate the confounding pattern and tests and indicate which terms will be falsely declared significant if certain assumptions are not met. The statistician must also determine which assumptions can be tested and which assumptions cannot and find the design best able to test the questionable assumptions. The statistician's job is to help the experimenter find the best compromise between experiment size, available information, necessary assumptions, and, ultimately, the statistical validity of the inference.

7.4 GENERATING THE LAYOUT SHEET

For the designs in all of the previous chapters, all allowable combinations of factor levels make up the layout sheet. Now, only a fraction of all combinations will be run. This section shows how to select that fraction based on the identity generator and modular arithmetic.

As usual, there are several ways to find the factor level combinations to run in the fractional design. They are all based on the use of modular arithmetic with the identity generators. There are some very clever methods for generating the factor level combinations when all factors are at two levels. Unfortunately, when all factors are not at two levels, new methods have to be learned. This results in much confusion. We prefer to use a slightly more complicated method in this chapter in order to save a lot of confusion in the next chapter.

In order to use modular arithmetic, we are going to label each level of each factor with either a 0 or a 1. Of course, it is a trivial matter to substitute actual low and high levels for 0 and 1 after the factor level combinations are generated. Let A, B, etc., ($= 0$ or 1) stand for the *level* of factors A, B, etc., as well as the factors themselves. The context should make it clear whether we are talking about the factor or the level of the factor. The factor level combinations making up the fractional design are generated from the identity generators by selecting those combinations whose values either sum to 0 or sum to 1 mod 2. A few examples show the simplicity of the procedure.

If there are three factors, A, B, and C, and the identity generator is $I = ABC$, the fraction to run consists of all combinations of factor levels satisfying

$$A + B + C = 0 \mod 2 \tag{7.4.1}$$

or all combinations satisfying

$$A + B + C = 1 \mod 2. \tag{7.4.2}$$

For all practical purposes, these two fractions are interchangeable. See Problem **7.8** for a discussion of the differences. We generally choose the primary block, the fraction corresponding to 0, although there is no overriding reason to do so. Since we are using mod 2 arithmetic, all even sums satisfy the first equation. Therefore, an even number of factors, either zero or two, must be 1. A set of factor level combinations corresponding to $I = ABC$ are

A	B	C	Σ	Σ mod 2
0	0	0	0	0
1	1	0	2	0
1	0	1	2	0
0	1	1	2	0

Notice that the sum Σ is always 0 mod 2, as required. If we had chosen the fraction corresponding to 1 mod 2, the opposite fraction would have been generated:

A	B	C	Σ	Σ mod 2
1	0	0	1	1
0	1	0	1	1
0	0	1	1	1
1	1	1	3	1

When there are two identity generators, there are two equations to set up. Since each equation could take the value 0 or 1 mod 2, there are four equivalent fractions. For example, if I = $ABCD = CDEF$ in a six factor fractional design, the primary block satisfies

$$\begin{aligned} A + B + C + D &= 0 \mod 2, \text{ and} \\ C + D + E + F &= 0 \mod 2. \end{aligned} \tag{7.4.3}$$

The other three possible fractions take the values 1 and 0, 0 and 1, and 1 and 1; all mod 2. The factor level combinations, $2^{6-2} = 16$ of them, satisfying the primary equations are given in Table 7.4.1. Notice that, in all cases, the sum of A through D is even and the sum of C through F is even, implying both sums are 0 mod 2. The combinations can be chosen by trial and error or a comprehensive search, but, after studying the combinations a little while, one can see the pattern used to come up with the combinations. We know there must be 16 combinations.

TABLE 7.4.1

FACTOR LEVEL COMBINATIONS FOR A 2^{6-2} DESIGN

A	B	C	D	E	F	Σ_A^D	Σ_C^F	A	B	C	D	E	F	Σ_A^D	Σ_C^F
0	0	0	0	0	0	0	0	1	0	0	1	0	1	2	2
0	0	0	0	1	1	0	2	1	0	0	1	1	0	2	2
0	0	1	1	0	0	2	2	1	0	1	0	0	1	2	2
0	0	1	1	1	1	2	4	1	0	1	0	1	0	2	2
0	1	0	1	0	1	2	2	1	1	0	0	0	0	2	0
0	1	0	1	1	0	2	2	1	1	0	0	1	1	2	2
0	1	1	0	0	1	2	2	1	1	1	1	0	0	4	2
0	1	1	0	1	0	2	2	1	1	1	1	1	1	4	4

Undoubtably, you noticed that the procedure was easy when one generator was present and much harder when two generators were present.

This observation is indeed correct. It is quite difficult to use this method on highly fractionated designs, even with the use of a computer. For example, a 2_{III}^{14-10} design will use 10 equations to select 16 combinations out of a possible 16,384 combinations. This is literally impossible by hand. If there are 20 factors or more, even computers will have difficulty searching through all possible combinations.

Fortunately, through a trick known as the *basic factor technique*, we can easily generate a fractional design when there is a high degree of fractionation.

The secret to the basic factor technique is to properly select the basic factors. Then, the generated design is a full factorial design in the basic factors and the remaining levels are determined by each of the equations. Thus, a 2^{k-p} design will have $k - p$ basic factors and the levels for the remaining p factors will be determined by these basic factors. The basic factors are determined by the rule given below.

RULE FOR FINDING BASIC FACTORS IN FRACTIONAL DESIGNS

- Start with a list of all factors in the experiment.
- Look at the first identity generator. If there is a factor in this generator not appearing in any other identity generator, remove it from the list. Otherwise, remove any factor in this generator from the list.
- Look at the next identity generator. If there is a factor remaining in the list that appears in this generator and not in any other generator, remove it from the list. Otherwise, remove any factor in this generator from the list.
- Repeat the previous step for every identity generator, removing one factor for each generator. If there are no factors both in the generator and the list, you have made an error and need to revisit a previous step.
- The remaining list is a list of basic factors for this fraction.

For the example given above, we start with the list $\{A,B,C,D,E,F\}$. Looking at the first generator, $ABCD$, we see that A and B are in this generator and not in the second generator. Therefore, we can remove either A or B from the list. Let's remove B leaving a list of $\{A,C,D,E,F\}$. Looking at the second generator, $CDEF$, E and F are candidates for removal since they are in the list and not in the first generator. Let's remove F from the list. We are left with a set of basic factors of $\{A,C,D,E\}$. Other sets are obviously possible.

Use the list of basic factors to generate the fractional design according to the following rule.

RULE FOR GENERATING A FRACTIONAL DESIGN USING BASIC FACTORS

- Generate a full factorial design using the basic factors.
- Use each of the equations to fill in values for the remaining factors.
- If there is more than one remaining factor in the equation, temporarily skip this equation until the values for the other remaining factors have been determined. This step is seldom necessary.

Following through on the example above, we start with a full factorial on the basic factors A, C, D, and E as given in Table 7.4.2.

TABLE 7.4.2

FULL FACTORIAL ON THE BASIC FACTORS A, C, D, AND E

A	B	C	D	E	F	Σ_A^D	Σ_C^F	A	B	C	D	E	F	Σ_A^D	Σ_C^F
0		0	0	0		0	0	1		0	0	0		1	0
0		0	0	1		0	1	1		0	0	1		1	1
0		0	1	0		1	1	1		0	1	0		2	1
0		0	1	1		1	2	1		0	1	1		2	2
0		1	0	0		1	1	1		1	0	0		2	1
0		1	0	1		1	2	1		1	0	1		2	2
0		1	1	0		2	2	1		1	1	0		3	2
0		1	1	1		2	3	1		1	1	1		3	3

To determine the value for B, we note that the sum of A, B, C, and D must be 0 mod 2, or the sum must be even. Therefore, if the sum of A, C, and D is odd, B must be 1. If the sum of A, C, and D is even, B must be 0. Similar reasoning indicates that F must be 1 if the sum of C, D, and E is odd and 0 if the sum is even. We included the sum in Table 7.4.2 for ease in calculating B and F. The resulting fraction is given in Table 7.4.3.

Although not in the same order, every combination in Table 7.4.1 appears in Table 7.4.3, and vice-versa. The two techniques are indeed equivalent.

A final technique that is often used in computers is to represent all factors in terms of the basic factors, run a complete factorial on the basic factors, and let the identity generators solve for the remaining factors mod 2. In this case, the generator $\mathrm{I} = ABCD$ solves to $B = ACD$ so that $B = A + C + D$ mod 2. This method guarantees the primary block will

TABLE 7.4.3

FACTOR LEVEL COMBINATIONS FOR A 2^{6-2} DESIGN
USING BASIC FACTORS $B = ACD$ MOD 2 AND $F = CDE$ MOD 2

A	B	C	D	E	F	Σ_A^D	Σ_C^F	A	B	C	D	E	F	Σ_A^D	Σ_C^F
0	0	0	0	0	0	0	0	1	1	0	0	0	0	2	0
0	0	0	0	1	1	0	2	1	1	0	0	1	1	2	2
0	1	0	1	0	1	2	2	1	0	0	1	0	1	2	2
0	1	0	1	1	0	2	2	1	0	0	1	1	0	2	2
0	1	1	0	0	1	2	2	1	0	1	0	0	1	2	2
0	1	1	0	1	0	2	2	1	0	1	0	1	0	2	2
0	0	1	1	0	0	2	2	1	1	1	1	0	0	4	2
0	0	1	1	1	1	2	4	1	1	1	1	1	1	4	4

be selected (*i.e.*, the sum will be 0 mod 2) and will quickly generate the factor level combinations for any set of identity generators. The results are given in Table 7.4.3.

7.5 ANOVA FOR FRACTIONAL DESIGNS

The rule for computing df given in Chapter **3** still applies to the fractional case. However, since every factor has two levels, every term in the model has 1 df and the rule is not worth reviewing.

Similarly, the rules for computing SS hold for fractional designs. However, one has to be very cautious because every combination does not exist. We must sum only over the combinations that have been run. The rule for hand calculating SS was written in such a way that no modification of the rule is necessary. Similarly, the rule for machine calculation holds *as it is written.* But, our standard practice of simplifying the summations *does not hold* since we are *not* summing over all possible levels. This may cause a problem with some computer packages. Be sure to verify the correctness before routine use.

To compute SS using a computer package, use one of two similar techniques, depending on the sophistication of the computer package. The most straight forward technique is the primary term technique. This requires a more sophisticated software package. A technique that always works, but is not as convenient, is the basic factor approach. Both techniques are similar in that they use one of the terms in a confounded group to represent the SS for the entire group. The statistician must realize this SS refers to the entire group and make the proper inferences.

Consider the primary factor technique first. We have previously defined the primary factor as the lowest ordered term in the confounded set. If your computer package lets you specify terms in the model, specify the model using only the primary term from each confounded set. The statistician must realize that the resulting SS is not due to the primary term alone but is the net result of all of the terms confounded with the primary term. It is the statistician's responsibility to properly interpret the results.

For the 2^{3-1} example given earlier, (7.2.1), merely enter the terms A, B, and C in the model. In this case, SS(A) is really SS(A &/or BC), an identification that is the statistician's responsibility.

For the 2^{6-2} example, (7.2.3), the model must include A through F, AB, AC, AD, AE, AF, CE, CF, ACE, and ACF. As in the previous paragraph, SS(A) is really SS(A &/or BCD &/or BEF &/or $ACDEF$), and so on.

Note that there is only one term entered for each confounded string. That is, do not include both AB and CD even though they have the same order because they are in the same confounded string. In general, it is wise to choose the primary term alphabetically among all terms having the lowest order in a given confounded string to keep from confusing the software program. To be more specific, most programs will recognize that SS(AC) equals SS(BD) but if BD is specified in the model along with ABC, SS(ABC) may not be correctly calculated. The reason is that a higher order interaction like ABC is defined in terms of all its lower order interactions, *e.g.*, AB, AC, and BC, so that these lower order interactions must be in the model. Always choosing the primary term alphabetically among all lowest order terms will guarantee all lower order interactions are included in the model.

If your computer package only recognizes full factorial designs or you are trying to analyze the data by hand, use the basic factor technique. This technique consists of entering only the basic factors and interactions into the computer and performing a full factorial analysis on the basic factors. Each confounded string will contain exactly one term made up of basic factors only and is representative of the entire string. When hand calculations have to be performed, this technique is often employed.

For the 2^{3-1} example, (7.2.1), any two factors can be basic factors so let us choose A and B. The representatives for each group are A, B, and AB. As before, SS(A) is really SS(A &/or BC) and SS(AB) is really SS(C &/or AB).

For the 2^{6-2} example, (7.2.3), arbitrary basic factors cannot be selected. For example if we chose A, B, C, and D as basic factors, we would need to find a term involving only A through D in every confounded string. We immediately run into problems with the string $E = ABF = CDF = ABCDE$, where every term contains either an E or an F. This is the very reason why basic factors cannot be arbitrarily chosen and the rule must be followed.

Going with our earlier choice for basic factors of A, C, D, and E gives the selective representatives of each confounded string, in the order of confounded strings given in (7.2.3), of A, ACD, C, D, E, CDE, CD, AC, AD, AE, $ACDE$, CE, DE, ACE, and ADE. Note that this is a complete factorial in the basic factors and each confounded string contains exactly one term defined in the basic factors alone. This full factorial is easily analyzed by computer packages or by hand. You must remember that, *e.g.*, SS(ACD) is really SS(B &/or ACD &/or AEF &/or $BCDEF$), and interpret the SS appropriately.

As in Chapter **3**, the MS values are simply the SS divided by the df and the F values are ratios of MS values. But since df are generally 1 and there are generally no tests, these are trivial statements.

MS and F values only become interesting when assumptions have been made. For example, it *may* (In this text we avoid any sweeping statements about interactions since every case is unique.) be reasonable to assume all three factor and higher order interactions are negligible in the 2^{6-2} example given in (7.2.3). Then, two of the confounded strings, namely $ACE = ADF = BCF = BDE$ and $ACF = ADE = BCE = BDF$, will be assumed negligible and can be pooled to form an error term with two df: MS(ε) = [SS(ACE &/or ADF &/or BCF &/or BDE) + SS(ACF &/or ADE &/or BCE &/or BDF)]/2. Of course, this pooling depends on the assumption that third order and higher terms are negligible and the resulting tests will be invalid if the assumptions do not hold.

7.6 SELECTING A FRACTIONAL DESIGN

So far we have indicated how to generate the confounding pattern, the factor level combinations, the EMS table, and how to analyze a set of data. However, all of these procedures depended on the selection of generators for the fractional design and we have never indicated how to select the generators. The best data analysis techniques in the world will not do much good if the design was poorly selected in the first place. In this section, we will show how to select generators to meet the individual needs of the experimenter. Three methods will be considered, trial and error selection of generators, the basic factor method, and table lookup of generators. We will briefly review each method.

Prior to selection of generators, one must decide which terms are of interest in the experiment. For example, if a team came up with a list of 25 factors that could potentially affect a process and the purpose of the experiment is only to pair down the list, one would probably decide to estimate the main effects only. On the other hand, if the number of factors was paired down to, say, eight, one may decide to estimate two or even three factor interactions as well as the main effects. In general, one must base the decision on the size of the required experiment and the knowledge of which interactions are likely to occur. This knowledge is not statistical in nature, but the statistician must be sure the client fully understands the meaning of the various interactions before the client's knowledge is translated into specific design requirements. This translation process takes practice and develops as the statistician gains experience.

Let us start with the trial and error method for choosing generators. This method is rapidly losing favor with the advent of computer programs and the fact that trial and error requires a lot of work. Nevertheless, we feel that knowledge of this method is invaluable because computers are not always available.

One starts the trial and error method by writing down all of the factors and interactions of interest. (We state this like it is a trivial procedure. In fact, the procedure usually takes considerable time with a client, explaining the meaning of interactions and questioning the client about the known behavior of the process.) The next higher power of two, say 2^r, is the suspected sample size. The number of generators needed is the number of factors minus the power of two in the suspected sample size; $q = p - r$ generators are needed to consider p factors in a fraction consisting of 2^r runs. There is no guarantee that a fractional design meeting all needs can be found in 2^r runs. In addition, there are typically many different sets of generators meeting the design requirements.

Actual selection of the generators is accomplished through trial and error. However, we can use the implied confounding in the list of required terms to help narrow down the search. A couple of tricks come to mind. One, if AB and CD are both in the required list of terms, $ABCD$ could not be a valid generator since it would force AB to be confounded with CD. Two, if there are 7 factors, the product of any two generators of length 5 will have length at most 4. Similar tricks are learned over time.

For example, suppose we had eight factors, A through H and wished to estimate all main effects and two factor interactions. We add the 28 two factor interactions to the 8 main effects to come up with 36 df at minimum. The next higher power of 2 is $2^6 = 64$ so we require at least 64 runs. Therefore, we must select $8 - 6 = 2$ generators, resulting in a 2^{8-2}

design. If any generator contained only 4 factors, a two factor interaction would be confounded with another two factor interaction. Therefore, each generator must have at least 5 factors. Since the product of the two generators is also a generator, the selected generators should have as few terms in common as possible. Therefore, we are motivated to try the extreme left and extreme right five factor interactions—$ABCDE$ and $DEFGH$. The implied generator is $ABCFGH$, containing at least five factors and resulting in a design that meets all requirements.

The reasoning used in the previous paragraph is typical of the reasoning employed in the trial and error method. With experience, one can get very good at the procedure. However, this procedure only works well when little fractionation occurs. When a lot of generators need to be selected, this method becomes extremely burdensome. The next method is very useful when there is a lot of fractionation and the resulting design is small. By small, we think that a 32 run design is a practical limit.

This technique is often called the *reverse basic factor method* or simply the *basic factor method*. Start by writing down a full factorial using the basic factors, usually the first factors in the list of factors to avoid confusion. Then, assign the remainder of the factors to the interactions involving the basic factors in such a way as to meet all requirements. If an assignment of a factor to a basic factor interaction results in a contradiction to the requirements, assign that factor to another basic factor interaction. Only the main effects need to be assigned to basic factor interactions. This assignment determines the identity generators and, as a side benefit, identifies the factor level combinations. One example easily illustrates the procedure.

For this example, suppose we wish to estimate the eight original factors A through H and the four interactions AB, BE, CF, and EF and hope to be able to use 16 runs. Since $16 = 2^4$, there are four basic factors, say A, B, C, and D. We usually choose the first factors alphabetically as the basic factors to avoid the confusion of having to relabel the basic factors differently than other factors. The first step is to write down the full factorial in the basic factors and assign the factors to the corresponding columns. Also place all of the interactions of the *assigned* basic factors as these columns will not be candidates for further assignments. The result is given below, with the basic terms on the bottom and the assigned original terms on the top.

original factors

A	*B*	*AB*	*C*				*D*							
A	*B*	*AB*	*C*	*AC*	*BC*	*ABC*	*D*	*AD*	*BD*	*ABD*	*CD*	*ACD*	*BCD*	*ABCD*

basic factors

This leaves plenty of basic factor terms unassigned. We wish to place the remaining factors in such a way that they are not assigned to any previously assigned column and the required interactions with the placed factors are not assigned to any previously assigned column. Let us start by trying to assign the original factor E to the basic factor term AC. Since we are interested in the BE interaction, we find that, in terms of the basic factor interactions, BE is equivalent to $B * AC = ABC$. This is an acceptable assignment since the ABC interaction among the basic factors is currently unassigned. We assign these two columns with the resulting assignments given below.

original factors

A	B	AB	C	E		BE	D							
A	B	AB	C	AC	BC	ABC	D	AD	BD	ABD	CD	ACD	BCD	$ABCD$

basic factors

Next, we try to assign the original factor F. This is trickier since there are two interactions involving F: CF and EF. We try assigning F to the empty basic factor interaction BC. Then, CF would correspond to $C * BC = B$, a column that is already assigned. This would *not* be a legitimate assignment.

Next we try assigning F to the basic factor interaction AD. We see that CF corresponds to the basic factor interaction ACD, unassigned, and EF corresponds to the basic factor interaction CD, also unassigned. We can then place G and H in any of the remaining columns since there are no interactions involving G or H. A final column assignment becomes

original factors

A	B	AB	C	E		BE	D	F			EF	CF	G	H
A	B	AB	C	AC	BC	ABC	D	AD	BD	ABD	CD	ACD	BCD	$ABCD$

basic factors

Since this is a $16 = 2^4$ run design in eight factors, we have a 2^{8-4} design and must have four generators. The four generators are determined by the assignment of the last four factors, E, F, G, and H. Since E was assigned to the basic factor interaction AC, we have $E = AC$ which corresponds to $\mathrm{I} = ACE$, the first generator. Likewise, $F = AD$ corresponds to $\mathrm{I} = ADF$, $G = BCD$ corresponds to $\mathrm{I} = BCDG$, and $H = ABCD$ corresponds to $\mathrm{I} = ABCDH$. This is a resolution III design.

As we can see from the column assignments given above, we get a resolution III design because we assigned one of the last four factors to a two factor interaction column. We should always try to assign the remaining factors to the highest order basic factor interactions possible. Had we assigned E to the ABD interaction column, F to the BCD interaction column, G to the BCD interaction column, and H to the $ABCD$ interaction column, we would have obtained a resolution IV design, generally a preferred design.

This method of assigning columns has an additional advantage: it is easy to generate the factor level combinations because the basic factors are already assigned, *e.g.*, A, B, C, and D are the basic factors. Start by assigning 0's and 1's to the basic factors using the standard order, Table 7.6.1.

TABLE 7.6.1

BASIC FACTOR LEVEL COMBINATIONS FOR A 2^{8-4}_{III} DESIGN

A	B	C	D	E	F	G	H	A	B	C	D	E	F	G	H
0	0	0	0					1	0	0	0				
0	0	0	1					1	0	0	1				
0	0	1	0					1	0	1	0				
0	0	1	1					1	0	1	1				
0	1	0	0					1	1	0	0				
0	1	0	1					1	1	0	1				
0	1	1	0					1	1	1	0				
0	1	1	1					1	1	1	1				

Now fill in the remainder of the columns using the techniques discussed in Section **7.4**. If we care to have the primary block as the fraction selected we will have to make sure $A+C+E=0$ mod 2, $A+D+F=0$ mod 2, $B+C+D+G=0$ mod 2, and $A+B+C+D+H=0$ mod 2. The easiest way is to let $E=A+C$ mod 2, $F=A+D$ mod 2, $G=B+C+D$ mod 2, and $H=A+B+C+D$ mod 2. The results are given in Table 7.6.2.

Many textbooks give all of the combinations for full factorials of size 8, 16, 32, and 64, corresponding to 3, 4, 5, and 6 basic factors. The remaining factors are assigned to columns of the full factorial on the basic factors. The values for the interaction columns get ignored if no factor is assigned to that column and become that factor's levels if a factor is assigned to that column. For example, the basic factor combinations for AB would already be filled out and the values for E $(= AC)$ would thus be assigned. Only a reordering of the columns is necessary to create the design. While this method saves some work on the part of the user, it

TABLE 7.6.2

FACTOR LEVEL COMBINATIONS FOR A 2_{III}^{8-4} DESIGN

A	B	C	D	E	F	G	H	A	B	C	D	E	F	G	H
0	0	0	0	0	0	0	0	1	0	0	0	1	1	0	1
0	0	0	1	0	1	1	1	1	0	0	1	1	0	1	0
0	0	1	0	1	0	1	1	1	0	1	0	0	1	1	0
0	0	1	1	1	1	0	0	1	0	1	1	0	0	0	1
0	1	0	0	0	0	1	1	1	1	0	0	1	1	0	0
0	1	0	1	0	1	0	0	1	1	0	1	1	0	1	1
0	1	1	0	1	0	0	0	1	1	1	0	0	1	1	1
0	1	1	1	1	1	1	1	1	1	1	1	0	0	0	0

requires tables for each power of 2. More seriously, we have seen the method misapplied so many times, by failing to order the columns, not selecting the correct columns, or trying to pick a level (high or low) from a column corresponding to an interaction, that we are hesitant to present it to anyone except statisticians. In addition, the values for the interaction columns depend on whether the combinations are generated using 0's and 1's and mod 2 arithmetic or using -1's and 1's and multiplication.

In addition to these two methods, many different tables of designs have been created. These tables are listed by resolution and categorized by the number of factors in the experiment. Appendices 14, 15, and 16 contain these designs for resolution V, IV, and III designs respectively and for five to 15 factors. These tables list the number of runs, the actual resolution (if a resolution V design requires the same number of runs as a resolution IV design, the resolution V design is listed even in the resolution IV table), the fraction, and the generators. To simplify the use of the tables, the generators are selected so the basic factors are the first factors alphabetically. This makes it easier to generate the layout sheet and slightly alter the design if necessary.

Use a resolution V design to estimate all main effects and two factor interactions. Use a resolution IV design to estimate the main effects without confounding them with any two factor interactions. Use a resolution III design when only main effects are of interest. With any of these designs, a few df not associated with main effects or two factor interactions may be left over. Use these as a crude check to see if any other interactions have an effect. This check will not prove there are no other interactions with significant effects, but does give some comfort in the assumptions.

Appendices 14, 15, and 16 are only useful for constructing general designs. If the experimenter has knowledge about specific interactions, these tables will be of no use. When such knowledge is present, the statistician must either hand calculate the generators or use some sort of computer package.

The tables in Appendices 14, 15, and 16 require considerable work but can be used to determine the entire confounding pattern as well as the design. For example, if a resolution V design for eight factors is required, we know from Appendix 14 that 64 runs are required and this is a 1/4 fraction. The generators are I = $BCDEFG$ and I = $ADEFH$ which implies a third generator of I = $ABCGH$. In standard order, the identity generator is I = $ABCGH$ = $ADEFH$ = $BCDEFG$. The complete confounding pattern is determined by mod 2 multiplication and given in Table 7.6.3. For example A is confounded with $BCGH$, $DEFH$, and $ABCDEFG$.

TABLE 7.6.3

CONFOUNDING PATTERN FOR THE 2_V^{8-2} DESIGN IN APPENDIX 14

A=BCGH=DEFH=ABCDEFG	CD=BEFG=ABDGH=ACEFH	ADG=BCDH=EFGH=ABCEF
B=ACGH=CDEFG=ABDEFH	CE=BDFG=ABEGH=ACDFH	AEG=BCEH=DFGH=ABCDF
C=ABGH=BDEFG=ACDEFH	CF=BDEG=ABFGH=ACDEH	AFG=BCFH=DEGH=ABCDE
D=AEFH=BCEFG=ABCDGH	CG=ABH=BDEF=ACDEFGH	BCD=EFG=ADGH=ABCEFH
E=ADFH=BCDFG=ABCEGH	CH=ABG=ACDEF=BDEFGH	BCE=DFG=AEGH=ABCDFH
F=ADEH=BCDEG=ABCFGH	DE=AFH=BCFG=ABCDEGH	BCF=DEG=AFGH=ABCDEH
G=ABCH=BCDEF=ADEFGH	DF=AEH=BCEG=ABCDFGH	BDE=CFG=ABFH=ACDEGH
H=ABCG=ADEF=BCDEFGH	DG=BCEF=ABCDH=AEFGH	BDF=CEG=ABEH=ACDFGH
AB=CGH=BDEFH=ACDEFG	DH=AEF=ABCDG=BCEFGH	BDG=CEF=ACDH=ABEFGH
AC=BGH=CDEFH=ABDEFG	EF=ADH=BCDG=ABCEFGH	BDH=ABEF=ACDG=CEFGH
AD=EFH=BCDGH=ABCEFG	EG=BCDF=ABCEH=ADFGH	BEF=CDG=ABDH=ACEFGH
AE=DFH=BCEGH=ABCDFG	EH=ADF=ABCEG=BCDFGH	BEG=CDF=ACEH=ABDFGH
AF=DEH=BCFGH=ABCDEG	FG=BCDE=ABCFH=ADEGH	BEH=ABDF=ACEG=CDFGH
AG=BCH=DEFGH=ABCDEF	FH=ADE=ABCFG=BCDEGH	BFG=CDE=ACFH=ABDEGH
AH=BCG=DEF=ABCDEFGH	GH=ABC=ADEFG=BCDEFH	BFH=ABDE=ACFG=CDEGH
BC=AGH=DEFG=ABCDEFH	ABD=BEFH=CDGH=ACEFG	CDH=ABDG=ACEF=BEFGH
BD=CEFG=ABEFH=ACDGH	ABE=BDFH=CEGH=ACDFG	CEH=ABEG=ACDF=BDFGH
BE=CDFG=ABDFH=ACEGH	ABF=BDEH=CFGH=ACDEG	CFH=ABFG=ACDE=BDEGH
BF=CDEG=ABDEH=ACFGH	ACD=BDGH=CEFH=ABEFG	DGH=ABCD=AEFG=BCEFH
BG=ACH=CDEF=ABDEFGH	ACE=BEGH=CDFH=ABDFG	EGH=ABCE=ADFG=BCDFH
BH=ACG=ABDEF=CDEFGH	ACF=BFGH=CDEH=ABDEG	FGH=ABCF=ADEG=BCDEH

As indicated by the mere size of Table 7.6.3, the generation of the complete confounding pattern is a monumental task. This is a task best left to a computer as most people do not have the patience to complete such a table and there is a high probability of making a mistake.

By construction of the designs in the appendices, the basic factors are the first factors in the experiment, *e.g.*, A through F in the 2_{IV}^{8-2} design. Furthermore, each generator in the appendix defines a new factor in terms of the basic factors. This is the quickest way to generate the factor level combinations as long as the primary block will do. For example, the first generator is $I = BCDEFG$ so $G = BCDEF$ and we can let $G = B+C+D+E+F$ mod 2. In a similar manner, $H = A+D+E+F$ mod 2. We write a complete factorial in terms of A through F and use the two identities to compute the values for G and H. The combinations are given in Table 7.6.4.

The final method for creating fractional designs is through the use of computers. As is to be expected, computer packages range in sophistication. Many are basic tabulations similar to Appendices 14 through 16. These packages require, as input, the number of factors and the desired resolution. The most sophisticated program we have seen to date is PROC FACTEX in the SAS system. This program requires, as input, the factors, a list of estimable terms, and a list of non-negligible terms. The program then performs a comprehensive search using a method like Franklin (1985), much like the basic factor technique given earlier, to find the smallest design that does not confound any of the estimable terms with themselves, does not confound the estimable terms with the non-negligible terms, but can confound non-negligible terms with themselves. This is the natural extension of the resolution principle given earlier. For example, a resolution III design would have all main effects in the estimable list and the non-negligible list empty. A resolution IV design would have all main effects in the estimable set and all two factor interactions in the non-negligible set. The Franklin method is more general since the sets can contain specific terms, not necessarily all terms of a certain order.

A difficulty with the SAS system is that it does not accept generators as input. It takes considerable effort to make the system choose the proper generators and the proper fraction in order to reproduce a given design. The procedure is illustrated in Problem **7.19** for those interested.

It is difficult to illustrate the use of computer packages as there are so many of them currently on the market. Our suggestion is to use whatever package is available to you and check the results with the examples worked in Section **7.8**.

TABLE 7.6.4

FACTOR LEVEL COMBINATIONS FOR THE 2_V^{8-2} DESIGN IN APPENDIX 14

A	B	C	D	E	F	G	H	A	B	C	D	E	F	G	H
0	0	0	0	0	0	0	0	1	0	0	0	0	0	0	1
0	0	0	0	0	1	1	1	1	0	0	0	0	1	1	0
0	0	0	0	1	0	1	1	1	0	0	0	1	0	1	0
0	0	0	0	1	1	0	0	1	0	0	0	1	1	0	1
0	0	0	1	0	0	1	1	1	0	0	1	0	0	1	0
0	0	0	1	0	1	0	0	1	0	0	1	0	1	0	1
0	0	0	1	1	0	0	0	1	0	0	1	1	0	0	1
0	0	0	1	1	1	1	1	1	0	0	1	1	1	1	0
0	0	1	0	0	0	1	0	1	0	1	0	0	0	1	1
0	0	1	0	0	1	0	1	1	0	1	0	0	1	0	0
0	0	1	0	1	0	0	1	1	0	1	0	1	0	0	0
0	0	1	0	1	1	1	0	1	0	1	0	1	1	1	1
0	0	1	1	0	0	0	1	1	0	1	1	0	0	0	0
0	0	1	1	0	1	1	0	1	0	1	1	0	1	1	1
0	0	1	1	1	0	1	0	1	0	1	1	1	0	1	1
0	0	1	1	1	1	0	1	1	0	1	1	1	1	0	0
0	1	0	0	0	0	1	0	1	1	0	0	0	0	1	1
0	1	0	0	0	1	0	1	1	1	0	0	0	1	0	0
0	1	0	0	1	0	0	1	1	1	0	0	1	0	0	0
0	1	0	0	1	1	1	0	1	1	0	0	1	1	1	1
0	1	0	1	0	0	0	1	1	1	0	1	0	0	0	0
0	1	0	1	0	1	1	0	1	1	0	1	0	1	1	1
0	1	0	1	1	0	1	0	1	1	0	1	1	0	1	1
0	1	0	1	1	1	0	1	1	1	0	1	1	1	0	0
0	1	1	0	0	0	0	0	1	1	1	0	0	0	0	1
0	1	1	0	0	1	1	1	1	1	1	0	0	1	1	0
0	1	1	0	1	0	1	1	1	1	1	0	1	0	1	0
0	1	1	0	1	1	0	0	1	1	1	0	1	1	0	1
0	1	1	1	0	0	1	1	1	1	1	1	0	0	1	0
0	1	1	1	0	1	0	0	1	1	1	1	0	1	0	1
0	1	1	1	1	0	0	0	1	1	1	1	1	0	0	1
0	1	1	1	1	1	1	1	1	1	1	1	1	1	1	0

7.7 RESTRICTION ERRORS AND BLOCKING

Most textbooks devote an entire chapter to the concept of running experiments in homogeneous blocks. As a result, many students feel that blocking and fractionation are separate concepts. In this text, we simply treat blocks as a factor in the experiment that contains a restriction error due to the constrained run order. The concept of fractionation and the concept of restricted run order are separate, but can occur in the same experiment. We will demonstrate simple restriction errors in this chapter. More complicated restriction errors will be illustrated in the Chapter **8**.

Since we have already covered each concept individually, we will use a couple of examples to illustrate how to use the concepts together.

For this example, there are five fixed factors, A through E, with interest in all main effects and two factor interactions and with the property that it is very expensive to change levels of factor A. Since it is expensive to change levels of A, we will select a level of A and run all those combinations before running the combinations associated with the other level of A. We recognize this as a restriction error on A, the restriction error concept being the same whether we run all factor level combinations or some fraction of the possible combinations.

Using Appendix 14, we see there are 16 runs in a 2_V^{5-1} experiment, the basic factors are A through D, $\mathrm{I} = ABCDE$, and a design is generated by $E = A + B + C + D$ mod 2. We randomly pick a level of A and then randomize the eight combinations involving that level before selecting the other level of A. A typical layout sheet is given in Table 7.7.1.

TABLE 7.7.1

LAYOUT SHEET FOR A 2_V^{5-1}, RESTRICTION ERROR ON A

run #	A	B	C	D	E	run #	A	B	C	D	E
1	1	0	1	1	1	9	0	0	0	0	0
2	1	1	1	0	1	10	0	0	1	1	0
3	1	0	0	1	0	11	0	0	1	0	1
4	1	0	0	0	1	12	0	1	0	1	0
5	1	0	1	0	0	13	0	1	1	0	0
6	1	1	0	1	1	14	0	1	0	0	1
7	1	1	1	1	0	15	0	1	1	1	1
8	1	1	0	0	0	16	0	0	0	1	1

The computation of the ANOVA table is straightforward. The final result is given in Table 7.7.2. You should try to reproduce this table to make sure you still understand how to compute EMS for fractional designs. If you have any trouble reproducing this table, go back and review Section **7.3**.

As is obvious from Table 7.7.2, there are no tests in this design because there are no df for error. We must either redesign this experiment to get tests, assume some interactions negligible, or use half normal plots to pool some terms after the data has been collected. The problem with half normal plots is that we do not know ahead of time what we can expect from the experiment.

TABLE 7.7.2

ANOVA FOR A 2_V^{5-1} DESIGN, RESTRICTION ERROR ON A

Source	df	EMS
A &/or $BCDE$	1	$\sigma^2 + \Phi(BCDE) + 8\sigma_\delta^2 + 8\Phi(A)$
δ_A	0	—
B &/or $ACDE$	1	$\sigma^2 + \Phi(ACDE) + 8\Phi(B)$
C &/or $ABDE$	1	$\sigma^2 + \Phi(ABDE) + 8\Phi(C)$
D &/or $ABCE$	1	$\sigma^2 + \Phi(ABCE) + 8\Phi(D)$
E &/or $ABCD$	1	$\sigma^2 + \Phi(ABCD) + 8\Phi(E)$
AB &/or CDE	1	$\sigma^2 + 2\Phi(CDE) + 4\Phi(AB)$
AC &/or BDE	1	$\sigma^2 + 2\Phi(BDE) + 4\Phi(AC)$
AD &/or BCE	1	$\sigma^2 + 2\Phi(BCE) + 4\Phi(AD)$
AE &/or BCD	1	$\sigma^2 + 2\Phi(BCD) + 4\Phi(AE)$
BC &/or ADE	1	$\sigma^2 + 2\Phi(ADE) + 4\Phi(BC)$
BD &/or ACE	1	$\sigma^2 + 2\Phi(ACE) + 4\Phi(BD)$
BE &/or ACD	1	$\sigma^2 + 2\Phi(ACD) + 4\Phi(BE)$
CD &/or ABE	1	$\sigma^2 + 2\Phi(ABE) + 4\Phi(CD)$
CE &/or ABD	1	$\sigma^2 + 2\Phi(ABD) + 4\Phi(CE)$
DE &/or ABC	1	$\sigma^2 + 2\Phi(ABC) + 4\Phi(DE)$
ε	0	—

The effect of the restriction error on the previous example, adding $8\sigma_\delta^2$ to EMS(A &/or $BCDE$), was very predictable. As this example will illustrate, using intuition to place restriction errors can easily lead you astray with fractional designs. We will again use the resolution V design on the factors A through E but A will be random and the restriction error will be on the ABC interaction. (That is, randomly select an ABC combination and randomly run all factor level combinations having this particular combination of ABC before randomly selecting another ABC combination.) The factor level combinations are the same as in the previous example, with a different run order reflecting the new restrictions on randomization. However, the EMS table, Table 7.7.3, is completely different from Table 7.7.2. This is a result of A being random and the different placement of the restriction error. Note that, even without assumptions, there are significant only tests on the fixed factors B through E. This is true despite the restriction error appearing in the EMS for A, B, C, AB, AC, BC, and DE and despite the confounding caused by fractionating the design. The restriction error for DE is a result of the confounding with ABC. One must always be on the lookout for tests on main effects despite the confounding that is present in fractional designs.

TABLE 7.7.3

ANOVA FOR 2_V^{5-1}, A RANDOM, RESTRICTION ERROR ON ABC

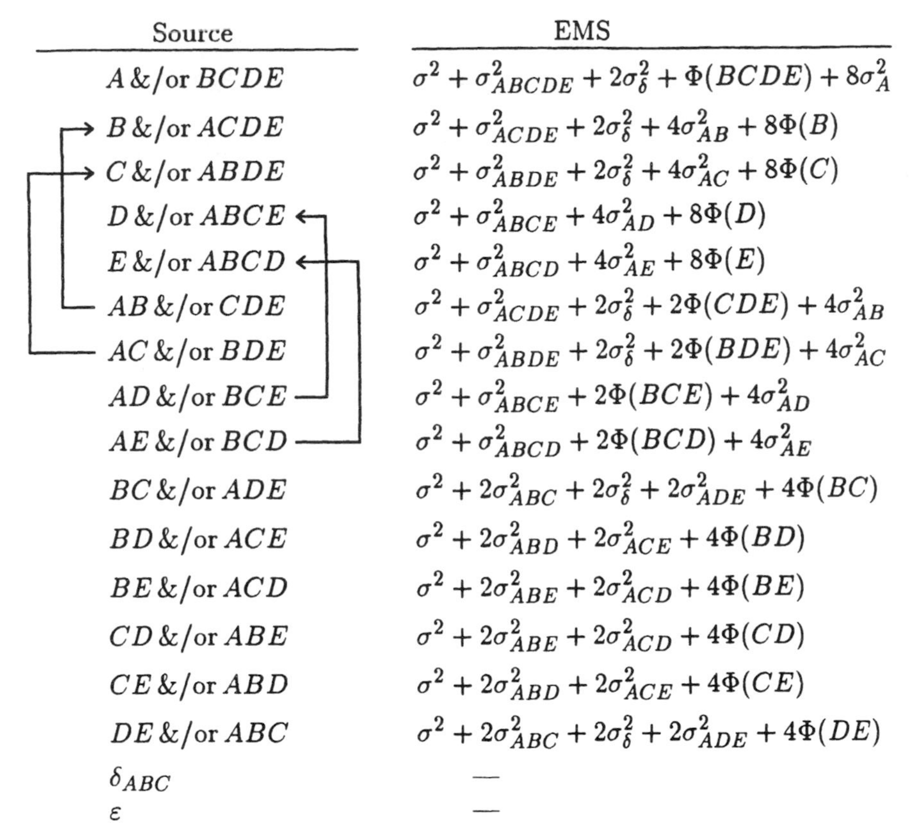

Source	EMS
A &/or $BCDE$	$\sigma^2 + \sigma^2_{ABCDE} + 2\sigma^2_\delta + \Phi(BCDE) + 8\sigma^2_A$
B &/or $ACDE$	$\sigma^2 + \sigma^2_{ACDE} + 2\sigma^2_\delta + 4\sigma^2_{AB} + 8\Phi(B)$
C &/or $ABDE$	$\sigma^2 + \sigma^2_{ABDE} + 2\sigma^2_\delta + 4\sigma^2_{AC} + 8\Phi(C)$
D &/or $ABCE$	$\sigma^2 + \sigma^2_{ABCE} + 4\sigma^2_{AD} + 8\Phi(D)$
E &/or $ABCD$	$\sigma^2 + \sigma^2_{ABCD} + 4\sigma^2_{AE} + 8\Phi(E)$
AB &/or CDE	$\sigma^2 + \sigma^2_{ACDE} + 2\sigma^2_\delta + 2\Phi(CDE) + 4\sigma^2_{AB}$
AC &/or BDE	$\sigma^2 + \sigma^2_{ABDE} + 2\sigma^2_\delta + 2\Phi(BDE) + 4\sigma^2_{AC}$
AD &/or BCE	$\sigma^2 + \sigma^2_{ABCE} + 2\Phi(BCE) + 4\sigma^2_{AD}$
AE &/or BCD	$\sigma^2 + \sigma^2_{ABCD} + 2\Phi(BCD) + 4\sigma^2_{AE}$
BC &/or ADE	$\sigma^2 + 2\sigma^2_{ABC} + 2\sigma^2_\delta + 2\sigma^2_{ADE} + 4\Phi(BC)$
BD &/or ACE	$\sigma^2 + 2\sigma^2_{ABD} + 2\sigma^2_{ACE} + 4\Phi(BD)$
BE &/or ACD	$\sigma^2 + 2\sigma^2_{ABE} + 2\sigma^2_{ACD} + 4\Phi(BE)$
CD &/or ABE	$\sigma^2 + 2\sigma^2_{ABE} + 2\sigma^2_{ACD} + 4\Phi(CD)$
CE &/or ABD	$\sigma^2 + 2\sigma^2_{ABD} + 2\sigma^2_{ACE} + 4\Phi(CE)$
DE &/or ABC	$\sigma^2 + 2\sigma^2_{ABC} + 2\sigma^2_\delta + 2\sigma^2_{ADE} + 4\Phi(DE)$
δ_{ABC}	—
ε	—

Note: tests are significant only tests on the main effects, detectable difference $\Delta = 3.67$, LARGE, and all df = 1.

7.8 EXAMPLES

Example 1. An experiment was performed in a laboratory to determine what factors influence the film build on painted panels. A total of eight fixed factors were selected for study: booth humidity (H), substrate temperature (S), fluid flow rate (R), target distance (D), booth temperature (T), base coat temperature (B), atomizing air pressure (A), and fan air pressure (F). Information was required on the main effects and the AF and HT interactions, which were thought to be important. Not much more was known about the remaining interactions. After considerable discussion, it was decided that the remaining two factor interactions could be important but it was unlikely that any three factor interaction was important. Therefore, a design was required that did not confound any of the main effects or the AF or HT interaction with any other two factor interaction but could confound the remaining two factor interactions among themselves.

Appendix 15 indicates that a resolution IV design requires 16 runs. We need to do better than a resolution IV design since neither of AF or HT can be confounded with any two factor interactions. It was hoped that these requirements could still be met in a 16 run design. Unfortunately, the assignments were impossible and a 32 run design was required. (This is a long hard task by hand assignment methods. We had access to a computer package capable of finding the smallest design meeting our requirements.) A set of identity generators for this 2_{IV}^{8-3} design is I $= RDTB = HSDTA = SRTF$, making the first five factors the basic factors and expressing the last three factors in terms of the first five. The confounding pattern, up to third order interactions, is given in Table 7.8.1. Even though the higher order interactions are not displayed, we know that each primary term is confounded with seven other terms. We see that all main effects, AF, and HT are confounded with third and higher order interactions.

TABLE 7.8.1

CONFOUNDING FOR EXAMPLE 1, UP TO THIRD ORDER TERMS

H	$HT = SDA = BAF$	$RD = TB = HAF$
$S = RTF = DBF$	$HB = SRA = TAF$	$RB = DT = HSA$
$R = STF = DTB$	$HA = SRB = SDT$	$RA = HSB = HDF$
$D = SBF = RTB$	$= RDF = TBF$	$DA = HST = HRF$
$T = SRF = RDB$	$HF = RDA = TBA$	$TA = HSD = HBF$
$B = SDF = RDT$	$SR = TF = HBA$	$BA = HSR = HTF$
A	$SD = BF = HTA$	$AF = HRD = HTB$
$F = SRT = SDB$	$ST = RF = HDA$	$HSF = HRT = HDB$
$HS = RBA = DTA$	$SB = DF = HRA$	$SRD = STB = RBF$
$HR = SBA = DAF$	$SA = HRB = HDT$	$= DTF$
$HD = STA = RAF$	$SF = RT = DB$	$SAF = RTA = DBA$

The factor level combinations are given, along with the collected data, in Table 7.8.3. The experimenters felt comfortable in assuming all three factor and higher order interactions were negligible. This led to the ANOVA table given in Table 7.8.2. Notice that the three degrees of freedom associated with third order and higher order interactions are pooled to test the remaining confounded terms. As a test on the assumption that three factor and higher order interactions are negligible, each of the remaining three factor interactions can be pooled to test the third three factor interaction. While not a proof that all three factor interactions are negligible, this test does offer some evidence.

The detectable sizes for the tests are also given in Table 7.8.2. Since we pooled the three terms corresponding to third and higher order interactions, there are three df for the tests. The tests for each of the individual third order terms is formed by the other third order terms, so have two df. In general, there is little that can be done about the detectability, Δ, in fractional designs. We cannot change levels of any factors since the fraction depends on the fact that all factors have two levels. And the size of the experiment was chosen to minimally meet the experimenter's needs. This does not imply the detectability is unimportant in fractional designs. It is still as important as ever and should be calculated. Rather, the experimenter should select actual levels of the factors to reflect the detectability in the design or should redesign the experiment to estimate more terms, forcing a larger experiment and smaller detectability. In the case of selecting actual levels, the levels can be close together if Δ is small but must be spread quite far apart when Δ is large. Spreading the actual level quite far apart may decrease the usefulness of the experiment.

The levels of the factors were chosen in accordance with the detectability (spread far enough apart to hopefully cause a change of 1.25 or 1.77 σ in the response variable), the factor level combinations were generated and randomized, and the experiment was run. The results of the experiment are given in Table 7.8.3.

The first step in the analysis is a quick plot of the raw data to verify that no spurious points were entered into the computer. The raw data plot is given in Figure 7.2. With even a quick glance, it is obvious that no data point is way out of line with the other data points. Hopefully, you have access to a friendly computer package that generates these plots as they take about 15 minutes to draw by hand.

The next step is a formal analysis of the data. Several techniques have been given earlier in this chapter. Any of these techniques should give the results in Table 7.8.4. All terms are tested with the three pooled third order interaction terms. Only three of the main effects were significant, H, R, and D with R being highly significant. Since there are no repeats, many people follow the formal analysis with the half normal plot of the mean squares given in Figure 7.3. As you can see, the half normal plots also indicate that H, R, and D are the most significant effects. An ad-

TABLE 7.8.2

EXAMPLE 1, IGNORING 4TH & HIGHER ORDER INTERACTIONS

Source	df	Δ	EMS
→ H	1	1.25	$\sigma^2 + 16\Phi(H)$
→ S	1	1.25	$\sigma^2 + 4\Phi(RTF) + 4\Phi(DBF) + 16\Phi(S)$
→ R	1	1.25	$\sigma^2 + 4\Phi(STF) + 4\Phi(DTB) + 16\Phi(R)$
→ D	1	1.25	$\sigma^2 + 4\Phi(SBF) + 4\Phi(RTB) + 16\Phi(D)$
→ T	1	1.25	$\sigma^2 + 4\Phi(SRF) + 4\Phi(RDB) + 16\Phi(T)$
→ B	1	1.25	$\sigma^2 + 4\Phi(SDF) + 4\Phi(RDT) + 16\Phi(B)$
→ A	1	1.25	$\sigma^2 + 16\Phi(A)$
→ F	1	1.25	$\sigma^2 + 4\Phi(SRT) + 4\Phi(SDB) + 16\Phi(F)$
→ HS	1	1.77	$\sigma^2 + 4\Phi(RBA) + 4\Phi(DTA) + 8\Phi(HS)$
→ HR	1	1.77	$\sigma^2 + 4\Phi(SBA) + 4\Phi(DAF) + 8\Phi(HR)$
→ HD	1	1.77	$\sigma^2 + 4\Phi(STA) + 4\Phi(RAF) + 8\Phi(HD)$
→ HT	1	1.77	$\sigma^2 + 4\Phi(SDA) + 4\Phi(BAF) + 8\Phi(HT)$
→ HB	1	1.77	$\sigma^2 + 4\Phi(SRA) + 4\Phi(TAF) + 8\Phi(HB)$
→ HA	1	1.77	$\sigma^2 + 4\Phi(SRB) + 4\Phi(SDT) + 4\Phi(RDF) + 4\Phi(TBF) + 8\Phi(HA)$
→ HF	1	1.77	$\sigma^2 + 4\Phi(RDA) + 4\Phi(TBA) + 8\Phi(HF)$
→ SR &/or TF	1	1.77	$\sigma^2 + 4\Phi(HBA) + 8\Phi(SR) + 8\Phi(TF)$
→ SD &/or BF	1	1.77	$\sigma^2 + 4\Phi(HTA) + 8\Phi(SD) + 8\Phi(BF)$
→ ST &/or RF	1	1.77	$\sigma^2 + 4\Phi(HDA) + 8\Phi(ST) + 8\Phi(RF)$
→ SB &/or DF	1	1.77	$\sigma^2 + 4\Phi(HRA) + 8\Phi(SB) + 8\Phi(DF)$
→ SA	1	1.77	$\sigma^2 + 4\Phi(HRB) + 4\Phi(HDT) + 8\Phi(SA)$
→ SF &/or RT &/or DB	1	1.77	$\sigma^2 + 8\Phi(SF) + 8\Phi(RT) + 8\Phi(DB)$
→ RD &/or TB	1	1.77	$\sigma^2 + 4\Phi(HAF) + 8\Phi(RD) + 8\Phi(TB)$
→ RB &/or DT	1	1.77	$\sigma^2 + 4\Phi(HSA) + 8\Phi(RB) + 8\Phi(DT)$
→ RA	1	1.77	$\sigma^2 + 4\Phi(HSB) + 4\Phi(HDF) + 8\Phi(RA)$
→ DA	1	1.77	$\sigma^2 + 4\Phi(HST) + 4\Phi(HRF) + 8\Phi(DA)$
→ TA	1	1.77	$\sigma^2 + 4\Phi(HSD) + 4\Phi(HBF) + 8\Phi(TA)$
→ BA	1	1.77	$\sigma^2 + 4\Phi(HSR) + 4\Phi(HTF) + 8\Phi(BA)$
→ AF	1	1.77	$\sigma^2 + 4\Phi(HRD) + 4\Phi(HTB) + 8\Phi(AF)$
— Third order terms	1	3.40	$\sigma^2 + 4\Phi(HSF) + 4\Phi(HRT) + 4\Phi(HDB)$
— Third order terms	1	3.40	$\sigma^2 + 4\Phi(SAF) + 4\Phi(RTA) + 4\Phi(DBA)$
— Third order terms	1	3.40	$\sigma^2 + 4\Phi(SRD) + 4\Phi(STB) + 4\Phi(RBF) + 4\Phi(DTF)$
ε	0	—	—

Note: all three factor and higher order interactions are assumed negligible.

TABLE 7.8.3

FACTOR LEVEL COMBINATIONS AND RESPONSES FOR EXAMPLE 1.

H	S	R	D	T	B	A	F	Film Build
70	100	0	12	70	65	50	50	0.15
50	100	0	15	90	65	40	40	0.16
70	70	0	12	70	65	40	40	0.19
50	70	20	12	70	85	50	50	0.38
70	100	20	15	90	85	50	50	0.22
70	70	20	12	90	65	50	40	0.35
50	100	20	12	90	65	50	50	0.30
50	70	20	15	90	85	50	40	0.26
50	100	20	12	70	85	40	40	0.30
50	70	0	12	90	85	40	50	0.15
70	70	20	15	70	65	50	50	0.27
70	100	0	15	90	65	50	40	0.16
70	70	20	15	90	85	40	40	0.28
70	100	0	15	70	85	40	50	0.16
50	100	20	15	90	85	40	50	0.27
70	100	20	15	70	65	40	40	0.27
50	100	20	15	70	65	50	40	0.28
50	70	0	15	70	85	40	40	0.22
50	100	0	15	70	85	50	50	0.20
70	100	20	12	70	85	50	40	0.32
70	70	0	15	90	65	40	50	0.13
50	70	0	15	90	65	50	50	0.16
50	70	0	12	70	65	50	40	0.34
50	70	20	12	90	65	40	40	0.35
70	70	20	12	70	85	40	50	0.28
50	100	0	12	90	85	50	40	0.21
50	100	0	12	70	65	40	50	0.20
70	100	20	12	90	65	40	50	0.27
70	70	0	12	90	85	50	50	0.14
50	70	20	15	70	65	40	50	0.27
70	70	0	15	70	85	50	40	0.15
70	100	0	12	90	85	40	40	0.15

vantage of the half normal plot is that it did not require any assumptions to indicate the most important effects. A disadvantage is that one cannot assess significance from half normal plots.

As is usual with fractionated designs, many terms were insignificant. To increase the detectability of the tests, the sometimes pooling rules were used to generate Table 7.8.5. As you can see, all main effects except B and A and a couple of interactions turned out to be significant. These became significant because the pooling added 13 df to the denominator and drastically decreased the detectable size (increased the sensitivity) of the tests.

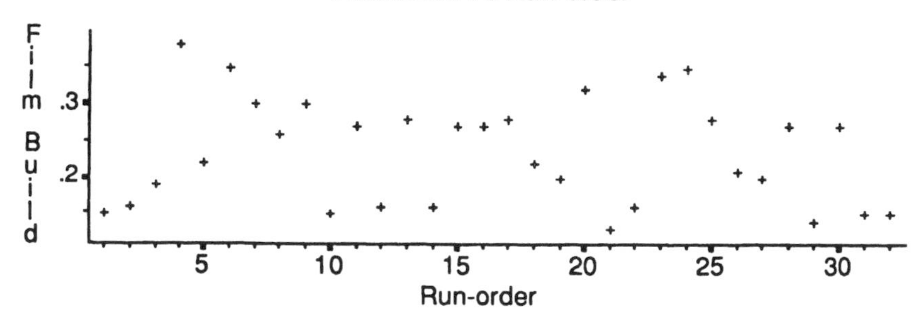

Figure 7.2 Raw Data Plot of Film Build by Run Order.

TABLE 7.8.4

ANOVA FOR EXAMPLE 1, 3 FACTOR INTERACTIONS ASSUMED NEGLIGIBLE, IGNORING 4TH & HIGHER ORDER INTERACTIONS

Term	df	MS	F_{calc}	p-value	Term	df	MS	F_{calc}	p-value
H	1	.0098	14.52*	.0318	SR &/or TF	1	.0005	0.17	.7105
S	1	.0028	4.17	.1339	SD &/or BF	1	.0021	3.13	.1750
R	1	.1012	150.00**	.0012	ST &/or RF	1	.0006	0.91	.4111
D	1	.0120	17.80*	.0244	SB &/or DF	1	.0018	2.67	.2001
T	1	.0055	8.17	.0647	SA	1	.0005	0.67	.4740
B	1	.0008	1.18	.3359	SF, RT &/or DB	1	.0025	3.63	.1529
A	1	.0018	2.67	.2010	RD &/or TB	1	.0018	2.67	.2010
F	1	.0061	8.96	.0580	RB &/or DT	1	.0001	0.17	.7105
HS	1	.0005	0.67	.4740	RA	1	.0001	0.17	.7105
HR	1	.0021	3.13	.1750	DA	1	.0041	6.00	.0917
HD	1	.0013	1.85	.2668	TA	1	.0008	1.19	.3559
HT	1	.0018	2.67	.2010	BA	1	.0003	0.46	.5451
HB	1	.0000	0.02	.9004	AF	1	.0001	0.17	.7105
HA	1	.0010	1.50	.3081	3rd order terms	1	.0010	2.00	.2929
HF	1	.0001	0.17	.7105	3rd order terms	1	.0000	0.00	1.000
					3rd order terms	1	.0010	2.00	.2929

* Significant at the .05 level. ** Significant at the .01 level.

The interactions that turned out to be significant are SF &/or RT &/or DB &/or higher order terms and DA &/or higher order terms. We can compare these to the interactions thought to be important before the experiment was run: AF and HT. The two sets are completely different.

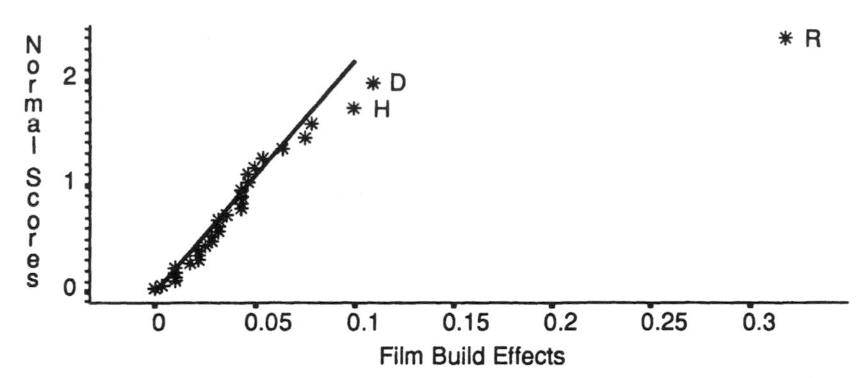

Figure 7.3 Half Normal Effects Plot with Largest Effects Labeled.

TABLE 7.8.5

ANOVA FOR EXAMPLE 1, SOMETIMES POOLING RULES APPLIED

Term	df	MS	F_{calc}	p-value	Term	df	MS	F_{calc}	p-value
H	1	.0098	20.04**	.0004	HR	1	.0021	4.32*	.0290
S	1	.0028	5.75*	.0290	HT	1	.0018	3.68	.0731
R	1	.1013	207.03**	.0000	SD &/or BF	1	.0021	4.32	.0541
D	1	.0121	24.56**	.0001	SB &/or DF	1	.0018	3.68	.0731
T	1	.0055	11.27**	.0040	SF, RT &/or DB	1	.0025	5.01*	.0398
B	1	.0080	1.64	.2191	RD &/or TB	1	.0018	3.68	.0731
A	1	.0018	3.68	.0731	DA	1	.0041	8.28*	.0109
F	1	.0061	12.37**	.0029	ε_{pooled}	16	.0005	—	—

* Significant at the .05 level. ** Significant at the .01 level.

Our experience indicates that this is the case the majority of the time an experiment is run. For this reason, we feel it is the statistician's responsibility to try to influence the client into running a larger experiment to test more and higher order interactions.

To get an idea of the relative contribution of each term in the model, we compute the variance components and percent contribution. These are given in Table 7.8.6. This is almost an ideal situation. Of the eight original factors, R alone accounts for almost 60% of the variation. R, D, and H account for over 70% of the variation. The error term, ε, accounts for only 5% of the variation. This means virtually all of the important factors have been considered in the experiment. Since only one interaction has more than a 2% contribution, the relation between the factors and the response is likely to be simple.

TABLE 7.8.6

VARIANCE COMPONENTS AND % CONTRIBUTION FOR EXAMPLE 1

Term	Estimate	% Contribution
$\Phi(R)$	0.0630	59%
$\Phi(D)$	0.0072	7%
$\Phi(H)$	0.0058	5%
ε	0.0049	5%
$\Phi(DA)$	0.0045	4%
$\Phi(F)$	0.0035	3%
$\Phi(T)$	0.0031	3%
$\Phi(SF) + \Phi(RT) + \Phi(DB)$	0.0025	2%
$\Phi(HR)$	0.0020	2%
$\Phi(SD) + \Phi(BF)$	0.0020	2%
$\Phi(HT)$	0.0016	2%
$\Phi(SB) + \Phi(DF)$	0.0016	2%
$\Phi(RD) + \Phi(TB)$	0.0016	2%
$\Phi(S)$	0.0015	1%
$\Phi(A)$	0.0008	1%
$\Phi(B)$	0.0002	0%

We can get a similar impression by looking at main effects plots in Figure 7.4 and interaction plots in Figure 7.5. All 95% confidence limits are included in the plots. For example, the slope of the main effect R is the largest, indicating R has the largest effect. When looking at Figure 7.5, keep in mind that the interaction is really the difference in slopes of the two lines, not the crossing point (if they cross at all). This makes it quite difficult to compare the effect of an interaction to the effect of a main effect. With sufficient experience, one can learn how to compare the differences at the endpoints of the interaction plots with the main effects plots in order to get a good idea of the relative contribution of each term in the model.

Example 2. Robust Design. A new idea, fast becoming popular within industry, is referred to as *robust design.* The idea behind the principle is to label the factors affecting a product or process as either control or

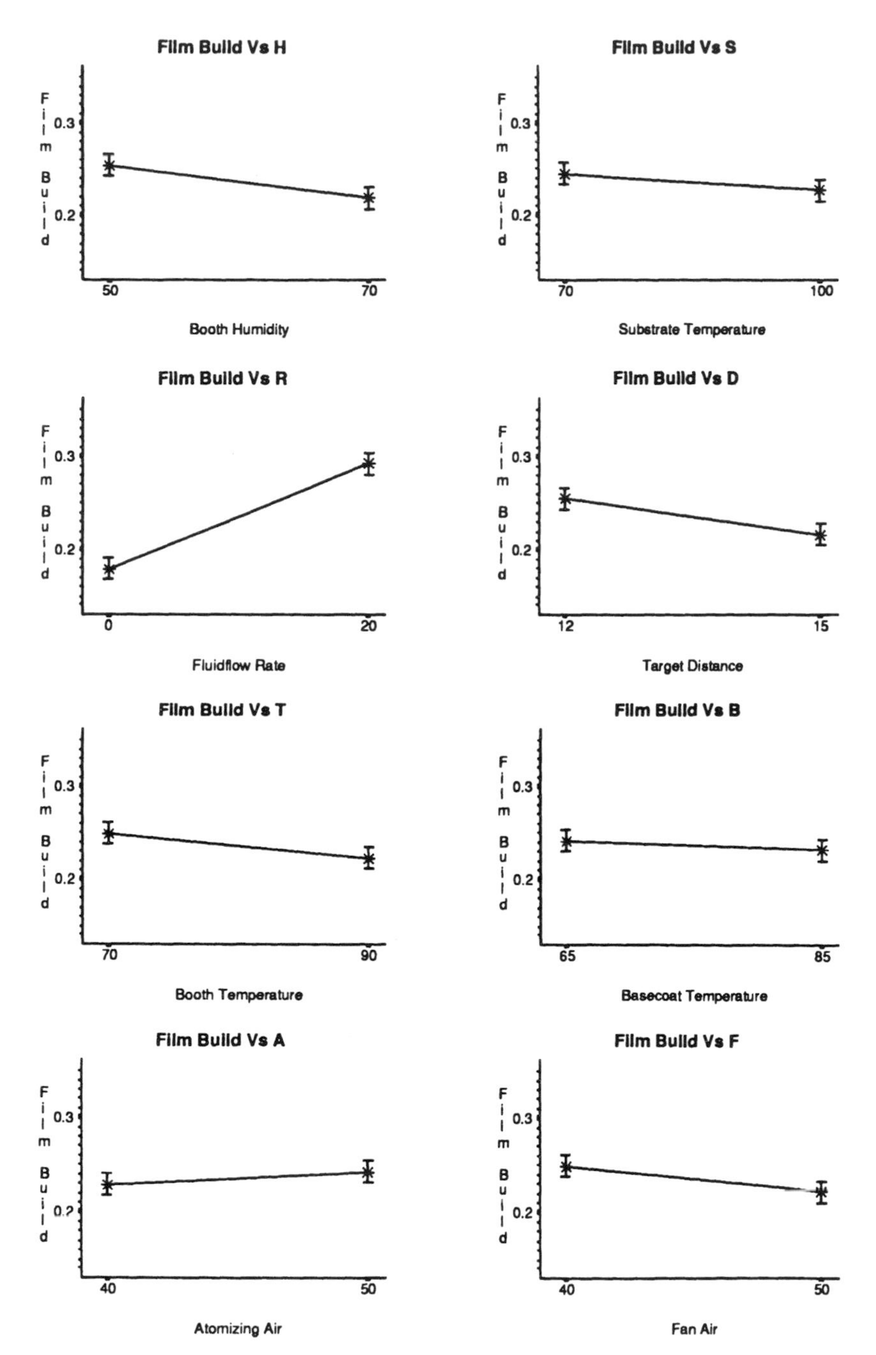

Figure 7.4 Main effects plots showing relative contribution of effects.

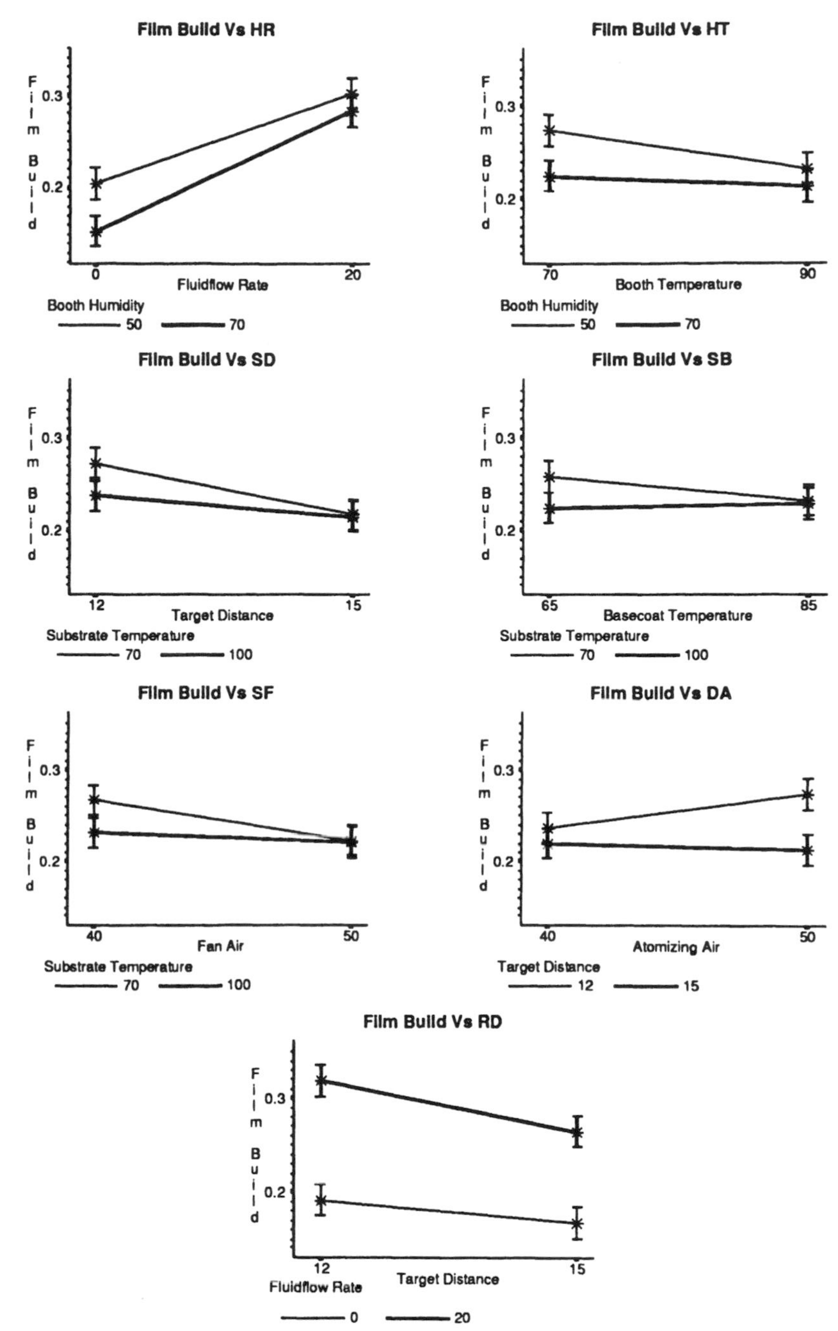

Figure 7.5 Interaction plots showing relative contribution of effects.

noise. Control factors are directly controllable by the factory, things such as part thickness, machine tool pressure, line speed, *etc.* Noise factors either cannot be controlled by the factory or are too expensive to be controlled by the factory; for example actual part thickness (which can only be controlled by expensive 100% sorting), temperature, humidity, customer usage, *etc.* The idea is to find the control settings that minimize the influence of the noise factors, thereby assuring the product behaves as consistently as possible despite the values of the factors that cannot be controlled.

We will illustrate the principle with a relatively simple example and develop more detail through Problems **7.25** through **7.28** at the end of this chapter. These should be pursued at the reader's option and all have the label *robust design*.

The intermediate shaft steering column, Figure 7.6, connects the steering wheel to the power steering motor. If the yoke and the tube become loose, undesirable play is detected in the steering wheel. If the yoke and the tube are too tight, the assembly may not come apart as desired in a crash or for repairs. The part is assembled by slipping the tube over the yolk and crimping the end of the tube into the pockets.

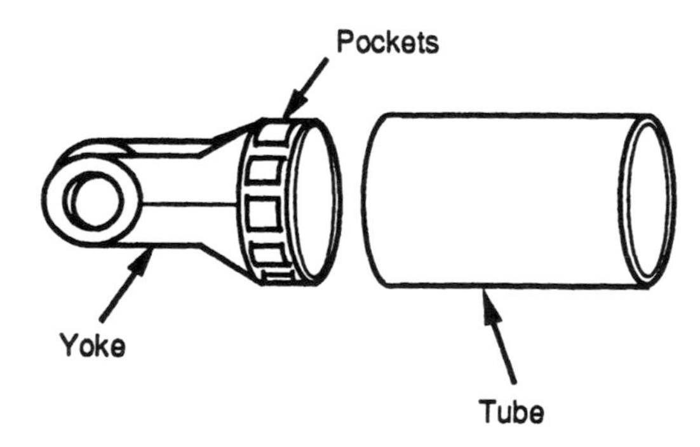

Figure 7.6 Intermediate shaft steering column.

A torque to separate of 30 is considered ideal. Common practice is to specify lower and upper specification limits around the target. Anywhere within the spec limits is considered acceptable, implying the loss function on the left side of Figure 7.7. A more realistic loss function is the quadratic loss function given on the right side of Figure 7.7. This implies any deviation from the target has an associated loss.

The functional form of the loss function is $K(Y-T)^2$ for some constant K, target value T, and response variable Y. Deviations from the target

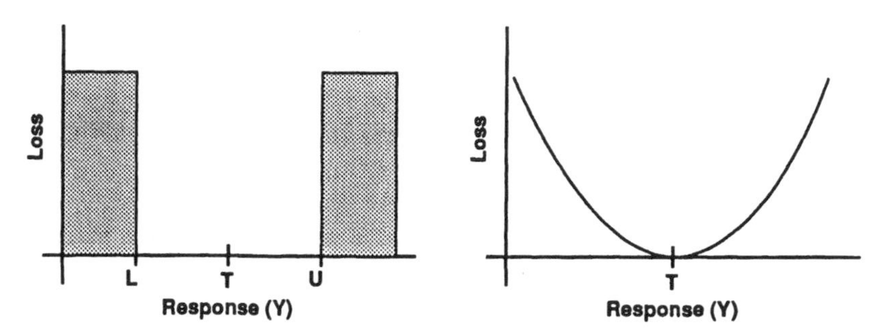

Figure 7.7 Comparison of specification and quadratic loss functions. L = lower spec limit, U = upper spec limit, T = target value

are caused by the noise variables so we wish to minimize the expected loss $\mathrm{E}[K(Y-T)^2]$ where the expectation is taken over all possible values of the noise variables. K plays no role in the minimization so will be taken as 1. The most robust design is the control factor combination that minimizes loss because bigger deviations from target have bigger losses.

The control factors are pocket depth, D; yoke concentricity, C; tube length, L; and power setting, P. (Refer to Figure 7.6) These are all fixed and can be set in the factory. The noise factors are clearance, C, and line voltage, V, also fixed. Clearance refers to the actual difference between the inside diameter of the tube and the outside diameter of the yolk. Line voltage will depend on what other machines in the factory are simultaneously firing. These noise factors are very difficult to control during daily operations but need to be controlled while running the experiment.

To keep this experiment simple to design and easy to understand, we will fractionate the control factors and run every combination of the noise factors in conjunction with the selected control factor combinations. More complicated situations will be discussed in the problems. Since there are 2 noise factors, 4 experimental runs are required to compute the loss. To keep the experiment to a reasonable size, only 8 control factor combinations will be run. The obvious choice for the identity generator is $\mathrm{I} = DCLP$, resulting in a 2_{IV}^{4-1} design. Among the control factors, main effects will be confounded with three factor interactions and two factor interactions will be confounded with two factor interactions:

$$
\begin{array}{ll}
D = CLP & DC = LP \\
C = DLP & DL = CP \\
L = CDP & DP = CL \\
P = DCL &
\end{array}
$$

A total of $8 \times 4 = 32$ (8 control combinations $\times$ 4 for the noise combinations) will be required in the experiment.

There are no degrees of freedom left for testing any of the terms in the model. We will use half normal plots to find insignificant terms that can be pooled to form an error term. This can only be done after the data has been collected so the tests and detectability are unknown prior to the collection of the data. There is no guarantee that tests will be available at all.

The basic data is given in Table 7.8.7, not in standard format. However, the purpose of the experiment is not to model the basic data, but to model and minimize the expected loss function, the expectation being taken across the noise factors. To stabilize the variance, it is better to model the ln of the loss function. For this purpose, the data has been arranged in rows of control factor combinations and the expected loss across the noise factor combinations computed for each control factor combination. This is the basic data used in the analysis. As you can see, there are essentially 8 data points in this experiment, despite having collected data on 32 observations.

TABLE 7.8.7

TORQUE DATA FOR THE INTERMEDIATE SHAFT STEERING COLUMN

CONTROL FACTORS				NOISE COMBINATIONS						
Pocket Depth	Yoke Conc.	Tube Length	Power Set	Clear.: Volt.:	0 0	0 1	1 0	1 1	Ave. Loss	ln Loss
0	0	0	1		23.4	33.4	6.7	17.8	186.7	5.23
0	0	1	0		34.7	21.7	17.9	6.3	199.8	5.30
0	1	0	0		33.8	22.3	18.1	7.0	186.1	5.23
0	1	1	1		22.6	34.1	7.1	16.9	191.9	5.26
1	0	0	0		26.9	17.0	42.6	30.7	84.5	4.44
1	0	1	1		16.5	28.6	33.5	43.1	92.0	4.52
1	1	0	1		14.9	28.5	33.0	42.6	99.5	4.60
1	1	1	0		28.1	14.6	43.6	32.4	107.9	4.68

To analyze the ln Loss data, the basic factor method will be used with the hand calculations method of Chapter **3**. We will use D, C, and L as the basic factors. In essence, we are ignoring P and analyzing a 2^3 experiment. The basic setup and calculations are given in Table 7.8.8.

The calculations in Table 7.8.8 give the following sums of squares: $\text{SS}(D) = (21.02^2 + 18.24^2)/4 - \text{CT} = 0.996$, $\text{SS}(C) = (19.49^2 + 19.77^2)/4 -$

TABLE 7.8.8

ANALYSIS OF LN LOSS USING BASIC FACTOR METHOD

		Pocket Depth D 0		Pocket Depth D 1	
		Yoke Concen. C		Yoke Concen. C	
		0	1	0	1
Tube Length	0	5.23	5.23	4.44	4.60
L	1	5.30	5.26	4.52	4.68

D	C	L	$D*C$ Cells	
$T_{0\cdot\cdot} = 21.02$	$T_{\cdot 0 \cdot} = 19.49$	$T_{\cdot\cdot 0} = 19.50$	$T_{00\cdot} = 10.53$	$T_{10\cdot} = 8.96$
$T_{1\cdot\cdot} = 18.24$	$T_{\cdot 1 \cdot} = 19.77$	$T_{\cdot\cdot 1} = 19.76$	$T_{01\cdot} = 10.49$	$T_{11\cdot} = 9.28$

$D*L$ Cells		$C*L$ Cells		
				$T_{\cdots} = 39.26$
$T_{0\cdot 0} = 10.46$	$T_{1\cdot 0} = 9.04$	$T_{\cdot 00} = 9.67$	$T_{\cdot 10} = 9.83$	$CT = 192.66845$
$T_{0\cdot 1} = 10.56$	$T_{1\cdot 1} = 9.20$	$T_{\cdot 01} = 9.82$	$T_{\cdot 11} = 9.94$	$\Sigma y^2 = 193.6698$

$CT = 0.010$, $SS(L) = (19.50^2 + 19.76^2)/4 - CT = 0.008$, $SS(D*C) = (10.53^2 + 10.49^2 + 8.96^2 + 9.28^2)/2 - SS(D) - SS(C) - CT = 0.016$, $SS(D*L) = (10.46^2 + 10.56^2 + 9.04^2 + 9.20^2)/2 - SS(D) - SS(L) - CT = 0.000$, $SS(C*L) = (9.67^2 + 9.82^2 + 9.83^2 + 9.94^2)/2 - SS(C) - SS(L) - CT = 0.000$, and $SS(D*C*L) = \Sigma y^2 - SS(D) - SS(C) - SS(L) - SS(D*C) - SS(D*L) - SS(C*L) - CT = 0.000$. Of course, these are the sums of squares associated with the basic factors, not the terms in the original experiment. To link these sums of squares to the original experiment we have to go back to the confounding pattern. Each string will contain one and only one basic factor term. The resulting ANOVA is given in Table 7.8.9.

As was known prior to the collection of the data, no tests are available on any of the confounded terms. One could, say, assume all two factor interactions are zero and pool them to test the main effect terms, but this can only be done with some additional expert knowledge of the system. To further make assumptions difficult, these assumptions need to be applied to the response, the loss computed across the noise factors, not the original torque data. It is much harder to have knowledge of the

TABLE 7.8.9

ANOVA FOR LN LOSS IN EXAMPLE 2

Source	df	MS
D &/or CLP	1	0.996
C &/or DLP	1	0.010
L &/or DCP	1	0.008
P &/or DCL	1	0.000
DC &/or LP	1	0.016
DL &/or CP	1	0.000
DP &/or CL	1	0.000

loss function than of the original data, so assumptions are an extremely difficult to make.

Rather than make assumptions, we recommend the use of the half normal plot in Figure 7.8 to guide the pooling of terms in the model. Half normal plots are always subject to individual interpretation. In this example, the fitted line could be placed through the three lowest points or the six lowest points, depending on individual preference. We chose to fit the three lowest points.

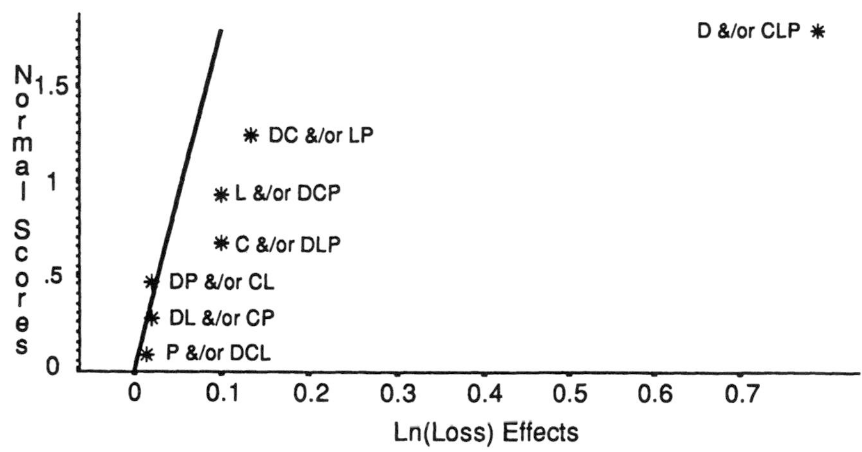

Figure 7.8 Half Normal Effects Plot of Ln(Loss) for Example 2.

According to the half normal plot, the terms that are candidates for pooling are, starting from the bottom of the plot, P, DCL, DL, CP, DP, and CL. We include all confounded terms in this list since a combined effect that is insignificant implies each individual term is insignificant. This gives us three df for error if all terms can be pooled. Unfortunately, CLP is confounded with D and this combined term is clearly important. By our sometimes pooling rules, the two factor interactions CL and CP as well as the main effect P cannot be pooled unless CLP can be pooled.

Rather than be stymied by the confounding pattern, we are going to step out on a limb for analyzing this experiment. Since we know P is insignificant, it is *reasonable* to believe all interactions involving P are also insignificant. Note that we have no evidence to support this claim and such a claim cannot be substantiated statistically without further experimentation. Under such a belief, the last three terms can be pooled, all interactions involving P dropped, and we are left with the simplified model

$$\ln(\text{Loss})_{ijk\ell} = \mu + D_i + C_j + DC_{ij} + L_k + \varepsilon_{\ell(ijk)}$$

and the ANOVA in Table 7.8.10.

TABLE 7.8.10

ANOVA FOR LN LOSS IN EXAMPLE 2 AFTER POOLING

Source	df	MS	F_{calc}	p-value
→ D	1	0.966	3409.6	0.000**
→ C	1	0.010	34.6	0.010*
→ L	1	0.008	29.8	0.012*
→ DC	1	0.016	57.2	0.005**
ε	3	0.000	—	—

* Significant at the .05 level

** Significant at the .01 level

Of course, one needs to temper the results given in Table 7.8.10 because one always artificially inflates the F values by purposely selecting the lowest MS values to pool. Nevertheless, these terms are the most important terms in the model.

The real goal of the experiment is to select the combination of control factors that minimizes the loss and therefore is least sensitive to variation in the noise factors. One can use graphical methods to select the best combination but, as we have seen earlier, this can be tricky in some circumstances. Instead, we will use the formal methods of Chapter **3** to predict the ln(Loss) for each of the possible 32 design points. Obviously, since P is not in the mathematical model, the prediction is the same for both high and low values of P. The estimates of the parameters in the model are $\hat{\mu} = 4.9075$, $\hat{D}_0 = 0.3475$, $\hat{D}_1 = -0.3475$, $\hat{C}_0 = -0.035$, $\hat{C}_1 = 0.035$, $\hat{L}_0 = -0.0325$, $\hat{L}_1 = 0.0325$, $\hat{DC}_{00} = 0.045$, $\hat{DC}_{01} = -0.045$, $\hat{DC}_{10} = -0.045$, and $\hat{DC}_{11} = 0.045$. (If you have forgotten how to do this, guess who needs to go back and review Chapter **3**.) The prediction for any given combination ijk of the ln(Loss) is given by $\ln(\text{Loss}_{ijk}) = \hat{\mu} + \hat{D}_i + \hat{C}_j + \hat{L}_j + \hat{DC}_{ij}$ with 95% confidence band given by $\sqrt{MSE}\, t_{.025} = 0.0536$ for all combinations. All predictions are given in Table 7.8.11, where we have merged $P = 0$ and $P = 1$ since they give identical predictions..

TABLE 7.8.11

PREDICTIONS FOR LN LOSS FOR EXAMPLE 2

D	C	L	P	Prediction	95% c.i.
1	0	0	0 or 1	4.415	(4.361,4.470)
1	0	1	0 or 1	4.513	(4.459,4.567)
1	1	0	0 or 1	4.608	(4.554,4.661)
1	1	1	0 or 1	4.673	(4.619,4.726)
0	1	0	0 or 1	5.210	(5.156,5.264)
0	0	0	0 or 1	5.233	(5.179,5.286)
0	1	1	0 or 1	5.278	(5.224,5.331)
0	0	1	0 or 1	5.298	(5.244,5.351)

The combination that minimizes loss and is therefore least sensitive to variation caused by the noise factors is $D = 1$, $C = 0$, and $L = 0$. It does not matter what the value of P is so the cheapest level (lowest power level) of this factor should be chosen. It is interesting to note that there is not a symmetry between factors C and L for choosing the best level even though there is a symmetry in mean values and MS values. This is caused by the interactions, significant for DC but not significant for DL.

The observed value for ln(Loss) for $D = 1$, $C = 0$, $L = 0$, and $P = 0$ is 4.44, well within the 95% confidence interval.

Example 3. An experiment was run to better understand a transmission control system. A team of engineers gathered and came up with 11 factors thought to influence the smoothness of the transmission. As usual, they were interested in running the smallest experiment possible to gather all of the required information about the factors.

A complete factorial with 11 factors at two settings and no repeats of any combination requires 2048 runs. Each run requires assembling a new transmission so 2048 runs is absolutely out of the question. A two step procedure was recommended with the first step consisting of a fractional design. The tables in Appendices 14, 15, and 16 were consulted to get an idea of the trade-offs between sample size and resolution of the design. With 11 factors, a resolution III design (Appendix 16) requires 16 runs. A resolution IV design (Appendix 15) requires 32 runs. A resolution V design (Appendix 14) requires 128 runs.

A full explanation of the meaning and significance of resolution was presented to the team. Transmissions are very complicated devices and past experience has indicated that many two factor interactions are likely to exist. If a resolution III design was run, two factor interactions would be confounded with main effects. As a result, a significant interaction would probably be misinterpreted as a significant main effect, too many main effects would be declared significant, and we would not be able to pare down the number of factors for a possible second experiment. Even though there would only be 16 runs, the 16 runs would be a wasted effort.

A resolution IV, 32 run design would probably be a better design. However, the engineers had considerable knowledge about the transmission that is not reflected in a general resolution IV design. It was felt that this information could be captured and used to create a design better than the one given in Appendix 15. This sort of information is highly valued by the statistician, who must translate engineering knowledge into statistical requirements. Proper communications about the meaning of an interaction is critical.

The factors under study, labeled A through K respectively, are oil temperature, piston stroke, restriction after stroke, hydraulic flow, air entrapment, leak rate, restriction before stroke, solenoid tolerance, input speed, relief pressure, and supply pressure. Since the actual factors are not of primary interest to the statistician, we will work with the letters A through K and not the factor names. Using knowledge of how the control system functions, the following interactions were quite likely to be important: $B * C$, $B * F$, $B * G$, $C * F$, $C * G$, $F * G$, and $G * H$. In addition, the remainder of the two factor interactions could be significant since a transmission is a complicated device. The 11 factors plus the 7 interactions sum to 18 df so there was good hope that a 32 run design could be found that did not confound any other two factor interaction with the 18 main effects and 7 interactions of primary interest.

Since all seven interactions involve the five factors B, C, F, G, and H, these five factors were selected as the basic factors. A full factorial in these factors ($2^5 = 32$) was written down and all main effects and interactions involving the basic factors were assigned. These are given below, split into two rows to fit across the page.

original factors	*B*	*C*	*BC*	*F*	*BF*	*CF*		*G*	*BG*	*CG*		*FG*				*H*	
basic factors	*B*	*C*	*BC*	*F*	*BF*	*CF*	*BCF*	*G*	*BG*	*CG*	*BCG*	*FG*	*BFG*	*CFG*	*BCFG*	*H*	*BH*

original factors							*GH*							
basic factors	*CH*	*BCH*	*FH*	*BFH*	*CFH*	*BCFH*	*GH*	*BGH*	*CGH*	*BCGH*	*FGH*	*BFGH*	*CFGH*	*BCFGH*

The next step is to assign the remaining 7 factors to open columns without violating any of the requirements. We must not have any two factor interaction confounded with a main effect or with any of the two factor interactions already assigned. We cannot assign any remaining main effect to a column containing a two factor interaction or a two factor interaction will be confounded with a main effect. In addition, we cannot assign any remaining main effect to a three factor interaction containing any of the already assigned two factor interactions or a two factor interaction will be confounded with an important two factor interaction. For example, *A* cannot be assigned to the *BCF* column or the *AF* interaction will be confounded with *CF* and we will be unable to estimate *CF* cleanly of any other two factor interaction. The unavailable columns are indicated below by the word *no*. This allows only 6 open columns from which to assign 7 factors and shows that no 32 run design meets all the requirements.

original factors	*B*	*C*	*BC*	*F*	*BF*	*CF*	*no*	*G*	*BG*	*CG*	*no*	*FG*	*no*	*no*		*H*	*no*
basic factors	*B*	*C*	*BC*	*F*	*BF*	*CF*	*BCF*	*G*	*BG*	*CG*	*BCG*	*FG*	*BFG*	*CFG*	*BCFG*	*H*	*BH*

original factors	*no*	*no*	*no*	*no*	*no*		*GH*	*no*	*no*		*no*			
basic factors	*CH*	*BCH*	*FH*	*BFH*	*CFH*	*BCFH*	*GH*	*BGH*	*CGH*	*BCGH*	*FGH*	*BFGH*	*CFGH*	*BCFGH*

A 64 run design meeting all criteria was run using the PROC FACTEX procedure in the SAS system. This design assigns 18 df to confounded strings of order 3 and higher. Under the assumption that all three factor

TABLE 7.8.12

EXAMPLE 3, ASSUMING 3RD ORDER INTERACTIONS NEGLIGIBLE

Source	df	Δ	EMS
A ←	1	1.83	$\sigma^2 + 8\sigma_\gamma^2 + 32\Phi(A)$
→ B	1	0.61	$\sigma^2 + 32\Phi(B)$
C* **	1	—	$\sigma^2 + 8\sigma_\gamma^2 + 16\sigma_\delta^2 + 32\Phi(C)$
D* **	1	—	$\sigma^2 + 8\sigma_\gamma^2 + 16\sigma_\delta^2 + 32\Phi(D)$
→ E	1	0.61	$\sigma^2 + 32\Phi(E)$
→ F	1	0.61	$\sigma^2 + 32\Phi(F)$
→ G	1	0.61	$\sigma^2 + 32\Phi(G)$
→ H	1	0.61	$\sigma^2 + 32\Phi(H)$
→ I	1	0.61	$\sigma^2 + 32\Phi(I)$
→ J	1	0.61	$\sigma^2 + 32\Phi(J)$
→ K	1	0.61	$\sigma^2 + 32\Phi(K)$
AC ←	1	2.59	$\sigma^2 + 8\sigma_\gamma^2 + 16\Phi(AC)$
AD, EK, GI ←	1	2.59	$\sigma^2 + 8\sigma_\gamma^2 + 16\Phi(AD, EK, GI)$
→ BC	1	0.86	$\sigma^2 + 16\Phi(BC)$
→ BF	1	0.86	$\sigma^2 + 16\Phi(BF)$
→ BG	1	0.86	$\sigma^2 + 16\Phi(BG)$
CD**	1	—	$\sigma^2 + 8\sigma_\gamma^2 + 16\sigma_\delta^2 + 16\Phi(CD)$
δ_{CD}	0	—	—
→ CF	1	0.86	$\sigma^2 + 16\Phi(CF)$
→ CG	1	0.86	$\sigma^2 + 16\Phi(CG)$
→ FG	1	0.86	$\sigma^2 + 16\Phi(FG)$
→ GH	1	0.86	$\sigma^2 + 16\Phi(GH)$
→ (24) 2 f.i.	1	0.86	$\sigma^2 + 16\Phi(2\text{ f.i.})$
ACD —	1	—	$\sigma^2 + 8\sigma_\gamma^2 + 8\Phi(ACD)$
γ_{ACD}	0	—	—
— pooled 3 f.i.	17	—	σ^2

* Significant Only test using CD **Insignificant Only test using ACD

interactions and higher are negligible, this provides 18 df for tests and a Δ of 0.61 for main effects and 0.86 for the interesting two factor interactions. The remaining two factor interactions are confounded among themselves and the remaining three factor interactions can test each of the individual three factor interactions as some justification on the assumptions.

This is an excellent design from the statistician's point of view. However, it does require 64 runs, a large design from the client's point of view. Fortunately, the dangers of a 32 run design along with the persuasion of the statistician (not to mention the availability of a summer employee to work on the project) convinced the team to consider the larger design.

While considering the larger design, the team mentioned that the experiment could be run fairly fast if complete randomization was not used. The hardest two factors to change are factors C and D, approximately equal in difficulty. The next most difficult factor to change is A. Thus, the statistician added a restriction error to the CD interaction and to the ACD interaction. The result was a design that changed levels of C and D at most four times and, within each block of 16 runs having the same levels for C and D, changed levels of A only once. The ANOVA for this final design is given in Table 7.8.12. Each of the interactions of interest is tested by the pooled 3 factor interactions. The remaining 2 factor interactions, some of which are confounded with other 2 factor interactions, are also tested by the pooled 3 factor interactions. The effects of the restriction errors are to change some direct tests to approximate tests and remove 1 df from the pooled error term.

Thus, with a minimum amount of redesign, a design was created that gave good statistical properties, was run in a convenient fashion, and met financial and time constraints.

PROBLEMS

7.1 *In a two-staged experimental approach consisting of a screening experiment followed by a full factorial experiment on the most important factors, explain why bad assumptions in the screening experiment lead to larger factorial experiments on the most important factors.*

7.2 *Compute the following. (a) 17 mod 2. (b) 17 mod 3. (c) 17 mod 5. (d) 17,484 mod 196.*

7.3 *Expand* $(I{+}ABC)(I{+}CDE)(IBEF)$ *to verify that you get* $I = ABC = CDE = ABDE = BEF = ACEF = BCDF = ADF$.

7.4 *Write the confounding pattern in standard order for the* 2^{6-3}_{III} *design of Section* **7.2** *having generators of* ABC, CDE, *and* BEF.

7.5 *Find the EMS for the* 2^{3-1}_{III} *design given in equation (7.1.2) when (a) all 3* *factors are fixed, (b) factors A and B are random and factor C is fixed, and (c) all 3 factors are random.*

7.6 *Repeat Problem* **7.5** *assuming there are two repeats of each combination. When and how can repeats help?*

7.8 *Suppose a 2^{5-2} design is created using the generators $I = ABCD$ and $I = BCE$. If all factors are fixed, the rule for EMS will yield a term $(1/4)\Phi(ABCDE)$ in the EMS for $ABCDE$. Use the definition of $\Phi(\cdot)$ given in Chapter* **3** *to write out the exact expression for $(1/4)\Phi(ABCDE)$. Now use first principle to figure out the exact quadratic that is to be tested. The exact quadratic does not have the coefficient (1/4) but contains only 1/4 of the terms in $\Phi(ABCDE)$. Since the term effects are interchangeable, the two expressions represent the same effect.*

7.8 *Write down the factor level combinations for (7.4.1) and (7.4.2). Use the formula for SS in Section* **7.5** *to compute SS(A) and SS(BC) for both (7.4.1) and (7.4.2). The formulae should agree, showing the confounding is the same in both designs. Now go back to the principles given in Chapter* **3** *and estimate the effect of $A(= \hat{A}_1 - \hat{A}_2)$ and the effect of $BC(= \widehat{BC}_{11} - \widehat{BC}_{12} - \widehat{BC}_{21} + \widehat{BC}_{22})$ for (7.4.1) and (7.4.2). One pair should agree and one pair should have opposite signs. This is the difference between the two designs and the reason some authors talk about negative confounding or negative aliasing.*

7.9 *Find the 16 combinations corresponding to the 1,1 fraction of (7.4.3) rather than the 0,0 fraction.*

7.10 *Find a set of basic factors other than $\{A, C, D, E\}$ for (7.4.3) and verify that the factor level combinations agree with Table 7.4.1.*

7.10 *For the basic factors you chose in Problem* **7.9***, write out the terms you would use if calculating SS by hand or if you only have access to a computer package that handles full factorial designs.*

7.12 *Mathematically, why does the method used to generate Table 7.4.3 guarantee selection of the primary block 0,0?*

7.13 *Suppose we had 7 factors, A through G, and we wished to estimate all main effects and AB, BC, and EG without confounding them with any other 2 factor interactions. Use trial and error to select a set of identity generators to minimize the size of the experiment.*

7.14 *Use the reverse basic factor technique to find identity generators in Problem* **7.13** *above.*

7.15 *The example in Section* **7.6** *assigned the remaining factors from the left to the right. Try assigning the remaining factors from the right to the left and give the resolution of the resulting design.*

7.16 *A suggestion of Greenfield (1977) is to choose the basic factors according to which factors appear most often in the interactions. Rework the reverse* *basic factor example in Section* **7.6** *using A, E, and F as the basic factors. Compare the confusion of using different basic factors to the simplicity of assigning the remaining factors.*

7.17 *Use the appendices to find the generators, confoundings, and factor level combinations for a 2_{IV}^{7-2} design.*

7.18 *Use the appendices to find the factor level combinations for a 2_{III}^{15-11} design.*

7.19 *Suppose one is trying to force PROC FACTEX in the SAS system to reproduce a fractional design that has already been created. With knowledge of how the Franklin (1985) algorithm works, this is not an impossible task. The first factors entered into SAS become the basic factors. Then, the next factor is assigned to the first available column starting from the right. The next factor is assigned to the next available column from the right. An so on. So to reproduce a given fraction, place the factors in the following order: (1) the basic factors, (2) assignments of the remaining factors as they appear from the right to the left. Put all unassigned basic factor combinations in the NONNEGLIGIBLE set in SAS. Try this on the second example given in Section* **7.6**.

7.20 *Run a Duncan's test on the HR interaction in the pooled model for Example 1 given in Table 7.8.5.*

7.21 *For Problem* **7.5**(a)*, find the primary block and use this to theoretically derive $\Phi(BC)$. Compare this to the definition of $\Phi(BC)$ given in Chapter* **3**.

7.22 *Compute the ANOVA table corresponding to Table 7.7.3 if A were fixed.*

7.23 *Crossed Designs. In Example 2, a design was generated for the control factors and a separate design was generated for the noise factors. Then, every noise combination was run with every control combination (and vice-versa). This is called crossing two designs. The confounding pattern for crossing two designs is found by combining each of the individual identity generators in the usual fashion. Suppose $I = ABCD$ and $I = EFG$ are two designs on 4 and 3 factors respectively and the two designs are crossed. Then, A is confounded with BCD from the first identity, with $AEFG$ from the second identity, and with $BCDEFG$ from the product of the two identities. Give the complete identity generator and the confounding pattern for the rest of the main effects.*

7.24 *Why would one even consider crossing two designs? Give two different reasons.*

7.25 *Robust Design. In Example 2, a resolution IV design was crossed (see Problem* **7.23***) with a full factorial, resulting in a resolution IV design. If all 6 factors, 4 control and 2 noise, were combined, a 32 run resolution VI design could be run. Obviously, a resolution VI design is superior to a resolution IV design. Create the resolution VI design and write all combinations* *in a format like Table 7.8.7. Discuss some possible disadvantages of the resolution VI design.*

7.26 *Robust Design. As we saw in Problem* **7.23**, *the combined design has twice as many control factor combinations but only half the noise combinations for each control combination. In some sense, there are twice as many data points (twice the control combinations) but each data point is only worth half as much (half the noise combinations). We can approximate the EMS for robustness as defined across the noise combinations for the combined design by treating the control factor combinations like a full factorial (since all control combinations exist) but adjusting the coefficients as if this were a half fraction (since only half the noise combinations exist). For a half fraction, recall that we multiply all coefficients by 1/2 and replace any coefficient less than 1 by 1. The EMS for robustness for the crossed design is simply a 2_{IV}^{4-1} design since all noise combinations are run. Assuming 3 factor and higher order interactions are negligible, compare the tests and detectability, Δ, for robustness in the combined and the crossed designs.*

7.27 *Robust Design. For the combined design given in Problem* **7.25**, *a different analysis approach must be taken since the loss computed across rows are not comparable. That is, the loss computed across noise combinations 0,1 and 1,0 is not expected to be equivalent to the loss computed across noise combinations 0,0 and 1,1. In order to make the loss across rows comparable, first model the raw data as a function of all factors, both noise and control, and predict the missing values. Then the loss across rows will be comparable. Actually, two approaches are possible. One is to use the model of the torque data to predict only the missing values, treat these missing values as actual data, compute the loss for each control combination, and model the loss as in Example 2. A second approach is to predict all control and noise combinations and calculate the loss based on the predictions. No modeling of the loss is necessary. Comment on the two approaches, giving your opinion as which is better under which circumstances. You will have to use your intuition.*

7.28 *Robust Design. Two methods for modeling data from a combined design are given in Problem* **7.27**. *A third method exists: calculate the mean and standard deviation across the noise combinations, separately model the mean and the ln standard deviation, and combine these two models to predict loss. This third method works with just filling in the missing values with predicted values as well as using all predicted values. Comment on how you think these methods stack up against the other two methods.*

7.29 *Create a layout sheet appropriate for the design given in Example 3.*

7.30 *Assuming a completely randomized 64 run design in Example 3, write the EMS table and compare it to Table 7.8.12.*

7.31 *In an experiment on the wear and life of gears, there were 15 factors, each fixed and having two levels. Only four factor and higher order interactions* *could be assumed negligible. Design a good experiment and show the ANOVA table. There is no need to list each individual main effect and interaction since all main effects and interactions of the same order will have the same df and detectability.*

7.32 *In a design having 11 factors, what is the minimal number of runs required to get information on all main effects and two factor interactions if we can assume all three factor and higher order interactions are negligible? Give the entire identity generator.*

7.33 *There are six factors, A through F, in an experiment. It is thought that A could interact with all other factors so all two factor interactions involving A must be estimable. In addition, factor E may interact with factors C and D. It is thought that the remainder of the two factor and higher order interactions are negligible. Can a 16 run experiment be constructed meeting all needs? If so, give the generators.*

7.34 *There are six factors, A through F, in an experiment. It is thought that the following interactions may be important: AB, CE, CF, DE, DF, and EF, one fewer interaction than required in Problem* **7.32**. *Again assuming the remainder of the two factor and the higher order interactions negligible, can a 16 run design be constructed? If so, give the generators.*

7.35 *If there are seven factors, all fixed and at two levels, use ANOVA to compare a resolution III design having two repeats to a resolution IV design. When would you use each design?*

7.36 *An experiment on the lost foam metal casting process was designed, run, and then the data was brought to a statistician. The experimenter found the design in some book somewhere, but no longer recalled where. He had no idea what an identity generator was, but remembered that complete randomization was used. Each casting was broken apart and the total fold area (a type of defect) was measured. Figure out what design was used, make some reasonable assumptions about interactions, and use the data to recommend a combination of factor levels that minimizes fold area. What is the predicted minimum fold area?*

FACTORS							FOLD AREA	
Pattern Coating	Metal Temp.	Mg/Sr Level	Susp. Oxide	Vacuum Level	Sand Perm.	Sand Temp.	Casting 1	Casting 2
0	0	0	0	0	0	0	101	184
0	0	0	1	1	1	1	265	392
0	1	1	0	0	1	1	170	75
0	1	1	1	1	0	0	276	302
1	0	1	0	1	0	1	621	931
1	0	1	1	0	1	0	490	341
1	1	0	0	1	1	0	582	346
1	1	0	1	0	0	1	480	455

7.37 *An experiment was run on the effect of the "A" pillar, located in front and to the left of the driver of an automobile, on the perceived visibility of the driver. The four factors varied were the width of the "A" pillar, the shape of the "A" pillar, the color of the "A" pillar, and cowl height. The cowl height is essentially the height of the dashboard. A 1/2 fraction was run since every combination requires costly modification of the vehicle. A total of 36 potential customers rated each of the 8 combinations on a scale of 1 for very good visibility to 4 for very poor visibility. (a) Comment on the use of 36 potential customers using detectability, Δ, calculations along with an estimate of the variability in a four point scale. (b) A computer generated ANOVA table and the average responses are given below. What can be concluded? Be very careful about stating assumptions and give the inference space.*

Source	df	SS	MS	F	p Value
Width	1	72.000	72.000	93.33	0.0001**
Shape	1	13.3472	13.3472	30.83	0.0001**
Color	1	0.0139	0.0139	0.10	0.7567
Cowl Height	1	6.7222	6.7222	15.92	0.0003**
Width*Shape	1	3.1250	3.1250	5.65	0.0231*
Width*Color	1	0.3472	0.3472	1.04	0.3142
Shape*Color	1	0.2222	0.2222	0.40	.05294
Subjects	35	31.777	0.9079		
Subjects*Width	35	27.0000	0.7714		
Subjects*Shape	35	15.1527	0.4329		
Subjects*Color	35	4.9861	0.1425		
Subjects*Cowl Height	35	14.7777	0.4222		
Subjects*Width*Shape	35	19.3750	0.5536		
Subjects*Width*Color	35	11.6528	0.3329		
Subjects*Shape*Color	35	19.2777	0.5508		

Width	Ave.	Shape	Ave.	Color	Ave.	Cowl Ht.	Ave.
Small	2.03	Flat	2.31	Dk. Blue	2.53	Low	2.38
Large	3.03	Bulged	2.74	Lt. Gray	2.52	High	2.68

Width/Shape Inter.	Ave.	Width/Shape Inter.	Ave.
Small/Flat	1.71	Small/Bulged	2.35
Large/Flat	2.92	Large/Bulged	3.14

7.38 *A problem facing products made by forming sheet metal is springback, the tendency of the part to return to its normal shape after it has been bent. If the springback were consistent, you could simply build in more bend and let it spring back to the desired shape. The problem is to find the* *combination of factor levels that minimizes the variance. The factors in the experiment are supplier, S, Gage of the Steel, G, Type of Steel, T, Die Type, D, and punch speed, P. The experiment was run using two different coils, C, to see if the results depend on coils. Coils are very difficult to change so all of one coil is done before starting on the next. You can assume the SP interaction is negligible, along with three factor and higher order interactions. Find the fractional design used in the experiment and model the data appropriately. What recommendations would you make in order to minimize springback variability? Carefully state any assumptions you make along the way and be sure to pool appropriately.*

Coil 1						Coil 2					
S	G	T	D	P	ln(s^2)	S	G	T	D	P	ln(s^2)
0	0	0	0	0	1.76	1	0	0	1	0	1.38
1	1	1	0	1	0.38	0	0	1	0	1	1.39
0	1	0	1	0	0.59	1	1	0	0	0	1.13
0	0	1	1	0	0.90	1	1	1	1	0	0.14
1	0	1	1	1	1.00	1	0	1	0	0	1.46
1	1	0	1	1	0.62	0	0	0	1	1	0.90
0	1	1	0	0	0.40	0	1	0	0	1	1.14
1	0	0	0	1	1.11	0	1	1	1	1	0.06

7.39 *In Problem* **7.38** *we said nothing about the way the data that went into the ln(s^2) calculation was collected. What would be the difference in interpretation if the parts were taken back to back versus complete randomization within a coil? What would be the difference between basing the calculations on 5 data points versus 25 data points? Then what is σ_ε^2? Even though we have learned new techniques, never, ever forget good statistical practice.*

7.40 *Blocking. A full factorial was desired on five fixed factors, each factor having 2 levels. This requires 32 runs. Unfortunately, only 16 runs can be made in one day and it is felt that there will be more variation from day to day than within a day. Define the error term as the within day variation and represent the day to day variation as a random factor (often called Block). Design a 32 run experiment and identify all confoundings and tests.*

7.41 *An experiment on printed wire board stenciling was run consisting of 9 factors each fixed at 2 levels. The factors, labeled A through I for convenience, are A = Material, B = Age of Material, C = Viscosity, D = Cure Before Thinning, E = Thickness of Application, F = Airset Time Before Oven Cure, G = Oven Temperature, H = Cure Time in Oven, and I = Board Cleanliness. In addition, 10 interactions, AB, BG, BH, CD, CE, CF, DF, FG, FH, and GH were thought to possibly exist. The identity generators $ACDH$, $BCDFG$, $DEFH$, and $DIGH$ were used to generate the following 32 run design. The exact run order was completely* *randomized. (a) What is the resolution of the design? (b) The measured response variable is the number of defective boards out of 10. Analyze the data below and give recommendations for ways to minimize the number of defects. (c) Do you see any problems with the data? If so, what?*

A	B	C	D	E	F	G	H	I	# def.	A	B	C	D	E	F	G	H	I	# def.
0	0	0	0	0	0	0	0	0	10	0	0	0	1	0	0	1	1	1	0
0	0	0	0	1	1	1	0	1	10	0	0	0	1	1	1	0	1	0	0
0	0	1	0	1	0	1	1	0	10	0	0	1	1	1	0	0	0	1	9
0	0	1	0	0	1	0	1	1	9	0	0	1	1	0	1	1	0	0	7
1	0	0	0	1	0	0	1	1	0	1	0	0	1	1	0	1	0	0	0
1	0	0	0	0	1	1	1	0	0	1	0	0	1	0	1	0	0	1	0
1	0	1	0	0	0	1	0	1	0	1	0	1	1	0	0	0	1	0	0
1	0	1	0	1	1	0	0	0	0	1	0	1	1	1	1	1	1	1	0
0	1	0	0	0	0	1	0	1	0	0	1	0	1	0	0	0	1	0	0
0	1	0	0	1	1	0	0	0	2	0	1	0	1	1	1	1	1	1	4
0	1	1	0	1	0	0	1	1	5	0	1	1	1	1	0	1	0	0	5
0	1	1	0	0	1	1	1	0	2	0	1	1	1	0	1	0	0	1	2
1	1	0	0	1	0	1	1	0	0	1	1	0	1	1	0	0	0	1	0
1	1	0	0	0	1	0	1	1	0	1	1	0	1	0	1	1	0	0	0
1	1	1	0	0	0	0	0	0	0	1	1	1	1	0	0	1	1	1	0
1	1	1	0	1	1	1	0	1	0	1	1	1	1	1	1	0	1	0	0

7.42 *An experiment was run to help maximize engine mount bonding strength. A 2^{7-2} design with identity generators $I = ABCDEF$ and $I = BCEG$ was created and completely randomized. The data is given below with the factors fixed, labeled A through G, and with levels labeled 1 and 2 for convenience. Analyze the data, pool to simplify the model, and find the highest predicted combination. You can assume all 3 factor and higher order interactions are negligible as long as you test each 3 factor interaction with the remainder of the other 3 factor interactions.*

A	B	C	D	E	F	G	Strength	A	B	C	D	E	F	G	Strength
1	1	2	1	1	2	2	11	2	2	2	2	1	1	1	14
2	1	1	2	1	1	1	18	2	2	1	2	2	1	1	5
1	1	1	1	1	1	1	27	2	2	1	1	1	1	2	6
1	1	2	1	2	1	1	19	1	1	1	1	2	2	2	23
2	1	2	2	1	2	2	5	1	1	2	2	2	2	1	15
2	2	2	1	2	1	2	11	1	2	2	2	1	2	1	14
1	2	1	1	2	1	1	19	2	1	2	2	2	1	1	7
1	2	2	1	2	2	2	17	1	1	1	2	2	1	2	30
2	1	1	2	2	2	2	19	1	2	1	2	2	2	1	16
2	1	2	1	2	2	1	5	1	2	2	1	1	1	1	22
2	1	1	1	2	1	2	19	1	1	2	2	1	1	2	20
2	1	1	1	1	2	1	13	2	2	2	1	1	2	1	4
2	1	2	1	1	1	2	11	1	2	1	1	1	2	2	13
2	2	2	2	2	2	2	5	2	2	1	1	2	2	1	5
1	1	1	2	1	2	1	27	1	2	1	2	1	1	2	21
2	2	1	2	1	2	2	8	1	2	2	2	2	1	2	24

7.43 *An experiment on manual window regulators (the window crank) has 5 factors, the Pinion Bushing, the Sector Combination, the Sector Bushing, the Sector Support, and the Clutch Housing, all fixed with 2 levels chosen. (a) Create a resolution III design. (b) Create a resolution IV design that does not confound either of the Sector Bushing/Sector Support or the Pinion Bushing/Sector Support interactions with any other of the 2 factor interactions. (c) Assuming we test the main effects with the pooled 2 factor interactions, compute the detectability, Δ, for each design. (d) Give your recommendations and cautions with each design.*

7.44 *An initial design dealing with the seam alignments on car seats had 25 factors involving the location of the various trim pieces, cushions, frame, and suspension, the possible tacking positions, the type of seat design, the thickness of the padding, the accent width, and whether this is a driver's or passenger's seat. Label the factors A through Y and find a resolution III design useful for finding the most significant main effects.*

7.45 *A crossover design was mentioned in Problems* **4.29** *and* **5.42**. *We can finally understand the complexities of that design. Identify the fractionation, write the model, analyze, and interpret the data given below.*

	Order 1							Order 2					
	Person							Person					
	1	2	3	4	5	6		7	8	9	10	11	12
Drug 2:	51.9	35.1	38.6	36.1	34.6	39.7	Drug 1:	50.8	41.1	39.1	35.7	33.7	31.2
Drug 1:	58.5	60.4	50.4	58.7	64.8	54.9	Drug 2:	62.2	51.4	50.7	51.1	51.4	45.1

7.46 *Censored Data. An experiment was performed to shed light on the effect of certain factors on cycles to failure in a milling process. Since it can take considerable time to cycle until failure, a highly fractionated design was run and tests were suspended (stopped) at 10,000 cycles if they had not previously failed. The data is given below. Analyze this data and make your best prediction as to which combination of factor levels maximizes cycles to failure. Note, there are 2 unusual circumstances to this problem. One, the data is likely to be heterogeneous with the variation growing as the mean grows. Two, 3 of the 8 data points are suspended. All we know is that the true value exceeds 10,000. To handle circumstance one, work with the ln of the data, not the data itself. To handle circumstance two, use the EM algorithm. Start with the value of 10,000 for these three data points. Fit the data and predict the mean and variance of the three suspended points. Using the predicted mean and variance, compute the expected cycles to failure conditional on the value exceeding 10,000. (You can assume normality of the ln's and will need calculus to solve the integral.) Now replace 10,000 with its expected value and iterate the entire process until it converges. It usually takes only a few iterations for the process to converge.*

D	E	A	B	C	Cycles
1	1	1	1	1	1,456
1	1	2	2	2	10,000*
1	2	1	1	2	676
1	2	2	2	1	8,718
2	1	1	2	2	3,441
2	1	2	1	1	5,092
2	2	1	2	1	10,000*
2	2	2	1	2	10,000*

*Suspended item

BIBLIOGRAPHY FOR CHAPTER 7

Anderson, V. L. and McLean, R. A. (1974). *Design of Experiments A Realistic Approach*, New York: Marcel Dekker. (Chapter 10)

Bennett, C. A. and Franklin, N. L. (1954). *Statistical Analysis in Chemistry and the Chemical Industry*, New York: John Wiley & Sons. (Chapters 7, 8)

Box, G. E. P. and Hunter, J. S. (1961). "The 2^{k-p} fractional factorial designs, part 1," *Technometrics*, **3**, 311–351.

Box, G. E. P. and Hunter, J. S. (1961). "The 2^{k-p} fractional factorial designs, part 2," *Technometrics*, **3**, 449–458.

Box, G. E. P., Hunter, W. G. and Hunter, J. S. (1978). *Statistics for Experimenters An Introduction to Design, Data Analysis, and Model Building*, New York: John Wiley & Sons. (Chapters 12, 13)

Brownlee, K. A., Kelly, B. K. and Loraine, P. K. (1948). "Fractional replication arrangements," *Biometrika*, **35**, 268–282.

Daniel, C. (1976). *Applications of Statistics to Industrial Experimentation*, New York: John Wiley & Sons. (Chapters 11, 12)

Das, M. N. and Giri, N. C. (1986). *Design and Analysis of Experiments*, 2nd ed., New York: John Wiley & Sons.

Finney, D. J. (1945). "Fractional replication of factorial arrangements," *Annals of Eugenics*, **12**, 291–301.

Franklin, M. F. (1985). "Selecting defining contrasts and confounded effects in p^{n-m} factorial experiments," *Technometrics*, **27**, 165–172.

Franklin, M. F. and Bailey, R. A. (1977). "Selection of defining contrasts and confounded effects in two-level experiments," *Applied Statistics*, **26**, 321-326.

Greenfield, A. A. (1976). "Selection of defining contrasts in two-level experiments," *Applied Statistics*, **25**, 64–67.

Greenfield, A. A. (1978). "Selection of defining contrasts in two-level experiments, a modification," *Applied Statistics*, **27**, 78.

Hicks, C. R. (1982). *Fundamental Concepts in the Design of Experiments*, New York: John Wiley & Sons. (Chapter 15)

John, P. M. W. (1971). *Statistical Design and Analysis of Experiments*, New York: The Macmillan Co. (Chapter 8)

Johnson, N. L. and Leone, F. C. (1977). *Statistics and Experimental Design in Engineering and the Physical Sciences*, 2nd ed., New York: John Wiley & Sons. (Chapter 15)

Lochner, R. H. and Matar, J. E. (1990). *Designing for Quality, An Introduction to the Best of Taguchi and Western Methods of Statistical Experimental Design*, Milwaukee, WI: ASQC Quality Press. (Chapters 4,6,9)

McLean, R. A. and Anderson, V. L. (1984). *Applied Factorial and Fractional Designs*, New York: Marcel Dekker. (Chapters 2, 5)

Montgomery, D. C. (1991). *Design and Analysis of Experiments*, 2nd ed., New York: John Wiley & Sons. (Chapter 9)

Montgomery, D. C. (1991). "Using fractional factorial designs for robust design Process Development," *Quality Engineering*, **3**, 193–205.

Ostle, B. and Malone, L. C. (1988). *Statistics in Research*, 4th ed., Ames, Iowa: Iowa State University Press.

Peng, K. C. (1967). *The Design and Analysis of Scientific Experiments*, Reading, MA: Addison-Wesley. (Chapter 7)

Shoemaker, A. C., Tsui, K. L. and Wu, C. F. J. (1991). "Economical experimentation methods for robust design," *Technometrics*, **33**, 415–427.

Steel, R. G. D. and Torrie, J. H. (1980). *Principles and Procedures of Statistics—A Biometrical Approach*, 2nd ed., New York: McGraw-Hill. (Chapter ?)

Taguchi, G. (1986). *Introduction to Quality Engineering*, White Plains, NY: Kraus International Publications. (Chapters 6–8)

Taguchi, G. (1988). *System of Experimental Designs*, White Plains, NY: Kraus International Publications. (Chapters 6, 7, 17)

Winer, B. J. (1971). *Statistical Principles in Experimental Design*, 2nd ed., New York: McGraw-Hill. (Chapter 8)

SOLUTIONS TO SELECTED PROBLEMS

7.1 *If one makes the assumption that an interaction is negligible and, in fact, it is not, then the main effect (or lowest order interaction) confounded with that interaction will be declared significant even though it may not be significant. This makes the ensuing full factorial larger because there are more factors to consider.*

7.4
$$I = ABC = ADF = BEF = CDE = ABDE = ACEF = BCDF$$
$$A = BC = DF = BDE = CEF = ACDE = ABEF = ABCDF$$
$$B = AC = EF = ADE = CDF = ABDF = BCDE = ABCEF$$
$$C = AB = DE = AEF = BDF = ACDF = BCEF = ABCDE$$
$$D = AF = CE = ABE = BCF = ABCD = BDEF = ACDEF$$
$$E = BF = CD = ABD = ACF = ABCE = ADEF = BCDEF$$
$$F = AD = BE = ACE = BCD = ABCF = CDEF = ABDEF$$
$$AE = BD = CF = ABF = ACD = BCE = DEF = ABCDEF$$

7.5 *The EMS for, respectively, (a), (b), and (c) are given below.*

Term	EMS for (a)
A &/or BC	$\sigma^2 + \Phi(BC) + 2\Phi(A)$
B &/or AC	$\sigma^2 + \Phi(AC) + 2\Phi(B)$
C &/or AB	$\sigma^2 + \Phi(AB) + 2\Phi(C)$

Term	EMS for (b)
A &/or BC	$\sigma^2 + (1/2)\sigma^2_{ABC} + \sigma^2_{BC} + \sigma^2_{AB} + 2\sigma^2_A$
B &/or AC	$\sigma^2 + (1/2)\sigma^2_{ABC} + \sigma^2_{AC} + \sigma^2_{AB} + 2\sigma^2_B$
C &/or AB	$\sigma^2 + (1/2)\sigma^2_{ABC} + \sigma^2_{BC} + \sigma^2_{AC} + \sigma^2_{AB} + 2\Phi(C)$

Term	EMS for (c)
A &/or BC	$\sigma^2 + \sigma^2_{ABC} + \sigma^2_{BC} + \sigma^2_{AC} + \sigma^2_{AB} + 2\sigma^2_A$
B &/or AC	$\sigma^2 + \sigma^2_{ABC} + \sigma^2_{BC} + \sigma^2_{AC} + \sigma^2_{AB} + 2\sigma^2_B$
C &/or AB	$\sigma^2 + \sigma^2_{ABC} + \sigma^2_{BC} + \sigma^2_{AC} + \sigma^2_{AB} + 2\sigma^2_C$

7.13 *It is first hoped that a 16 run design will do the trick since a resolution IV design requires 16 runs. Unfortunately, the extra requirement of having to estimate three interactions without confounding them with any other two factor interactions makes a 32 run design necessary. Since this is a 1/4 fraction, 2 identity generators are required. We need to choose these generators to contain at least 4 factors (don't forget the implied generator must also contain at least 4 factors). Because there are 7 factors, the product of any two 5 factor generators is at most a 4 factor generator. This 4 factor generator cannot contain any of the 2 factor interactions of interest or the interaction of interest will be confounded with another two factor interaction. A set of generators that works is* $I = ABDEG$ *and* $I = BCDEF$ *which also gives* $I = ACFG$. *The crucial step is that the four factor interaction not contain any of* AB, BC, *or* EG. *Other generators are obviously possible.*

7.17 *The primary generators are* $I = BCDE$, $I = ACDF$, *and* $I = ABDG$. *The confoundings, in standard order, are:*

$$I = ABDG = ABEF = ACDF = ACEG = BCDE = BCFG = DEFG$$
$$A = BDG = BEF = CDF = CEG = ABCDE = ABCFG = ADEFG$$
$$B = ADG = AEF = CDE = CFG = ABCDF = ABCEG = ADEFG$$
$$C = ADF = AEG = BDE = BFG = ABCDG = ABCEF = CDEFG$$
$$D = ABG = ACF = BCE = EFG = ABDEF = ACDEG = BCDFG$$
$$E = ABF = ACG = BCD = DFG = ABDEG = ACDEF = BCDEG$$
$$F = ABE = ACD = BCG = DEG = ABDFG = ACEFG = BCDEF$$
$$G = ABD = ACE = BCF = DEG = ABEFG = ACDFG = BCDEG$$
$$AB = DG = EF = ACDE = ACFG = BCDF = BCEG = ABDEFG$$
$$AC = DF = EG = ABDE = ABFG = BCDG = BCEF = ACDEFG$$
$$AD = BG = CF = ABCE = AEFG = BDEF = CDEG = ABCDEF$$
$$AE = BF = CG = ABCD = ADFG = BDEG = CDEF = ABCEFG$$
$$AF = BE = CD = ABCG = ADEG = BDFG = CEFG = ABCDEF$$

$$AG = BD = CE = ABCF = ADEF = BEFG = CDFG = ABCDEG$$
$$BC = DE = FG = ABDF = ABEG = ACDG = ACEF = BCDEFG$$

The factor level combinations, labeled 0 and 1, can be found by writing a full factorial on A through D and letting $E = B + C + D \bmod 2$, $F = A + C + D \bmod 2$, and $G = A + B + D \bmod 2$.

A	B	C	D	E	F	G
0	0	0	0	0	0	0
0	0	0	1	1	1	1
0	0	1	0	1	1	0
0	0	1	1	0	0	1
0	1	0	0	1	0	1
0	1	0	1	0	1	0
0	1	1	0	0	1	1
0	1	1	1	1	0	0
1	0	0	0	1	1	1
1	0	0	1	0	0	0
1	0	1	0	0	0	1
1	0	1	1	1	1	0
1	1	0	0	0	1	0
1	1	0	1	1	0	1
1	1	1	0	1	0	0
1	1	1	1	0	1	1

7.33 *Yes. A set of generators that works is $I = ABCDE$ and $I = ACDF$. Of course, others are possible. Our assignment is*

original factors														
A	B	AB	C	AC		DE	D	AD		CE	AF	F	AE	E
A	B	AB	C	AC	BC	ABC	D	AD	BD	ABD	CD	ACD	BCD	$ABCD$
basic factors														

7.35 *At first glance, it appears that repeating a resolution III design is superior (better detectability) to a resolution IV design as long as one is interested only in the main effects. However, two factor interactions can cause one to falsely conclude a main effect is significant in the resolution III design. This will not occur in the resolution IV design which is capable of checking two factor interactions. If one is willing to assume all two factor interactions have negligible effect, the two factor interactions can be pooled in the resolution IV design and the power is the same as the resolution III design. The resolution IV design will have some checks on the assumptions, the resolution III design will not. Finally, the resolution III design has a true estimate of error while the resolution IV design must use a three factor interaction for error. All else being equal, I prefer the resolution IV design because it gives the same power under the assumption that all two factor* *interactions have negligible effect and gives some checks on the assumptions. If repeats are cheap and each different factor level combination is expensive, the resolutions III design is better.*

ANOVAs for Problem **7.35**

2^{7-4}_{III} Design, 2 Repeats

Term	df	Δ
A &/or BG &/or CF &/or DE	1	1.31
B &/or AG &/or CE &/or DF	1	1.31
C &/or AF &/or BE &/or DG	1	1.31
D &/or AE &/or BF &/or CG	1	1.31
E &/or AD &/or BC &/or FG	1	1.31
F &/or AC &/or BD &/or EG	1	1.31
G &/or AB &/or CD &/or EF	1	1.31
error	8	

2^{7-3}_{IV} Design

Term	df	Δ
A	1	3.67
B	1	3.67
C	1	3.67
D	1	3.67
E	1	3.67
F	1	3.67
G	1	3.67
AB &/or DG &/or EF	1	5.18
AC &/or DF &/or DF	1	5.18
AD &/or BG &/or CF	1	5.18
AE &/or BF &/or CG	1	5.18
AF &/or BE &/or CD	1	5.18
AG &/or BD &/or CE	1	5.18
BC &/or DE &/or FG	1	5.18
3 factor interactions	1	

7.42 *There is no absolute correct method to proceed but we analyzed the data as follows: Test each three factor interaction with the remainder of the other three factor interactions. It is reasonable to assume all three factor interactions except ABG are negligible and these are pooled to form the error term. Next apply the "sometimes pooling" rules to pool AC, AD &/or FG, AE, BD, BE &/or CG, and BF. The ANOVA is given on the next page. Finding the highest predicted value is very difficult in general. We used the computer to predict all 128 possible combinations and then sorted them to get the values displayed below the ANOVA table. One can usually find the highest or lowest combination using a graphical approach. Start by looking at the significant effects. Since A, B, C, and F are highly significant, it is likely that the significant confounded interactions are actually AF, BC, and CF. Next plot the main effects and these three interactions, displayed on the following two pages. Look first at the interaction plots to check for consistency. We are lucky because the selections are consistent; the highest combination is A=1, B=1, C=1, and F=1. Use the main effects plots to select the remaining levels, D=2, E=1, and G=2. E and G are insignificant so all combinations of E and G should have similar predictions. The graphical technique does very well, picking out the second best combination. Since the standard error for prediction is 1.87 and the difference between the best and second best is 1, one need not be concerned about selecting the second best.*

ANOVA for Problem **7.42**

Source	df	SS	MS	F_{calc}	p value
A	1	830.281	830.281	372.71	0.0000**
B	1	132.031	132.031	59.27	0.0000**
C	1	132.031	132.031	59.27	0.0000**
D	1	16.531	16.531	7.42	0.0165*
E	1	0.781	0.781	0.35	0.5362
F	1	166.531	166.531	74.76	0.0000**
G	1	5.281	5.281	2.37	0.1459
AB	1	5.281	5.281	2.37	0.1459
AF &/or DG	1	11.281	11.281	5.06	0.0410*
AG &/or DF	1	5.281	5.281	2.37	0.1459
BC &/or EG	1	318.781	318.781	143.10	0.0000**
BG &/or CE	1	0.031	0.031	0.01	0.9074
CD	1	7.031	7.031	3.16	0.0974
CF	1	30.031	30.031	13.48	0.0025**
DE	1	9.031	9.031	4.05	0.0637
EF	1	7.031	7.031	3.16	0.0974
ABG &/or ⋯	1	9.031	9.031	4.05	0.0637
pooled error	14	31.187	2.228		

*Significant at the 0.05 level **Significant at the 0.01 level

Predicted Values for Strength

A	B	C	D	E	F	G	Strength	Std Err
1	1	1	2	1	1	1	31.125	1.87
1	1	1	2	1	1	2	30.125	1.87
1	1	1	2	2	1	1	29.438	1.87
1	1	1	2	2	1	2	28.438	1.87
1	1	1	1	2	1	1	28.125	1.87
1	1	1	1	1	1	1	27.688	1.87
1	1	1	1	2	1	2	27.125	1.87
1	1	1	1	1	1	2	26.688	1.87

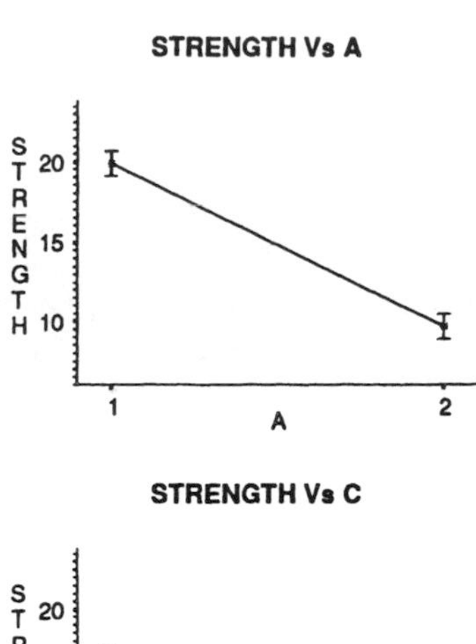
STRENGTH Vs A
STRENGTH
20
15
10
1
2
A

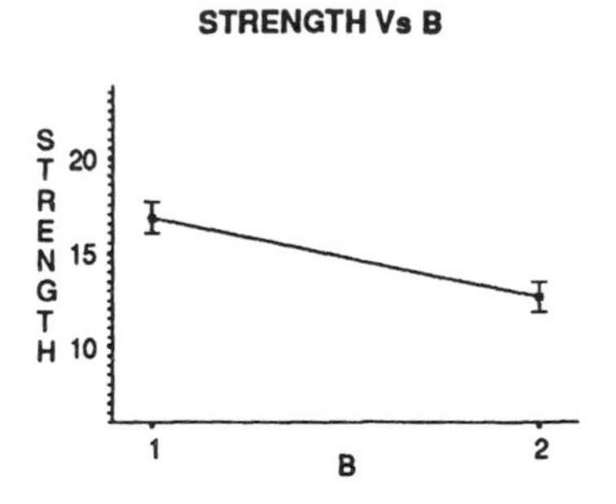
STRENGTH Vs B
STRENGTH
20
15
10
1
2
B

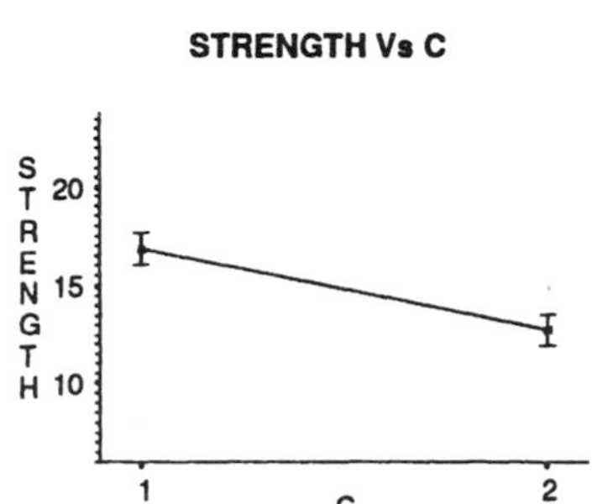
STRENGTH Vs C
STRENGTH
20
15
10
1
2
C

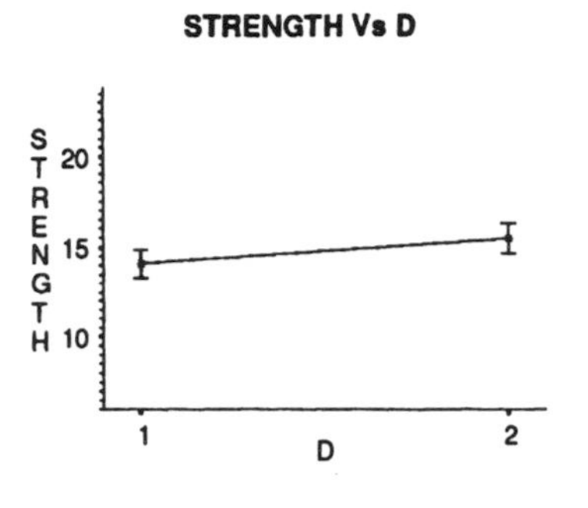
STRENGTH Vs D
STRENGTH
20
15
10
1
2
D

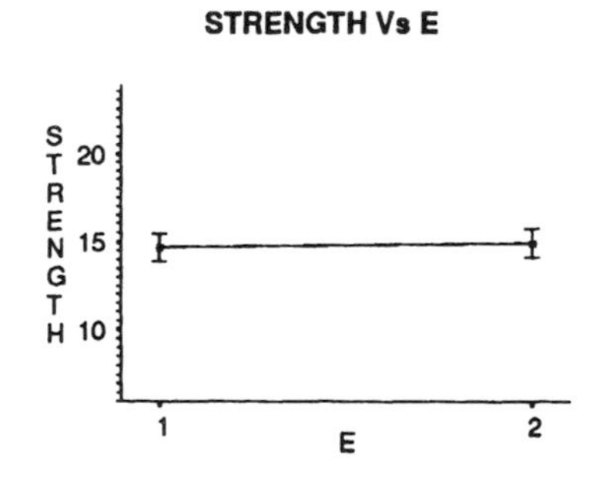
STRENGTH Vs E
STRENGTH
20
15
10
1
2
E

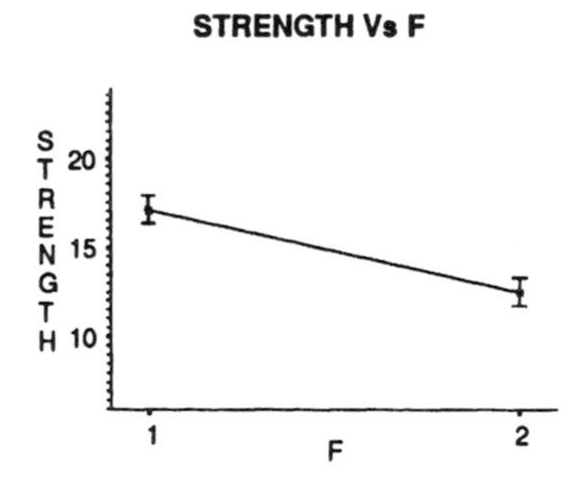
STRENGTH Vs F
STRENGTH
20
15
10
1
2
F

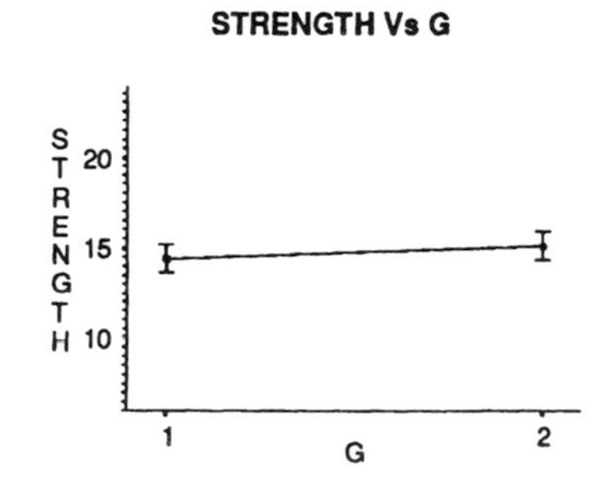
STRENGTH Vs G
STRENGTH
20
15
10
1
2
G

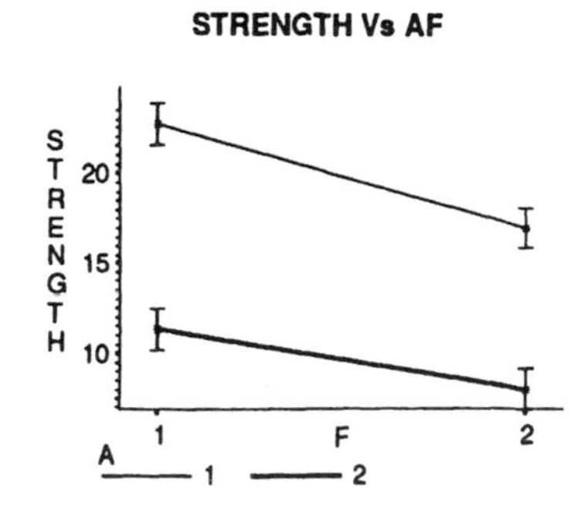
STRENGTH Vs AF
STRENGTH
20
15
10
1
2
F
A
1
2

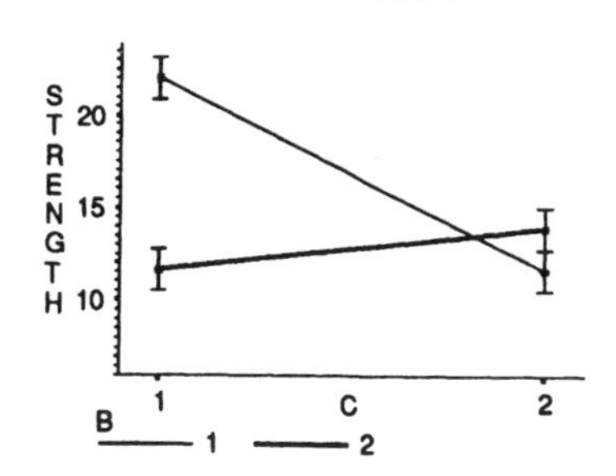
STRENGTH Vs BC
STRENGTH
20
15
10
1
C
2
B
1
2

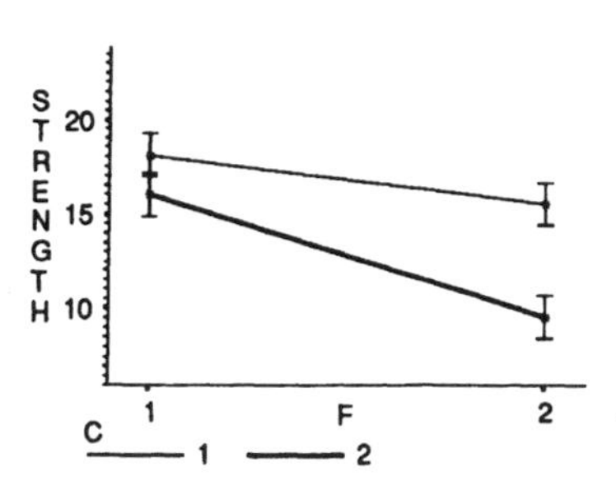
STRENGTH Vs CF
STRENGTH
20
15
10
1
F
2
C
1
2

8

Other Fractional Designs

8.0 INTRODUCTION

The previous chapter covered fractional factorial design for all factors at two levels. In this chapter, we extend those results to factors at prime levels. For practical purposes, most fractionation is done on factors containing two or three levels. After covering prime level fractionation we consider powers of primes. Powers of primes are extremely useful when the experiment must be run in blocks, that is, when the number of factor level combinations that can be run under similar conditions is limited. Mixed level designs follow the power of primes designs. We then give orthogonal main effect designs. These are useful when runs are so expensive that we can only hope to estimate main effects, no interactions. Finally, blocking and some irregular fractional plans are discussed.

The primary emphasis in this chapter will be on creating the appropriate designs. Some emphasis will be placed on the computation of mean squares, primarily through the use of computers. Little emphasis will be placed on the df, EMS, and F tests as these should be thoroughly familiar to the reader by now.

8.1 PRIME LEVEL FRACTIONS

Prime level fractions refer to experiments performed when all factors have the same number of levels and that number is prime. In Chapter 7, all factors had two levels. In the section, all factors will have three, five, seven, and so on levels. Obviously, since fractionation is performed to reduce the size of the experiment, the prime number three will be more important than five, which will be more important than seven, and so on.

There are two fundamental differences between this section and Chapter 7 and it is important to understand both. First, the arithmetic mod two must be replaced with arithmetic mod the prime number. This is pretty much straightforward. The second difference is that interaction terms must be broken into pieces. This needs elaboration.

When working with two level factors, all main effects *and* interactions had the same df: 1. When working with three level factors, main effects will have 2 df, two factor interactions will have 4 df, three factor interactions will have 8 df, and so on. Since the df do not match, it is impossible to completely confound main effects with interactions and interactions with other interactions not of the same degree.

To handle this problem, the interactions are broken into pieces that have the same df as the main effects. That is, for the prime number three, two factor interactions must be broken into two orthogonal pieces having 2 df each, three factor interactions must be broken into four orthogonal pieces having 2 df each, and so on. Then the individual pieces can be confounded with other pieces or with main effects.

From a practical point of view, the pieces should be viewed simply as pieces of the interaction, nothing more. All pieces should be treated as containing identical information about the interaction and the df and SS for the pieces will add up to df and SS for the interaction from which they came. If a piece of an interaction is confounded with, say a main effect, and another piece is confounded with higher order interactions, the estimate of the remaining piece is to be taken as an estimate of the entire interaction, just with fewer df than expected. We will also say that the main effect is partially confounded with the interaction. Thus, if the piece of the interaction not confounded with the main effect is negligible, it is reasonable to assume the piece confounded with the main effect is also negligible. Such reasoning allows more complex analysis of prime level fractions than is possible with two level fractions.

Using the notation developed in the previous chapter, we will label the factors A, B, C, etc., having levels 0, 1, 2, $\cdots$, $p-1$ for any prime p. The pieces of the interactions will be denoted with lower case letters to distinguish them from the interactions themselves. All pieces will have $p-1$ df and all confounding will involve lower case letters.

8.1.1 Three Level Fractions

For example, if $p = 3$, the factor A would have 2 df and be represented by the piece a. The factor B would have 2 df and be represented by the piece b. The interaction AB would have 4 df and must be represented by two pieces. We will use the convention that the first letter of every piece must have exponent 1. The pieces of the AB interaction would then be ab and ab^2, each with 2 df. Continuing along the same lines, the pieces of the ABC interaction would be abc, abc^2, ab^2c, and ab^2c^2, each with 2 df.

Notice that the piece a really represents a and a^2, each with 1 df, b really represents b and b^2, each with 1 df, ab really represents ab and a^2b^2, each with 1 df, and ab^2 really represents ab^2 and $a^2b^4 = a^2b$ where the exponent on b was simplified mod 3. If desired, we could break everything down to single level df and work with these pieces. But the complication added by all the confoundings is not worth the simplification gained by working with 1 df.

However, we must always keep in mind that the pieces *must have the exponent of the first letter equal to 1.* Thus, if we run across a piece like a^2bc^2, we must multiply it by itself (square it, cube it, *etc.*) mod 3 until the first exponent is 1: $a^2bc^2 = a^4b^2c^4 = ab^2c$.

We are now in a position to consider an example. Suppose there are four factors, A, B, C, and D, all at three levels. A complete factorial requires 81 runs, even if each combination were run only once. We would like to be able to reduce this to $27 = 3^{4-1}$ runs. It is natural to use the highest order interaction as the identity generator. However, the highest order interaction $ABCD$ has 16 df so 8 separate pieces exist. Any of these pieces could be selected and we arbitrarily select ab^2cd^2. The match between the original terms and their pieces is given in Table 8.1.1.

If we sum the df in Table 8.1.1, we find there are 40 pieces each having 2 df. This accounts for the 80 df in an 81 run full factorial experiment. When we run a 1/3 fraction in 27 runs, there are only 26 df for estimating the pieces. Since there are 2 df associated with each piece, a total of 13 confounded strings can be estimated. This implies each confounded string contains 3 different pieces and accounts for 39 of the 40 pieces. The 40th piece is the confounded with the identity generator. This is just like the 2 factor case, a 1/2 fraction confounds 2 terms, a 1/4 fraction confounds 4 terms, and so on. Here, a 1/3 fraction confounds 3 pieces, a 1/9 fraction confounds 9 pieces, and so on.

The question is, what happened to the 2 df for the piece confounded with the identity, having only 1 df? The answer is that both df's get confounded with the identity. Therefore, for the identity, and the identity only, we include the square of the piece. This yields the confounding pattern in Table 8.1.2. We show the intermediate steps for a and b only.

Notice that all of the pieces are accounted for once and only once and that the term was sometimes squared so the first letter always has exponent one. This is a resolution IV design so can be abbreviated as a

TABLE 8.1.1

PIECES OF A 3^4 DESIGN USED FOR CONFOUNDING

Source	df	Pieces	df	Source	df	Pieces	df
A	2	a	2	CD	4	cd	2
B	2	b	2			cd^2	2
AB	4	ab	2	ACD	8	acd	2
		ab^2	2			acd^2	2
C	2	c	2			ac^2d	2
AC	4	ac	2			ac^2d^2	2
		ac^2	2	BCD	8	bcd	2
BC	4	bc	2			bcd^2	2
		bc^2	2			bc^2d	2
ABC	8	abc	2			bc^2d^2	2
		abc^2	2	$ABCD$	16	$abcd$	2
		ab^2c	2			$abcd^2$	2
		ab^2c^2	2			abc^2d	2
D	2	d	2			abc^2d^2	2
AD	4	ad	2			ab^2cd	2
		ad^2	2			ab^2cd^2	2
BD	4	bd	2			ab^2c^2d	2
		bd^2	2			$ab^2c^2d^2$	2
ABD	8	abd	2				
		abd^2	2				
		ab^2d	2				
		ab^2d^2	2				

3_{IV}^{4-1} design. This can be seen by looking at the length of the smallest piece in the identity generator or by looking at the lowest order of the pieces confounded with the main effects.

You will notice that we continually refer to the pieces of the interactions rather than the interactions themselves. For example, we see that pieces of the ABC interaction are confounded with D and pieces of the AD, BD, and CD interactions. This means, if the ABC interaction happened to be significant, we could see a significance in any or all of D, AD, BD, or CD.

To create the ANOVA table, combine the pieces together keeping careful track of the df. For the main effects, this is straightforward. Notice, however, that pieces of the AB, AC, and AD interaction are listed first in two different alias strings. These should therefore have 4 df.

TABLE 8.1.2

CONFOUNDING PATTERN FOR 3_{IV}^{4-1} WITH I = ab^2cd^2

$$
\begin{aligned}
I &= ab^2cd^2 = a^2bc^2d \\
a &= a^2b^2cd^2 = a^4b^4c^2d^4 = abc^2d \\
&= a^3bc^2d = bc^2d \\
b &= ab^3cd^2 = acd^2 \\
&= a^2b^2c^2d = abcd^2 \\
c &= ab^2c^2d^2 = ab^2d^2 \\
d &= ab^2c = ab^2cd \\
ab &= ac^2d = bcd^2 \\
ab^2 &= ab^2c^2d = cd^2 \\
ac &= abcd = bd \\
ac^2 &= abd = bcd \\
ad &= abc^2 = bc^2d^2 \\
ad^2 &= abc^2d^2 = bc^2 \\
bc &= ac^2d^2 = abd^2 \\
bd^2 &= acd = abc \\
cd &= ab^2c^2 = ab^2d
\end{aligned}
$$

Pieces of the BC, BD, and CD interactions only appear first in one alias string so should have only 2 df. The final result is given in Table 8.1.3. Note that this is order dependent.

Table 8.1.3, along with the corresponding SS and MS will be obtained if you analyze the data using type I sums of squares with the following model:

$$Y = A + B + C + D + AB + AC + AD + BC + BD + CD + \varepsilon.$$

Both the SS and the df will sum to the total. But, type I sum of squares depends on the order of the terms in the model. And the order was chosen as a result of an arbitrary alphabetic labeling convention adopted in the previous chapter. If, in the confounded string involving the pieces ad^2 and bc^2, bc^2 were placed first, then BC would have 4 df and AD would have 2 df. This corresponds to using type I SS on the mathematical model

$$Y = A + B + C + D + AB + AC + BC + AD + BD + CD + \varepsilon.$$

TABLE 8.1.3

ANOVA FOR A 3_{IV}^{4-1} DESIGN USING TYPE I SS

Source	df
A &/or BCD &/or $ABCD$	2
B &/or ACD &/or $ABCD$	2
C &/or ABD &/or $ABCD$	2
D &/or ABC &/or $ABCD$	2
AB &/or CD &/or ACD &/or BCD &/or $ABCD$	4
AC &/or BD &/or ABD &/or BCD &/or $ABCD$	4
AD &/or BC &/or ABC &/or BCD &/or $ABCD$	4
BC &/or ABD &/or ACD	2
BD &/or ABD &/or ACD	2
CD &/or ABC &/or ACD	2

The reason for the difference is that type I SS are computed sequentially. That is, type I estimates the effect of a term given all previous terms. So the 2 df for the string confounding AD and BC will go to whichever term is listed first in the model.

If one uses type III SS, the results are not order dependent. However, all two factor interactions will have 2 df and the sum of the individual df will not equal the total df. This is because type III estimates the effect of any specific term in the model given *all other* terms in the model. Thus, the estimate of AB will already have the piece confounded with CD removed—leaving only 2 df.

Given this confusion, which method of estimating effects and df is correct? Fortunately, we do not really need to answer that question. When we design a resolution IV experiment, we believe main effects are of primary interest and are not to be confounded with any two factor interactions. The two factor interactions themselves are not of interest. For a resolution IV design, the SS for all main effects are the same for all types of SS and all orderings of the model. This is because main effects are *orthogonal* to all of the two factor interactions. This property was built into the design by the method of construction.

When there is more than one identity generator creating a design, the complete identity generator is formed by squaring each individual generator and forming all possible cross products. That is, if X and Y are two generators, the complete identity generator consists of $I = X = X^2 = Y = Y^2 = XY = XY^2 = X^2Y = X^2Y^2$. This gives a 1/9th fraction and every term will be confounded with 8 other terms because there are 8 terms associated with the identity I.

Specifically, if the generators were ab^2ce and ad^2ef^2, the complete identity generator would be

$$\begin{aligned} \mathrm{I} &= ab^2ce = a^2bc^2e^2 = ad^2ef^2 = a^2de^2f \\ &= a^2b^2cd^2e^2f^2 = b^2cdf = bc^2d^2f^2 = abc^2def, \end{aligned}$$

a resolution IV design. As can be seen by this simple example, the confounding gets quite involved quite quickly and is probably a task best suited for a computer.

Given the identity generators, the actual design points can be found using modular arithmetic as was done in Chapter 7. That is, for factors A, B, C, and D, whose levels take the values 0, 1, or 2, if the identity generator is ab^2cd^2, the primary design is the set of all values satisfying

$$A + 2B + C + 2D = 0 \text{ mod } 3.$$

This can be solved by trial and error or by taking A, B, and C as the basic factors, writing down all 27 combinations of A, B, and C, and letting

$$D = \begin{cases} 0, & \text{if } A + 2B + C = 0 \text{ mod } 3 \\ 1, & \text{if } A + 2B + C = 1 \text{ mod } 3 \\ 2, & \text{if } A + 2B + C = 2 \text{ mod } 3 \end{cases}$$

These values for D make the sum 0 mod 3. Again this is a good task for a computer and packages are becoming more readily available.

Finding the generators that will meet a given design criteria can be done using predetermined tables, existing computer programs, or using the basic factor technique given in Chapter 7. A list of 41 designs ranging from a 3_{IV}^{4-1} to a 3_{V}^{10-5} was generated by the National Bureau of Standards (1961) and is given in McLean and Anderson (1984). Proc FACTEX in the SAS system is a good example of a computer program that can handle any prime fractionation. However, when no computer is available, one can use the basic factor technique from Chapter 7. This must be performed on the pieces of the interactions and all of the pieces of the basic factors must be written out.

As an example of the basic factor technique, suppose there are 7 factors, A–G, all at 3 levels, and we wish to estimate the main effects as well as the interactions AB, AC, and AD. The main effects require 14 df and the interactions require 12 df, for a total of 26. Therefore, there is hope that a 27 run 3^{7-4} design can be used. We start by writing down a full factorial on the pieces of the basic factors A, B, and C. To the pieces of the basic factors, we assign the main effects and appropriate pieces of the original factors A, B, and C.

original pieces

a	b	ab	ab^2	c	ac	ac^2						
a	b	ab	ab^2	c	ac	ac^2	bc	bc^2	abc	abc^2	ab^2c	ab^2c^2

basic pieces

We next place d in the right most column and compute where ad and ad^2 will fall. They do not conflict with any other assignment.

original pieces

a	b	ab	ab^2	c	ac	ac^2	ad^2		ad			d
a	b	ab	ab^2	c	ac	ac^2	bc	bc^2	abc	abc^2	ab^2c	ab^2c^2

basic pieces

As a final step, we assign the remaining main effects to the open columns.

original pieces

a	b	ab	ab^2	c	ac	ac^2	ad^2	g	ad	f	e	d
a	b	ab	ab^2	c	ac	ac^2	bc	bc^2	abc	abc^2	ab^2c	ab^2c^2

basic pieces

This gives one assignment that satisfies the desired criteria. Of course, there are many other assignments that work.

The placement of the main effects again determines the identity generators. The right most column shows $d = ab^2c^2$. Multiplying each side of the equation by d^2 (not d because we are using mod 3 arithmetic) shows that $\mathrm{I} = ab^2c^2d^2$. We also find $\mathrm{I} = ab^2ce^2$, $\mathrm{I} = abc^2f^2$, and $\mathrm{I} = bc^2g^2$. The confounding pattern is very complicated.

8.1.2 General Prime Level Fractions

Fractionation of prime levels greater than 3 is used less often but poses no special problems. Let p be the prime number of levels each factor contains. Then, all arithmetic is done mod p, each piece will have $p - 1$ df, and a two factor interaction would consist of the pieces ab, ab^2, ab^3, $\cdots$, and ab^{p-1}. If X were the identity generator, the complete identity generator would consist of $\mathrm{I} = X = X^2 = \cdots = X^{p-1}$ and in a p^{n-m} design, each confounded string would contain p^m pieces.

We leave it to the reader to fill in all of the details if the need should ever arise. Simply follow the pattern for $p = 3$.

Fractional designs that have had enough popularity to earn special names are the *Latin square design* and the *Graeco-Latin square design*. The Latin square, Table 8.1.4a, is a p_{III}^{3-1} design formed by a generator like I = abc and the Graeco-Latin square, Table 8.1.4b, is a p_{III}^{4-2} design formed by generators like I = abc and I = ab^2d. The names come from the agricultural literature where the first two factors are rows and columns and are feared to have an effect and the third and/or fourth factors are treatments labeled using Latin and Latin and Greek letters respectively. Usually there are restriction errors on the randomization in these designs.

TABLE 8.1.4

LATIN (5_{III}^{3-1}) AND GRAECO-LATIN (5_{III}^{4-2}) SQUARE DESIGNS

		A 1	2	3	4	5
B	1	1	5	4	3	2
	2	5	4	3	2	1
	3	4	3	2	1	5
	4	3	2	1	5	4
	5	2	1	5	4	3

table contains C

		A 1	2	3	4	5
B	1	1,1	5,4	4,2	3,5	2,3
	2	5,5	4,3	3,1	2,4	1,2
	3	4,4	3,2	2,5	1,3	5,1
	4	3,3	2,1	1,4	5,2	4,5
	5	2,2	1,5	5,3	4,1	3,4

table contains C, D

8.2 POWERS OF PRIMES

Factors that are powers of primes can be fractionated using a simple trick. As is often the case, the trick is most easily demonstrated using an example.

Suppose an experiment consists of an eight level factor and four different two level factors. To fractionate this experiment, we must realize that eight is a power of two, $8 = 2^3$, and represent the eight level factor as three different two level *pseudo* factors. If we label the eight level factor A, we can label the three pseudo factors A_1, A_2, and A_3, and associate the original factor with the pseudo factors as in Table 8.2.1. There is nothing unique about this association, any will do.

The 7 df for A are associated with the pseudo factors *and* their interactions. Thus, each of A_1, A_2, A_3, A_1A_2, A_1A_3, A_2A_3, and $A_1A_2A_3$ actually represent one df for the main effect A. We then create the design

TABLE 8.2.1

AN 8 LEVEL FACTOR AS 3 PSEUDO FACTORS

A	A_1	A_2	A_3
0	0	0	0
1	0	0	1
2	0	1	0
3	0	1	1
4	1	0	0
5	1	0	1
6	1	1	0
7	1	1	1

using the pseudo factors, realizing that the interactions of the pseudo factors are really main effects for A, and substitute the levels of the original for the levels of the pseudo factors.

A typical question for this set up may be: can a resolution IV design be created for the 8×2^4 experiment in 16 runs? The main effects alone require 11 df so things seem doubtful. However, using Appendix 15, a resolution IV design exists for 7 factors, the 3 pseudo factors and the 4 two level factors, each at two levels. Perhaps there is a chance.

Label the factors A, B, C, D, and E and label the pseudo factors A_1, A_2, and A_3. We must be able to estimate all of the interactions involving the A_i as well as the main effects B, C, D, and E and, furthermore, cannot have any main effects confounded with any two factor interaction. A cannot be used as a basic factor since it has 8 levels, so we will use B, C, D, and E as the basic factors. Start by assigning the first four "factors", A_1, A_2, A_3, and B to the basic factors.

original factors

A_1	A_2	A_1A_2	A_3	A_1A_3	A_2A_3	$A_1A_2A_3$	B	A_1B	A_2B	A_1A_2B	A_3B	A_1A_3B	A_2A_3B	$A_1A_2A_3B$
B	C	BC	D	BD	CD	BCD	E	BE	CE	BCE	DE	BDE	CDE	$BCDE$

basic factors

Since all interactions involving any or all of the A_i's are really part of the main effect A, an equivalent representation is

original factors

A	A	A	A	A	A	A	B	AB	AB	AB	AB	AB	AB	AB
B	C	BC	D	BD	CD	BCD	E	BE	CE	BCE	DE	BDE	CDE	$BCDE$

basic factors

All columns are assigned to main effects or two factor interactions so no more main effects can be assigned without violating the resolution IV criteria. We conclude that it is not possible to create a resolution IV design in 16 runs.

Next we try to get a resolution IV design in 32 runs. We can still use the 16 run assignment given previously and add 16 more columns. We must use one of the pseudo factors as a basic factor. To avoid confusion, let us label this factor F. We retain the previous assignments and assign the original factor C to the next basic factor to get the following assignments:

original factors

C	AC	AC	AC	AC	AC	AC	AC	BC							
F	BF	CF	BCF	DF	BDF	CDF	$BCDF$	EF	BEF	CEF	$BCEF$	DEF	$BDEF$	$CDEF$	$BCDEF$

basic factors

No matter where we place the original factor D, the two factor interactions between A and D will fill the remainder of the unassigned columns. It will then be impossible to place the original factor E without confounding it with a main effect or two factor interaction.

Thus, 64 runs are required to get a resolution IV design on five factors with one factor at 8 levels and the remaining four at 2 levels. This size is unacceptable in many experiments.

Perhaps, if we relax the resolution IV requirement by assuming all interactions involving A are negligible, we could still get a 16 run experiment. We start with the usual assignment, not assigning any interaction with A because we are assuming interactions with A are negligible. (Of course, this has to be justified using other expertise about the process being tested.)

original factors

A	A	A	A	A	A	A	B							
B	C	BC	D	BD	CD	BCD	E	BE	CE	BCE	DE	BDE	CDE	$BCDE$

basic factors

No matter where we assign the original factor C, the BC interaction will be confounded with a piece of the main effect A. As a matter of fact, it is impossible to assign the original factors in such a way that even one main effect will not be confounded with some two factor interaction. A 32 run design is required.

To demonstrate how to generate the factor level combinations with a design having pseudo factors, let us accept the resolution III design resulting from assigning the original factor C to the basic factor interaction CDE, D to the basic factor interaction BDE, and E to the basic factor interaction BCE. We must go back to the pseudo factors to create the design. That is, recall that basic factor B is really pseudo factor A_1, basic factor C is really pseudo factor A_2, and basic factor D is really pseudo factor A_3.

original pseudo factors

A_1	A_2	A_1A_2	A_3	A_1A_3	A_2A_3	$A_1A_2A_3$	B			E		D		C
B	C	BC	D	BD	CD	BCD	E	BE	CE	BCE	DE	BDE	CDE	$BCDE$

basic factors

The identity generators are

$$C = A_2A_3B$$
$$D = A_1A_3B$$
$$E = A_1A_2B$$

which creates the factor level combinations through the equations

$$C = A_2 + A_3 + B \bmod 2,$$
$$D = A_1 + A_3 + B \bmod 2, \text{ and}$$
$$E = A_1 + A_2 + B \bmod 2,$$

where the levels of the factors take the values 0 or 1. Create a full factorial in A_1, A_2, A_3, and B, solve for C, D, and E using the equations given above, and substitute the value for A from the values for A_1, A_2, and A_3. The final design is given in Table 8.2.3. The reader must make sure the entire process is fully understood since we will cover one example only.

The confounding pattern is determined using the techniques of the previous section. The only difference is that the pseudo factors make up the pieces used for confounding. Everything else is identical. For the example above, the identity generators would be

$$\begin{aligned} \mathrm{I} &= A_2A_3BC \\ &= A_1A_3BD = A_1A_2CD \\ &= A_1A_2BE = A_1A_3CE = A_2A_3DE = BCDE. \end{aligned}$$

This implies that, for example, B is confounded with pieces of the AC, AD, $ABCD$, AE, $ABCE$, $ABDE$, and the CDE interaction. C is confounded with pieces of the AB, $ABCD$, AD, $ABCE$, AE, $ACDE$, and the BDE interaction. And so forth.

Using the alphabetic notation, type I SS would assign 7 df to A, 1 df to B, C, D, and E, and 4 df to the AB interaction. The analysis usually assumes all interactions negligible and tests the main effects with the 4

TABLE 8.2.3

8×2^4 RESOLUTION III DESIGN

A_1	A_2	A_3	A	B	C	D	E
0	0	0	0	0	0	0	0
0	0	0	0	1	1	1	1
0	0	1	1	0	1	1	0
0	0	1	1	1	0	0	1
0	1	0	2	0	1	0	1
0	1	0	2	1	0	1	0
0	1	1	3	0	0	1	1
0	1	1	3	1	1	0	0
1	0	0	4	0	0	1	1
1	0	0	4	1	1	0	0
1	0	1	5	0	1	0	1
1	0	1	5	1	0	1	0
1	1	0	6	0	1	1	0
1	1	0	6	1	0	0	1
1	1	1	7	0	0	0	0
1	1	1	7	1	1	1	1

TABLE 8.2.4

A 9 LEVEL FACTOR AS 2 PSEUDO FACTORS

A	A_1	A_2
0	0	0
1	0	1
2	0	2
3	1	0
4	1	1
5	1	2
6	2	0
7	2	1
8	2	2

df that were assigned to AB. This may or may not make sense from a practical point of view.

The tricks in this section work for all powers of primes. For example, a nine level factor can be represented by 2 three level pseudo factors as illustrated in Table 8.2.4.

8.3 MIXED LEVEL DESIGNS

Fractions of different prime levels can be constructed by working with each prime individually and crossing the resulting designs. Crossing two designs consists of running the entire second design with each design point of the first design. While the procedure is relatively easy to understand, the designs are not very desirable because they tend to have large run sizes.

Consider a resolution III design with 9 factors; 5 with two levels, 3 with three levels, and 1 with six levels. Label the two level factors A–E, the three level factors F–H, and the six level factor J. (We skip I to avoid confusion with the identity I.) Break the six level factor into a two and a three level factor as in Table 8.3.1.

TABLE 8.3.1

A 6 LEVEL FACTOR AS 2 AND 3 LEVEL PSEUDO FACTORS

J	J_1	J_2
0	0	0
1	0	1
2	0	2
3	1	0
4	1	1
5	1	2

Now, collect the two level factors and create a resolution III design. With 6 factors, Appendix 16 gives the identity generators I $= ABCD = BCE = ACJ_1$. The resulting design has 8 runs and is given in Table 8.3.2.

TABLE 8.3.2

A 2_{III}^{6-3} DESIGN

A	B	C	D	E	J_1
0	0	0	0	0	0
0	0	1	1	1	1
0	1	0	1	1	0
0	1	1	0	0	1
1	0	0	1	0	1
1	0	1	0	1	0
1	1	0	0	1	1
1	1	1	1	0	0

Next, collect the 4 three level factors and create a resolution III design. A set of identity generators that work are I $= FG^2H = FGJ_2$, yielding the 9 run design in Table 8.3.3.

TABLE 8.3.3

A 3^{4-2}_{III} DESIGN

F	G	H	J_2
0	0	0	0
0	1	1	2
0	2	2	1
1	0	2	2
1	1	0	1
1	2	1	0
2	0	1	1
2	1	2	0
2	2	0	2

The crossed design will have $8 \times 9 = 72$ runs and is constructed by running the entire 3^{4-2}_{III} design with every run of the 2^{6-3}_{III} design (or vice-versa). This design is given in Table 8.3.4. It will have resolution III because crossed designs have resolution equal to the smallest resolution of its individual components. We can call this a 1/72 replicate of a $2^5 \times 3^3 \times 6$ design having resolution III.

The confounding pattern for crossing two designs is extremely lengthy but straightforward. The complete identity generator for the crossed design consists of cross multiplying the complete identity generator for each individual design. Each term in the complete identity generator for the two level fraction has 1 df. Each term in the complete identity generator for the three level fraction has 2 df. Each term in the complete identity generator for the crossing will have $1 \times 2 = 2$ df. In this case, the complete identity generator is

$$\text{From the } 2^{6-3}_{III}, \quad \text{I} = \begin{cases} ABCD \\ BCE = ADE \\ ACJ_1 = BDJ_1 = ABEJ_1 = CDEJ_1 \end{cases}$$

$$\text{From the } 3^{4-2}_{III}, \quad \text{I} = \begin{cases} FG^2H = F^2GH^2 \\ FGJ_2 = F^2G^2J_2^2 = F^2HJ_2 = \cdots = FH^2J_2^2 \end{cases}$$

TABLE 8.3.4

A 72 RUN RESOLUTION III 1/72 REPLICATE OF A $2^5 \times 3^3 \times 6$ DESIGN

A	B	C	D	E	F	G	H	J_1	J_2	J
0	0	0	0	0	0	0	0	0	0	0
0	0	0	0	0	0	1	1	0	2	2
0	0	0	0	0	0	2	2	0	1	1
0	0	0	0	0	1	0	2	0	2	2
0	0	0	0	0	1	1	0	0	1	1
0	0	0	0	0	1	2	1	0	0	0
0	0	0	0	0	2	0	1	0	1	1
0	0	0	0	0	2	1	2	0	0	0
0	0	0	0	0	2	2	0	0	2	2
0	0	1	1	1	0	0	0	1	0	3
0	0	1	1	1	0	1	1	1	2	5
0	0	1	1	1	0	2	2	1	1	4
0	0	1	1	1	1	0	2	1	2	5
0	0	1	1	1	1	1	0	1	1	4
0	0	1	1	1	1	2	1	1	0	3
0	0	1	1	1	2	0	1	1	1	4
0	0	1	1	1	2	1	2	1	0	3
0	0	1	1	1	2	2	0	1	2	5
0	1	0	1	1	0	0	0	0	0	0
0	1	0	1	1	0	1	1	0	2	2
0	1	0	1	1	0	2	2	0	1	1
0	1	0	1	1	1	0	2	0	2	2
0	1	0	1	1	1	1	0	0	1	1
0	1	0	1	1	1	2	1	0	0	0
0	1	0	1	1	2	0	1	0	1	1
0	1	0	1	1	2	1	2	0	0	0
0	1	0	1	1	2	2	0	0	2	2
0	1	1	0	0	0	0	0	1	0	3
0	1	1	0	0	0	1	1	1	2	5
0	1	1	0	0	0	2	2	1	1	4
0	1	1	0	0	1	0	2	1	2	5
0	1	1	0	0	1	1	0	1	1	4
0	1	1	0	0	1	2	1	1	0	3
0	1	1	0	0	2	0	1	1	1	4
0	1	1	0	0	2	1	2	1	0	3
0	1	1	0	0	2	2	0	1	2	5
1	0	0	1	0	0	0	0	1	0	3
1	0	0	1	0	0	1	1	1	2	5
1	0	0	1	0	0	2	2	1	1	4
1	0	0	1	0	1	0	2	1	2	5
1	0	0	1	0	1	1	0	1	1	4
1	0	0	1	0	1	2	1	1	0	3
1	0	0	1	0	2	0	1	1	1	4
1	0	0	1	0	2	1	2	1	0	3
1	0	0	1	0	2	2	0	1	2	5
1	0	1	0	1	0	0	0	0	0	0
1	0	1	0	1	0	1	1	0	2	2
1	0	1	0	1	0	2	2	0	1	1
1	0	1	0	1	1	0	2	0	2	2
1	0	1	0	1	1	1	0	0	1	1
1	0	1	0	1	1	2	1	0	0	0
1	0	1	0	1	2	0	1	0	1	1
1	0	1	0	1	2	1	2	0	0	0
1	0	1	0	1	2	2	0	0	2	2
1	1	0	0	1	0	0	0	1	0	3
1	1	0	0	1	0	1	1	1	2	5
1	1	0	0	1	0	2	2	1	1	4
1	1	0	0	1	1	0	2	1	2	5
1	1	0	0	1	1	1	0	1	1	4
1	1	0	0	1	1	2	1	1	0	3
1	1	0	0	1	2	0	1	1	1	4
1	1	0	0	1	2	1	2	1	0	3
1	1	0	0	1	2	2	0	1	2	5
1	1	1	1	0	0	0	0	0	0	0
1	1	1	1	0	0	1	1	0	2	2
1	1	1	1	0	0	2	2	0	1	1
1	1	1	1	0	1	0	2	0	2	2
1	1	1	1	0	1	1	0	0	1	1
1	1	1	1	0	1	2	1	0	0	0
1	1	1	1	0	2	0	1	0	1	1
1	1	1	1	0	2	1	2	0	0	0
1	1	1	1	0	2	2	0	0	2	2

$$\text{From crossing,}\quad I = \begin{cases} ABCDFG^2H = ABCDF^2GH^2 \\ ABCDFGJ_2 = \cdots = ABCDFH^2J_2^2 \\ BCEFG^2H = ADEF^2GH^2 \\ BCEFGJ_2 = \cdots = ADEFH^2J_2^2 \\ ACJ_1FG^2H = \cdots = CDEJ_1F^2GH^2 \\ ACJ_1FGJ_2 = \cdots = CDEJ_1FH^2J_2^2. \end{cases}$$

In computing the actual confounding, we must remember that quantities like ACJ_1FGJ_2 simply represent a piece of the $ACFGJ$ interaction. This piece will have 2 df because we multiply the ACJ_1 part having 1 df by the FGJ_2 part having 2 df. Also note that all of the interactions between terms estimable in the first design and terms estimable in the second design will be estimable in the crossed design.

From a practical point of view, 72 runs to estimate the 16 df for main effects is not very efficient. This is a big disadvantage of mixed designs that will be addressed later in this chapter.

Even though this design turned out to be quite large, the idea of crossing designs should not be dismissed. Crossing designs will often result in cheaper experiments when some factor combinations are expensive and others are inexpensive. We illustrate with an example.

Suppose we are interested in designing a transmission that has smooth shift points regardless of the temperature, speed, or load. There are five different *design factors* under study, labeled A–E. The three *running condition factors*, temperature, speed, and load, are labeled F, G, and H respectively. All factors have two levels and a resolution V design is required. From Appendix 14, a 2_V^{8-2} experiment requires 64 runs. In examining the design, it is discovered that there are 32 different *design factor* combinations. This is undesirable because each different *design factor* combination requires construction of an expensive one-of-a-kind transmission. Changing temperature, speed, and load is a simple and inexpensive proposition. Can a cheaper design be constructed?

The key to this problem is to realize we want to minimize the number of combinations of the factors A through E, not the number of combinations of the factors A through H. Again, from Appendix 14, a resolution V design on factors A through E requires 16 runs. A resolution V design on the factors F through H requires a full factorial design, 8 runs. The cross product of the two designs will be a resolution V design and will require $16 \times 8 = 128$ runs. This is twice as many runs, but only half as many expensive transmissions. From the standpoint of expense, the 128 run cross product design is actually cheaper than the 64 run design.

From a statistical point of view, both are resolution V designs. From a practical point of view, the 128 run design may be better.

8.4 ORTHOGONAL MAIN EFFECT DESIGNS

When the size of the design is of utmost importance and there are mixed levels for the factors, orthogonal main effect designs, OMEDs, are appropriate. These designs need not be balanced, *i.e.*, there need not be the same number of runs for each level of a factor, but the main effects are orthogonal. This means the estimate of one main effect is not influenced by the other main effects. Usually the confounding pattern is so complicated that even two factor interactions are not considered. These designs should only be run under the strong assumption that *all* interactions are negligible. A trick that can be used when there are a few interactions will be given later in this section.

Appendix 17 gives a list of orthogonal main effect designs for factors having 2 to 6 levels and experiment sizes up to 50. These have been generated using methods given in Dey (1985) and Wang and Wu (1991). A tabulation is given first followed by the designs themselves. The tabulation gives all known (as of the time of this book) minimal sized designs up to 6 levels and 50 runs. Whenever possible, designs have been combined to save space and association or substitution tables provided.

Even though specific designs have been listed, Appendix 17 can be used for all designs (up to 6 levels and 50 runs) using the following rule.

RULE FOR FINDING THE SMALLEST OMED

- If any factor exceeds 6 levels, Appendix 17 cannot be used.
- For the required design, compute the number of factors greater than or equal to 6, 5, 4, 3, and 2 levels.
- Starting from the top of Appendix 17 and working toward the bottom, select the first tabulated design that meets or exceeds requirements for all of the levels 6, 5, 4, 3, and 2.
- The design is listed on the appropriate page. Assign the factors to the tabulated design from left to right, largest levels to smallest levels. If the tabulated design does not have a column exactly matching the required number of levels, use a column with a greater number of levels.
- It is acceptable to leave columns unassigned.
- When assigning a factor to a column having extra levels, simply repeat one or more of the assigned levels. The design will still be orthogonal.

Suppose we wish to study the main effects in an experiment that contains 1 four level factor, 2 three level factors, and 15 two level factors, *i.e.*, a $4 \times 3^2 \times 2^{15}$ design. The number of factors $\geq$ 6, 5, 4, 3, and 2 are, respectively, 0, 0, 1, 3, and 18. Going down the rows in Appendix 17, we find that a $4^3 \times 2^{22}$ design in 32 runs, having number of factors $\geq$ 6, 5, 4, 3, and 2 equal to 0, 0, 3, 3, and 25 respectively, is the first design to meet the requirements. In the displayed 2^{31} design, we make columns 1–3, 4–6, and 7-9 into four level columns using the association at the bottom of the page. We assign the first four level column to the four level factor and the next 2 four level columns to the three level factors, relabeling the fourth level to, say, the second level of the three level factors. We will end up with the last 7 columns of the 2^{31} design unassigned so we will simply ignore them. The resulting design is given in Table 8.4.1. It goes without saying that the design should be randomized.

TABLE 8.4.1

A $4 \times 3^2 \times 2^{15}$ ORTHOGONAL MAIN EFFECT DESIGN IN 32 RUNS

4	3	3	2	2	2	2	2	2	2	2	2	2	2	2	2	2	2
0	0	0	1	1	0	0	1	1	1	0	1	1	0	1	0	1	1
0	0	1	0	0	0	1	0	1	0	1	1	0	1	1	1	0	1
0	1	1	0	1	1	1	1	0	1	0	1	1	1	0	0	0	0
0	1	2	1	0	1	0	0	0	0	1	1	0	0	0	1	1	0
0	2	1	1	1	0	0	1	1	0	0	0	0	0	0	1	1	0
0	2	2	0	0	0	1	0	1	1	1	0	1	1	0	0	0	0
0	1	0	0	1	1	1	1	0	0	0	0	0	1	1	1	0	1
0	1	1	1	0	1	0	0	0	1	1	0	1	0	1	0	1	1
1	0	1	1	0	1	1	1	0	1	1	0	0	0	0	1	0	1
1	0	2	0	1	1	0	0	0	0	0	0	1	1	0	0	1	1
1	1	0	0	0	0	0	1	1	1	1	0	0	1	1	1	1	0
1	1	1	1	1	0	1	0	1	0	0	0	1	0	1	0	0	0
1	2	0	1	0	1	1	1	0	0	1	1	1	0	1	0	0	0
1	2	1	0	1	1	0	0	0	1	0	1	0	1	1	1	1	0
1	1	1	0	0	0	0	1	1	0	1	1	1	1	0	0	1	1
1	1	2	1	1	0	1	0	1	1	0	1	0	0	0	1	0	1
2	0	1	1	1	0	0	0	0	1	1	0	0	1	1	0	0	0
2	0	2	0	0	0	1	1	0	0	0	0	1	0	1	1	1	0
2	1	0	0	1	1	1	0	1	1	1	0	0	0	0	0	1	1
2	1	1	1	0	1	0	1	1	0	0	0	1	1	0	1	0	1
2	2	0	1	1	0	0	0	0	0	1	1	1	1	0	1	0	1
2	2	1	0	0	0	1	1	0	1	0	1	0	0	0	0	1	1
2	1	1	0	1	1	1	0	1	0	1	1	1	0	1	1	1	0
2	1	2	1	0	1	0	1	1	1	0	1	0	1	1	0	0	0
3	0	0	1	0	1	1	0	1	1	0	1	1	1	0	1	1	0
3	0	1	0	1	1	0	1	1	0	1	1	0	0	0	0	0	0
3	1	1	0	0	0	0	0	0	1	0	1	1	0	1	1	0	1
3	1	2	1	1	0	1	1	0	0	1	1	0	1	1	0	1	1
3	2	1	1	0	1	1	0	1	0	0	0	0	1	1	0	1	1
3	2	2	0	1	1	0	1	1	1	1	0	1	0	1	1	0	1
3	1	0	0	0	0	0	0	0	0	0	0	0	0	0	0	0	0
3	1	1	1	1	0	1	1	0	1	1	0	1	1	0	1	1	0

The df that go into the error term will come from three sources: the df left listed in the appendix, the number of repeated levels, and the unassigned columns. There are 0 df left listed in the appendix, 2 factors have repeated 1 level for $2 \times 1 = 2$ df, and there are 7 unassigned two level factors. Thus, the df for error sums to 9 and with the df for the main effects accounts for all 31 df.

As a check that you understand the entire process, you should independently follow the rules and compare your results to those in Table 8.4.1.

8.5 OTHER MIXED LEVEL DESIGNS

Procedures for developing resolution IV and resolution V mixed level designs are given in Dey (1985) along with a table indicating which resolution IV designs are known to exist. These will not be given.

McLean and Anderson (1984) present approximately 40 fractional factorial designs dealing with factors having two and three levels. These designs estimate all main effects and two factor interactions, are nearly orthogonal, and were originally published by the National Bureau of Standards (1961). The designs are created by combining fractions of the two level factors with fractions of the three level factors. Using the example in the National Bureau of Standards pamphlet (1961), consider a $2^3 \times 3^2$ experiment. One cannot fractionate either the two level or the three level pieces without confounding main effects with two factor interactions. So, break the 2^3 design into two 1/2 fractions, R_1 and R_2, corresponding to $A+B+C = 0 \bmod 2$ and $A+B+C = 1 \bmod 2$ and break the 3^2 design into three 1/3 fractions, S_1, S_2, and S_3, corresponding to $D+2E = 0, 1,$ and 2 mod 3. Then cross R_1 with S_1, R_2 with S_2, and R_2 again with S_3. This results in the $4 \times 3 + 4 \times 3 + 4 \times 3 = 36$ run design derived in Table 8.5.1. This design is not quite orthogonal due to the last crossing. Exact orthogonality in a resolution V design requires a full factorial consisting of 72 runs.

A trick to be used when mixed level designs are required and a few interactions need to be estimated is to alter orthogonal main effect designs by combining the main effects and interactions into main effects having the product of the levels of the original main effects. For example, the AB interaction for two level factors A and B can be estimated by combining A and B into a four level factor, say C, to be used in an orthogonal main effect design. When the data is collected, the 3 df for C must be broken down into the two main effect df and the interaction df. This is basically the reverse of the pseudo factor technique introduced in Section **8.2**.

TABLE 8.5.1

DERIVING A $2^3 \times 3^2$ DESIGN IN 36 RUNS

A 2^3 in 2 blocks

R_1			R_2		
A	B	C	A	B	C
0	0	0	1	1	1
1	1	0	1	0	0
1	0	1	0	1	0
0	1	1	0	0	1

A 3^2 in 3 blocks

S_1		S_2		S_3	
D	E	D	E	D	E
0	0	1	0	2	0
1	2	0	1	0	2
2	1	2	2	1	1

The associated $2^3 \times 3^2$ design in 36 runs

R_1S_1					R_2S_2					R_2S_3				
A	B	C	D	E	A	B	C	D	E	A	B	C	D	E
0	0	0	0	0	1	1	1	1	0	1	1	1	2	2
0	0	0	1	2	1	1	1	0	1	1	1	1	0	0
0	0	0	2	1	1	1	1	2	2	1	1	1	1	1
1	1	0	0	0	1	0	0	1	0	1	0	0	2	2
1	1	0	1	2	1	0	0	0	1	1	0	0	0	0
1	1	0	2	1	1	0	0	2	2	1	0	0	1	1
1	0	1	0	0	0	1	0	1	0	0	1	0	2	2
1	0	1	1	2	0	1	0	0	1	0	1	0	0	0
1	0	1	2	1	0	1	0	2	2	0	1	0	1	1
0	1	1	0	0	0	0	1	1	0	0	0	1	2	2
0	1	1	1	2	0	0	1	0	1	0	0	1	0	0
0	1	1	2	1	0	0	1	2	2	0	0	1	1	1

8.6 BLOCKING

The concept of blocking has played a sufficiently important role in design of experiments that we decided to include a special section on it. Most of the material presented here has already appeared in this book, in one chapter or another. Here we bring it together explicitly.

We also introduce two other types of designs that have not previously appeared in this book. We let the interested reader pursue these further.

The idea behind blocking is that the entire experiment cannot be run under the same homogeneous conditions. This restriction may be due to the amount of material available, the amount of space available, the amount of time available, the amount of test equipment available, or any other similar limitation. If the experimenter failed to recognize the homogeneity and completely randomized, the error would contain both the within block variation and the between block variation (as well as the interaction if it exists). If blocks are included as a factor in the experiment, the error would only contain the within block variation as the between blocks variation is accounted for by the factor blocks. This should lead to a smaller error term and more precise tests. The experiment must therefore be altered to reflect the limitation.

The historic approach to this problem is to design the experiment first and then try to figure out a way to run it in blocks that meet the homogeneity condition. The formulation may go as follows. I have 8 factors, all at two levels. Based on Appendix 15, I can run a resolution IV experiment in 16 runs. However, I can only run 8 treatment combinations under homogeneous conditions. Can I still use a 16 run experiment without confounding blocks with any main effect?

The standard approach to solving this problem is to add an additional generator to the four generators given in the appendix. The design points that are associated with the 0 value for this generator give one block and the design points associated with the 1 value give the other block. The confounded string for this additional generator gives the terms that will be confounded with blocks.

To be more specific, the identity generators for the 2_{IV}^{8-4} design in Appendix 15 are I = $BCDE$, I = $ACDF$, I = $ABDG$, and I = $ABCH$. If we make the blocking generator AE, the set of design points for which $A+E = 0$ mod 2 forms one block of 8 runs and the set of design points for which $A+E = 1$ mod 2 forms the other block, as in Table 8.6.1. Blocks are confounded with AE, BF, CG, DH, $ABCD$, $ABGH$, $ACFH$, $ADFG$, $BCEH$, $BDEG$, $CDEF$, $EFGH$, $ABCDEF$, $ABDEFH$, $ACDEGH$, and $BCDFGH$. Since blocks are not confounded with any main effects, we conclude that we can run a resolution IV design for 8 factors in two blocks of eight runs each.

The approach taken in this book is to explicitly recognize blocks as a factor and design the experiment with full understanding of this additional factor. The rationale for recognizing blocks as a factor goes as follows. Data taken within a block is more homogeneous and therefore represents the source of error from which we wish to draw inferences. Each block will have a different mean effect so blocks influence the response variable. This makes blocks a factor, and it is generally random. Since the block size is 8 and we hope to use 16 runs, blocks must have 2 levels. In addition, all of block 1 is completed before block 2 is started so there is a restriction error on the factor blocks.

TABLE 8.6.1

A 2^{8-4}_{IV} Design in 2 Blocks of 8 Each

Block 1 (A+E=0 mod 2)								Block 2 (A+E=1 mod 2)							
A	B	C	D	E	F	G	H	A	B	C	D	E	F	G	H
0	0	0	0	0	0	0	0	0	0	0	1	1	1	1	0
0	0	1	1	0	0	0	1	0	0	1	0	1	1	1	1
0	1	0	1	0	1	0	1	0	1	0	0	1	0	1	1
0	1	1	0	0	1	0	0	0	1	1	1	1	0	1	0
1	0	0	1	1	0	0	1	1	0	0	0	0	1	1	1
1	0	1	0	1	0	0	0	1	0	1	1	0	1	1	0
1	1	0	0	1	1	0	0	1	1	0	1	0	0	1	0
1	1	1	1	1	1	0	1	1	1	1	0	0	0	1	1

We can now reformulate the question stated above as follows. I have 8 factors all at two levels. However, I can only run 8 treatment combinations under homogeneous conditions. Since I cannot create a resolution III design with 8 runs, I will need more than one block, making this a 9 factor experiment. What is the best tradeoff between the size of the experiment, directly controlled by the number of levels of the factor blocks, and the amount of information to be obtained in the experiment? If block has two levels, the experiment will consist of 16 runs. If block has three levels, the experiment will have 24 runs. If block has four levels, the experiment will have 32 runs. And so on. If block has two levels, we ask the following question. In 16 runs, can I estimate the main effects for 9 two level factors without confounding them with any of the 28 two factor interactions of the first 8 factors? If block has three levels, we ask the following question. In 24 runs, can I estimate the main effects for 8 two level factors and 1 three level without confounding them with any of the 28 two factor interactions of the two level factors. And so on.

As we discovered in an earlier section of this chapter, fractionation works best when all the factors are at the same prime or power of the same prime. So, for practical purposes, we should consider a 16 or a 32 run experiment, having 2 or $4 = 2^2$ levels for blocks. We must also remember that there is a restriction error on the factor called block.

It appears that we have a contradiction in this example. If we add the factor $\mathcal{B}$, for blocks, we end up with 9 factors and find in Appendix 15 that 32 runs are required for a resolution IV design. Yet, the previous design appeared to be resolution IV. But, as soon as we recognized blocks as a factor, the assignment $\mathcal{B} = AE$ shows us that the above design is not resolution IV.

The proof comes from a paper by Lorenzen and Wincek (1992). Their proof shows that the 8 factor design generated by I $= BCDE$, I $=$ $ACDF$, I $= ABDG$, and I $= ABCH$ and with blocking generator AE is identical to the 9 factor design generated by I $= BCDE$, I $= ACDF$, I $= ABDG$, I $= ABCH$, and I $= AE\mathcal{B}$. This 9 factor design is given in Table 8.6.2. Compare Table 8.6.2 with Table 8.6.1 to see that they are equivalent. The last generator for the 9 factor design shows that this is a resolution III design, not a resolution IV design. Specifically, A is confounded with $E\mathcal{B}$, E is confounded with $A\mathcal{B}$, and $\mathcal{B}$ is confounded with AE. For this very reason, we insist that blocks be treated as another factor in the experiment and have not treated blocks in the standard fashion. If one wishes to assume all interactions with blocks are negligible, then one may have some justification for calling this a resolution IV design.

TABLE 8.6.2

A 2^{9-5}_{III} DESIGN SORTED BY $\mathcal{B}$, IDENTICAL TO TABLE 8.6.1

A	B	C	D	E	F	G	H	$\mathcal{B}$	A	B	C	D	E	F	G	H	$\mathcal{B}$
0	0	0	0	0	0	0	0	0	0	0	0	1	1	1	1	0	1
0	0	1	1	0	0	0	1	0	0	0	1	0	1	1	1	1	1
0	1	0	1	0	1	0	1	0	0	1	0	0	1	0	1	1	1
0	1	1	0	0	1	0	0	0	0	1	1	1	1	0	1	0	1
1	0	0	1	1	0	0	1	0	1	0	0	0	0	1	1	1	1
1	0	1	0	1	0	0	0	0	1	0	1	1	0	1	1	0	1
1	1	0	0	1	1	0	0	0	1	1	0	1	0	0	1	0	1
1	1	1	1	1	1	0	1	0	1	1	1	0	0	0	1	1	1

Example 1. Consider a 2^5_{IV} design in four blocks of size 8 each. Label the five factors A–E. The word "blocks" tips us off that there are really six factors. We restate the problem as requiring a $2^5 \times 4$ design in 32 runs with a restriction error on the four level factor. Let $\mathcal{B}$ be the four level block factor with pseudo factors $\mathcal{B}_1$ and $\mathcal{B}_2$ both at 2 levels. Some identity generators that work are I $= ABD\mathcal{B}_1$ and I $= ACE\mathcal{B}_2$. Assuming three factor and higher order interactions are negligible, these generators yield the ANOVA in Table 8.6.3 and the confounding pattern in Table 8.6.4. If one is willing to assume all interactions involving blocks are negligible,

TABLE 8.6.3

ANOVA FOR A $2^5 \times 4$ DESIGN IN 32 RUNS

Source	df
$\mathcal{B}_i$ &/or ABD &/or ACE &/or $BCDE$	3
restriction error $\delta_{j(i)}$	0
A_k	1
B_ℓ	1
C_m	1
D_n	1
E_o	1
2 factor interactions of A–E	10
2 factor interactions involving $\mathcal{B}$	9
error (assuming three factor interactions are negligible)	4
Total	31

the 9 df for two factor interactions involving $\mathcal{B}$ can be pooled with the existing 4 df for error.

This is a very good design. The detectable size, using 4 df for the error is 1.10 for the main effects and 2.40 for the two factor interactions. If we pool the two factor interactions involving blocks into error, the detectable size drops to 0.88 and 1.24 respectively.

From Table 8.6.4, we see that the block factor, the main effects A–E, and the interactions BC, BE, CD, and DE are not confounded with any other two factor interactions. If there are a few interactions of interest, we can possibly estimate them by relabeling the factors so the interactions of interest correspond to BC, BE, CD, or DE. Even assuming three factor and higher order interaction are negligible, it is impossible to estimate the remaining two factor interactions as they are confounded among themselves.

Example 2. Consider a 36 run $2^2 \times 3^2$ factorial experiment that cannot be completed under homogeneous conditions. Is it better to run 9 blocks of 4, 6 blocks of 6, or 4 blocks of 9?

Consider the first alternative, 9 blocks of 4. Blocks will have 9 levels and we need a 1/9 replicate of a $2^2 \times 3^2 \times 9$ design. Break the 9 level factor into its pseudo factors, label the two level factors A and B and label the three level factors and pseudo factors C, D, $\mathcal{B}_1$, and $\mathcal{B}_2$. Working with

TABLE 8.6.4

CONFOUNDING FOR A $2^5 \times 4$ DESIGN IN 32 RUNS

$$\mathcal{B} = ABD = ACE = ABD\mathcal{B} = ACE\mathcal{B} = BCDE = BCDE\mathcal{B}$$
$$A = BD\mathcal{B} = CE\mathcal{B} = ABCDE\mathcal{B}$$
$$B = AD\mathcal{B} = CDE\mathcal{B} = ABCE\mathcal{B}$$
$$C = AE\mathcal{B} = BDE\mathcal{B} = ABCD\mathcal{B}$$
$$D = AB\mathcal{B} = BCE\mathcal{B} = ACDE\mathcal{B}$$
$$E = AC\mathcal{B} = BCD\mathcal{B} = ABDE\mathcal{B}$$
$$AB = D\mathcal{B} = BCE\mathcal{B} = ACDE\mathcal{B}$$
$$AC = E\mathcal{B} = BCD\mathcal{B} = ABDE\mathcal{B}$$
$$AD = B\mathcal{B} = CDE\mathcal{B} = ABCE\mathcal{B}$$
$$AE = C\mathcal{B} = BDE\mathcal{B} = ABCD\mathcal{B}$$
$$BC = DE\mathcal{B} = ABE\mathcal{B} = ACD\mathcal{B}$$
$$BD = A\mathcal{B} = CE\mathcal{B} = ABCDE\mathcal{B}$$
$$BE = CD\mathcal{B} = ADE\mathcal{B} = ABC\mathcal{B}$$
$$CD = BE\mathcal{B} = ABC\mathcal{B} = ADE\mathcal{B}$$
$$CE = A\mathcal{B} = BD\mathcal{B} = ABCDE\mathcal{B}$$
$$DE = BC\mathcal{B} = ABE\mathcal{B} = ACD\mathcal{B}$$
$$A\mathcal{B} = BD\mathcal{B} = CE\mathcal{B} = ABCDE$$
$$B\mathcal{B} = AD\mathcal{B} = CDE = ABCE = CDE\mathcal{B} = ABCE\mathcal{B}$$
$$C\mathcal{B} = AE\mathcal{B} = BDE = ABCD = BDE\mathcal{B} = ABCD\mathcal{B}$$
$$D\mathcal{B} = AB\mathcal{B} = BCE = ACDE = BCE\mathcal{B} = ACDE\mathcal{B}$$
$$E\mathcal{B} = AC\mathcal{B} = BCD = ABDE = BCD\mathcal{B} = ABDE\mathcal{B}$$
$$BC\mathcal{B} = ABE = ACD = DE\mathcal{B} = ABE\mathcal{B} = ACD\mathcal{B}$$
$$CD\mathcal{B} = ABC = ADE = BE\mathcal{B} = ABC\mathcal{B} = ADE\mathcal{B}$$

the three level factors and the three level pseudo factors representing blocks, we need a 3^{4-2} fraction on C, D, $\mathcal{B}_1$, and $\mathcal{B}_2$. Let C and D be basic factors. From Section **8.1**, we must work with the pieces of the interactions. One possible assignments is

original factors

C	D	$\mathcal{B}_1$	$\mathcal{B}_2$
c	d	cd	cd^2

basic factor pieces .

Since the main effect for the 9 level factor called block also includes $\mathcal{B}_1\mathcal{B}_2 = (cd)(cd^2) = C$ and $\mathcal{B}_1\mathcal{B}_2^2 = (cd)(cd^2)^2 = D$, we find both main effects C and D confounded with blocks. This is a poor design.

Consider next 6 blocks of 6 runs each. Blocks will have 6 levels and we need a 1/6 rep of a $2^2 \times 3^2 \times 6$ design. Again break blocks into pseudo factors $\mathcal{B}_1$ having two levels and $\mathcal{B}_2$ having three levels. We will need a 1/2 replicate of A, B, and $\mathcal{B}_1$ and a 1/3 replicate of C, D, and $\mathcal{B}_2$. We can let the following assignments be made, corresponding to the identity generators $\mathrm{I} = AB\mathcal{B}_1$ for the two level factors and $\mathrm{I} = BC^2\mathcal{B}_2$ for the three level factors.

original factors			original factors			
A	B	$\mathcal{B}_1$	C	D		$\mathcal{B}_2$
A	B	AB	C	D	CD	CD^2
basic factors			basic factors			

In both cases you obtain a resolution III design. Under the assumption that blocks do not interact with any of the factors A–D, you will get estimates of A and B from the two level fraction, estimates of C, D and 2 df for CD from the three level fraction, and estimates of AC, AD, BC, and BD from the cross product. This is a good design as long as the estimate of AB is not crucial.

Finally consider 4 blocks of 9 each. Break the four level factor called blocks into two pseudo factors having two levels each. We need a 2^{4-2} design that estimates A, B, $\mathcal{B}_1$, $\mathcal{B}_2$, and $\mathcal{B}_1\mathcal{B}_2$.

original factors		
A	B	
A	B	AB
basic factors		

It is impossible to assign 3 effects to one column. Again this is a poor design since main effects must be confounded with blocks.

Example 3. Another special type of blocked designs is known as the *balanced incomplete block design* or simply the BIBD. We will give a simple case and let the interested reader pursue the bibliography further. These designs are appropriate when, similar to the 4 blocks example above, the blocks are so small that no fractional design will possibly fit. The difference is that experiment size is not limited because many blocks can exist.

Suppose there are two factors of interest, each having two levels, and we wish to estimate main effects and the interaction. We can run any number of blocks but the block size is only two. If two blocks are used, we have a 2^{3-1} design in A, B, and $\mathcal{B}$. If we take A and B as the basic factors, there are three possible assignments of $\mathcal{B}$: to the basic factor interaction AB, to the basic factor B, or to the basic factor A.

original factors

$$\begin{array}{ccc} \mathcal{B}? & \mathcal{B}? & \mathcal{B}? \\ A & B & AB \end{array}$$

basic factors

These assignments give the three identity generators I $= AB\mathcal{B}$, I $= B\mathcal{B}$, and I $= A\mathcal{B}$. For each of these assignments, blocks will be confounded with a term of interest, AB, B, and A respectively. The secret is to run all three designs, using the data in the second and third design to estimate AB, using the data in the first and third design to estimate B, and using the data in the first and second design to estimate A.

Each separate design is called a replicate, $\mathcal{R}$, and blocks are nested within replicates. The layout of the design is given in Figure 8.1 and the ANOVA is given in Table 8.6.5. The combinations within each box must be completed before starting on the next box so there is a restriction error on blocks.

Replicate	I				II				III			
Block	1		2		3		4		5		6	
	A	B	A	B	A	B	A	B	A	B	A	B
	0	0	0	1	0	0	0	1	0	0	1	0
	1	1	1	0	1	0	1	1	0	1	1	1

Figure 8.1 BIBD for 2 factors in blocks of size 2. Replicates correspond to I $= AB\mathcal{B}$, I $= B\mathcal{B}$, and I $= A\mathcal{B}$.

The df for the terms in this model follow the rules of Chapter **3** once it is recognized that blocks are nested within replicates. The 3 df for error come from the fact that two different replicates provide independent estimates for each of A, B, and AB. The total df is 11, since there are 12 runs. The SS for $\mathcal{R}$ and $\mathcal{B}$ follow the rules in Chapter **3**. However, the SS for A, B, and AB must be modified to only include the replicates for which those effects can be estimated. That is, SS(A) must only use the

TABLE 8.6.4

ANOVA FOR THE BIBD IN FIGURE 8.1

Source	df
$\mathcal{R}_i$	2
$\mathcal{B}_{j(i)}$	3
$\delta_{(ij)}$	0
→ A_k	1
→ B_ℓ	1
→ $AB_{k\ell}$	1
$\varepsilon_{(ijk\ell)}$	3
Total	11

data in replicates 1 and 2, SS(B) must only use the data in replicates 1 and 3, and SS(AB) must only use the data in replicates 2 and 3. These can be computed from the rules in Chapter **3** as long as you use the proper data and "pretend" there are only two replicates.

From a computational point of view, use type I SS with replicates and blocks listed in the model before A, B, and AB. Of course, this analysis assumes all interactions of replicates and blocks with A and B are negligible.

The confounding pattern and EMS for the balanced incomplete block design are beyond this text. No simple rules exist when designs are added, as the 3 replicates were in this example, only when they are crossed.

Example 4 A clinic was trying to compare the effects of 3 different drugs. To control the huge person to person variability, a crossover design (see Chapter **7**) was to be conducted. Each person was to receive one drug, given a washout period, then receive a second drug, again given a washout period, and then receive the third drug. There are $6 = 3!$ different orders to receive the three drugs. Two people are randomly chosen for each order.

The factors in this experiment are orders, O_i, $i = 1, \cdots, 6$; person nested within order, $P_{j(i)}$, $j = 1, 2$; trial, T_k, $k = 1, 2, 3$ indicating whether this was the first, second, or third drug received; and drug, D_ℓ, $\ell = 1, 2, 3$. A typical layout for the crossover design is given in Figure 8.2. This is constructed by breaking the 6 level factor O into a two level pseudo factor O_1 and a three level pseudo factor O_2. A 3^{3-1}_{III} Latin Square design using $I = O_2TD$ generates the first three columns. Since we do not wish to repeat the same orders, the last three columns must be orthogonal to the

3_{III}^{3-1} Latin Square design generated by $I = O_2TD$, say the 3_{III}^{3-1} Latin Square design generated by $I = O_2TD^2$. (Any different three factor piece would work since the pieces are all orthogonal to each other.) The modular labeling used to generate this design, 0, 1, and 2, has been replaced with 1, 2, and 3 respectively to increase the readability of the figure.

	Order	1	2	3	4	5	6
	Person	1,2	3,4	5,6	7,8	9,10	11,12
Trial	1	D1	D3	D2	D1	D2	D3
	2	D3	D2	D1	D2	D3	D1
	3	D2	D1	D3	D3	D1	D2

Figure 8.2. Crossover design for 3 drugs, D1, D2, and D3.

There is a restriction error on Person but the order of each person should be randomized. Some of the confounding can be understood by looking at the first three and the second three orders separately. Again, the complete confounding pattern is very complicated.

Example 5. Lattice Design. A lattice design is appropriate when there is only one factor of interest, that factor has many levels, and the number of runs that can be run under homogeneous conditions is small. We give one simple example and let the interested reader pursue these designs further using the references at the end of the chapter.

Suppose a factor A had 25 different levels but an experiment could only be run in blocks of size 5. How should the design be constructed?

Note first that A can be broken into two pseudo factors, A_1 and A_2 each having 5 levels. For a block size of 5, running all levels of A will result in 5 blocks, *i.e.*, $\mathcal{B} = 5$. Since the 20 df associated with the pseudo factor interaction A_1A_2 are really main effects for A, there is no way to fractionate the design without confounding blocks with a main effect.

One possible way to estimate all levels of A is to run a balanced incomplete block design with 6 replicates. Each replicate will consist of the 25 levels of A broken into 5 blocks of 5 different levels each. Specifically, the 6 replicates would correspond to $I = A_1\mathcal{B}$, $I = A_2\mathcal{B}$, $I = A_1A_2\mathcal{B}$, $I = A_1A_2^2\mathcal{B}$, $I = A_1A_2^3\mathcal{B}$, and $I = A_1A_2^4\mathcal{B}$. Under the assumption that blocks do not interact with treatments, 5 of the 6 replicates would be used to estimate each piece of the A_1A_2 design. The association between

A_1, A_2, and A would have to be used to get estimates of each individual level of A.

The problem with the balanced incomplete design is that it requires 150 runs, 25 for each of the 6 fractions indicated above. If one used any two of the 6 fractions given above, say $I = A_1A_2\mathcal{B}$ and $I = A_1A_2^2\mathcal{B}$, one could get estimates of A_1A_2 from the second replicate, of $A_1A_2^2$ from the first replicate, and of the remaining pieces of A from both replicates. This is called the simple lattice. The construction is given in Figure 8.3. Substituting for A, the actual design is given in Figure 8.4. The ANOVA is given in Table 8.6.5.

$I = A_1A_2\mathcal{B}$

		$\mathcal{B}$ 0	1	2	3	4
	0	0	4	3	2	1
	1	4	3	2	1	0
A_2	2	3	2	1	0	4
	3	2	1	0	4	3
	4	1	0	4	3	2

$I = A_1A_2^2\mathcal{B}$

		$\mathcal{B}$ 0	1	2	3	4
	0	0	4	3	2	1
	1	3	2	1	0	4
A_2	2	1	0	4	3	2
	3	4	3	2	1	0
	4	2	1	0	4	3

Figure 8.3 Construction of the Lattice Design in Example 5. Boxes contain levels for A_1.

Replication $\mathcal{R}$

I: Blocks $\mathcal{B}$ 1	2	3	4	5	II: Blocks $\mathcal{B}$ 6	7	8	9	10
1	5	4	3	2	1	5	4	3	2
10	9	8	7	6	9	8	7	6	10
14	13	12	11	15	12	11	15	14	13
18	17	16	20	19	20	19	18	17	16
22	21	25	24	23	23	22	21	25	24

Figure 8.4 A Lattice Design for Example 5. Boxes contain A.

TABLE 8.6.5

ANOVA FOR A SIMPLE LATTICE DESIGN

Source	df
Replication $\mathcal{R}$	1
Blocks(Reps) $\mathcal{B}$	8
restriction(Blocks) δ	0
Treatments	24
R×T = error	16
Total	49

PROBLEMS

8.1 *Assuming 3 factor and higher order interactions are negligible, construct the design combinations, the ANOVA table, and the complete confounding pattern in standard order for a 3_{III}^{5-2} design using $I = abcd$ and $I = ad^2e$.*

8.2 *a) Use the basic factor technique to find a set of identity generators for 6 three level fixed factors, labeled A through F, when the main effects and the BC, BE, and CE interactions are of interest. b) After discussion with the subject area experts, it is reasonable to assume AB, CD, EF, and all three factor and higher order interactions are negligible. Use this information to find a set of identity generators using the basic factor technique.*

8.3 *Find the design and complete confounding pattern for a 5_{III}^{3-1} design using the identity generator $I = ab^2c^3$.*

8.4 *Assuming two factor and higher order interactions are negligible in Problem* **8.3**, *give the ANOVA table including detectable sizes. If factors A and B were random rather than fixed, what would the new ANOVA table look like? Use the same assumptions for both designs.*

8.5 *Create a Latin square design for 3 factors each at 5 levels, i.e., create a 5_{III}^{3-1} design. Use the identity generator $I = ab^4c^3$. Display the design in the Latin square format as in Table 8.1.4a.*

8.6 *Create a 3_{III}^{4-2} Graeco-Latin square design for 4 factors each at 3 levels using the identities $I = abc$ and $I = ab^2d^2$. Write the design using the format in Table 8.1.4b. The following data was collected. Analyze this data and draw conclusions. Make sure your assumptions are stated and your reasoning is carefully documented.*

A	B	C	D	y	A	B	C	D	y	A	B	C	D	y
1	1	1	1	10.7	2	1	3	3	13.9	3	1	2	2	11.2
1	2	3	2	15.0	2	2	2	1	15.1	3	2	1	3	9.6
1	3	2	3	16.2	2	3	1	2	13.0	3	3	3	1	10.9

8.7 *Compute the ANOVA table associated with the Graeco-Latin square design given in Figure 8.1.4b. State all of your assumptions and be sure to include the df and detectable sizes.*

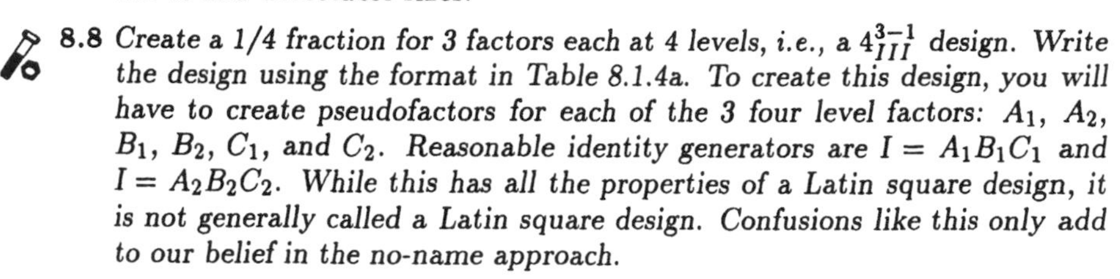

8.8 *Create a 1/4 fraction for 3 factors each at 4 levels, i.e., a* 4_{III}^{3-1} *design. Write the design using the format in Table 8.1.4a. To create this design, you will have to create pseudofactors for each of the 3 four level factors:* A_1, A_2, B_1, B_2, C_1, *and* C_2. *Reasonable identity generators are* $I = A_1B_1C_1$ *and* $I = A_2B_2C_2$. *While this has all the properties of a Latin square design, it is not generally called a Latin square design. Confusions like this only add to our belief in the no-name approach.*

8.9 *Repeat Problem* **8.8** *using the identity generators* $I = A_1A_2B_1B_2C_1$ *and* $I = A_1B_1B_2C_2$. *Determine the confounding pattern to see what is wrong with this set of generators. Now display the design using the format in Table 8.1.4a to visualize the problem with the confounding pattern.*

8.10 *Create a design for a* $8 \times 4 \times 2^5$ *experiment. Assuming all factors are fixed, all interactions involving the 8 level factor are negligible, all interactions involving the 4 level factor are negligible, and all third and higher order interactions are negligible, write the ANOVA for your design. Be sure to include all of the two level interactions among the two level factors in your ANOVA table.*

8.11 *Can one more two level factor be included in the above example without increasing the size of the experiment? Questions like this should always be asked when fractional designs are considered.*

8.12 *If 27 runs are available in an experiment involving a nine level factor and 4 different three level factors, can any interactions involving the 9 level factor be considered? Why or why not?*

8.13 *For the first example in Section* **8.3**, *displayed in Table 8.3.4, find the smallest orthogonal main effect design possible using Appendix 17. Compare the two designs.*

8.14 *Let us revisit the transmission example of Section* **8.3**. *Use Appendix 14 to construct the 64 run design. Verify that all 32 combinations of levels of factors A through E exist.*

8.15 *It is obvious from the description in the transmission example of Section* **8.3** *that, in the 64 run design, only 32 transmissions are going to be built. This implies a restriction error. Where does the restriction error belong? What is the meaning of the restriction error? This is also an object lesson for the reader. Even when considering extremely sophisticated designs, we must never lose sight of the fundamentals.*

8.16 *Create the 128 run design for the transmission example of Section* **8.3** *to verify that there are only 16 combinations of factors A through E. Again, as in Problem* **8.15**, *there will most likely be 16 different transmissions built. Identify and interpret the meaning of the restriction error. How* *does this added information influence your decision as to which design to run?*

8.17 *Give the complete confounding pattern for the 128 run transmission example of Section* **8.3**.

8.18 *A squeeze casting experiment involves 5 factors at two levels, 7 factors at three levels, and 1 factor at four levels. Create the smallest orthogonal main effect design.*

8.19 *In the squeeze casting experiment given in Problem* **8.18**, *it is feared that several interactions may be important. One interaction involves 2 two level factors and the other interaction involves a different two level factor and a three level factor. By how much does this increase the size of the experiment? You need not create this design.*

8.20 *A large screening experiment to study the effect of design and manufacturing factors on the acoustics of a muffler system was required. A design team had already met and determined that there were 17 factors in all, 2 of them having three levels and 15 having two levels. Each run requires hand construction of a muffler so run size is critical. The team is willing to discuss requirements. What suggestions should be made?*

8.21 *Example 1 in Section* **8.6** *considers a 2^5_V design in 4 blocks of size 8 (which we know as a 4×2^5 design in 32 runs). Suppose the block size is only 4 and we are still willing to assume all two factor interactions involving blocks are negligible. To estimate all main effects and all 10 two factor interactions, how many blocks are needed? (Caution, this is a tricky question.)*

8.22 *Suppose we relax the conditions in Problem* **8.21** *so that main effects can be confounded with two factor interactions, i.e., a resolution IV design, but we still require a block size of 4. How many blocks are needed? (This is not a tricky question.)*

8.23 *Suppose for the second example of Section* **8.6** *that all interactions of the $2^2 \times 3^2$ design must be estimated but the block size is only 4. The number of blocks is not a problem. Design an experiment to estimate all effects of interest, state any assumptions, and indicate how to analyze the data.*

8.24 *Find a 32 run design for an 8×2^5 experiment. Assume all two factor interactions involving the eight level factor are negligible. Do not confound any of the main effects with the remaining two factor interactions.*

8.25 *Give the ANOVA, including detectable size Δ, for the three treatment crossover design, Example 4 in Section* **8.6**.

8.26 *Problem* **8.25** *shows that the three treatment crossover design, Example 4 in Section* **8.6**, *is an orthogonal main effect design in 36 runs. Using Appendix 17, can a smaller design be created? If yes, give the design. If no, explain why not.*

8.27 *An experiment conducted by a social scientist considered the effect of several factors on citizen attitude. The three factors considered are A, schools in the university, arts, science, and engineering; B, home environment, low, medium, and high; and C, scholastic index, low, medium, and high. All levels are labeled 0, 1, and 2. The score is the average of 5 students. The test was administered by 1 instructor in 1 room. The room held 45 students so the tests were conducted over 3 days. The results are given below. Analyze the experiment and draw conclusions. Comment on the design and the analysis method, giving some alternative designs.*

Day 1				Day 2				Day 3			
A	B	C	Score	A	B	C	Score	A	B	C	Score
0	0	0	174	0	0	2	193	0	0	1	186
1	0	1	183	1	0	0	207	1	0	2	175
2	0	2	185	2	0	1	205	2	0	0	191
0	1	2	163	0	1	1	164	0	1	0	184
1	1	0	168	1	1	2	176	1	1	1	193
2	1	1	213	2	1	0	216	2	1	2	215
0	2	1	175	0	2	0	187	0	2	2	197
1	2	2	186	1	2	1	168	1	2	0	182
2	2	0	201	2	2	2	221	2	2	1	207

8.28 *In a farm cardiac study, the effects of certain farm tasks upon heart action were studied. The variable measured was oxygen uptake when men carried one or two pails at six different heights using three different loads. Three men were used in the experiment and the complete factorial would have been a $3 \times 2 \times 6 \times 3$ experiment. Each man could perform only six tasks in any one day and 18 days were allowed for the experiment. All two factor interactions not involving men are of interest and three factor and higher order interactions can be assumed negligible. Men are considered random but the other three factors are taken as fixed. a) Design a good experiment showing the layout of the design pictorially. b) Give the ANOVA table for your design. c) Describe the randomization scheme.*

8.29 *An experiment on automotive paint finishes dealt with 7 factors each at two levels. The factors are A = primer/surfacer supplier, B = primer/surfacer film build, C=primer/surfacer sanding technique, D = base coat/clear coat supplier, E = base coat film build, F = clear coat film build, and G = final coat sanding coat. All factors are coded as 0 or 1. A 2^{7-2} fraction was run in two blocks of 16 treatment combinations each. The identity generators are $F = ABCE$, $G = ACD$, and $\mathcal{B} = AC$. A measured variable was DOI, standing for distinctness of image, a measure of the sharpness of the reflection. Assuming AC, DG, all two factor interactions involving blocks,* *and all three factor interactions are negligible, analyze this experiment and make recommendations to the experimenter.*

Block 1

A	B	C	D	E	F	G	DOI	A	B	C	D	E	F	G	DOI
0	0	0	0	0	0	0	56.5	1	0	1	0	0	0	0	90.7
0	1	0	0	1	0	0	56.8	1	1	1	0	1	0	0	88.3
0	1	0	0	0	1	0	62.6	1	1	1	0	0	1	0	96.7
0	0	0	0	1	1	0	66.1	1	0	1	0	1	1	0	89.8
1	0	1	1	0	0	1	109.0	0	0	0	1	0	0	1	79.6
1	1	1	1	1	0	1	104.8	0	1	0	1	1	0	1	79.1
1	1	1	1	0	1	1	113.8	0	1	0	1	0	1	1	78.8
1	0	1	1	1	1	1	106.0	0	0	0	1	1	1	1	78.4

Block 2

A	B	C	D	E	F	G	DOI	A	B	C	D	E	F	G	DOI
0	1	1	1	0	0	0	81.1	1	1	0	1	0	0	0	79.1
0	0	1	1	1	0	0	78.7	1	0	0	1	1	0	0	77.7
0	0	1	1	0	1	0	79.9	1	0	0	1	0	1	0	79.2
0	1	1	1	1	1	0	79.3	1	1	0	1	1	1	0	80.0
1	1	0	0	0	0	1	80.1	0	1	1	0	0	0	1	97.1
1	0	0	0	1	0	1	83.3	0	0	1	0	1	0	1	89.0
1	0	0	0	0	1	1	85.8	0	0	1	0	0	1	1	87.2
1	1	0	0	1	1	1	80.0	0	1	1	0	1	1	1	88.4

8.30 *The following data was collected for the BIBD in Example 3 of Section* **8.6**. *Analyze the data and draw conclusions. Be sure you state your assumptions.*

Replicate	Block	A	B	y
I	1	0	0	15.6
I	1	1	1	14.0
I	2	0	1	11.2
I	2	1	0	16.3
II	3	0	0	24.9
II	3	1	0	26.3
II	4	0	1	23.1
II	4	1	1	25.0
III	5	0	0	22.6
III	5	0	1	19.8
III	6	1	0	29.6
III	6	1	1	26.5

8.31 *Based on Figure* **8.3**, *one could conclude that lattice designs are constructed using Latin square designs. Follow the procedure outlined in Example 5* *using the blocks generated by $I = A_1\mathcal{B}$ and $I = A_2\mathcal{B}$. These are obviously not Latin squares. Yet, when carried to the end, as in Figure **8.4**, it is clear that this is a lattice design having the same ANOVA as Table 8.6.5.*

BIBLIOGRAPHY FOR CHAPTER 8

Anderson, V. L. and McLean, R. A. (1974). *Design of Experiments A Realistic Approach*, New York: Marcel Dekker. (Chapters 8–12)

Box, G. E. P. and Hunter, J. S. (1961). "The 2^{k-p} fractional factorial designs, part 1," *Technometrics*, **3**, 311–351.

Box, G. E. P. and Hunter, J. S. (1961). "The 2^{k-p} fractional factorial designs, part 2," *Technometrics*, **3**, 449–458.

Box, G. E. P., Hunter, W. G. and Hunter, J. S. (1978). *Statistics for Experimenters An Introduction to Design, Data Analysis, and Model Building*, New York: John Wiley & Sons. (Chapters 12, 13)

Cheng, C.-S. (1989). "Some orthogonal main-effect plans for asymmetrical factorials," *Technometrics*, **31**, 475–477.

Daniel, C. (1976). *Applications of Statistics to Industrial Experimentation*, New York: John Wiley & Sons. (Chapters 10, 12–14)

Das, M. N. and Giri, N. C. (1986). *Design and Analysis of Experiments*, 2nd ed., New York: John Wiley & Sons. (Chapters 4–6, 9–12)

Dey, A. (1985). *Orthogonal Fractional Factorial Designs*, New York: John Wiley & Sons

Franklin, M. F. (1985). "Selecting defining contrasts and confounded effects in p^{n-m} factorial experiments," *Technometrics*, **27**, 165–172.

Franklin, M. F. and Bailey, R. A. (1977). "Selection of defining contrasts and confounded effects in two-level experiments," *Applied Statistics*, **26**, 321-326.

Hedayat, A. and Wallis, W. d. (1978). "Hadamard Matrices and their applications," *The Annals of Statistics*, **6**, 1184–1238.

Ghosh, S., editor. (1990). *Statistical Design and Analysis of Industrial Experiments*, New York: Marcel Dekker.

Hicks, C. R. (1982). *Fundamental Concepts in the Design of Experiments*, New York: John Wiley & Sons. (Chapters 12, 14, 15)

John, P. M. W. (1971). *Statistical Design and Analysis of Experiments*, New York: The Macmillan Co. (Chapters 6–9, 11-14)

Johnson, N. L. and Leone, F. C. (1977). *Statistics and Experimental Design in Engineering and the Physical Sciences*, 2nd ed., New York: John Wiley & Sons. (Chapters 14, 15)

Lamacraft, R. R. and Hall, W. B. (1982). "Tables of cyclic incomplete block designs," *Australian Journal of Statistics*, **24**, 350–360.

Lin, P. K. H. (1986). "Using the Chinese Remainder Theorem in constructing confounded designs for mixed factorial experiments," *Communications in Statistics–Theory and Methods*, **15**, 1389–1398.

Lin, P. K. H. (1987). "A simple method for constructing confounded designs for mixed factorial experiments," *Communications in Statistics–Theory and Methods*, **16**, 407–419.

Lin, P. K. H. (1987). "Confounding in mixed factorial experiments through isomorphisms," *Communications in Statistics–Theory and Methods*, **16**, 421–429.

Lochner, R. H. and Matar, J. E. (1990). *Designing for Quality, An Introduction to the Best of Taguchi and Western Methods of Statistical Experimental Design*, Milwaukee, WI: ASQC Quality Press. (Chapters 4, 7)

Lorenzen, T. J. and Wincek, M. A. (1992). "Blocking is simply fractionation," *Research Publication GMR-7709*: Mathematics Department, General Motors Research Laboratories, Warren, MI 48090-9055.

McLean, R. A. and Anderson, V. L. (1984). *Applied Factorial and Fractional Designs*, New York: Marcel Dekker.

Monod, H. and Bailey, R. A. (1992). "Pseudofactors: normal use to improve design and facilitate analysis," *Applied Statistics*, **41**, 317–336.

Montgomery, D. C. (1991). *Design and Analysis of Experiments*, 2nd ed., New York: John Wiley & Sons. (Chapters 5, 6, 12)

Nigam, A. K. and Gupta, V. K. (1985). "Construction of orthogonal main-effect plans using Hadamard Matrices," *Technometrics*, **27**, 37–40.

Ogawa, J. (1974). *Statistical Theory of the Analysis of Experimental Designs*, New York: Marcel Dekker

Peng, K. C. (1967). *The Design and Analysis of Scientific Experiments*, Reading, MA: Addison-Wesley. (Chapters 6, 7)

Plackett, R. L. and Burman, J. P. (1946). "The design of multifactorial experiments," *Biometrika*, **33**, 328–332.

Street, A. P. and Street, D. J. (1987). *Combinatorics of Experimental Design*, Oxford: Clarendon Press.

Taguchi, G. (1986). *Introduction to Quality Engineering*, White Plains, NY: Kraus International Publications. (Chapters 6–8)

Taguchi, G. (1988). *System of Experimental Designs*, White Plains, NY: Kraus International Publications. (Chapters 6, 7, 17)

Voss, D. T. (1986). "On Generalizations of the classical method of confounding to asymmetric factorial experiments," *Communications in Statistics: Theory and Methods*, **15**, 1299–1314.

Wang, J. C. and Wu, C. F. J. (1991). "An approach to the construction of asymmetrical orthogonal arrays," *Journal of the American Statistical Association*, **86**, 450–456.

Winer, B. J. (1971). *Statistical Principles in Experimental Design*, 2nd ed., New York: McGraw-Hill. (Chapter 8)

SOLUTIONS TO SELECTED PROBLEMS

8.1 *The design combinations, ANOVA, and confounding are*

A	*B*	*C*	*D*	*E*	*A*	*B*	*C*	*D*	*E*	*A*	*B*	*C*	*D*	*E*
1	1	1	1	1	2	1	1	3	2	3	1	1	2	3
1	1	2	3	3	2	1	2	2	1	3	1	2	1	2
1	1	3	2	2	2	1	3	1	3	3	1	3	3	1
1	2	1	3	3	2	2	1	2	1	3	2	1	1	2
1	2	2	2	2	2	2	2	1	3	3	2	2	3	1
1	2	3	1	1	2	2	3	3	2	3	2	3	2	3
1	3	1	2	2	2	3	1	1	3	3	3	1	3	1
1	3	2	1	1	2	3	2	3	2	3	3	2	2	3
1	3	3	3	3	2	3	3	2	1	3	3	3	1	2

ANOVA Term	df	Confounding Pattern
A	2	*DE*, *ADE*, *BCD*, *BCE*, *ABCD*, *ABCE*, *ABCDE*
B	2	*ACD*, *ACE*, *CDE*, *ABCD*, *ABCE*, *ABDE*, *BCDE*
C	2	*ABD*, *ABE*, *BDE*, *ABCD*, *ABCE*, *ACDE*, *BCDE*
D	2	*AE*, *ABC*, *ADE*, *BCE*, *ABCD*, *BCDE*, *ABCDE*
E	2	*AD*, *ABC*, *ADE*, *BCD*, *ABCE*, *BCDE*, *ABCDE*
AB	4	*CD*, *CE*, *ACD*, *ACE*, *BCD*, *BCE*, *BDE*, *ABCD*, *ABCE*, *ABDE*, *ACDE*, *ABCDE*
AC	4	*BD*, *BE*, *ABD*, *ABE*, *BCD*, *BCE*, *CDE*, *ABCD* *ABCE*, *ABDE*, *ACDE*, *ABCDE*
AD	2	*AE*, *BC*, *DE*, *ABCD*, *ABCE*, *BCDE*, *ABCDE*
BC	2	*ABD*, *ABE*, *ACD*, *ACE*, *BDE*, *CDE*, *ABCDE*
BD	2	*CE*, *ABC*, *ABE*, *ACD*, *ABDE*, *ACDE*, *BCDE*, *ABCDE*
BE	2	*CD*, *ABC*, *ABD*, *ACE*, *ABDE*, *ACDE*, *BCDE*, *ABCDE*

8.2 *a) Using B, C, and E as the basic factors leads to the following identity generators: $I = abc^2e^2$, $I = bc^2de^2$, and $I = bce^2f$. This is a resolution III design since the first identity times the second identity squared is acd. b) This is a tough one. By making the assignments $I = abce$, $I = bce^2d$, and $I = bc^2ef$, one finds that the unassigned column corresponds to ab, cd, ef, and higher order interactions. By the assumptions, there will be a single df that will serve as an error term. We know of no way other than trial and error to make this assignment.*

8.6 *To analyze this data, notice first that there are no degrees of freedom for error even assuming all two factor and higher order interactions are negligible. Therefore we use the half-normal effects plot given below to conclude A and C have the largest effects. We assume all interactions are small compared to the main effects and pool B and D to get an upper bound on the true error. (It is a true estimate if B and D have no effect.) The F-value for A is 8.515 with a p-value of .036. The F-value for C is 5.437 with a p-value of .072. Ths error is used to get confidence bounds in the main effects plots for A and C. We conclude that the third level of A and probably the first level of C are significantly lower than the other two levels. This is not a very good design.*

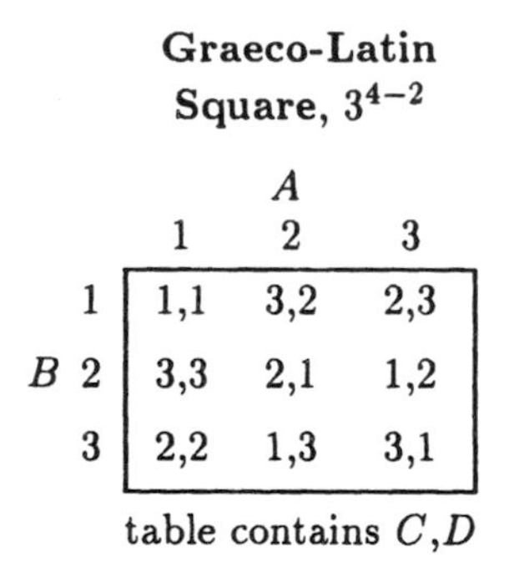

Graeco-Latin Square, 3^{4-2}

		A 1	2	3
	1	1,1	3,2	2,3
B	2	3,3	2,1	1,2
	3	2,2	1,3	3,1

table contains C,D

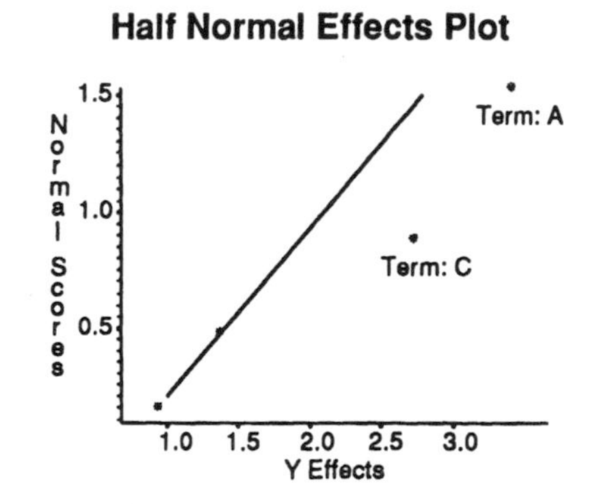

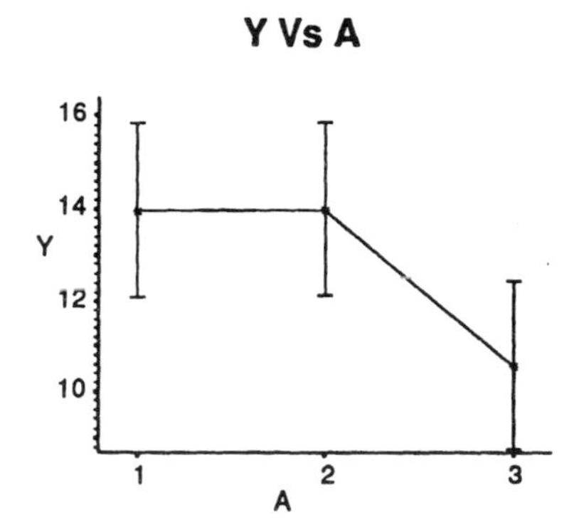

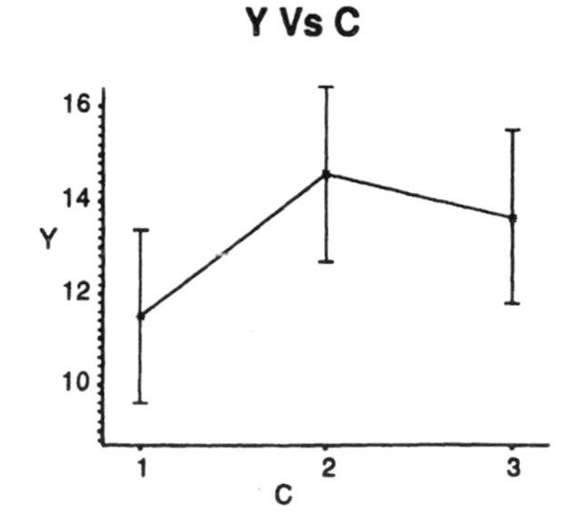

8.8 *Write a full factorial in the two level factors A_1, A_2, B_1, and B_2 and solve for C_1 and C_2 using $C_1 = A_1 + B_1$ mod 2 and $C_2 = A_2 + B_2$ mod 2 as in Chapter* **7**. *Writing the design in the format of Table 8.1.4a gives the following design. If you labeled the four level factors differently, your table may look different.*

		A			
		1	2	3	4
B	1	1	2	3	4
	2	2	1	4	3
	3	3	4	1	2
	4	4	3	2	1

table contains C

8.13 *A $6 \times 4^3 \times 2^{19}$ design in 32 runs will do the trick. This design has the advantage of requiring less than half as many runs. It's disadvantages are that it is unbalanced, that no interactions can be estimated, and that the confounding pattern is unknown. Assuming two factor and higher order interactions are negligible, both designs have plenty of df for estimating error.*

8.20 *The smallest orthogonal main effect design is a $4^2 \times 2^{25}$ design in 32 runs. There will be 12 df for error. Assuming all factors are fixed, this gives detectable sizes of approximately 0.88 for all factors. This is a very small detectability for a screening experiment. Make the following suggestions. 1) You can add up to 7 or 8 more factors to the study. This will not increase the size of the experiment and will still keep the detectability under 1.25. 2) Eliminate 2 of the two level factors to reduce the size of the design to 24. This still leaves 6 df for error with a detectable size of about 1.13. 3) Change the three level factors to two level factors, if it makes sense to do so. Then a 2^{19} design in 20 runs is possible. This leaves 2 df for error with a detectable size of approximately 2.15, still not bad. 4) Change the three level factors to two level factors and reduce the number of factors to 14. In 16 runs, this leaves 1 df for error with a detectable size of 3.67. This is our least favorite option.*

8.21 *If we insist on using balanced blocks as in Example 1, there are no designs that can be run in blocks of size 4. However, we can combine the ideas in Examples 3 and 5 to get a resolution V design in 8 blocks of size 4. Appendix 14 shows that a resolution V design in 5 factors requires 16 runs. This is 4 blocks of size 4 so we denote blocks by $\mathcal{B}_1$ and $\mathcal{B}_2$ for a total of 7 two level factors. We use the 16 run resolution IV design for 7 factors given in Appendix 15 to get the complete identity generator: $I = BCDE$, $I = ACD\mathcal{B}_1 = ABE\mathcal{B}_1$, and $I = ABD\mathcal{B}_2 = ACE\mathcal{B}_2 =$*

$BC\mathcal{B}_1\mathcal{B}_2 = DE\mathcal{B}_1\mathcal{B}_2$. *Assuming interactions with blocks and three factor interactions are negligible, we can estimate all main effects and two factor interactions except* BC *and* DE. *To get estimates on* BC *and* DE, *use the same design after relabeling the factors. One possibility is* $I = ABCE$, $I = DBC\mathcal{B}_1 = \cdots$, *and* $I = DAC\mathcal{B}_2 = \cdots$. *This will confound two different two factor interactions,* AB *and* CE, *with each other and with blocks. The combination of these two designs estimates all main effects and all two factor interactions not involving blocks in 8 blocks of 4 each. You should try to confound interactions of lesser importance with blocks since only half as much information is available for these terms.*

8.22 *This can be accomplished in 4 blocks of 4 each using the 7 factor resolution IV design in Appendix 15. Letting* $\mathcal{B}_1$ *and* $\mathcal{B}_2$ *represent the four level block factor, the identity generators are* $I = BCDE$, $I = ACD\mathcal{B}_1$, *and* $I = ABD\mathcal{B}_2$. *As usual, there will be a restriction error on blocks.*

9

Response Surface Designs

9.0 INTRODUCTION

The purpose of this chapter is to introduce the reader to the concept of response surface methodology and give some response surface designs. This is merely an introduction as many good books are available for further study. Some are listed in the bibliography at the end of this chapter.

As in the previous chapters of this book, there are two distinct parts to response surface methodology, the design of the experiment and the analysis of the data. We will only consider the design in this chapter. Most readers should already be familiar with regression techniques necessary to analyze the data.

It is extremely important to note the change in meaning of the math model in this chapter. The math model in this chapter is a predictive equation, predicting the value of the response variable for given levels of the factors. This implies that all factors in this chapter are fixed and quantitative. In Section **9.4**, we will see how to handle a few fixed qualitative factors.

Response surface methodology does not handle random factors.

The rules of the previous eight chapters do not generally apply to this chapter, even though many of the thought processes and some of the designs are the same. Specific designs have been created for specific models and these designs are listed in the appropriate section. There is no general method for creating designs based on the form of the mathematical model. For each particular set of circumstances, a specific design is suggested. There are very few redesign options available for each design and this will not be the focus of the chapter.

9.1 LINEAR MODELS

The simplest equation describing the relation between a response variable and the factors is the linear model. Suppose Y is the response variable and $X_1, X_2, \cdots, X_n$ are the factors. The linear response surface model takes the form

$$Y = \beta_0 + \beta_1 X_1 + \cdots + \beta_n X_n + \varepsilon, \tag{9.1.1}$$

where $\beta_0, \beta_1, \cdots, \beta_n$ are unknown coefficients giving the linear relation between X_i and Y. A picture of the fitted surface for two factors is given in Figure 9.1.

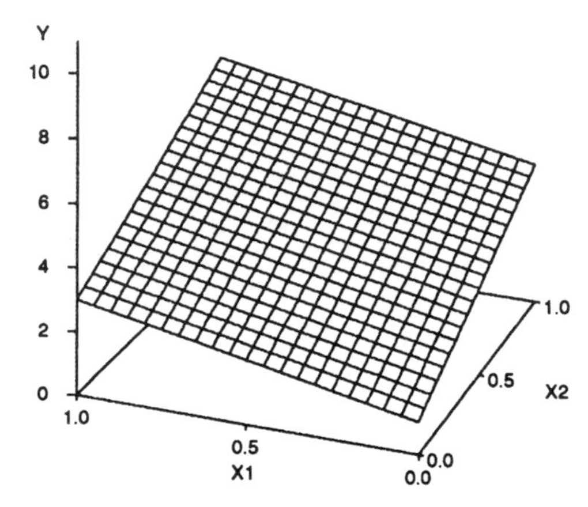

Figure 9.1 Representation of a linear model, 2 factors.

To create a design for model (9.1.1), notice first that each factor is related to the response in a linear fashion, *i.e.*, $\beta_i X_i$. Notice second that X_i does not occur in the equation with any other factor so all factors independently affect the response.

A linear equation is determined by two points so each factor must have at least two levels. Exactly two levels at the extremes of the factor space is optimal. Since each factor independently affects the response, it is not necessary to estimate any interactions. This describes the orthogonal main effect designs of Chapter 8 where each factor has two levels. Therefore, one can simply use Appendix 17 whenever a linear response model is desired.

Engineers were interested in increasing the force to initiate lateral slip in a riveted brake assembly. A team of experts decided to study the

effect of 8 variables, 4 variables dealing with the design of the brake and 4 variables dealing with the assembly process. The design factors were X_1 = shoe surface finish, X_2 = counter bore depth, X_3 = rivet length, and X_4 = rivet hardness. The four process variables were X_5 = insert spring force, X_6 = anvil and hammer force, X_7 = anvil and hammer age, and X_8 = riveting order. The linear response surface takes the form

$$y = \beta_0+\beta_1X_1 + \beta_2X_2 + \beta_3X_3 + \beta_4X_4+ \beta_5X_5 + \beta_6X_6 + \beta_7X_7 + \beta_8X_8 + \varepsilon. \tag{9.1.2}$$

High and low values for each of the 8 factors are, respectively, 50 and 200 rku, 0.06" and 0.09", 0.30" and 0.32", 100 and 200 bhu, 10 and 20 lbs, 500 and 650 lbs, 0 and 90 days, and 1 and 6. All factors are fixed and quantitative. A sketch of the assembly is given in Figure 9.2. We use Appendix 17 to find the design for an eight factor linear response surface, e.g., eight factors at two levels each. The design, as well as the collected data, is given in Table 9.1.1.

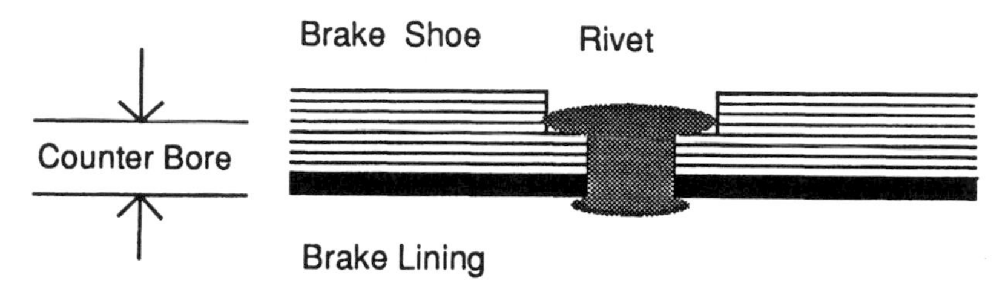

Figure 9.2 Cutaway view of a riveted brake assembly.

The analysis for this experiment is as follows. There are 8 df for the model and 3 df for error. MS(model) = 2810.8 while MS(error) = 491.4. This gives an overall F of 5.72 with a p-value of 0.09. This is not very good, probably because there were so few df for error. (A value of 10 df for error is often recommended.) The correlation coefficient squared, R^2, is 0.938 and the root mean square error is 22.17. We get the following coefficients with their associated standard errors in parentheses: $\beta_0 = 772.17$ (209.9), $\beta_1 = 0.70$ (0.26), $\beta_2 = 1094$ (427), $\beta_3 = (-2425)$ (640), $\beta_4 = 0.245$ (0.128), $\beta_5 = 3.95$ (1.28), $\beta_6 = -0.14$ (0.09), $\beta_7 = -0.076$ (0.14), and $\beta_8 = 2.63$ (2.56). The predictive equation for the force to initiate lateral slip, y, is given by

$$y = 772.17+0.70X_1 + 1094X_2 - 2425X_3 + 0.245X_4+ 3.95X_5 - 0.14X_6 - 0.076X_7 + 2.63X_8. \tag{9.1.3}$$

TABLE 9.1.1

DESIGN AND DATA, LINEAR MODEL, RIVETED BRAKE ASSEMBLY

shoe surface (rku)	counter bore depth (in)	rivet length (in)	rivet hardness (bhu)	insert spring force (lbs)	anvil & hammer force (lbs)	anvil & hammer age (days)	start with rivet #	force to init. lateral slip
100	0.06	0.32	100	10	500	90	6	139
100	0.09	0.30	200	10	500	0	6	260
50	0.09	0.32	100	20	500	0	1	165
100	0.06	0.32	200	10	650	0	1	114
100	0.09	0.30	200	20	500	90	1	254
100	0.09	0.32	100	20	650	0	6	193
50	0.09	0.32	200	10	650	90	1	140
50	0.06	0.32	200	20	500	90	6	159
50	0.06	0.30	200	20	650	0	6	202
100	0.06	0.30	100	20	650	90	1	201
50	0.09	0.30	100	10	650	90	6	142
50	0.06	0.30	100	10	500	0	1	142

A better model would be found by pooling the insignificant factors X_7 and X_8. The process of simplifying the model is often called parsimony.

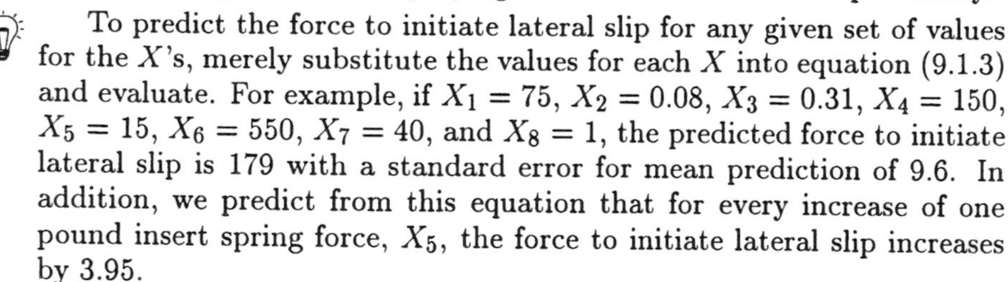

To predict the force to initiate lateral slip for any given set of values for the X's, merely substitute the values for each X into equation (9.1.3) and evaluate. For example, if $X_1 = 75$, $X_2 = 0.08$, $X_3 = 0.31$, $X_4 = 150$, $X_5 = 15$, $X_6 = 550$, $X_7 = 40$, and $X_8 = 1$, the predicted force to initiate lateral slip is 179 with a standard error for mean prediction of 9.6. In addition, we predict from this equation that for every increase of one pound insert spring force, X_5, the force to initiate lateral slip increases by 3.95.

Of course, this is not the end of the analysis because we need to address issues such as the goodness of the fit, which coefficients are significantly different from zero, the accuracy of the coefficients and the predictions, and the combination of X's yielding the highest force to lateral slip. We will not cover regression concepts in this brief overview of response surface designs.

9.2 INTERACTION MODELS

Sometimes it is known that the linear response surface model is not quite adequate because a few factors are thought to interact with each other. This is handled in the model by adding a term consisting of the product of the two factors that interact. Of course, the design must be modified to be able to handle this interaction. A picture of the fitted surface for two factors with an interaction is given in Figure 9.3.

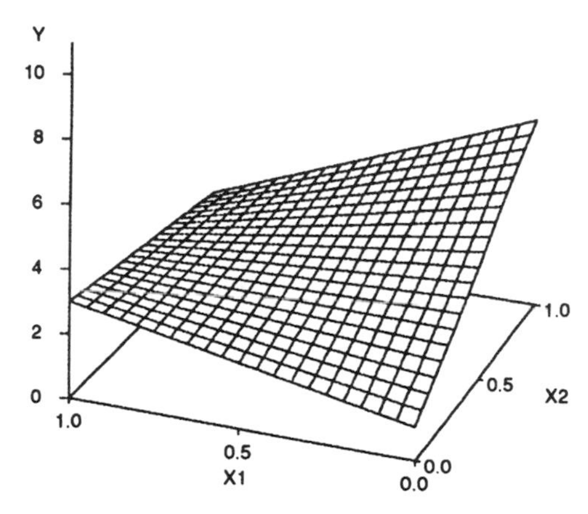

Figure 9.3 Representation of a linear model + interaction, 2 factors.

Suppose in the previous example that the effect of anvil and hammer force is thought to depend on the rivet hardness. We add the X_4X_6 interaction term to the linear model given in (9.1.2) to obtain the response surface model:

$$\begin{aligned} y = \beta_0 + &\beta_1X_1 + \beta_2X_2 + \beta_3X_3 + \beta_4X_4 + \\ &\beta_5X_5 + \beta_6X_6 + \beta_7X_7 + \beta_8X_8 + \beta_{46}X_4X_6 + \varepsilon. \end{aligned} \tag{9.2.1}$$

β_{46} is the unknown coefficient associated with the interaction of X_4 and X_6. β_{46} is read as beta four six, not beta forty-six.

We get a design for this model by using the tricks in Chapter 8. In this case, we represent X_4, X_6, and X_4X_6 as a single 4 level factor. We then look for a 4×2^6 orthogonal main effect design in Appendix 17. The minimum size is 16 which comes from a 4×2^{12} design. The final result is given in Table 9.2.1.

To create Table 9.2.1 from the $2^{15} = 4 \times 2^{12}$ design given in Appendix 17, assign X_4 to column 1, X_6 to column 2, and the remaining X's to columns 4 through 9. You should pause now and reproduce Table 9.2.1 to make sure you understand the process. If you have difficulty, go back and review Chapter 8.

TABLE 9.2.1

LINEAR MODEL + 1 INTERACTION, RIVETED BRAKE ASSEMBLY

shoe surface (rku)	counter bore depth (in)	rivet length (in)	rivet hardness (bhu)	insert spring force (lbs)	anvil & hammer force (lbs)	anvil & hammer age (days)	start with rivet #
50	0.06	0.30	100	20	500	90	1
50	0.09	0.32	100	20	500	0	6
100	0.06	0.32	100	10	500	0	1
100	0.09	0.30	100	10	500	90	6
50	0.06	0.30	100	20	650	0	6
50	0.09	0.32	100	20	650	90	1
100	0.06	0.32	100	10	650	90	6
100	0.09	0.30	100	10	650	0	1
50	0.06	0.30	200	10	500	0	1
50	0.09	0.32	200	10	500	90	6
100	0.06	0.32	200	20	500	90	1
100	0.09	0.30	200	20	500	0	6
50	0.06	0.30	200	10	650	0	6
50	0.09	0.32	200	10	650	90	1
100	0.06	0.32	200	20	650	90	6
100	0.09	0.30	200	20	650	0	1

9.3 QUADRATIC MODELS

A reasonable model to consider when there is no knowledge about a response curve is the quadratic response model. It takes the form

$$\begin{aligned} Y = \beta_0 + \beta_1 X_1 + \cdots \beta_n X_n + \beta_{11} X_1^2 + \cdots + \beta_{nn} X_n^2 + \\ \beta_{12} X_1 X_2 + \cdots + \beta_{1n} X_1 X_n + \\ \beta_{23} X_2 X_3 + \cdots + \beta_{n-1\,n} X_{n-1} X_n + \varepsilon, \end{aligned} \tag{9.3.1}$$

consisting of all linear terms, all squared terms, and all linear by linear interactions of the factors. A picture of a fitted quadratic response curve is given in Figure 9.4.

9.3.1 Fixed Designs

A quadratic equation is determined by three points so one possibility for designing for equation (9.3.1) is to use three level factorials. Unfortunately, three level factorials provide for an estimate of all interactions of

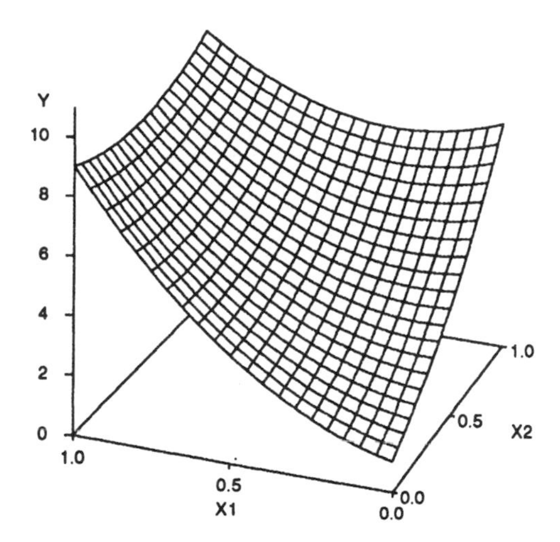

Figure 9.4 Representation of a quadratic model in 2 factors.

linear by quadratic and quadratic by quadratic terms as well as the linear by linear terms in equation (9.3.1). This tends to make the experiment size unacceptably high. For example, with as few as four factors, 81 runs are required.

Box and Wilson (1951) created an alternate class of designs known as the *central composite design*. This set of designs consists of two separate components. The first component is a full or fractional two level experiment designed to estimate all of the linear and interaction terms. The second components consists of center points and star or axial points used to estimate the quadratic terms. These components are illustrated in Figure 9.5.

Recode the factor values so –1 represents the lowest value of interest, 1 represents the highest value of interest, and 0 represents the center of the region of interest. Use Appendix 14 to create the fractional factorial component for the design. Note that Appendix 14 creates resolution V fractions using the coding 0 and 1. The value 0 must be recoded as –1 for use in the central composite design. The center point has the value 0 for all coded factors. The star points take the value $\pm\alpha$ for one of the coded factors and 0 for all of the other coded factors. The value α will be discussed later.

The coded design for a central composite design with 3 factors is given in Table 9.3.1. The first eight rows consist of the 2^3 full factorial design and are recoded appropriately. The ninth row is the center point.

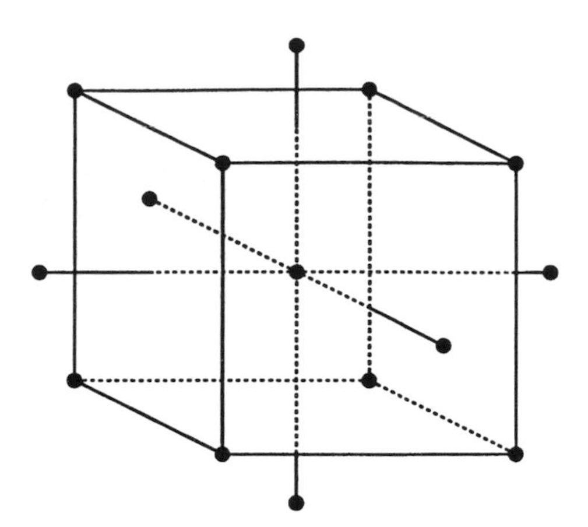

Figure 9.5 Representation of a central composite design in 3 factors.

TABLE 9.3.1

A CENTRAL COMPOSITE DESIGN FOR 3 FACTORS.

X_1	X_2	X_3
-1	-1	-1
-1	-1	1
-1	1	-1
-1	1	1
1	-1	-1
1	-1	1
1	1	-1
1	1	1
0	0	0
$-\alpha$	0	0
α	0	0
0	$-\alpha$	0
0	α	0
0	0	$-\alpha$
0	0	α

Rows 10 through 15 represent the star points. Notice that a total of 15 runs are required. This compares to $27 = 3^3$ for the full factorial three level design. Table 9.3.2 shows the size comparison between the central composite design with 1 center point and the resolution V three level fractional factorial design for $n = 2, \cdots, 8$. Considerable efficiency is gained using the central composite design.

TABLE 9.3.2

SIZE OF DESIGN TO ESTIMATE QUADRATIC MODEL

Number of Factors	Number of Coefficients	3 Level Fraction	Central Composite	:	2 Level Fraction	Star + Center
2	6	9	9	=	4	5
3	10	27	15	=	8	7
4	15	81	25	=	16	9
5	21	81	27	=	16	11
6	28	243	45	=	32	13
7	36	243	79	=	64	15
8	45	243	81	=	64	17

The difference between the number of coefficients to be estimated and the size of the design in Table 9.3.2 is the df for error. However, the error represents both true error and *lack of fit* error. Lack of fit error is the inadequacy of the quadratic model to represent the true model, whatever that model may be. To get an estimate of true error, at least one design point must be replicated. This true error can then be used to estimate lack of fit error.

Typically the center point is replicated to get an estimate of true error. Alternatively, one could replicate the fractional design or replicate the star points to get an estimate of true error. If one is interested in increasing the accuracy of the model, one could increase the resolution of the fractional design, say from resolution V to resolution VI or even to a full factorial. These are all possible redesign options and the experimenter has to make tradeoffs between the increased sensitivity of the design and the increased size of the experiment.

Another design or redesign option is the choice of the constant α. If simplicity is of utmost importance, choose $\alpha = 1$. This is called a face centered central composite design because the star points will be located in the center of each face of the cube. Only three different values for each factor need be studied. The disadvantages of this design are that the quadratic terms are not estimated with as much accuracy as the linear terms and the estimates of the linear and quadratic are correlated.

Another property that is sometimes desirable is *rotatability*. A design is rotatable if the variance of the prediction depends only on the distance from the center of the design, not the direction. Such designs are particularly important for drilling or punching operations where distance from the target is the crucial measurement. A rotatable design is obtained by letting $\alpha = \sqrt[4]{F}$, where F represents the number of runs in the fractional factorial portion of the central composite, listed as a column in Table 9.3.2. The corresponding α for two through eight factors is given below.

Values of α for Rotatable Central Composite Design

factors:	2	3	4	5	6	7	8
α:	1.414	1.682	2.000	2.000	2.378	2.828	2.828

An advantage of the fractional factorial three level design over the central composite design is orthogonality. That is, the estimates of the coefficients are not influenced by the values of the other coefficients. By properly choosing α, the central composite design can also be made orthogonal. The solution depends on the size of the two level fraction, which we label F, and the size of the star + center part of the central composite design, T. Both F and T are listed in Table 9.2.3 for central composite designs with one center point. Let $Q = (\sqrt{F+T} - \sqrt{F})^2$. Then, $\alpha = \sqrt[4]{QF/4}$. Unlike rotatability, the choice of α to make the design orthogonal will depend on the number of center points used. The α corresponding to the designs in Table 9.3.2 are given below. If more than one center point is used, use the formula given above.

Values of α for Orthogonal Central Composite Designs, 1 Center Point

factors:	2	3	4	5	6	7	8
α:	1.000	1.215	1.414	1.547	1.724	1.885	2.000

Since orthogonality depends on the number of center points but rotatability does not, it is possible to find the number of center points required to make the α for orthogonality equal to the α for rotatability. Unfortunately, the number of center points must be an integer so exact equality is not always possible. The values given below show the number of center point and the α necessary to obtain exact orthogonality with as near rotatability as possible.

Number of Centerpoints and α for Orthogonality and Near Rotatability

factors:	2	3	4	5	6	7	8
# CP:	8	9	12	10	15	22	20
α:	1.414	1.668	2.000	2.000	2.393	2.828	2.828

A final decision with central composite designs must be made when you scale the design from [–1,1] to the region of experimentation and α

exceeds 1. There are two choices, let the factorial portion correspond to the region of experimentation so the star points extend beyond the the original region of experimentation or scale the design so the star points just touch the limits of experimentation. The decision is non-statistical. The wider the region of experimentation, the more accurate the estimates of the coefficients. However, if the response at the star points is not representative of the response inside the design region, the entire equation is suspect.

It is generally true that the extreme points of a region of interest will give the best statistical accuracy for estimating a response surface. However, in practice, the extreme conditions may not truly represent the behavior of the response. For example, in an automobile engine, the extremes of spark advance and fuel retardation will cause knock and not represent the behavior of the engine under no knock conditions. In some cases, the engine won't even run. To handle extreme point problems such as this, Box and Behnken (1960) created a series of designs that avoid the extremes in all variables. For three factors, the design is illustrated in Figure 9.6.

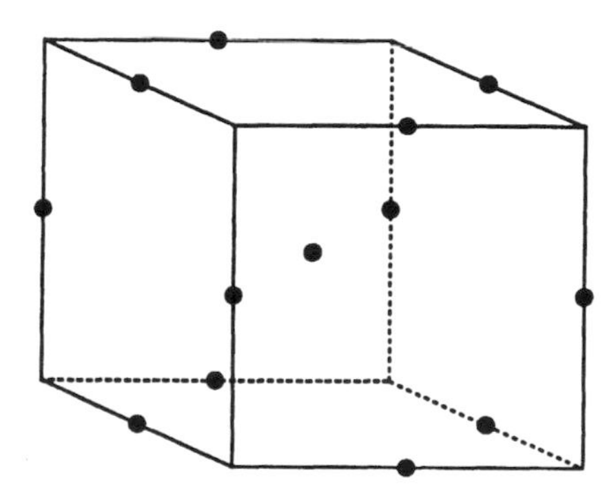

Figure 9.6 Representation of a Box-Behnken design in 3 factors.

The Box-Behnken designs are constructed by combining two level factorial designs on some of the factors with center points on the remaining factors. At least one center points for all factors should be added to the design. For three, four, or five factor designs, two factors are chosen to be at the extreme values, and the remaining factors are fixed at their center values. Each selection of two factors produces four design points, say ± 1 for the two selected factors and 0 for the remaining factors. Create the four design points for each combination of two factors and add one center point. For three factors, there are $\binom{3}{2} = 3$ ways to choose two factors, yielding a design size of $3 \times 4 + 1 = 13$. The design points are given in Table 9.3.3.

TABLE 9.3.3

A BOX-BEHNKEN DESIGN FOR 3 FACTORS.

X_1	X_2	X_3
1	1	0
1	−1	0
−1	1	0
−1	−1	0
1	0	1
1	0	−1
−1	0	1
−1	0	−1
0	1	1
0	1	−1
0	−1	1
0	−1	−1
0	0	0

For four factors, there are six ways to choose two factors so the design size is 25. For five factors, there are ten ways to choose two factors for a total design size of 41. For six and seven factor designs, three factors are selected at a time and all eight combinations of extreme values are run. To keep the size of the design down, not all combinations of three columns are run. For a six factor design, the columns selected are (1,2,4), (2,3,5), (3,4,6), (4,5,1), (5,6,2), and (6,1,3) for a total design size of 49. The triplets were not sorted to see the pattern used. For seven factors, the columns selected are (1,2,4), (1,3,5), (1,6,7), (2,3,6), (2,5,7), (3,4,7), and (4,5,6) for a total design size of 57. Note that, for both six and seven factor designs, each factor is at its extremes exactly three times.

9.3.2 Sequential Designs

The central composite design can be run in several blocks. The purpose of this is to provide an analysis of the data before all of the runs are completed. We will consider the central composite design in two blocks and in three blocks. Of course, as in Chapter 8, blocks must be a factor considered in the analysis of the data.

When using two blocks, it is natural to separate a central composite design into a block containing the two level fraction and a block containing the star points, corresponding to the linear and linear by linear components in one block and the quadratic components in the other block. To be able to estimate the block effect, at least 2 center points must be run in each block. The formula for the star value α which yields rotatability is given in the previous section since the value is unaffected by the number of center points. However, the formula for orthogonality changes to $\alpha = \sqrt{n(1 + \mathrm{CP}_f/F)/(1 + \mathrm{CP}_s/S)}$, where n is the number of factors in the design, F is the number of runs in the fractional block given in Table 9.3.2, CP_f is the number of center points in the fractional block, S is the number of star points ($S = 2n$), and CP_s is the number of center points in the star block. For the minimum recommended 2 center points in each design, the values for α necessary for orthogonality are given below.

Values of α for 2 Blocks, 2 Centerpoints Each

factors:	2	3	4	5	6	7	8
α:	1.414	1.789	2.108	2.309	2.567	2.785	2.954

To obtain a design that is orthogonal and as nearly rotatable as possible, use the number of center points and α values given in Table 9.3.4. The design for an orthogonal and nearly rotatable central composite design for three factors is given in Table 9.3.5.

TABLE 9.3.4

NUMBER OF CENTERPOINTS AND α FOR ORTHOGONALITY AND NEAR ROTATABILITY USING 2 BLOCKS

factors:	2	3	4	5	6	7	8
number of CP_f:	2	6	4	8	8	4	8
number of CP_s:	2	4	2	2	2	3	2
α:	1.414	1.690	2.000	2.000	2.366	2.828	2.828

A two staged analysis can be used with central composite designs in two blocks. Run the fractional factorial (and associated center points) block first. At this point you can estimate all linear as well as all linear by linear terms. The center points are used to get an estimate of *lack of fit* as well as an estimate of *true error*. The estimate of lack of fit will have 1 df. The estimate of true error will have $\mathrm{CP}_f - 1$ df and form the denominator for all tests.

TABLE 9.3.5

A CENTRAL COMPOSITE DESIGN FOR 3 FACTORS IN 2 BLOCKS

BLOCK 1			BLOCK 2		
X_1	X_2	X_3	X_1	X_2	X_3
-1	-1	-1	-1.69	0	0
-1	-1	1	1.69	0	0
-1	1	-1	0	-1.69	0
-1	1	1	0	1.69	0
1	-1	-1	0	0	-1.69
1	-1	1	0	0	1.69
1	1	-1	0	0	0
1	1	1	0	0	0
0	0	0	0	0	0
0	0	0	0	0	0
0	0	0			
0	0	0			
0	0	0			
0	0	0			

If the lack of fit test is significant, then run the second block and fit the full quadratic model. One should use a large value for the significance level for testing lack of fit since since the test is not very powerful. One must also remember to include a block factor in the analysis since something may have changed between the running of the first block and the running of the second block.

Some central composite designs can also be run in three blocks. The first block corresponds to the linear components in the model. The second block corresponds to the linear by linear components. The third block corresponds to the quadratic components. Again, there will be a factor called blocks that must be included in the model.

In order to run a central composite design in three blocks, start with the two block design just given. The first block is a resolution V fractional factorial that estimates all linear effects and all linear by linear interactions. It may be possible to split this block into a resolution III block estimating linear effects only and the remainder of the block estimating linear by linear interactions. In other words, we can run the central composite design in three blocks if it is possible to add a generator to the resolution V fraction and end up with a resolution III design. The number of center points given in the two block design get split equally.

We illustrate the procedure with a three factor central composite design. A full factorial design is used on the corner points and there are 6 center points in the first block of a two block design. Break the 2^3 full factorial into two blocks using the generator I = $X_1X_2X_3$, where the level -1 is replaced by 0. The first block corresponds to $X_1 + X_2 + X_3 = 0$ mod 2 and the second block corresponds to $X_1 + X_2 + X_3 = 1$ mod 2. Three center points go with each block. After replacing 0 with -1, the design is as given in Table 9.3.6.

TABLE 9.3.6

A CENTRAL COMPOSITE DESIGN FOR 3 FACTORS IN 3 BLOCKS

BLOCK 1			BLOCK 2			BLOCK 3		
X_1	X_2	X_3	X_1	X_2	X_3	X_1	X_2	X_3
-1	-1	-1	-1	-1	1	-1.69	0	0
-1	1	1	-1	1	-1	1.69	0	0
1	-1	1	1	-1	-1	0	-1.69	0
1	1	-1	1	1	1	0	1.69	0
0	0	0	0	0	0	0	0	-1.69
0	0	0	0	0	0	0	0	1.69
0	0	0	0	0	0	0	0	0
						0	0	0
						0	0	0
						0	0	0

For the four factor design, the 2^4 factorial piece can be broken into two blocks using the identity generator I = $X_1X_2X_3X_4$, where the -1 level must be replaced with a 0 for purposes of finding the combinations as in the example above. For the five factor design and the eight factor design, it is impossible to break the resolution V fraction into two resolution III pieces. Therefore, three block central composite designs do not exist for five or eight factors. For six and seven factor designs, use the block generator I = $X_1X_2X_3$ to create three block designs.

In analyzing the data from a three block central composite design, run block 1 first. Use the data from this block to estimate the linear effects, the lack of fit, and the true error. If there is a significant lack of fit, run block 2. Use the combined data to estimate the linear terms, the linear by linear terms, the block effect, the lack of fit, and the true error. If there is still a significant lack of fit, run block 3 to estimate the full quadratic model. Using this sequential approach, fewer data points are collected when a simpler model adequately describes the data.

Either the two or the three block approach can be used to find a maximum (or minimum) value. We will illustrate the procedure with the three block design but the idea is the same for the two block design. Start with a given region and use block 1 to fit a linear model. Use the results of this experiment to move to a new region in the direction of most improvement. This method is called *steepest ascent*. Now use block 1 and a linear model in the new region. When the linear model becomes insignificant and the lack of fit term becomes large, you know you are approaching the region containing the maximum. Do not change the region but run block 2. The analysis will indicate whether you need to move in a direction other than horizontally or vertically. When the linear plus linear by linear interaction model becomes insignificant and the lack of fit significant, it is time to use block 3 to estimate the curvature and find the maximum. The details are left to the interested reader.

9.4 POLYNOMIAL MODELS

In addition to the second order model previously discussed, general polynomial models can be used to approximate response surfaces. Polynomial models can have any power for any factor as well as any interaction of powers of factors. For example, if there were three factors in an experiment, X_1, X_2, and X_3, and X_1 was thought to be linear, X_2 cubic, X_3 quadratic, and there were thought to be simple linear by linear interactions among the three factors, the polynomial model would take the form

$$Y = \beta_0 + \beta_1 X_1 + \beta_2 X_2 + \beta_3 X_3 + \beta_{22} X_2^2 + \beta_{222} X_2^3 + \\ \beta_{33} X_3^2 + \beta_{12} X_1 X_2 + \beta_{13} X_1 X_3 + \beta_{23} X_2 X_3 + \varepsilon.$$

As indicated earlier, we are using the notation β_{222} to represent the unknown coefficient associated with the X_2^3 term. Determination of the form of the model is a non-statistical task but a good consultant should be able to guide the experimenter.

There are several different methods for designing experiments when polynomial designs are used. These are often referred to as "alphabet optimality" since each method is referred to by a letter of the alphabet. The most popular method is the D-optimal method. This method minimizes the overall variance of the estimated coefficients β_i. Other methods include A-optimality—minimizing the sum of the variances of the estimated coefficients, E-optimality—minimizing the maximum variance of the estimated coefficients, G-optimality—minimizing the maximum variance of the predicted responses, and V-optimality—minimizing the average variance of the predicted responses. Computer search routines are necessary to find the optimum and are becoming more readily available.

To use these methods, the polynomial model must be specified as well as the size of the design and a discrete set of candidate points from which the actual design points will be selected. The minimal number of candidate levels for any factor is one more than the maximum power of that factor in the polynomial model. In the previous model, the minimal number of candidate points for X_1 is 2, for X_2 is 4, and for X_3 is 3. Our experience has been that the minimal number of candidate levels is usually sufficient.

In an automotive safety experiment on the combined safety of air bags and shoulder harnesses, three factors were considered: the belt length L at 118, 124, and 128 inches; the vent size V at 20, 35, and 45 mm; and the belt stretch S at 6% and 12%. Belts at length 118 inches with 6% stretch were not available for experimentation. Due to limited sled time and cost, only 10 runs can be made. The head injury criteria, HIC, will be measured. The polynomial approximation relating L, V, and S to HIC is thought to take the form

$$\text{HIC} = \beta_0 + \beta_1 L + \beta_2 V + \beta_3 S + \beta_{11} L^2 + \beta_{22} V^2 + \beta_{12} LV + \beta_{23} VS + \varepsilon.$$

The candidate region for this experiment consists of all combinations of levels of L, V, and S except those combinations that have both $L = 118$ and $S = 6\%$. There are a total of 15 combinations available for the design. The 10 points selected as being D-optimal are given in Table 9.4.1. Since there are many local minima, different computer algorithms may lead to different D-optimal designs. For all practical purposes, these different designs can be treated as if they were identical.

We can also create models that will account for qualitative factors, factors that do not take numeric values. This is accomplished through the use of *indicator factors*. An indicator factor is a quantitative factor that takes the values 0 or 1. For a qualitative factor having n levels, there will be $n - 1$ associated indicator factors.

Suppose there were three different manufacturers of air bags in the previous example and the experimenters were interested in the effect of manufacturer. As a qualitative factor, M cannot be directly incorporated into the model. To incorporate it into the model we first create two indicator factors I_1 and I_2 as illustrated in Table 9.4.2.

Then the model with manufacturer included takes the form

$$\begin{aligned}\text{HIC} = \beta_0 &+ \beta_1 L + \beta_2 V + \beta_3 S + \beta_{11} L^2 + \beta_{22} V^2 + \\ &\beta_{12} LV + \beta_{23} VS + \beta_4 I_1 + \beta_5 I_2 + \varepsilon.\end{aligned}$$

TABLE 9.4.1

D-OPTIMAL DESIGN FOR THE AUTOMOTIVE SAFETY EXPERIMENT

L (in)	V (mm)	S (%)
118	20	12
118	45	12
124	20	6
124	35	12
124	45	6
128	20	6
128	20	12
128	35	6
128	45	6
128	45	12

TABLE 9.4.2

REPRESENTING A QUALITATIVE FACTOR AS INDICATOR FACTORS

Manufacturer	I_1	I_2
A	0	0
B	1	0
C	0	1

Both I_1 and I_2 take the value 0 when using an air bag from manufacturer A. The factor I_1 takes the value 1 if you are using an air bag from manufacturer B. And the factor I_2 takes the value 1 if you are using an air bag from manufacturer C. It is impossible for both I_1 and I_2 to take the value 1 in the same factor level combination.

Manufacturer A is taken as the baseline in this model. The factor I_1 represents the difference between manufacturer B and manufacturer A. The factor I_2 represents the difference between manufacturer C and manufacturer A. If I_1 is significant, then we say manufacturer B differs from manufacturer A. If I_2 is significant, then we say manufacturer C differs from manufacturer A. To test whether manufacturer B differs from C, we must test if $\beta_4 = \beta_5$.

To create a design for the automotive safety experiment with the extra factor M, the candidate space grows to 45 points, the 15 candidate points given earlier with each of the conditions $I_1 = I_2 = 0$; $I_1 = 1,\ I_2 = 0$; and $I_1 = 0,\ I_2 = 1$. A D-optimal design for a 12 run experiment is given in Table 9.4.3.

TABLE 9.4.3

D-OPTIMAL DESIGN WITH THE QUALITATIVE FACTOR M

L (in)	V (mm)	S (%)	I_1	I_2	M
118	20	12	1	0	B
118	35	12	0	0	A
118	45	12	0	1	C
124	20	6	0	0	A
124	35	12	0	1	C
124	45	6	0	0	A
124	45	12	1	0	B
128	20	6	0	1	C
128	20	12	0	0	A
128	35	6	1	0	B
128	45	6	0	1	C
128	45	12	0	0	A

If a qualitative factor is thought to interact with a quantitative factor, then include every interaction involving the indicator factors and the quantitative factors. For example, if V was thought to interact with M, then include VI_1 and VI_2 in the model. If a qualitative factor is thought to interact with another qualitative factor, then include the interaction of every indicator factor from the first qualitative factor with every indicator factor from the second qualitative factor.

9.5 MIXTURE DESIGNS

In a mixture experiment, ingredients are mixed together in a certain proportion to make a product. Each of the ingredients is a factor in the experiment and the proportion of the ingredient is the level of the factor. A special property of mixture experiments that has to be dealt with is the fact that the sum of the levels of the factors must be 1. Since the constant β_0 is really $\beta_0 \times 1$, β_0 would be completely confounded with $\beta_1 + \cdots + \beta_n$. We must drop β_0 from the model. Thus, when we design a mixture experiment, we must design for a model without the intercept term and make sure that the sum of the levels of the factors is one.

Using the same line of reasoning, we do not include terms of the form $\beta_{ii}X_i^2$ in the quadratic model. The reasoning is as follows. If the model includes all second order terms, then $X_1^2 + X_1X_2 + \cdots + X_1X_n = X_1(X_1 + X_2 + \cdots + X_n) = X_1$ and the term X_1 is represented twice in the model. The same goes for X_i^3, X_i^4, and so on.

The intuition behind designing for mixture experiments comes from the fact that the restriction of summing to one effectively reduces the dimension of the problem. This is most easily seen in the three dimensional case given in Figure 9.7. Because $X_1 + X_2 + X_3$ has to equal 1, the three dimensional cube is effectively reduced to a two dimensional triangle. In the two dimensional triangle, the bottom line is $X_1 = 0$, the upper right hand line is $X_2 = 0$ and the upper left hand line is $X_3 = 0$.

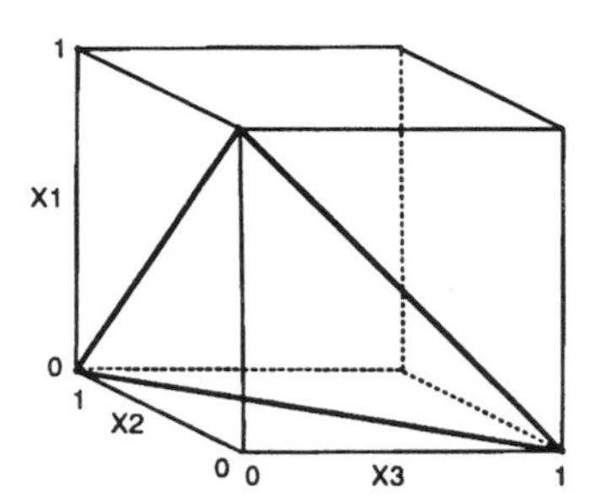

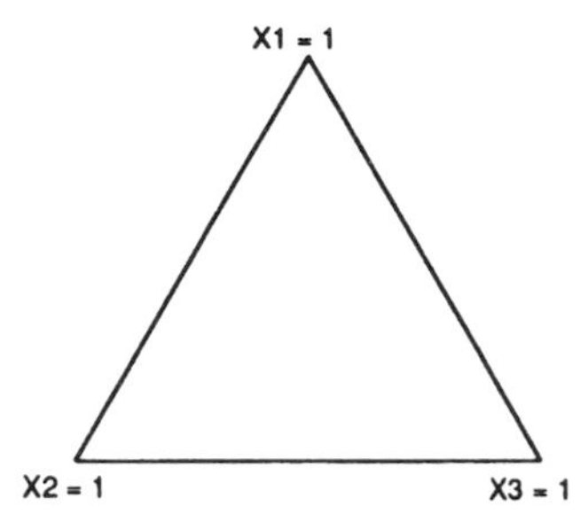

Figure 9.7 The three factor mixture space with 2 d representation.

The equation for a mixture model that corresponds to an n polynomial in n factors is

$$\begin{aligned} Y = &\beta_1 X_1 + \beta_2 X_2 + \beta_3 X_3 + \cdots + \beta_n X_n + \\ &\beta_{12} X_1 X_2 + \beta_{13} X_1 X_3 + \cdots \beta_{n\text{-}1\,n} X_{n\text{-}1} X_n + \\ &\beta_{123} X_1 X_2 X_3 + \cdots + \beta_{n\text{-}2\,n\text{-}1\,n} X_{n\text{-}2} X_{n\text{-}1} X_n + \\ &\vdots \\ &\beta_{12\cdots n} X_1 X_2 \cdots X_n + \varepsilon \end{aligned}$$

The design for the n polynomial is called a simplex centroid, Scheffé (1963). It consists of n points of the form $(1,0,0,\cdots,0)$, $\binom{n}{2}$ points of the form $(1/2,1/2,0,\cdots,0)$, $\binom{n}{3}$ points of the form $(1/3,1/3,1/3,\cdots,0)$, $\cdots$, and 1 point of the form $(1/n,1/n,1/n,\cdots,1/n)$. The simplex centroid for three factors is given in Figure 9.8.

For a linear model of the form

$$Y = \beta_1 X_1 + \beta_2 X_2 + \cdots + \beta_n X_n + \varepsilon$$

use the first set of points from the simplex centroid. That is, the design for a linear model in n factors is given by the factor level combinations $(1,0,0,\cdots,0)$, $(0,1,0,\cdots,0)$, $\cdots$, and $(0,0,0,\cdots,1)$.

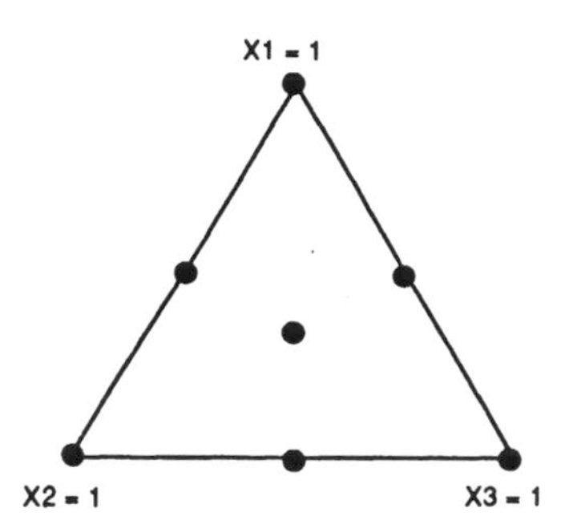

Figure 9.8 A simplex centroid in 3 factors.

For a quadratic model taking the form

$$Y = \beta_1 X_1 + \beta_2 X_2 + \cdots + \beta_n X_n + \beta_{12} X_1 X_2 + \cdots + \beta_{n-1n} X_{n-1} X_n + \varepsilon,$$

use the first and second set of points from the simplex centroid.

The extension to higher order models uses what is known as the simplex lattice designs, Scheffé (1958). For a cubic model, use every combination of levels involving 0, 1/3, 2/3, and 1 for each factor that sums to 1. For a quartic model, use every combination of levels involving 0, 1/4, 1/2, 3/4, and 1 that sums to one. And so on. Representations of the cubic and quartic simplex lattice designs in three factors are shown in Figure 9.9.

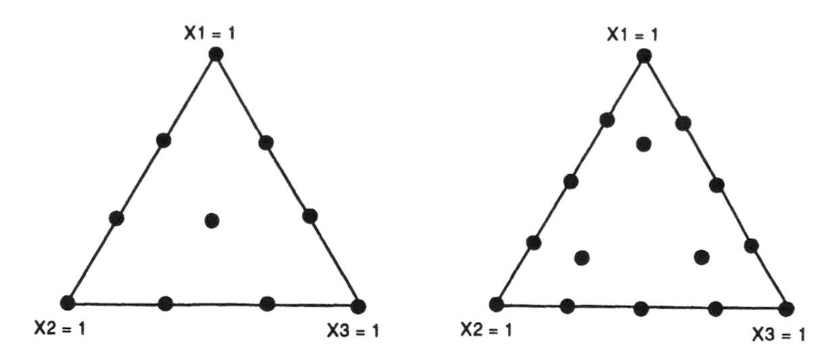

Figure 9.9 A simplex lattice for cubic and quartic models, 3 factors

Often times there are realistic constraints on the factors of the form $0 \leq \ell_i \leq X_i \leq u_i \leq 1$, for lower bounds ℓ_i and upper bounds u_i on the factor levels to be used in the experiment. These are handled using the extreme vertices design, McLean and Anderson (1966). If there are n factors, the first step is to find all of the admissible vertices. Write down all possible factor level combinations using the lower and upper constraints, ℓ_i and u_i, for all factors but one. If the level that makes the total sum to 1 falls within the bounds for that factor, the combination is an extreme vertex. Do this for every factor. If a simple linear model is to be used, the extreme vertices represent the design. Note that duplicates may occur and should be counted as one vertex. There may be more vertices than factors so some use only a subset of the vertices.

For a quadratic model with 3 factors, add the centers of the lines connecting the vertices and the overall centroid to the extreme vertices to create the design. The center of each line is found by averaging the factor levels for each pair of vertices that have the same upper or lower bound for one factor. Note that duplicates are again possible and should be used only once. The overall centroid is the average of all extreme vertices. Typically there will be more centers of lines than necessary so some people use only the centers associated with the longest lines.

For a quadratic model with more than 3 factors, the centroids of the 2-dimensional faces and the overall centroid must be found. Each centroid of a 2-dimensional face is found by averaging the levels for the remaining factors for each set of vertices that agree on $n - 3$ of the levels. Again, the overall centroid is the average of all extreme vertices.

This extends to higher order models with 3-dimensional faces found by averaging vertices that agree on $n - 4$ of the levels.

A mixture experiment on 3 factors had the following constraints:

$$0.2 \leq X_1 \leq 0.6 \qquad 0.1 \leq X_2 \leq 0.4 \qquad 0.1 \leq X_3 \leq 0.5 \; .$$

To find the extreme vertices, use the limits on X_1 and X_2 first, then on X_1 and X_3, and finally on X_2 and X_3. The procedure is illustrated below, *na* indicating that the combination is not admissible.

X_1	X_2	X_3	X_1	X_2	X_3	X_1	X_2	X_3
.20	.10	*na*	.20	*na*	.10	*na*	.10	.10
.20	.40	.40	.20	.30	.50	.40	.10	.50
.60	.10	.30	.60	.30	.10	.50	.40	.10
.60	.40	*na*	.60	*na*	.50	*na*	.40	.50

The center of each line is found by averaging the 2 extreme vertices that have $X_1 = .2$, the 2 that have $X_1 = .6$, the 2 that have $X_2 = .1$, the 2 that have $X_2 = .4$, the 2 that have $X_3 = .1$, and the 2 that have $X_3 = .5$.

We do not average the 2 points that have $X_2 = .3$ since .3 is not a lower or upper bound for X_2. (The overall centroid is $X_1 = .417$, $X_2 = .267$, and $X_3 = .317$. The designs for linear and quadratic models are listed in Table 9.5.1 and displayed in Figure 9.10.

TABLE 9.5.1

LINEAR & QUADRATIC DESIGNS, CONSTRAINED MIXTURE EXAMPLE

Linear Points			Quadratic Points		
X_1	X_2	X_3	X_1	X_2	X_3
.20	.40	.40	.20	.35	.45
.60	.10	.30	.60	.20	.20
.20	.30	.50	.35	.40	.25
.60	.30	.10	.40	.30	.30
.40	.10	.50	.50	.10	.40
.50	.40	.10	.30	.20	.50
			.417	.267	.317

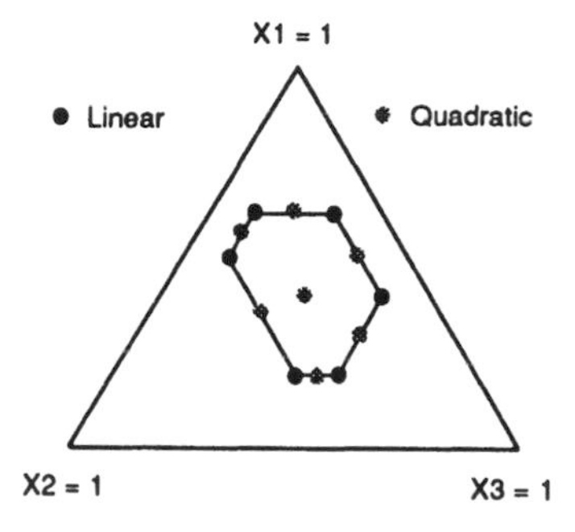

Figure 9.10 Linear & quadratic designs, constrained mixture example.

In manufacturing a particular type of flare, the chemical constituents are magnesium X_1, sodium nitrate X_2, strontium nitrate X_3, and binder X_4. Engineering experience has indicated that the following constraints on a proportion by weight basis should be utilized:

$$0.40 \leq X_1 \leq 0.60$$
$$0.10 \leq X_2 \leq 0.50$$
$$0.10 \leq X_3 \leq 0.50$$
$$0.03 \leq X_4 \leq 0.08\ .$$

The problem is to design for a quadratic model in order to find the treatment combination which gives maximum illumination.

The admissible extreme vertices are found by fixing three of the factors at their extremes and checking if the fourth is admissible. There are four ways to fix three of the factors and $8 = 2^3$ extremes within each. These will be used to estimate the linear portion of the model.

X_1	X_2	X_3	X_4
0.40	0.10	0.10	*na*
0.40	0.10	0.50	*na*
0.40	0.50	0.10	*na*
0.40	0.50	0.50	*na*
0.60	0.10	0.10	*na*
0.60	0.10	0.50	*na*
0.60	0.50	0.10	*na*
0.60	0.50	0.50	*na*

X_1	X_2	X_3	X_4
0.40	0.10	0.47	0.03
0.40	0.10	0.42	0.08
0.40	0.50	*na*	0.03
0.40	0.50	*na*	0.08
0.60	0.10	0.27	0.03
0.60	0.10	0.22	0.08
0.60	0.50	*na*	0.03
0.60	0.50	*na*	0.08

X_1	X_2	X_3	X_4
0.40	0.47	0.10	0.03
0.40	0.42	0.10	0.08
0.40	*na*	0.50	0.03
0.40	*na*	0.50	0.08
0.60	0.27	0.10	0.03
0.60	0.22	0.10	0.08
0.60	*na*	0.50	0.03
0.60	*na*	0.50	0.08

X_1	X_2	X_3	X_4
na	0.10	0.10	0.03
na	0.10	0.10	0.08
na	0.10	0.50	0.03
na	0.10	0.50	0.08
na	0.50	0.10	0.03
na	0.50	0.10	0.08
na	0.50	0.50	0.03
na	0.50	0.50	0.08

In order to complete the design, one must determine the centroids of the 2-dimensional faces. These are made up of the averages of all vertices that agree on $4 - 3 = 1$ of the factor levels. These centroids, along with the overall centroid which is the average of all vertices, are given below. The extreme vertices plus these centroids constitute the entire design.

X_1	X_2	X_3	X_4	Factor Level
0.40	0.2725	0.2725	0.055	$X_1 = 0.40$
0.60	0.1725	0.1725	0.055	$X_1 = 0.60$
0.50	0.1000	0.3450	0.055	$X_2 = 0.10$
0.50	0.3450	0.1000	0.055	$X_3 = 0.10$
0.50	0.2350	0.2350	0.030	$X_4 = 0.03$
0.50	0.2100	0.2100	0.080	$X_4 = 0.08$
0.50	0.2225	0.2225	0.055	centroid

PROBLEMS

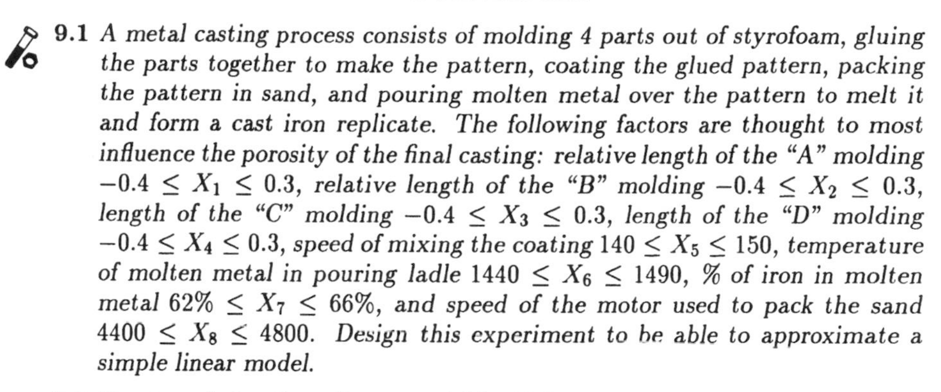

9.1 *A metal casting process consists of molding 4 parts out of styrofoam, gluing the parts together to make the pattern, coating the glued pattern, packing the pattern in sand, and pouring molten metal over the pattern to melt it and form a cast iron replicate. The following factors are thought to most influence the porosity of the final casting: relative length of the "A" molding* $-0.4 \leq X_1 \leq 0.3$, *relative length of the "B" molding* $-0.4 \leq X_2 \leq 0.3$, *length of the "C" molding* $-0.4 \leq X_3 \leq 0.3$, *length of the "D" molding* $-0.4 \leq X_4 \leq 0.3$, *speed of mixing the coating* $140 \leq X_5 \leq 150$, *temperature of molten metal in pouring ladle* $1440 \leq X_6 \leq 1490$, *% of iron in molten metal* $62\% \leq X_7 \leq 66\%$, *and speed of the motor used to pack the sand* $4400 \leq X_8 \leq 4800$. *Design this experiment to be able to approximate a simple linear model.*

9.2 *Create a design for a linear model in 10 factors.*

9.3 *Create a design for a linear model plus one interaction* (X_1X_2) *in 10 factors.*

9.4 *Create a design for a linear model plus three interactions* (X_1X_2, X_3X_4, *and* X_5X_6) *in 10 factors.*

9.5 *In the construction of front wheel drive boot seals, used to hold the grease in the axle joints and hold the dirt out, elongation of the seal is critical to performance. Five factors were studied in the experiment: blow pressure in psi* $80 \leq X_1 \leq 95$, *die temperature* $182° \leq X_2 \leq 198°$, *program setting in psi* $2 \leq X_3 \leq 4$, *screw speed in rpm* $57 \leq X_4 \leq 62$, *and pre-blow pressure in psi* $55 \leq X_5 \leq 80$. *Design a face centered central composite design for fitting a second order polynomial.*

9.6 *Design an orthogonal central composite for the experiment described in Problem* **9.5**. *Assume the limits refer to the fractional factorial piece of the central composite design.*

9.7 *Design an orthogonal central composite for the experiment described in Problem* **9.5**. *Assume the limits are absolute limits so must equal the star points, not the fractional factorial points.*

9.8 *For a central composite design with 6 factors and five center points, what value of* α *is needed for orthogonality? What value of* α *is needed for rotatability?*

9.9 *Create a Box-Behnken design for 4 factors. Label the factors* X_1 *through* X_4 *and the levels -1, 0, and 1.*

9.10 *Create a Box-Behnken design for 5 factors. Label the factors* X_1 *through* X_5 *and the levels -1, 0, and 1.*

9.11 *Create a Box-Behnken design for 6 factors. Label the factors* X_1 *through* X_6 *and the levels -1, 0, and 1.*

9.12 *Create a Box-Behnken design for 7 factors. Label the factors X_1 through X_7 and the levels –1, 0, and 1.*

9.13 *Create an orthogonal and nearly rotatable design for a central composite design in four factors that is to be run in two blocks. Do likewise for a central composite design to be run in three blocks.*

9.14 *Write down the mathematical model associated with both designs in Problem* **9.13**. *Discuss restriction errors as they apply to the designs.*

9.15 *In a weapons evaluation section of a large corporation, an experiment was needed to investigate factors affecting firing time of a weapon. The factors and their lower and upper limits were*

Primary igniter	*5 to 15 mg*
Primary initiator	*5 to 15 mg*
Packing pressure	*12,000 to 28,000 psi*
Amount of igniter	*80 to 120 mg*
Delay column pressure	*20,000 to 40,000 psi*

Eventually an equation of the response surface is desired in which linear by linear and quadratic terms may be important. The number of runs is to be held to a minimum. Design an experiment showing the levels of each factor to be used and explain why you used this particular design.

9.16 *A model of the aging process of an emissions system is desired to help design for stricter automotive emission standards. The following factors are to be considered in the experiment: converter, fuel rail, egr valve, throttle body, front and rear oxygen sensor. For all of these factors, parts are available that have been aged 5, 25, and 50 thousand miles. All factors are thought to contain curvature so all quadratic terms must be in the model. No interactions are thought to be important. There are only 3 oxygen sensors available, aged at 5, 25, and 50 thousand miles, and they can be used either in the front or the rear, but not at the same time. Therefore, no test is possible if the same aged sensor is in the front and the rear. A total of 25 tests can be run. Create a D-optimal design for this experiment.*

9.17 *Use D-optimality with the levels in the range [–1,1] for 4 factors, with factors 1 and 2 quadratic, factors 3 and 4 linear and all linear interactions included. Use 10 more runs than the number of unknown coefficients.*

9.18 *Create a D-optimal design that compares the quadratic effects of the same two factors from three different suppliers. Assume the levels for the two factors fall in the range [–1,1]. For the same number of runs, how does the D-optimal design compare to a face centered central composite used at each supplier?*

9.19 *What is a reasonable model for 4 factors each at three levels when the fourth factor is qualitative? Define the terms in your model.*

9.20 *Write the simplex lattice design for a mixture problem for a quadratic model in 4 factors.*

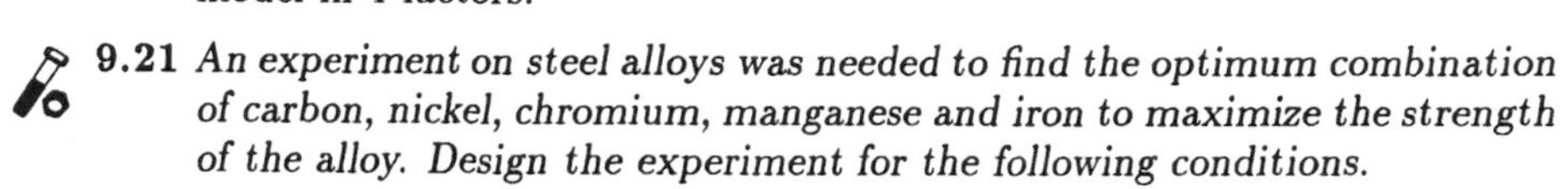

9.21 *An experiment on steel alloys was needed to find the optimum combination of carbon, nickel, chromium, manganese and iron to maximize the strength of the alloy. Design the experiment for the following conditions.*

	Low	*High*
Carbon	*0.0004*	*0.0010*
Nickel	*0.0800*	*0.1200*
Chromiun	*0.1200*	*0.2000*
Manganese	*0.0050*	*0.0200*
Iron	*fixed at 0.7000*	

9.22 *Explain how one could run mixture designs in blocks. Write down a model for the 3 factor case corresponding to the way you set up the blocks.*

9.23 *In mixture experiments involving fertilizer, both the mixture and the amount of fertilizer will have an influence. How can you design for experiments such as this?*

BIBLIOGRAPHY FOR CHAPTER 9

Anderson, V. L. and McLean, R. A. (1974). *Design of Experiments A Realistic Approach*, New York: Marcel Dekker. (Chapter 13)

Box, G. E. P. and Behnken, D. W. (1960). "Some new three-level designs for the study of quantitative variables," *Technometrics*, **2**, 455–475.

Box, G. E. P. and Draper, N. R. (1987). *Empirical Model Building and Response Surfaces*, New York: John Wiley.

Box, G. E. P. and Hunter, J. S. (1957). "Multifactor experimental designs for exploring response surfaces," *Annals of Mathematical Statistics*, **8**, 195–241.

Box, G. E. P. and Wilson, K. B. (1951). "On the experimental attainment of optimum conditions," *Journal of the Royal Statistical Society, B*, **13**, 1–45.

Cornell, J. A. (1973). "Experiments with mixtures: a review," *Technometrics*, **15**, 437–455.

Cornell, J. A. (1979). "Experiments with mixtures: an update and bibliography," *Technometrics*, **21**, 95–106.

Cornell, J. A. (1981). *Experiments with Mixtures: Designs, Models, and the Analysis of Mixture Data*, New York: John Wiley & Sons.

Ghosh, S., editor. (1990). *Statistical Design and Analysis of Industrial Experiments*, New York: Marcel Dekker.

Gorman, J. W. and Hinman, J. E. (1962). "Simplex-lattice designs for multicomponent systems," *Technometrics*, **4**, 463–487.

Khuri, A. I. and Cornell, J. A. (1987). *Response Surfaces: Designs and Analysis*, New York: Marcel Dekker.

McLean, R. A. and Anderson, V. L. (1966). "Extreme vertices design of mixture experiments," *Technometrics*, **8**, 447–454.

Myers, R. H. (1976). *Response Surface Methodology*, Blacksburg, VA: Author.

Myers, R. H., Khuri, A. I., and Carter, W. H. (1989). "Response surface methodology: 1966-1988," *Technometrics*, **31**, 137–157.

Scheffé, H. (1958). "Experiments with mixtures," *Journal of the Royal Statistical Society, B*, **20**, 344–360.

Scheffé, H. (1963). "The simplex centroid design for experiments with mixtures," *Journal of the Royal Statistical Society, B*, **25**, 235–263.

Snee, R. D. (1975). "Experimental designs for quadratic models in constrained mixture spaces," *Technometrics*, **17**, 149–159.

Snee, R. D. and Marquardt, D. W. (1974). "Extreme vertices designs for linear mixture models," *Technometrics*, **16**, 399–408.

SOLUTIONS TO SELECTED PROBLEMS

9.1 *Design for a metal casting process using Appendix 17, not randomized.*

X_1	X_2	X_3	X_4	X_5	X_6	X_7	X_8
0.3	-0.4	0.3	-0.4	140	1440	66%	4800
0.3	0.3	-0.4	0.3	140	1440	62%	4800
-0.4	0.3	0.3	-0.4	150	1440	62%	4400
0.3	-0.4	0.3	0.3	140	1490	62%	4400
0.3	0.3	-0.4	0.3	150	1440	66%	4400
0.3	0.3	0.3	-0.4	150	1490	62%	4800
-0.4	0.3	0.3	0.3	140	1490	66%	4400
-0.4	-0.4	0.3	0.3	150	1440	66%	4800
-0.4	-0.4	-0.4	0.3	150	1490	62%	4800
0.3	-0.4	-0.4	-0.4	150	1490	66%	4400
-0.4	0.3	-0.4	-0.4	140	1490	66%	4800
-0.4	-0.4	-0.4	-0.4	140	1440	62%	4400

9.5 *Central composite design for front wheel drive boot problem, not randomized.*

X_1	X_2	X_3	X_4	X_5
80	182°	2	57	55
80	182°	2	62	80
80	182°	4	57	80
80	182°	4	62	55
80	198°	2	57	80
80	198°	2	62	55
80	198°	4	57	55
80	198°	4	62	80
95	182°	2	57	80
95	182°	2	62	55
95	182°	4	57	55
95	182°	4	62	80
95	198°	2	57	55
95	198°	2	62	80
95	198°	4	57	80
95	198°	4	62	55

X_1	X_2	X_3	X_4	X_5
80	190°	3	59.5	67.5
95	190°	3	59.5	67.5
87.5	182°	3	59.5	67.5
87.5	198°	3	59.5	67.5
87.5	190°	2	59.5	67.5
87.5	190°	4	59.5	67.5
87.5	190°	3	57	67.5
87.5	190°	3	80	67.5
87.5	190°	3	59.5	55
87.5	190°	3	59.5	80
87.5	190°	3	59.5	67.5

9.8 *The α for orthogonality is 1.949. The α for rotatability is 2.378.*

9.9 *A Box-Behnken design for 4 factors, not randomized.*

X_1	X_2	X_3	X_4
1	1	0	0
1	-1	0	0
-1	1	0	0
-1	-1	0	0
1	0	1	0
1	0	-1	0
-1	0	1	0
-1	0	-1	0
1	0	0	1
1	0	0	-1
-1	0	0	1
-1	0	0	-1

X_1	X_2	X_3	X_4
0	1	1	0
0	1	-1	0
0	-1	1	0
0	-1	-1	0
0	1	0	1
0	1	0	-1
0	-1	0	1
0	-1	0	-1
0	0	1	1
0	0	1	-1
0	0	-1	1
0	0	-1	-1
0	0	0	0

9.14 *The key thing to note in this problem is that blocks is a qualitative factor having 2 or 3 levels depending on the number of blocks used. Define an indicator function I_2 that takes the value 1 if the treatment level comes* *from block 2 and is 0 otherwise. Do likewise for I_3 and block 3. The model for 2 blocks is*

$$\begin{aligned} Y = \beta_0 + \beta_{I_2} I_2 + \delta + \beta_1 X_1 + \beta_2 X_2 + \beta_3 X_3 + \beta_4 X_4 + \beta_{11} X_1^2 + \\ \beta_{22} X_2^2 + \beta_{33} X_3^2 + \beta_{44} X_4^2 + \beta_{12} X_1 X_2 + \beta_{13} X_1 X_3 + \\ \beta_{14} X_1 X_4 + \beta_{23} X_2 X_3 + \beta_{24} X_2 X_4 + \beta_{34} X_3 X_4 + \varepsilon. \end{aligned}$$

The model for 3 blocks is

$$\begin{aligned} Y = \beta_0 + \beta_{I_2} I_2 + \delta + \beta_{I_3} I_3 + \gamma + \beta_1 X_1 + \beta_2 X_2 + \beta_3 X_3 + \beta_4 X_4 + \\ \beta_{11} X_1^2 + \beta_{22} X_2^2 + \beta_{33} X_3^2 + \beta_{44} X_4^2 + \beta_{12} X_1 X_2 + \beta_{13} X_1 X_3 + \\ \beta_{14} X_1 X_4 + \beta_{23} X_2 X_3 + \beta_{24} X_2 X_4 + \beta_{34} X_3 X_4 + \varepsilon. \end{aligned}$$

There are restriction errors associated with both I_2 and I_3 because one block is completed before the next is started. However, I_2 and I_3 are placed in the model to account for anything that changed between the first block and the second, including the restriction error effect. Therefore, one could drop the restriction error without any loss of meaning in the model.

9.16 *A D-optimal design for the emissions system problem. Your answer may not agree with this answer depending on the algorithm used.*

C	R	E	T	F	R
5	5	5	5	50	25
5	5	5	25	5	50
5	25	5	25	25	5
5	25	50	5	50	5
5	25	50	50	5	25
5	50	25	5	5	25
5	50	25	50	50	5
5	50	50	25	25	50
25	5	5	50	25	50
25	5	25	25	25	5
25	5	50	5	5	50
25	25	25	5	50	25
25	25	25	25	5	50

C	R	E	T	F	R
25	25	50	50	25	5
25	50	5	5	25	5
25	50	5	50	5	25
25	50	50	25	50	25
50	5	25	5	25	5
50	5	50	25	5	25
50	5	50	50	50	5
50	25	5	5	5	50
50	25	5	25	50	25
50	25	25	50	25	50
50	50	5	25	50	5
50	50	50	5	25	50

9.21 *We use a quadratic model to find a maximum. The extreme vertices are in the left column, appropriate for the linear portion. The centroids of the 2 dimensional faces and the overall centroid are in the right column, appropriate for the quadratic portion.*

C	Ni	Cr	Mn
.0004	.0800	.2000	.0196
.0004	.0800	.1996	.0200
.0004	.0946	.2000	.0050
.0004	.1200	.1596	.0200
.0004	.1200	.1746	.0050
.0010	.0800	.1990	.0200
.0010	.0800	.2000	.0190
.0010	.0940	.2000	.0005
.0010	.1200	.1590	.0200
.0010	.1200	.1740	.0005

C	Ni	Cr	Mn
.0001	.0988	.1864	.0138
.0004	.0989	.1868	.0139
.0007	.0800	.1997	.0197
.0007	.0872	.2000	.0122
.0007	.1000	.1793	.0200
.0007	.1072	.1872	.0050
.0007	.1200	.1668	.0125
.0007	.0989	.1866	.0139

Appendices

APPENDIX 1

A Table of Random Numbers

	1-5	6-10	11-15	16-20	21-25	26-30	31-35	36-40	41-45	46-50	51-55
1	13850	97278	13196	36822	73035	23373	85007	77708	61140	49077	30435
2	98513	23371	75397	77638	38265	01902	05355	48230	75638	40435	82480
3	12902	49790	47102	95353	77813	31610	90046	30113	51610	20872	38750
4	22840	04331	87875	89070	89799	14061	38510	64474	06955	15296	45521
5	85823	09558	05668	66273	44640	46397	12222	56828	87061	13258	52463
6	43650	98872	36771	83227	02839	70884	98103	78792	07367	16561	32600
7	85339	64465	23027	27956	36547	27965	54647	41028	86008	00037	62191
8	43499	29700	45747	49769	34935	76683	80446	63945	89419	11644	59663
9	20580	94685	50466	97019	20260	59845	97809	63570	77011	78504	94569
10	53254	91665	49093	94386	66655	97532	68138	87304	24313	26087	54083
11	64878	12640	07711	39019	91509	72396	44739	62101	55124	47626	93761
12	24382	79057	15336	43848	03650	17442	55948	40736	48190	63029	23861
13	24447	90523	30025	84006	71037	23939	88699	80520	03472	93820	03041
14	94713	82346	55545	70048	03932	42721	09440	78852	14267	57769	46074
15	57867	00687	41386	78540	95467	97096	03194	45772	75279	61387	03891
16	36464	32501	13265	15070	00942	53956	54443	93996	60407	85430	59979
17	18390	12322	00241	00887	30164	72310	12781	93867	99110	50012	98764
18	25724	31123	29119	77481	14979	42553	44519	17868	55742	37051	76140
19	26331	82734	03676	22775	69355	56950	83931	79775	86114	99913	49882
20	17306	51914	74099	72226	45466	41378	56682	77302	51808	72526	20089
21	54982	34644	54900	77882	19604	42142	32927	42712	38823	63828	71038
22	28780	68472	25915	43835	50980	62775	85951	59120	76796	19532	32129
23	29301	41943	04840	66967	46012	21864	24200	24024	00986	45207	33460
24	78377	55802	43158	91842	10378	96619	11630	54972	10602	01660	10593
25	73004	96709	91140	78806	53249	65413	64903	39872	24218	90168	08523
26	06957	62400	10570	23123	32729	66470	72395	67806	59392	22555	74798
27	86143	36763	75439	47285	83635	72692	75209	99726	69728	46925	03620
28	59576	92668	72127	90351	96384	03032	32887	32249	99841	71217	53928
29	33956	83502	44254	74879	94983	55102	24609	56479	15670	18519	20013
30	42522	28267	31960	25699	05453	44760	20641	20019	46642	22217	19610
31	24002	14709	03651	62884	75983	15772	01111	83019	61859	08301	23020
32	93763	60499	70409	58726	15817	43009	43453	46117	35342	96355	81447
33	62472	56895	93509	93588	84533	11640	60623	20389	91072	27545	17530
34	23016	07282	87620	38276	86183	10219	19293	52976	89042	21390	15655
35	50178	16356	24568	36710	98908	10527	00621	93693	94334	07400	18287
36	48511	02179	83017	32369	25923	64485	32910	38550	77384	85331	55860
37	95439	91637	19391	28111	78388	81445	37916	80942	72410	48686	96937
38	56062	32140	15540	80546	30963	27281	80227	42739	40635	00087	95965
39	31802	73935	75131	62521	86477	12119	48156	18608	38815	88322	93737
40	40914	90967	13797	62012	21179	36240	39519	86564	94086	77097	00921

APPENDIX 2

Upper 90% Critical Values for the F Distribution — $F_{\nu_1,\nu_2}(.90)$

ν_2 \ ν_1	1	2	3	4	5	6	7	8	9	10	11	12	13	14	15	20	25	30	40	60	120	∞
1	.025	.117	.181	.220	.246	.265	.279	.289	.298	.304	.310	.315	.319	.322	.325	.336	.343	.347	.353	.358	.364	.370
2	.020	.111	.183	.231	.265	.289	.307	.321	.333	.342	.350	.356	.362	.367	.371	.386	.396	.402	.410	.418	.426	.434
3	.019	.109	.185	.239	.276	.304	.325	.342	.356	.367	.376	.384	.391	.396	.402	.420	.432	.439	.449	.459	.469	.480
4	.018	.108	.187	.243	.284	.314	.338	.356	.371	.384	.394	.403	.411	.418	.423	.445	.458	.467	.478	.490	.502	.514
5	.018	.108	.188	.247	.290	.322	.347	.367	.383	.397	.408	.418	.426	.433	.440	.463	.478	.488	.501	.514	.527	.541
6	.017	.107	.189	.249	.294	.327	.354	.375	.392	.406	.419	.429	.438	.446	.453	.478	.494	.505	.519	.533	.548	.564
7	.017	.107	.190	.251	.297	.332	.359	.381	.399	.414	.427	.438	.448	.456	.463	.490	.507	.519	.534	.550	.566	.564
8	.017	.107	.190	.253	.299	.335	.363	.386	.405	.421	.434	.446	.455	.464	.472	.500	.518	.531	.547	.563	.581	.599
9	.017	.107	.191	.254	.302	.338	.367	.390	.410	.426	.440	.452	.462	.471	.479	.509	.528	.541	.558	.575	.594	.613
10	.017	.106	.191	.255	.303	.340	.370	.394	.414	.431	.445	.457	.468	.477	.486	.516	.536	.550	.567	.586	.605	.625
11	.017	.106	.191	.256	.305	.343	.373	.397	.417	.434	.449	.462	.473	.482	.491	.523	.543	.557	.576	.595	.615	.637
12	.017	.106	.192	.257	.306	.344	.375	.400	.420	.438	.453	.466	.477	.487	.496	.528	.549	.564	.583	.603	.625	.647
13	.016	.106	.192	.257	.307	.346	.377	.402	.423	.441	.456	.469	.481	.491	.500	.533	.555	.570	.590	.611	.633	.656
14	.016	.106	.192	.258	.308	.347	.378	.404	.425	.443	.459	.472	.484	.494	.504	.538	.560	.576	.596	.618	.640	.665
15	.016	.106	.192	.258	.309	.348	.380	.406	.427	.446	.461	.475	.487	.498	.507	.542	.565	.581	.602	.624	.647	.672
16	.016	.106	.192	.259	.310	.349	.381	.407	.429	.448	.464	.478	.490	.500	.510	.546	.569	.585	.607	.629	.654	.680
17	.016	.106	.193	.259	.310	.350	.382	.409	.431	.450	.466	.480	.492	.503	.513	.549	.573	.589	.611	.635	.659	.686
18	.016	.106	.193	.260	.311	.351	.384	.410	.432	.451	.468	.482	.494	.505	.515	.552	.576	.593	.615	.639	.665	.693
19	.016	.106	.193	.260	.311	.352	.384	.411	.434	.453	.469	.484	.496	.508	.518	.555	.579	.597	.619	.644	.670	.698
20	.016	.106	.193	.260	.312	.353	.385	.412	.435	.454	.471	.486	.498	.510	.520	.557	.582	.600	.623	.648	.675	.704
21	.016	.106	.193	.260	.312	.353	.386	.413	.436	.456	.472	.487	.500	.511	.522	.560	.585	.603	.626	.652	.679	.709
22	.016	.106	.193	.261	.313	.354	.387	.414	.437	.457	.474	.489	.502	.513	.523	.562	.587	.606	.630	.655	.683	.714
23	.016	.106	.193	.261	.313	.354	.388	.415	.438	.458	.475	.490	.503	.515	.525	.564	.590	.608	.633	.659	.687	.718
24	.016	.106	.193	.261	.313	.355	.388	.416	.439	.459	.476	.491	.504	.516	.527	.566	.592	.611	.635	.662	.691	.723
25	.016	.106	.193	.261	.314	.355	.389	.417	.440	.460	.477	.492	.506	.517	.528	.568	.594	.613	.638	.665	.694	.727
26	.016	.106	.193	.261	.314	.356	.389	.417	.441	.461	.478	.493	.507	.519	.529	.570	.596	.615	.640	.668	.698	.731
28	.016	.106	.193	.262	.315	.356	.390	.419	.442	.462	.480	.495	.509	.521	.532	.573	.600	.619	.645	.673	.704	.738
30	.016	.106	.193	.262	.315	.357	.391	.420	.443	.464	.482	.497	.511	.523	.534	.575	.603	.622	.649	.678	.710	.745
40	.016	.106	.194	.263	.317	.360	.394	.424	.448	.469	.487	.504	.518	.530	.542	.585	.615	.636	.664	.696	.731	.772
60	.016	.106	.194	.264	.318	.362	.398	.428	.453	.475	.494	.510	.525	.538	.550	.596	.628	.650	.682	.717	.757	.807
120	.016	.105	.194	.265	.320	.365	.401	.432	.458	.480	.500	.518	.533	.547	.560	.609	.642	.667	.702	.742	.791	.856
∞	.016	.105	.195	.266	.322	.367	.405	.436	.463	.486	.507	.525	.542	.556	.570	.622	.659	.687	.726	.774	.839	1.00

APPENDIX 2 (Continued)

Upper 25% Critical Values for the F Distribution — $F_{\nu_1,\nu_2}(.25)$

ν_2 \ ν_1	1	2	3	4	5	6	7	8	9	10	11	12	13	14	15	20	25	30	40	60	120	∞
1	5.83	7.50	8.20	8.58	8.82	8.98	9.10	9.19	9.26	9.32	9.37	9.41	9.44	9.47	9.49	9.58	9.63	9.67	9.71	9.76	9.80	9.85
2	2.57	3.00	3.15	3.23	3.28	3.31	3.34	3.35	3.37	3.38	3.39	3.39	3.40	3.41	3.41	3.43	3.44	3.44	3.45	3.46	3.47	3.48
3	2.02	2.28	2.36	2.39	2.41	2.42	2.43	2.44	2.44	2.44	2.45	2.45	2.45	2.45	2.46	2.46	2.46	2.47	2.47	2.47	2.47	2.47
4	1.81	2.00	2.05	2.06	2.07	2.08	2.08	2.08	2.08	2.08	2.08	2.08	2.08	2.08	2.08	2.08	2.08	2.08	2.08	2.08	2.08	2.08
5	1.69	1.85	1.88	1.89	1.89	1.89	1.89	1.89	1.89	1.89	1.89	1.89	1.89	1.89	1.89	1.88	1.88	1.88	1.88	1.87	1.87	1.87
6	1.62	1.76	1.78	1.79	1.79	1.78	1.78	1.78	1.77	1.77	1.77	1.77	1.77	1.76	1.76	1.76	1.75	1.75	1.75	1.74	1.74	1.74
7	1.57	1.70	1.72	1.72	1.71	1.71	1.70	1.70	1.69	1.69	1.69	1.68	1.68	1.68	1.68	1.67	1.67	1.66	1.66	1.65	1.65	1.65
8	1.54	1.66	1.67	1.66	1.66	1.65	1.64	1.64	1.63	1.63	1.63	1.62	1.62	1.62	1.62	1.61	1.60	1.60	1.59	1.59	1.58	1.58
9	1.51	1.62	1.63	1.63	1.62	1.61	1.60	1.60	1.59	1.59	1.58	1.58	1.58	1.57	1.57	1.56	1.55	1.55	1.54	1.54	1.53	1.53
10	1.49	1.60	1.60	1.59	1.59	1.58	1.57	1.56	1.56	1.55	1.55	1.54	1.54	1.54	1.53	1.52	1.52	1.51	1.51	1.50	1.49	1.48
11	1.47	1.58	1.58	1.57	1.56	1.55	1.54	1.53	1.53	1.52	1.52	1.51	1.51	1.51	1.50	1.49	1.49	1.48	1.47	1.47	1.46	1.45
12	1.46	1.56	1.56	1.55	1.54	1.53	1.52	1.51	1.51	1.50	1.49	1.49	1.49	1.48	1.48	1.47	1.46	1.45	1.45	1.44	1.43	1.42
13	1.45	1.55	1.55	1.53	1.52	1.51	1.50	1.49	1.49	1.48	1.47	1.47	1.47	1.46	1.46	1.45	1.44	1.43	1.42	1.42	1.41	1.40
14	1.44	1.53	1.53	1.52	1.51	1.50	1.49	1.48	1.47	1.46	1.46	1.45	1.45	1.44	1.44	1.43	1.42	1.41	1.41	1.40	1.39	1.38
15	1.43	1.52	1.52	1.51	1.49	1.48	1.47	1.46	1.46	1.45	1.44	1.44	1.43	1.43	1.43	1.41	1.40	1.40	1.39	1.38	1.37	1.36
16	1.42	1.51	1.51	1.50	1.48	1.47	1.46	1.45	1.44	1.44	1.43	1.43	1.42	1.42	1.41	1.40	1.39	1.38	1.37	1.36	1.35	1.34
17	1.42	1.51	1.50	1.49	1.47	1.46	1.45	1.44	1.43	1.43	1.42	1.41	1.41	1.41	1.40	1.39	1.38	1.37	1.36	1.35	1.34	1.33
18	1.41	1.50	1.49	1.48	1.46	1.45	1.44	1.43	1.42	1.42	1.41	1.40	1.40	1.40	1.39	1.38	1.37	1.36	1.35	1.34	1.33	1.32
19	1.41	1.49	1.49	1.47	1.46	1.44	1.43	1.42	1.41	1.41	1.40	1.40	1.39	1.39	1.38	1.37	1.36	1.35	1.34	1.33	1.32	1.30
20	1.40	1.49	1.48	1.47	1.45	1.44	1.43	1.42	1.41	1.40	1.39	1.39	1.38	1.38	1.37	1.36	1.35	1.34	1.33	1.32	1.31	1.29
21	1.40	1.48	1.48	1.46	1.44	1.43	1.42	1.41	1.40	1.39	1.39	1.38	1.37	1.37	1.37	1.35	1.34	1.33	1.32	1.31	1.30	1.28
22	1.40	1.48	1.47	1.45	1.44	1.42	1.41	1.40	1.39	1.39	1.38	1.37	1.37	1.36	1.36	1.34	1.33	1.32	1.31	1.30	1.29	1.28
23	1.39	1.47	1.47	1.45	1.43	1.42	1.41	1.40	1.39	1.38	1.37	1.37	1.36	1.36	1.35	1.34	1.33	1.32	1.31	1.30	1.28	1.27
24	1.39	1.47	1.46	1.44	1.43	1.41	1.40	1.39	1.38	1.37	1.37	1.36	1.36	1.35	1.35	1.33	1.32	1.31	1.30	1.29	1.28	1.26
25	1.39	1.47	1.46	1.44	1.42	1.41	1.40	1.39	1.38	1.37	1.36	1.36	1.35	1.35	1.34	1.33	1.32	1.31	1.29	1.28	1.27	1.25
26	1.38	1.46	1.45	1.44	1.42	1.41	1.39	1.38	1.37	1.37	1.36	1.35	1.35	1.34	1.34	1.32	1.31	1.30	1.29	1.28	1.26	1.25
28	1.38	1.46	1.45	1.43	1.41	1.40	1.39	1.38	1.37	1.36	1.35	1.34	1.34	1.33	1.33	1.31	1.30	1.29	1.28	1.27	1.25	1.24
30	1.38	1.45	1.44	1.42	1.41	1.39	1.38	1.37	1.36	1.35	1.34	1.34	1.33	1.33	1.32	1.30	1.29	1.28	1.27	1.26	1.24	1.23
40	1.36	1.44	1.42	1.40	1.39	1.37	1.36	1.35	1.34	1.33	1.32	1.31	1.31	1.30	1.30	1.28	1.26	1.25	1.24	1.22	1.21	1.19
60	1.35	1.42	1.41	1.38	1.37	1.35	1.33	1.32	1.31	1.30	1.29	1.29	1.28	1.27	1.27	1.25	1.23	1.22	1.21	1.19	1.17	1.15
120	1.34	1.40	1.39	1.37	1.35	1.33	1.31	1.30	1.29	1.28	1.27	1.26	1.25	1.25	1.24	1.22	1.20	1.19	1.18	1.16	1.13	1.10
∞	1.32	1.39	1.37	1.35	1.33	1.31	1.29	1.28	1.27	1.25	1.25	1.24	1.23	1.22	1.22	1.19	1.17	1.16	1.14	1.12	1.08	1.00

APPENDIX 2 (Continued)

Upper 10% Critical Values for the F Distribution — $F_{\nu_1,\nu_2}(.10)$

ν_2 \ ν_1	1	2	3	4	5	6	7	8	9	10	11	12	13	14	15	20	25	30	40	60	120	∞
1	39.9	49.5	53.6	55.8	57.2	58.2	58.9	59.4	59.9	60.2	60.5	60.7	60.9	61.1	61.2	61.7	62.1	62.2	62.5	62.8	63.1	63.3
2	8.53	9.00	9.16	9.24	9.29	9.33	9.35	9.37	9.38	9.39	9.40	9.41	9.41	9.42	9.42	9.44	9.45	9.46	9.47	9.47	9.48	9.49
3	5.54	5.46	5.39	5.34	5.31	5.28	5.27	5.25	5.24	5.23	5.22	5.22	5.21	5.20	5.20	5.19	5.17	5.17	5.16	5.15	5.14	5.13
4	4.54	4.32	4.19	4.11	4.05	4.01	3.98	3.95	3.94	3.92	3.91	3.90	3.89	3.88	3.87	3.84	3.83	3.82	3.80	3.79	3.78	3.76
5	4.06	3.78	3.62	3.52	3.45	3.40	3.37	3.34	3.32	3.30	3.28	3.27	3.26	3.25	3.24	3.21	3.19	3.17	3.16	3.14	3.12	3.10
6	3.78	3.46	3.29	3.18	3.11	3.05	3.01	2.98	2.96	2.94	2.92	2.90	2.89	2.88	2.87	2.84	2.81	2.80	2.78	2.76	2.74	2.72
7	3.59	3.26	3.07	2.96	2.88	2.83	2.79	2.75	2.72	2.70	2.68	2.67	2.65	2.64	2.63	2.59	2.57	2.56	2.54	2.51	2.49	2.47
8	3.46	3.11	2.92	2.81	2.73	2.67	2.62	2.59	2.56	2.54	2.52	2.50	2.49	2.48	2.46	2.42	2.40	2.38	2.36	2.34	2.32	2.29
9	3.36	3.01	2.81	2.69	2.61	2.55	2.51	2.47	2.44	2.42	2.40	2.38	2.36	2.35	2.34	2.30	2.27	2.25	2.23	2.21	2.18	2.16
10	3.29	2.92	2.73	2.61	2.52	2.46	2.41	2.38	2.35	2.32	2.30	2.28	2.27	2.26	2.24	2.20	2.17	2.16	2.13	2.11	2.08	2.06
11	3.23	2.86	2.66	2.54	2.45	2.39	2.34	2.30	2.27	2.25	2.23	2.21	2.19	2.18	2.17	2.12	2.10	2.08	2.05	2.03	2.00	1.97
12	3.18	2.81	2.61	2.48	2.39	2.33	2.28	2.24	2.21	2.19	2.17	2.15	2.13	2.12	2.10	2.06	2.03	2.01	1.99	1.96	1.93	1.90
13	3.14	2.76	2.56	2.43	2.35	2.28	2.23	2.20	2.16	2.14	2.12	2.10	2.08	2.07	2.05	2.01	1.98	1.96	1.93	1.90	1.88	1.85
14	3.10	2.73	2.52	2.39	2.31	2.24	2.19	2.15	2.12	2.10	2.07	2.05	2.04	2.02	2.01	1.96	1.93	1.91	1.89	1.86	1.83	1.80
15	3.07	2.70	2.49	2.36	2.27	2.21	2.16	2.12	2.09	2.06	2.04	2.02	2.00	1.99	1.97	1.92	1.89	1.87	1.85	1.82	1.79	1.76
16	3.05	2.67	2.46	2.33	2.24	2.18	2.13	2.09	2.06	2.03	2.01	1.99	1.97	1.95	1.94	1.89	1.86	1.84	1.81	1.78	1.75	1.72
17	3.03	2.64	2.44	2.31	2.22	2.15	2.10	2.06	2.03	2.00	1.98	1.96	1.94	1.93	1.91	1.86	1.83	1.81	1.78	1.75	1.72	1.69
18	3.01	2.62	2.42	2.29	2.20	2.13	2.08	2.04	2.00	1.98	1.95	1.93	1.92	1.90	1.89	1.84	1.80	1.78	1.75	1.72	1.69	1.66
19	2.99	2.61	2.40	2.27	2.18	2.11	2.06	2.02	1.98	1.96	1.93	1.91	1.89	1.88	1.86	1.81	1.78	1.76	1.73	1.70	1.67	1.63
20	2.97	2.59	2.38	2.25	2.16	2.09	2.04	2.00	1.96	1.94	1.91	1.89	1.87	1.86	1.84	1.79	1.76	1.74	1.71	1.68	1.64	1.61
21	2.96	2.57	2.36	2.23	2.14	2.08	2.02	1.98	1.95	1.92	1.90	1.87	1.86	1.84	1.83	1.78	1.74	1.72	1.69	1.66	1.62	1.59
22	2.95	2.56	2.35	2.22	2.13	2.06	2.01	1.97	1.93	1.90	1.88	1.86	1.84	1.83	1.81	1.76	1.73	1.70	1.67	1.64	1.60	1.57
23	2.94	2.55	2.34	2.21	2.11	2.05	1.99	1.95	1.92	1.89	1.87	1.84	1.83	1.81	1.80	1.74	1.71	1.69	1.66	1.62	1.59	1.55
24	2.93	2.54	2.33	2.19	2.10	2.04	1.98	1.94	1.91	1.88	1.85	1.83	1.81	1.80	1.78	1.73	1.70	1.67	1.64	1.61	1.57	1.53
25	2.92	2.53	2.32	2.18	2.09	2.02	1.97	1.93	1.89	1.87	1.84	1.82	1.80	1.79	1.77	1.72	1.68	1.66	1.63	1.59	1.56	1.52
26	2.91	2.52	2.31	2.17	2.08	2.01	1.96	1.92	1.88	1.86	1.83	1.81	1.79	1.77	1.76	1.71	1.67	1.65	1.61	1.58	1.54	1.50
28	2.89	2.50	2.29	2.16	2.06	2.00	1.94	1.90	1.87	1.84	1.81	1.79	1.77	1.75	1.74	1.69	1.65	1.63	1.59	1.56	1.52	1.48
30	2.88	2.49	2.28	2.14	2.05	1.98	1.93	1.88	1.85	1.82	1.79	1.77	1.75	1.74	1.72	1.67	1.63	1.61	1.57	1.54	1.50	1.46
40	2.84	2.44	2.23	2.09	2.00	1.93	1.87	1.83	1.79	1.76	1.74	1.71	1.69	1.68	1.66	1.61	1.57	1.54	1.51	1.47	1.42	1.38
60	2.79	2.39	2.18	2.04	1.95	1.87	1.82	1.77	1.74	1.71	1.68	1.66	1.64	1.62	1.60	1.54	1.50	1.48	1.44	1.40	1.35	1.29
120	2.75	2.35	2.13	1.99	1.90	1.82	1.77	1.72	1.68	1.65	1.63	1.60	1.58	1.56	1.54	1.48	1.44	1.41	1.37	1.32	1.26	1.19
∞	2.71	2.30	2.08	1.94	1.85	1.77	1.72	1.67	1.63	1.60	1.57	1.55	1.52	1.50	1.49	1.42	1.38	1.34	1.29	1.24	1.17	1.00

APPENDIX 2 (Continued)

Upper 5% Critical Values for the F Distribution — $F_{\nu_1,\nu_2}(.05)$

ν_2 \ ν_1	1	2	3	4	5	6	7	8	9	10	11	12	13	14	15	20	25	30	40	60	120	∞
1	161.	199.	216.	225.	230.	234.	237.	239.	241.	242.	243.	244.	245.	245.	246.	248.	249.	250.	251.	252.	253.	254.
2	18.5	19.0	19.2	19.2	19.3	19.3	19.4	19.4	19.4	19.4	19.4	19.4	19.4	19.4	19.4	19.4	19.5	19.5	19.5	19.5	19.5	19.5
3	10.1	9.55	9.28	9.12	9.01	8.94	8.89	8.85	8.81	8.79	8.76	8.74	8.73	8.72	8.70	8.66	8.63	8.62	8.59	8.57	8.55	8.53
4	7.71	6.94	6.59	6.39	6.26	6.16	6.09	6.04	6.00	5.97	5.94	5.91	5.89	5.87	5.86	5.80	5.77	5.74	5.72	5.69	5.66	5.63
5	6.61	5.79	5.41	5.19	5.05	4.95	4.88	4.82	4.77	4.73	4.70	4.68	4.66	4.64	4.62	4.56	4.52	4.50	4.46	4.43	4.40	4.36
6	5.99	5.14	4.76	4.53	4.39	4.28	4.21	4.15	4.10	4.06	4.03	4.00	3.98	3.96	3.94	3.87	3.84	3.81	3.77	3.74	3.70	3.67
7	5.59	4.74	4.35	4.12	3.97	3.87	3.79	3.73	3.68	3.64	3.60	3.57	3.55	3.53	3.51	3.44	3.40	3.38	3.34	3.31	3.27	3.23
8	5.32	4.46	4.07	3.84	3.69	3.58	3.50	3.44	3.39	3.35	3.31	3.28	3.26	3.24	3.22	3.15	3.11	3.08	3.04	3.00	2.97	2.93
9	5.12	4.26	3.86	3.63	3.48	3.37	3.29	3.23	3.18	3.14	3.10	3.07	3.05	3.03	3.01	2.94	2.89	2.86	2.83	2.79	2.75	2.71
10	4.96	4.10	3.71	3.48	3.33	3.22	3.14	3.07	3.02	2.98	2.94	2.91	2.89	2.86	2.85	2.77	2.73	2.70	2.66	2.62	2.58	2.54
11	4.84	3.98	3.59	3.36	3.20	3.09	3.01	2.95	2.90	2.85	2.82	2.79	2.76	2.74	2.72	2.65	2.60	2.57	2.53	2.49	2.45	2.40
12	4.75	3.89	3.49	3.26	3.11	3.00	2.91	2.85	2.80	2.75	2.72	2.69	2.66	2.64	2.62	2.54	2.50	2.47	2.43	2.38	2.34	2.30
13	4.67	3.81	3.41	3.18	3.03	2.92	2.83	2.77	2.71	2.67	2.63	2.60	2.58	2.55	2.53	2.46	2.41	2.38	2.34	2.30	2.25	2.21
14	4.60	3.74	3.34	3.11	2.96	2.85	2.76	2.70	2.65	2.60	2.57	2.53	2.51	2.48	2.46	2.39	2.34	2.31	2.27	2.22	2.18	2.13
15	4.54	3.68	3.29	3.06	2.90	2.79	2.71	2.64	2.59	2.54	2.51	2.48	2.45	2.42	2.40	2.33	2.28	2.25	2.20	2.16	2.11	2.07
16	4.49	3.63	3.24	3.01	2.85	2.74	2.66	2.59	2.54	2.49	2.46	2.42	2.40	2.37	2.35	2.28	2.23	2.19	2.15	2.11	2.06	2.01
17	4.45	3.59	3.20	2.96	2.81	2.70	2.61	2.55	2.49	2.45	2.41	2.38	2.35	2.33	2.31	2.23	2.18	2.15	2.10	2.06	2.01	1.96
18	4.41	3.55	3.16	2.93	2.77	2.66	2.58	2.51	2.46	2.41	2.37	2.34	2.31	2.29	2.27	2.19	2.14	2.11	2.06	2.02	1.97	1.92
19	4.38	3.52	3.13	2.90	2.74	2.63	2.54	2.48	2.42	2.38	2.34	2.31	2.28	2.26	2.23	2.16	2.11	2.07	2.03	1.98	1.93	1.88
20	4.35	3.49	3.10	2.87	2.71	2.60	2.51	2.45	2.39	2.35	2.31	2.28	2.25	2.22	2.20	2.12	2.07	2.04	1.99	1.95	1.90	1.84
21	4.32	3.47	3.07	2.84	2.68	2.57	2.49	2.42	2.37	2.32	2.28	2.25	2.22	2.20	2.18	2.10	2.05	2.01	1.96	1.92	1.87	1.81
22	4.30	3.44	3.05	2.82	2.66	2.55	2.46	2.40	2.34	2.30	2.26	2.23	2.20	2.17	2.15	2.07	2.02	1.98	1.94	1.89	1.84	1.78
23	4.28	3.42	3.03	2.80	2.64	2.53	2.44	2.37	2.32	2.27	2.24	2.20	2.18	2.15	2.13	2.05	2.00	1.96	1.91	1.86	1.81	1.76
24	4.26	3.40	3.01	2.78	2.62	2.51	2.42	2.36	2.30	2.25	2.22	2.18	2.15	2.13	2.11	2.03	1.97	1.94	1.89	1.84	1.79	1.73
25	4.24	3.39	2.99	2.76	2.60	2.49	2.40	2.34	2.28	2.24	2.20	2.16	2.14	2.11	2.09	2.01	1.96	1.92	1.87	1.82	1.77	1.71
26	4.23	3.37	2.98	2.74	2.59	2.47	2.39	2.32	2.27	2.22	2.18	2.15	2.12	2.09	2.07	1.99	1.94	1.90	1.85	1.80	1.75	1.69
28	4.20	3.34	2.95	2.71	2.56	2.45	2.36	2.29	2.24	2.19	2.15	2.12	2.09	2.06	2.04	1.96	1.91	1.87	1.82	1.77	1.71	1.65
30	4.17	3.32	2.92	2.69	2.53	2.42	2.33	2.27	2.21	2.16	2.13	2.09	2.06	2.04	2.01	1.93	1.88	1.84	1.79	1.74	1.68	1.62
40	4.08	3.23	2.84	2.61	2.45	2.34	2.25	2.18	2.12	2.08	2.04	2.00	1.97	1.95	1.92	1.84	1.78	1.74	1.69	1.64	1.58	1.51
60	4.00	3.15	2.76	2.53	2.37	2.25	2.17	2.10	2.04	1.99	1.95	1.92	1.89	1.86	1.84	1.75	1.69	1.65	1.59	1.53	1.47	1.39
120	3.92	3.07	2.68	2.45	2.29	2.18	2.09	2.02	1.96	1.91	1.87	1.83	1.80	1.78	1.75	1.66	1.60	1.55	1.50	1.43	1.35	1.25
∞	3.84	3.00	2.60	2.37	2.21	2.10	2.01	1.94	1.88	1.83	1.79	1.75	1.72	1.69	1.67	1.57	1.51	1.46	1.39	1.32	1.22	1.00

APPENDIX 2 (Continued)

Upper 1% Critical Values for the F Distribution — $F_{\nu_1,\nu_2}(.01)$

ν_2 \ ν_1	1	2	3	4	5	6	7	8	9	10	11	12	13	14	15	20	25	30	40	60	120	∞
1	4052.	5000.	5403.	5625.	5764.	5859.	5928.	5981.	6022.	6056.	6083.	6106.	6126.	6143.	6157.	6209.	6240.	6261.	6287.	6313.	6339.	6366.
2	98.5	99.0	99.2	99.2	99.3	99.3	99.4	99.4	99.4	99.4	99.4	99.4	99.4	99.4	99.4	99.4	99.5	99.5	99.5	99.5	99.5	99.5
3	34.1	30.8	29.5	28.7	28.2	27.9	27.7	27.5	27.3	27.2	27.1	27.0	27.0	26.9	26.9	26.7	26.6	26.5	26.4	26.3	26.2	26.1
4	21.2	18.0	16.7	16.0	15.5	15.2	15.0	14.8	14.7	14.5	14.5	14.4	14.3	14.2	14.2	14.0	13.9	13.8	13.7	13.7	13.6	13.5
5	16.3	13.3	12.1	11.4	11.0	10.7	10.5	10.3	10.2	10.1	9.96	9.89	9.83	9.77	9.72	9.55	9.45	9.38	9.30	9.20	9.11	9.02
6	13.7	10.9	9.78	9.15	8.75	8.47	8.26	8.10	7.98	7.87	7.79	7.72	7.66	7.60	7.56	7.40	7.29	7.23	7.15	7.06	6.97	6.88
7	12.2	9.55	8.45	7.85	7.46	7.19	6.99	6.84	6.72	6.62	6.54	6.47	6.41	6.36	6.31	6.16	6.06	5.99	5.91	5.82	5.74	5.65
8	11.3	8.65	7.59	7.01	6.63	6.37	6.18	6.03	5.91	5.81	5.73	5.67	5.61	5.56	5.52	5.36	5.26	5.20	5.12	5.03	4.95	4.86
9	10.6	8.02	6.99	6.42	6.06	5.80	5.61	5.47	5.35	5.26	5.18	5.11	5.05	5.01	4.96	4.81	4.71	4.65	4.57	4.48	4.40	4.31
10	10.0	7.56	6.55	5.99	5.64	5.39	5.20	5.06	4.94	4.85	4.77	4.71	4.65	4.60	4.56	4.41	4.31	4.25	4.17	4.08	4.00	3.91
11	9.65	7.21	6.22	5.67	5.32	5.07	4.89	4.74	4.63	4.54	4.46	4.40	4.34	4.29	4.25	4.10	4.00	3.94	3.86	3.78	3.69	3.60
12	9.33	6.93	5.95	5.41	5.06	4.82	4.64	4.50	4.39	4.30	4.22	4.16	4.10	4.05	4.01	3.86	3.76	3.70	3.62	3.54	3.45	3.36
13	9.07	6.70	5.74	5.21	4.86	4.62	4.44	4.30	4.19	4.10	4.02	3.96	3.91	3.86	3.82	3.66	3.57	3.51	3.43	3.34	3.25	3.17
14	8.86	6.51	5.56	5.04	4.69	4.46	4.28	4.14	4.03	3.94	3.86	3.80	3.75	3.70	3.66	3.51	3.41	3.35	3.27	3.18	3.09	3.00
15	8.68	6.36	5.42	4.89	4.56	4.32	4.14	4.00	3.89	3.80	3.73	3.67	3.61	3.56	3.52	3.37	3.28	3.21	3.13	3.05	2.96	2.87
16	8.53	6.23	5.29	4.77	4.44	4.20	4.03	3.89	3.78	3.69	3.62	3.55	3.50	3.45	3.41	3.26	3.16	3.10	3.02	2.93	2.84	2.75
17	8.40	6.11	5.18	4.67	4.34	4.10	3.93	3.79	3.68	3.59	3.52	3.46	3.40	3.35	3.31	3.16	3.07	3.00	2.92	2.83	2.75	2.65
18	8.29	6.01	5.09	4.58	4.25	4.01	3.84	3.71	3.60	3.51	3.43	3.37	3.32	3.27	3.23	3.08	2.98	2.92	2.84	2.75	2.66	2.57
19	8.18	5.93	5.01	4.50	4.17	3.94	3.77	3.63	3.52	3.43	3.36	3.30	3.24	3.19	3.15	3.00	2.91	2.84	2.76	2.67	2.58	2.49
20	8.10	5.85	4.94	4.43	4.10	3.87	3.70	3.56	3.46	3.37	3.29	3.23	3.18	3.13	3.09	2.94	2.84	2.78	2.69	2.61	2.52	2.42
21	8.02	5.78	4.87	4.37	4.04	3.81	3.64	3.51	3.40	3.31	3.24	3.17	3.12	3.07	3.03	2.88	2.78	2.72	2.64	2.55	2.46	2.36
22	7.95	5.72	4.82	4.31	3.99	3.76	3.59	3.45	3.35	3.26	3.18	3.12	3.07	3.02	2.98	2.83	2.73	2.67	2.58	2.50	2.40	2.31
23	7.88	5.66	4.76	4.26	3.94	3.71	3.54	3.41	3.30	3.21	3.14	3.07	3.02	2.97	2.93	2.78	2.69	2.62	2.54	2.45	2.35	2.26
24	7.82	5.61	4.72	4.22	3.90	3.67	3.50	3.36	3.26	3.17	3.09	3.03	2.98	2.93	2.89	2.74	2.64	2.58	2.49	2.40	2.31	2.21
25	7.77	5.57	4.68	4.18	3.85	3.63	3.46	3.32	3.22	3.13	3.06	2.99	2.94	2.89	2.85	2.70	2.60	2.54	2.45	2.36	2.27	2.17
26	7.72	5.53	4.64	4.14	3.82	3.59	3.42	3.29	3.18	3.09	3.02	2.96	2.90	2.86	2.81	2.66	2.57	2.50	2.42	2.33	2.23	2.13
28	7.64	5.45	4.57	4.07	3.75	3.53	3.36	3.23	3.12	3.03	2.96	2.90	2.84	2.79	2.75	2.60	2.51	2.44	2.35	2.26	2.17	2.06
30	7.56	5.39	4.51	4.02	3.70	3.47	3.30	3.17	3.07	2.98	2.91	2.84	2.79	2.74	2.70	2.55	2.45	2.39	2.30	2.21	2.11	2.01
40	7.31	5.18	4.31	3.83	3.51	3.29	3.12	2.99	2.89	2.80	2.73	2.66	2.61	2.56	2.52	2.37	2.27	2.20	2.11	2.02	1.92	1.80
60	7.08	4.98	4.13	3.65	3.34	3.12	2.95	2.82	2.72	2.63	2.56	2.50	2.44	2.39	2.35	2.20	2.10	2.03	1.94	1.84	1.73	1.60
120	6.85	4.79	3.95	3.48	3.17	2.96	2.79	2.66	2.56	2.47	2.40	2.34	2.28	2.23	2.19	2.03	1.93	1.86	1.76	1.66	1.53	1.38
∞	6.63	4.61	3.78	3.32	3.02	2.80	2.64	2.51	2.41	2.32	2.25	2.18	2.13	2.08	2.04	1.88	1.77	1.70	1.59	1.47	1.32	1.00

APPENDIX 2 (Continued)

Upper 0.1% Critical Values for the F Distribution — $F_{\nu_1,\nu_2}(.001)$

ν_2 \ ν_1	1	2	3	4	5	6	7	8	9	10	11	12	13	14	15	20	25	30	40	60	120	∞
1	4053*	5000*	5404*	5625*	5764*	5859*	5929*	5981*	6023*	6056*	6084*	6107*	6126*	6143*	6158*	6209*	6240*	6261*	6287*	6313*	6340*	6366*
2	999.	999.	999.	999.	999.	999.	999.	999.	999.	999.	999.	999.	999.	999.	999.	999.	999.	999.	999.	1000.	1000.	1000.
3	167.	149.	141.	137.	135.	133.	132.	131.	130.	129.	129.	128.	128.	128.	128.	127.	126.	125.	125.	124.	124.	124.
4	74.1	61.2	56.2	53.5	51.7	50.5	49.7	49.0	48.5	48.0	47.7	47.4	47.2	46.9	46.8	46.0	45.7	45.5	45.1	44.7	44.4	44.1
5	47.2	37.1	33.2	31.1	29.7	28.8	28.2	27.7	27.2	26.9	26.6	26.4	26.2	26.0	25.9	25.4	25.1	24.9	24.6	24.3	24.1	23.8
6	35.5	27.0	23.7	21.9	20.8	20.0	19.5	19.0	18.7	18.4	18.2	18.0	17.8	17.7	17.6	17.1	16.8	16.7	16.4	16.2	16.0	15.8
7	29.2	21.7	18.8	17.2	16.2	15.5	15.0	14.6	14.3	14.1	13.9	13.7	13.6	13.4	13.3	12.9	12.7	12.5	12.3	12.1	11.9	11.7
8	25.4	18.5	15.8	14.4	13.5	12.9	12.4	12.0	11.8	11.5	11.4	11.2	11.1	10.9	10.8	10.5	10.3	10.1	9.92	9.72	9.53	9.33
9	22.9	16.4	13.9	12.6	11.7	11.1	10.7	10.4	10.1	9.89	9.72	9.57	9.44	9.33	9.24	8.90	8.69	8.55	8.37	8.18	8.00	7.81
10	21.0	14.9	12.6	11.3	10.5	9.93	9.52	9.20	8.96	8.75	8.59	8.45	8.32	8.22	8.13	7.80	7.60	7.47	7.30	7.12	6.94	6.76
11	19.7	13.8	11.6	10.3	9.58	9.05	8.65	8.35	8.12	7.92	7.76	7.63	7.51	7.41	7.32	7.01	6.82	6.68	6.52	6.35	6.18	6.00
12	18.6	13.0	10.8	9.63	8.89	8.38	8.00	7.71	7.48	7.29	7.14	7.00	6.89	6.79	6.71	6.40	6.22	6.09	5.93	5.76	5.59	5.42
13	17.8	12.3	10.2	9.07	8.35	7.86	7.49	7.21	6.98	6.80	6.65	6.52	6.41	6.31	6.23	5.93	5.75	5.63	5.47	5.31	5.14	4.97
14	17.1	11.8	9.73	8.62	7.92	7.44	7.08	6.80	6.58	6.40	6.26	6.13	6.02	5.93	5.85	5.56	5.38	5.25	5.10	4.94	4.77	4.60
15	16.5	11.3	9.34	8.25	7.57	7.09	6.74	6.47	6.26	6.08	5.93	5.81	5.71	5.61	5.53	5.25	5.07	4.95	4.80	4.64	4.48	4.31
16	16.1	11.0	9.01	7.94	7.27	6.80	6.46	6.19	5.98	5.81	5.67	5.55	5.44	5.35	5.27	4.99	4.82	4.70	4.54	4.39	4.23	4.06
17	15.7	10.7	8.73	7.68	7.02	6.56	6.22	5.96	5.75	5.58	5.44	5.32	5.22	5.13	5.05	4.78	4.60	4.48	4.33	4.18	4.02	3.85
18	15.4	10.4	8.49	7.46	6.81	6.35	6.02	5.76	5.56	5.39	5.25	5.13	5.03	4.94	4.87	4.59	4.42	4.30	4.15	4.00	3.84	3.67
19	15.1	10.2	8.28	7.27	6.62	6.18	5.85	5.59	5.39	5.22	5.08	4.97	4.87	4.78	4.70	4.43	4.26	4.14	3.99	3.84	3.68	3.51
20	14.8	9.95	8.10	7.10	6.46	6.02	5.69	5.44	5.24	5.08	4.94	4.82	4.72	4.64	4.56	4.29	4.12	4.00	3.86	3.70	3.54	3.38
21	14.6	9.77	7.94	6.95	6.32	5.88	5.56	5.31	5.11	4.95	4.81	4.70	4.60	4.51	4.44	4.17	4.00	3.88	3.74	3.58	3.42	3.26
22	14.4	9.61	7.80	6.81	6.19	5.76	5.44	5.19	4.99	4.83	4.70	4.58	4.49	4.40	4.33	4.06	3.89	3.78	3.63	3.48	3.32	3.15
23	14.2	9.47	7.67	6.70	6.08	5.65	5.33	5.09	4.89	4.73	4.60	4.48	4.39	4.30	4.23	3.96	3.79	3.68	3.53	3.38	3.22	3.05
24	14.0	9.34	7.55	6.59	5.98	5.55	5.24	4.99	4.80	4.64	4.51	4.39	4.30	4.21	4.14	3.87	3.71	3.59	3.45	3.29	3.14	2.97
25	13.9	9.22	7.45	6.49	5.89	5.46	5.15	4.91	4.71	4.56	4.42	4.31	4.22	4.13	4.06	3.79	3.63	3.52	3.37	3.22	3.06	2.89
26	13.7	9.12	7.36	6.41	5.80	5.38	5.07	4.83	4.64	4.48	4.35	4.24	4.14	4.06	3.99	3.72	3.56	3.44	3.30	3.15	2.99	2.82
28	13.5	8.93	7.19	6.25	5.66	5.24	4.93	4.69	4.50	4.35	4.22	4.11	4.01	3.93	3.86	3.60	3.43	3.32	3.18	3.02	2.86	2.69
30	13.3	8.77	7.05	6.12	5.53	5.12	4.82	4.58	4.39	4.24	4.11	4.00	3.91	3.82	3.75	3.49	3.33	3.22	3.07	2.92	2.76	2.59
40	12.6	8.25	6.59	5.70	5.13	4.73	4.44	4.21	4.02	3.87	3.75	3.64	3.55	3.47	3.40	3.15	2.98	2.87	2.73	2.57	2.41	2.23
60	12.0	7.77	6.17	5.31	4.76	4.37	4.09	3.86	3.69	3.54	3.42	3.32	3.23	3.15	3.08	2.83	2.67	2.55	2.41	2.25	2.08	1.89
∞	10.8	6.91	5.42	4.62	4.10	3.74	3.50	3.27	3.12	2.98	2.86	2.74	2.67	2.59	2.51	2.27	2.11	1.99	1.84	1.66	1.45	1.00

* Multiply these entries by 100.

APPENDIX 3

Critical Values for the t Distribution — $t_{\alpha/2}$

df	$t_{.25}$	$t_{.10}$	$t_{.05}$	$t_{.025}$	$t_{.01}$	$t_{.005}$	$t_{.001}$	$t_{.0005}$
1	1.000	3.078	6.314	12.706	31.821	63.657	318.309	636.619
2	0.816	1.886	2.920	4.303	6.965	9.925	22.327	31.599
3	0.765	1.638	2.353	3.182	4.541	5.841	10.215	12.924
4	0.741	1.533	2.132	2.776	3.747	4.604	7.173	8.610
5	0.727	1.476	2.015	2.571	3.365	4.032	5.893	6.869
6	0.718	1.440	1.943	2.447	3.143	3.707	5.208	5.959
7	0.711	1.415	1.895	2.365	2.998	3.499	4.785	5.408
8	0.706	1.397	1.860	2.306	2.896	3.355	4.501	5.041
9	0.703	1.383	1.833	2.262	2.821	3.250	4.297	4.781
10	0.700	1.372	1.812	2.228	2.764	3.169	4.144	4.587
11	0.697	1.363	1.796	2.201	2.718	3.106	4.025	4.437
12	0.695	1.356	1.782	2.179	2.681	3.055	3.930	4.318
13	0.694	1.350	1.771	2.160	2.650	3.012	3.852	4.221
14	0.692	1.345	1.761	2.145	2.624	2.977	3.787	4.140
15	0.691	1.341	1.753	2.131	2.602	2.947	3.733	4.073
16	0.690	1.337	1.746	2.120	2.583	2.921	3.686	4.015
17	0.689	1.333	1.740	2.110	2.567	2.898	3.646	3.965
18	0.688	1.330	1.734	2.101	2.552	2.878	3.610	3.922
19	0.688	1.328	1.729	2.093	2.539	2.861	3.579	3.883
20	0.687	1.325	1.725	2.086	2.528	2.845	3.552	3.850
21	0.686	1.323	1.721	2.080	2.518	2.831	3.527	3.819
22	0.686	1.321	1.717	2.074	2.508	2.819	3.505	3.792
23	0.685	1.319	1.714	2.069	2.500	2.807	3.485	3.768
24	0.685	1.318	1.711	2.064	2.492	2.797	3.467	3.745
25	0.684	1.316	1.708	2.060	2.485	2.787	3.450	3.725
26	0.684	1.315	1.706	2.056	2.479	2.779	3.435	3.707
28	0.683	1.313	1.701	2.048	2.467	2.763	3.408	3.674
30	0.683	1.310	1.697	2.042	2.457	2.750	3.385	3.646
40	0.681	1.303	1.684	2.021	2.423	2.704	3.307	3.551
60	0.679	1.296	1.671	2.000	2.390	2.660	3.232	3.460
120	0.677	1.289	1.658	1.980	2.358	2.617	3.160	3.373
∞	0.674	1.282	1.645	1.960	2.326	2.576	3.090	3.291

APPENDIX 4

$\alpha = .10$ Critical Values for Duncan's Test

	Number of Means Spanned – k													
df	2	3	4	5	6	7	8	9	10	12	15	30	60	100
1	8.929	8.929	8.929	8.929	8.929	8.929	8.929	8.929	8.929	8.929	8.929	8.929	8.929	8.929
2	4.129	4.129	4.129	4.129	4.129	4.129	4.129	4.129	4.129	4.129	4.129	4.129	4.129	4.129
3	3.328	3.330	3.330	3.330	3.330	3.330	3.330	3.330	3.330	3.330	3.330	3.330	3.330	3.330
4	3.015	3.074	3.081	3.081	3.081	3.081	3.081	3.081	3.081	3.081	3.081	3.081	3.081	3.081
5	2.850	2.934	2.964	2.969	2.969	2.969	2.969	2.969	2.969	2.969	2.969	2.969	2.969	2.969
6	2.748	2.846	2.890	2.908	2.911	2.911	2.911	2.911	2.911	2.911	2.911	2.911	2.911	2.911
7	2.679	2.785	2.838	2.864	2.876	2.878	2.878	2.878	2.878	2.878	2.878	2.878	2.878	2.878
8	2.630	2.741	2.800	2.832	2.849	2.857	2.858	2.858	2.858	2.858	2.858	2.858	2.858	2.858
9	2.592	2.708	2.771	2.808	2.829	2.840	2.845	2.846	2.846	2.846	2.846	2.846	2.846	2.846
10	2.563	2.681	2.748	2.788	2.813	2.827	2.835	2.839	2.839	2.839	2.839	2.839	2.839	2.839
11	2.540	2.660	2.729	2.772	2.799	2.817	2.827	2.833	2.835	2.835	2.835	2.835	2.835	2.835
12	2.521	2.643	2.714	2.759	2.788	2.808	2.820	2.828	2.832	2.832	2.832	2.832	2.832	2.832
13	2.504	2.628	2.701	2.748	2.779	2.800	2.814	2.824	2.829	2.832	2.832	2.832	2.832	2.832
14	2.491	2.616	2.690	2.739	2.771	2.794	2.810	2.820	2.827	2.833	2.833	2.833	2.833	2.833
15	2.479	2.605	2.681	2.730	2.765	2.788	2.805	2.817	2.825	2.833	2.833	2.833	2.833	2.833
16	2.469	2.596	2.672	2.723	2.759	2.784	2.802	2.814	2.824	2.833	2.835	2.835	2.835	2.835
17	2.460	2.587	2.665	2.717	2.753	2.779	2.798	2.812	2.822	2.834	2.838	2.838	2.838	2.838
18	2.452	2.580	2.659	2.711	2.749	2.776	2.795	2.810	2.821	2.834	2.840	2.840	2.840	2.840
19	2.445	2.574	2.653	2.706	2.744	2.772	2.793	2.808	2.820	2.834	2.843	2.843	2.843	2.843
20	2.439	2.568	2.648	2.702	2.741	2.769	2.790	2.807	2.819	2.835	2.845	2.845	2.845	2.845
21	2.433	2.563	2.643	2.698	2.737	2.766	2.788	2.805	2.818	2.835	2.847	2.847	2.847	2.847
22	2.428	2.558	2.639	2.694	2.734	2.764	2.786	2.804	2.817	2.835	2.848	2.848	2.848	2.848
23	2.424	2.554	2.635	2.691	2.731	2.762	2.785	2.803	2.816	2.835	2.850	2.850	2.850	2.850
24	2.420	2.550	2.632	2.688	2.729	2.760	2.783	2.801	2.816	2.836	2.851	2.851	2.851	2.851
25	2.416	2.546	2.628	2.685	2.726	2.758	2.782	2.800	2.815	2.836	2.853	2.853	2.853	2.853
26	2.412	2.543	2.625	2.683	2.724	2.756	2.780	2.799	2.815	2.836	2.854	2.854	2.854	2.854
28	2.406	2.537	2.620	2.678	2.720	2.753	2.778	2.798	2.814	2.837	2.856	2.856	2.856	2.856
30	2.400	2.532	2.615	2.674	2.717	2.750	2.776	2.796	2.813	2.837	2.859	2.863	2.863	2.863
40	2.381	2.514	2.600	2.660	2.705	2.740	2.768	2.791	2.810	2.838	2.866	2.897	2.897	2.897
60	2.363	2.497	2.584	2.646	2.693	2.731	2.761	2.786	2.807	2.840	2.874	2.934	2.934	2.934
120	2.344	2.479	2.568	2.632	2.682	2.721	2.754	2.781	2.804	2.842	2.883	2.974	3.001	3.001
∞	2.326	2.462	2.552	2.619	2.670	2.712	2.746	2.776	2.801	2.844	2.892	3.019	3.113	3.163

APPENDIX 4 (Continued)

$\alpha = .05$ Critical Values for Duncan's Test

Number of Means Spanned – k

df	2	3	4	5	6	7	8	9	10	12	15	30	60	100
1	17.97	17.97	17.97	17.97	17.97	17.97	17.97	17.97	17.97	17.97	17.97	17.97	17.97	17.97
2	6.085	6.085	6.085	6.085	6.085	6.085	6.085	6.085	6.085	6.085	6.085	6.085	6.085	6.085
3	4.501	4.516	4.516	4.516	4.516	4.516	4.516	4.516	4.516	4.516	4.516	4.516	4.516	4.516
4	3.927	4.013	4.033	4.033	4.033	4.033	4.033	4.033	4.033	4.033	4.033	4.033	4.033	4.033
5	3.635	3.749	3.797	3.814	3.814	3.814	3.814	3.814	3.814	3.814	3.814	3.814	3.814	3.814
6	3.460	3.586	3.649	3.680	3.694	3.697	3.697	3.697	3.697	3.697	3.697	3.697	3.697	3.697
7	3.344	3.477	3.548	3.588	3.611	3.622	3.625	3.625	3.625	3.625	3.625	3.625	3.625	3.625
8	3.261	3.398	3.475	3.521	3.549	3.566	3.575	3.579	3.579	3.579	3.579	3.579	3.579	3.579
9	3.199	3.339	3.420	3.470	3.502	3.523	3.536	3.544	3.547	3.547	3.547	3.547	3.547	3.547
10	3.151	3.293	3.376	3.430	3.465	3.489	3.505	3.516	3.522	3.525	3.525	3.525	3.525	3.525
11	3.113	3.256	3.341	3.397	3.435	3.462	3.480	3.493	3.501	3.509	3.509	3.509	3.509	3.509
12	3.081	3.225	3.312	3.370	3.410	3.439	3.459	3.474	3.484	3.495	3.498	3.498	3.498	3.498
13	3.055	3.200	3.288	3.348	3.389	3.419	3.441	3.458	3.470	3.484	3.490	3.490	3.490	3.490
14	3.033	3.178	3.268	3.328	3.371	3.403	3.426	3.444	3.457	3.474	3.484	3.484	3.484	3.484
15	3.014	3.160	3.250	3.312	3.356	3.389	3.413	3.432	3.446	3.465	3.478	3.478	3.478	3.478
16	2.998	3.144	3.235	3.298	3.343	3.376	3.402	3.422	3.437	3.458	3.473	3.473	3.473	3.473
17	2.984	3.130	3.221	3.285	3.331	3.365	3.392	3.412	3.429	3.451	3.469	3.469	3.469	3.469
18	2.971	3.117	3.210	3.274	3.320	3.356	3.383	3.404	3.421	3.445	3.465	3.465	3.465	3.465
19	2.960	3.106	3.199	3.264	3.311	3.347	3.375	3.397	3.415	3.440	3.462	3.462	3.462	3.462
20	2.950	3.097	3.190	3.255	3.303	3.339	3.368	3.390	3.409	3.435	3.459	3.461	3.461	3.461
21	2.941	3.088	3.181	3.247	3.295	3.332	3.361	3.385	3.403	3.431	3.456	3.465	3.465	3.465
22	2.933	3.080	3.173	3.239	3.288	3.326	3.355	3.379	3.398	3.427	3.453	3.468	3.468	3.468
23	2.926	3.072	3.166	3.233	3.282	3.320	3.350	3.374	3.394	3.423	3.451	3.471	3.471	3.471
24	2.919	3.066	3.160	3.227	3.276	3.315	3.345	3.370	3.390	3.420	3.449	3.473	3.473	3.473
25	2.913	3.060	3.154	3.221	3.271	3.310	3.341	3.366	3.386	3.417	3.447	3.476	3.476	3.476
26	2.907	3.054	3.149	3.216	3.266	3.305	3.336	3.362	3.382	3.414	3.445	3.478	3.478	3.478
28	2.897	3.044	3.139	3.207	3.257	3.297	3.329	3.355	3.376	3.409	3.442	3.483	3.483	3.483
30	2.888	3.035	3.131	3.199	3.250	3.290	3.322	3.349	3.371	3.405	3.439	3.486	3.486	3.486
40	2.858	3.005	3.102	3.171	3.224	3.266	3.300	3.328	3.352	3.389	3.429	3.500	3.500	3.500
60	2.829	2.976	3.073	3.143	3.198	3.241	3.277	3.307	3.333	3.374	3.419	3.515	3.573	3.537
120	2.800	2.947	3.045	3.116	3.172	3.217	3.254	3.287	3.314	3.359	3.409	3.532	3.596	3.601
∞	2.772	2.918	3.017	3.089	3.146	3.193	3.232	3.265	3.294	3.343	3.339	3.550	3.668	3.735

APPENDIX 4 (Continued)

$\alpha = .01$ Critical Values for Duncan's Test

df	Number of Means Spanned – k													
	2	3	4	5	6	7	8	9	10	12	15	30	60	100
1	90.03	90.03	90.03	90.03	90.03	90.03	90.03	90.03	90.03	90.03	90.03	90.03	90.03	90.03
2	14.04	14.04	14.04	14.04	14.04	14.04	14.04	14.04	14.04	14.04	14.04	14.04	14.04	14.04
3	8.260	8.321	8.321	8.321	8.321	8.321	8.321	8.321	8.321	8.321	8.321	8.321	8.321	8.321
4	6.511	6.677	6.740	6.756	6.756	6.756	6.756	6.756	6.756	6.756	6.756	6.756	6.756	6.756
5	5.702	5.893	5.989	6.040	6.065	6.074	6.074	6.074	6.074	6.074	6.074	6.074	6.074	6.074
6	5.243	5.439	5.549	5.614	5.655	5.680	5.694	5.701	5.703	5.703	5.703	5.703	5.703	5.703
7	4.949	5.145	5.260	5.333	5.383	5.416	5.439	5.454	5.464	5.472	5.472	5.472	5.472	5.472
8	4.745	4.939	5.056	5.134	5.189	5.227	5.256	5.276	5.291	5.309	5.317	5.317	5.317	5.317
9	4.596	4.787	4.906	4.986	5.043	5.086	5.117	5.142	5.160	5.185	5.202	5.202	5.202	5.202
10	4.482	4.671	4.789	4.871	4.931	4.975	5.010	5.036	5.058	5.087	5.112	5.112	5.112	5.112
11	4.392	4.579	4.697	4.780	4.841	4.887	4.923	4.952	4.975	5.009	5.039	5.049	5.049	5.049
12	4.320	4.504	4.622	4.705	4.767	4.815	4.852	4.882	4.907	4.944	4.978	5.007	5.007	5.007
13	4.260	4.442	4.560	4.643	4.706	4.755	4.793	4.824	4.850	4.889	4.927	4.971	4.971	4.971
14	4.210	4.391	4.508	4.591	4.654	4.703	4.743	4.775	4.802	4.843	4.884	4.940	4.940	4.940
15	4.167	4.346	4.463	4.547	4.610	4.660	4.700	4.733	4.760	4.803	4.846	4.913	4.913	4.913
16	4.131	4.308	4.425	4.508	4.572	4.622	4.662	4.696	4.724	4.768	4.813	4.890	4.890	4.890
17	4.099	4.275	4.391	4.474	4.538	4.589	4.630	4.664	4.692	4.737	4.785	4.869	4.869	4.869
18	4.071	4.246	4.361	4.445	4.509	4.559	4.601	4.635	4.664	4.710	4.759	4.850	4.850	4.850
19	4.046	4.220	4.335	4.419	4.482	4.533	4.575	4.610	4.639	4.686	4.736	4.833	4.833	4.833
20	4.024	4.197	4.312	4.395	4.459	4.510	4.552	4.587	4.617	4.664	4.715	4.818	4.822	4.822
21	4.004	4.177	4.291	4.374	4.438	4.489	4.531	4.567	4.597	4.645	4.697	4.805	4.816	4.816
22	3.986	4.158	4.272	4.355	4.419	4.470	4.513	4.548	4.578	4.627	4.680	4.792	4.810	4.810
23	3.970	4.141	4.254	4.337	4.401	4.453	4.496	4.531	4.562	4.611	4.665	4.780	4.805	4.805
24	3.955	4.126	4.239	4.322	4.386	4.437	4.480	4.516	4.546	4.596	4.651	4.770	4.800	4.800
25	3.942	4.111	4.224	4.307	4.371	4.423	4.466	4.502	4.532	4.582	4.638	4.760	4.795	4.795
26	3.930	4.099	4.211	4.294	4.358	4.410	4.452	4.488	4.520	4.570	4.626	4.751	4.791	4.791
28	3.908	4.076	4.188	4.270	4.334	4.386	4.429	4.465	4.497	4.548	4.604	4.735	4.784	4.784
30	3.889	4.056	4.168	4.250	4.314	4.366	4.409	4.445	4.477	4.528	4.586	4.721	4.777	4.777
40	3.825	3.988	4.098	4.180	4.243	4.295	4.339	4.376	4.408	4.461	4.521	4.671	4.754	4.761
60	3.762	3.922	4.030	4.111	4.174	4.226	4.270	4.307	4.340	4.394	4.456	4.620	4.729	4.765
120	3.702	3.858	3.964	4.044	4.107	4.158	4.202	4.239	4.272	4.327	4.392	4.568	4.703	4.770
∞	3.643	3.796	3.900	3.978	4.040	4.091	4.135	4.172	4.205	4.261	4.327	4.514	4.675	4.776

APPENDIX 5

Coefficients of Orthogonal Polynomials

n	Polynomial	z_1	z_2	z_3	z_4	z_5	z_6	z_7	z_8	z_9	z_{10}	$\sum z_i^2$	λ
2	Linear	−1	1									2	2
3	Linear	−1	0	1								2	1
	Quadratic	1	−2	1								6	3
4	Linear	−3	−1	1	3							20	2
	Quadratic	1	−1	−1	1							4	1
	Cubic	−1	3	−3	1							20	10/3
5	Linear	−2	−1	0	1	2						10	1
	Quadratic	2	−1	−2	−1	2						14	1
	Cubic	−1	2	0	−2	1						10	5/6
	Quartic	1	−4	6	−4	1						70	35/12
6	Linear	−5	−3	−1	1	3	5					70	2
	Quadratic	5	−1	−4	−4	−1	5					84	3/2
	Cubic	−5	7	4	−4	−7	5					180	5/3
	Quartic	1	−3	2	2	−3	1					28	7/12
7	Linear	−3	−2	−1	0	1	2	3				28	1
	Quadratic	5	0	−3	−4	−3	0	5				84	1
	Cubic	−1	1	1	0	−1	−1	1				6	1/6
	Quartic	3	−7	1	6	1	−7	3				154	7/12
8	Linear	−7	−5	−3	−1	1	3	5	7			168	2
	Quadratic	7	1	−3	−5	−5	−3	1	7			168	1
	Cubic	−7	5	7	3	−3	−7	−5	7			264	2/3
	Quartic	7	−13	−3	9	9	−3	−13	7			616	7/12
	Quintic	−7	23	−17	−15	15	17	−23	7			2184	7/10
9	Linear	−4	−3	−2	−1	0	1	2	3	4		60	1
	Quadratic	28	7	−8	−17	−20	−17	−8	7	28		2772	3
	Cubic	−14	7	13	9	0	−9	−13	−7	14		990	5/6
	Quartic	14	−21	−11	9	18	9	−11	−21	14		2002	7/12
	Quintic	−4	11	−4	−9	0	9	4	−11	4		468	3/20
10	Linear	−9	−7	−5	−3	−1	1	3	5	7	9	330	2
	Quadratic	6	2	−1	−3	−4	−4	−3	−1	2	6	132	1/2
	Cubic	−42	14	35	31	12	−12	−31	−35	−14	42	8580	5/3
	Quartic	18	−22	−17	3	18	18	3	−17	−22	18	2860	5/12
	Quintic	−6	14	−1	−11	−6	6	11	1	−14	6	780	1/10

APPENDIX 6

Critical Values for the χ^2 Distribution — $\chi^2_{df}(\alpha)$

df	$\chi^2(.25)$	$\chi^2(.10)$	$\chi^2(.05)$	$\chi^2(.025)$	$\chi^2(.01)$	$\chi^2(.005)$	$\chi^2(.001)$
1	1.32	2.71	3.84	5.02	6.64	7.88	10.83
2	2.77	4.61	5.99	7.38	9.21	10.60	13.82
3	4.11	6.25	7.81	9.35	11.35	12.84	16.27
4	5.39	7.78	9.49	11.14	13.28	14.86	18.47
5	6.63	9.24	11.07	12.83	15.09	16.75	20.52
6	7.84	10.64	12.59	14.45	16.81	18.55	22.46
7	9.04	12.02	14.07	16.01	18.48	20.28	24.32
8	10.22	13.36	15.51	17.53	20.09	21.96	26.12
9	11.39	14.68	16.92	19.02	21.67	23.59	27.88
10	12.55	15.99	18.31	20.48	23.21	25.19	29.59
11	13.70	17.28	19.68	21.92	24.73	26.76	31.26
12	14.85	18.55	21.03	23.34	26.22	28.30	32.91
13	15.98	19.81	22.36	24.74	27.69	29.82	34.53
14	17.12	21.06	23.68	26.12	29.14	31.32	36.12
15	18.25	22.31	25.00	27.49	30.58	32.80	37.70
16	19.37	23.54	26.30	28.85	32.00	34.27	39.25
17	20.49	24.77	27.59	30.19	33.41	35.72	40.79
18	21.60	25.99	28.87	31.53	34.81	37.16	42.31
19	22.72	27.20	30.14	32.85	36.19	38.58	43.82
20	23.83	28.41	31.41	34.17	37.57	40.00	45.31
21	24.93	29.62	32.67	35.48	38.93	41.40	46.80
22	26.04	30.81	33.92	36.78	40.29	42.80	48.27
23	27.14	32.01	35.17	38.08	41.64	44.18	49.73
24	28.24	33.20	36.42	39.36	42.98	45.56	51.18
25	29.34	34.38	37.65	40.65	44.31	46.93	52.62
30	34.80	40.26	43.77	46.99	50.89	53.67	59.70
40	45.61	51.81	55.76	59.34	63.69	66.77	73.40
50	56.33	63.17	67.50	71.42	76.15	79.49	86.66
60	66.98	74.40	79.08	83.30	88.38	91.95	99.61
70	77.58	85.53	90.53	95.02	100.43	104.25	112.32
80	88.13	96.58	101.88	106.63	112.33	116.32	124.84
90	98.65	107.57	113.15	118.14	124.12	128.30	137.21
100	109.14	118.50	124.34	129.56	135.81	140.17	149.45

APPENDIX 7

Coefficients a_{n-i+1} for the W Test of Normality

i	n =	2	3	4	5	6	7	8	9	10
1		0.7071	0.7071	0.6872	0.6646	0.6431	0.6233	0.6052	0.5888	0.5739
2			0.0000	0.1668	0.2413	0.2806	0.3031	0.3164	0.3244	0.3290
3					0.0000	0.0875	0.1401	0.1743	0.1976	0.2141
4							0.0000	0.0561	0.0947	0.1224
5									0.0000	0.0399

i	11	12	13	14	15	16	17	18	19	20
1	0.5601	0.5475	0.5359	0.5251	0.5150	0.5056	0.4968	0.4886	0.4808	0.4734
2	0.3315	0.3325	0.3325	0.3318	0.3306	0.3290	0.3273	0.3253	0.3232	0.3211
3	0.2260	0.2347	0.2412	0.2460	0.2495	0.2521	0.2540	0.2552	0.2561	0.2565
4	0.1429	0.1586	0.1707	0.1803	0.1878	0.1939	0.1988	0.2027	0.2059	0.2085
5	0.0695	0.0922	0.1100	0.1240	0.1354	0.1447	0.1523	0.1587	0.1641	0.1686
6	0.0000	0.0303	0.0538	0.0727	0.0880	0.1005	0.1109	0.1197	0.1271	0.1334
7			0.0000	0.0240	0.0434	0.0593	0.0725	0.0837	0.0932	0.1013
8					0.0000	0.0196	0.0359	0.0496	0.0612	0.0712
9							0.0000	0.0164	0.0303	0.0422
10									0.0000	0.0140

i	21	22	23	24	25	26	27	28	29	30
1	0.4664	0.4598	0.4536	0.4476	0.4419	0.4364	0.4312	0.4262	0.4214	0.4168
2	0.3189	0.3167	0.3144	0.3122	0.3100	0.3078	0.3056	0.3035	0.3014	0.2993
3	0.2567	0.2566	0.2564	0.2560	0.2554	0.2548	0.2541	0.2533	0.2525	0.2516
4	0.2106	0.2122	0.2136	0.2146	0.2154	0.2160	0.2164	0.2167	0.2168	0.2169
5	0.1724	0.1756	0.1783	0.1806	0.1826	0.1842	0.1856	0.1868	0.1878	0.1886
6	0.1388	0.1435	0.1475	0.1510	0.1540	0.1567	0.1590	0.1610	0.1628	0.1643
7	0.1083	0.1144	0.1197	0.1243	0.1284	0.1320	0.1351	0.1380	0.1404	0.1427
8	0.0798	0.0873	0.0938	0.0997	0.1047	0.1092	0.1132	0.1167	0.1200	0.1228
9	0.0525	0.0615	0.0693	0.0759	0.0823	0.0878	0.0926	0.0969	0.1008	0.1044
10	0.0261	0.0366	0.0457	0.0540	0.0610	0.0673	0.0730	0.0781	0.0827	0.0869
11	0.0000	0.0121	0.0227	0.0321	0.0403	0.0476	0.0542	0.0601	0.0654	0.0702
12			0.0000	0.0107	0.0201	0.0284	0.0359	0.0426	0.0486	0.0541
13					0.0000	0.0094	0.0179	0.0254	0.0322	0.0383
14							0.0000	0.0084	0.0160	0.0229
15									0.0000	0.0076

Shapiro and Wilk (1965) used approximations for n>20. This table is exact.

APPENDIX 8

Critical Values for the W Test — W_α

n	$W_{.25}$	$W_{.10}$	$W_{.05}$	$W_{.025}$	$W_{.01}$	$W_{.001}$
2	1.000	1.000	1.000	1.000	1.000	1.000
3	0.853	0.793	0.772	0.761	0.754	0.750
4	0.860	0.799	0.761	0.726	0.692	0.650
5	0.867	0.813	0.776	0.742	0.700	0.624
6	0.876	0.828	0.793	0.760	0.718	0.625
7	0.885	0.841	0.809	0.778	0.739	0.643
8	0.892	0.852	0.822	0.793	0.755	0.667
9	0.899	0.862	0.835	0.808	0.773	0.680
10	0.906	0.871	0.845	0.820	0.784	0.698
11	0.911	0.878	0.854	0.830	0.797	0.718
12	0.916	0.885	0.862	0.839	0.808	0.730
13	0.920	0.891	0.869	0.847	0.819	0.746
14	0.924	0.896	0.876	0.855	0.827	0.756
15	0.927	0.901	0.882	0.862	0.836	0.768
16	0.931	0.906	0.887	0.869	0.843	0.777
17	0.933	0.910	0.892	0.874	0.850	0.784
18	0.936	0.914	0.897	0.880	0.857	0.793
19	0.939	0.917	0.901	0.885	0.863	0.806
20	0.941	0.920	0.904	0.889	0.867	0.810
21	0.943	0.923	0.908	0.892	0.871	0.815
22	0.945	0.925	0.911	0.896	0.875	0.822
23	0.946	0.928	0.914	0.899	0.880	0.830
24	0.948	0.930	0.917	0.903	0.885	0.836
25	0.950	0.932	0.919	0.906	0.888	0.839
26	0.951	0.934	0.922	0.910	0.891	0.843
27	0.952	0.937	0.925	0.912	0.895	0.848
28	0.954	0.938	0.926	0.914	0.898	0.853
29	0.955	0.940	0.928	0.917	0.900	0.857
30	0.956	0.941	0.930	0.919	0.903	0.864

Note: These values are based on the exact coefficients in Appendix 7.

APPENDIX 9

Critical Values for the Anderson-Darling Test — B^2_α

$B^2_{.25}$	$B^2_{.20}$	$B^2_{.15}$	$B^2_{.10}$	$B^2_{.05}$	$B^2_{.025}$	$B^2_{.01}$	$B^2_{.005}$
0.472	0.509	0.561	0.631	0.752	0.873	1.035	1.159

APPENDIX 10

Table of Minimal Detectable Differences $\Delta = \delta/\sigma$ for a One Fixed Factor Design, $\alpha = .05$, $\beta = .10$.

Number Repeats	Number of Fixed Treatment Combinations 2	3	4	5	6
2	6.796	6.548	6.395	6.333	6.317
3	3.589	3.838	3.967	4.065	4.149
4	2.767	3.010	3.148	3.251	3.337
5	2.348	2.568	2.698	2.795	2.876
6	2.081	2.280	2.401	2.492	2.567
7	1.890	2.073	2.186	2.271	2.341
8	1.745	1.915	2.020	2.100	2.166
10	1.534	1.684	1.778	1.850	1.910
12	1.385	1.521	1.607	1.673	1.727
14	1.273	1.398	1.478	1.539	1.589
16	1.185	1.301	1.375	1.432	1.479
18	1.112	1.222	1.292	1.345	1.390
20	1.052	1.155	1.222	1.273	1.315
22	1.000	1.099	1.162	1.210	1.251
24	0.956	1.050	1.110	1.157	1.195
26	0.917	1.007	1.065	1.109	1.146
28	0.882	0.969	1.025	1.068	1.103
30	0.851	0.935	0.989	1.030	1.065
40	0.734	0.806	0.852	0.888	0.918
60	0.597	0.655	0.693	0.722	0.747
80	0.516	0.566	0.599	0.624	0.645
100	0.461	0.506	0.535	0.558	0.577
200	0.325	0.357	0.377	0.393	0.407
500	0.205	0.225	0.238	0.248	0.257
1000	0.145	0.159	0.168	0.176	0.181

APPENDIX 11

Table of Minimal Detectable Differences $\Delta = \delta/\sigma$
for a One Random Factor Design, $\alpha = .05$, $\beta = .10$.

Number	Number of Random Treatment Combinations				
Repeats	2	3	4	5	6
2	21.394	6.577	4.136	3.165	2.640
3	11.964	3.957	2.605	2.052	1.746
4	9.321	3.120	2.074	1.645	1.407
5	7.938	2.667	1.779	1.415	1.213
6	7.046	2.370	1.584	1.262	1.083
7	6.405	2.156	1.443	1.151	0.988
8	5.915	1.992	1.334	1.064	0.915
10	5.203	1.752	1.175	0.938	0.806
12	4.701	1.583	1.062	0.848	0.729
14	4.321	1.455	0.976	0.780	0.671
16	4.021	1.354	0.909	0.726	0.625
18	3.776	1.272	0.853	0.682	0.587
20	3.571	1.203	0.807	0.645	0.555
22	3.397	1.144	0.768	0.614	0.528
24	3.245	1.093	0.733	0.586	0.505
26	3.113	1.048	0.703	0.562	0.484
28	2.995	1.008	0.677	0.541	0.466
30	2.890	0.973	0.653	0.522	0.450
40	2.492	0.839	0.563	0.450	0.388
60	2.026	0.682	0.458	0.366	0.315
80	1.751	0.589	0.396	0.317	0.273
100	1.564	0.526	0.353	0.283	0.244
200	1.103	0.371	0.249	0.199	0.172
500	0.697	0.234	0.157	0.126	0.108
1000	0.492	0.166	0.111	0.089	0.077

APPENDIX 12

Table of Minimal Detectable Differences $\Delta = \sqrt{\Phi(Y)}/\sigma$ for a Fixed Factor or Interaction Y in a General Design, $\alpha = .05$, $\beta = .10$. Divide Table Entry by $\sqrt{C}$, where C = Coefficient in EMS Preceding $\Phi(Y)$.

Den.	Numerator df								
df	1	2	3	4	5	6	10	20	50
1	10.368	7.331	5.986	5.184	4.637	4.233	3.279	2.318	1.466
2	6.795	6.711	5.986	5.184	4.637	4.233	3.279	2.318	1.466
3	5.014	4.631	4.475	4.389	4.336	4.233	3.279	2.318	1.466
4	4.395	3.901	3.692	3.577	3.501	3.450	3.279	2.318	1.466
5	4.091	3.538	3.301	3.167	3.079	3.018	2.886	2.318	1.466
6	3.913	3.323	3.068	2.921	2.825	2.758	2.609	2.318	1.466
7	3.795	3.183	2.914	2.758	2.656	2.583	2.423	2.287	1.466
8	3.712	3.084	2.805	2.642	2.535	2.458	2.288	2.142	1.466
10	3.604	2.953	2.661	2.489	2.375	2.292	2.107	1.944	1.466
12	3.536	2.871	2.571	2.392	2.272	2.186	1.989	1.814	1.466
14	3.490	2.814	2.508	2.325	2.202	2.112	1.907	1.721	1.466
16	3.455	2.774	2.463	2.276	2.150	2.058	1.847	1.652	1.466
18	3.429	2.742	2.428	2.239	2.111	2.016	1.800	1.598	1.449
20	3.409	2.718	2.401	2.210	2.079	1.984	1.762	1.554	1.399
22	3.393	2.698	2.379	2.186	2.054	1.957	1.732	1.519	1.357
24	3.379	2.682	2.361	2.166	2.033	1.936	1.707	1.489	1.322
26	3.369	2.669	2.345	2.150	2.016	1.917	1.686	1.464	1.292
28	3.358	2.657	2.333	2.136	2.001	1.901	1.667	1.442	1.266
30	3.350	2.647	2.322	2.124	1.988	1.888	1.652	1.423	1.242
40	3.322	2.613	2.284	2.082	1.944	1.841	1.597	1.355	1.159
60	3.295	2.580	2.246	2.041	1.900	1.794	1.542	1.287	1.070
80	3.281	2.564	2.227	2.022	1.879	1.771	1.515	1.252	1.022
100	3.273	2.554	2.216	2.009	1.866	1.758	1.498	1.231	0.993
200	3.257	2.534	2.195	1.986	1.840	1.731	1.466	1.187	0.930
500	3.248	2.523	2.182	1.972	1.825	1.714	1.446	1.161	0.889
1000	3.245	2.519	2.178	1.967	1.820	1.709	1.440	1.152	0.875

APPENDIX 13

Table of Minimal Detectable Differences $\Delta = \sigma_Y / \sigma$ for a Random Factor or Interaction Y in a General Design, $\alpha = .05$, $\beta = .10$. Divide Table Entry by $\sqrt{C}$, where C = Coefficient in EMS Preceding σ_Y^2.

Den.	Numerator df								
df	1	2	3	4	5	6	10	20	50
1	80.218	41.231	34.549	31.932	30.554	29.707	28.171	27.143	26.573
2	30.255	13.038	10.182	9.068	8.481	8.121	7.465	7.025	6.780
3	23.276	9.301	7.001	6.100	5.624	5.330	4.792	4.429	4.225
4	20.722	7.949	5.849	5.024	4.585	4.313	3.813	3.471	3.278
5	19.423	7.264	5.265	4.476	4.054	3.793	3.308	2.973	2.782
6	18.641	6.853	4.913	4.145	3.733	3.476	2.998	2.665	2.473
7	18.121	6.580	4.679	3.924	3.518	3.264	2.789	2.455	2.260
8	17.750	6.385	4.512	3.765	3.364	3.111	2.638	2.301	2.104
10	17.258	6.126	4.289	3.554	3.157	2.907	2.433	2.091	1.887
12	16.946	5.963	4.148	3.420	3.025	2.775	2.300	1.953	1.742
14	16.731	5.850	4.050	3.327	2.933	2.684	2.207	1.854	1.637
16	16.574	5.767	3.979	3.259	2.866	2.617	2.137	1.780	1.557
18	16.454	5.704	3.924	3.206	2.814	2.565	2.084	1.723	1.494
20	16.360	5.655	3.881	3.165	2.774	2.524	2.041	1.676	1.443
22	16.284	5.615	3.846	3.132	2.741	2.491	2.007	1.638	1.400
24	16.221	5.582	3.818	3.104	2.713	2.464	1.978	1.606	1.364
26	16.168	5.554	3.794	3.081	2.690	2.440	1.953	1.579	1.333
28	16.123	5.530	3.773	3.061	2.671	2.421	1.933	1.556	1.306
30	16.084	5.510	3.755	3.044	2.654	2.404	1.915	1.535	1.282
40	15.950	5.440	3.694	2.985	2.595	2.344	1.851	1.463	1.196
60	15.819	5.371	3.634	2.927	2.537	2.286	1.788	1.390	1.105
80	15.755	5.337	3.605	2.899	2.508	2.257	1.757	1.352	1.056
100	15.716	5.317	3.587	2.882	2.491	2.240	1.738	1.329	1.025
200	15.640	5.277	3.552	2.848	2.457	2.205	1.700	1.282	0.960
500	15.595	5.253	3.531	2.828	2.437	2.185	1.677	1.254	0.919
1000	15.580	5.246	3.524	2.821	2.431	2.178	1.670	1.244	0.904

APPENDIX 14

Resolution V Two Level Fractional Factorial Designs

# factors	factors	# runs	resolution	fraction	generators
5	$A-E$	16	V	1/2	$I = ABCDE$
6	$A-F$	32	VI	1/2	$I = ABCDEF$
7	$A-G$	64	VII	1/2	$I = ABCDEFG$
8	$A-H$	64	V	1/4	$I = BCDEFG$
					$= ADEFH$
9	$A-I$	128	VI	1/4	$I = CDEFGH$
					$= ABEFGI$
10	$A-J$	128	V	1/8	$I = ABCDEFGH$
					$= DEFGI$
					$= BCFGJ$
11	$A-K$	128	V	1/16	$I = ABCDEFGH$
					$= DEFGI$
					$= BCFGJ$
					$= ACEGK$
12	$A-L$	256	VI	1/16	$I = BCDEFGHI$
					$= AEFGHJ$
					$= ACDGHK$
					$= ABDFHL$
13	$A-M$	256	V	1/32	$I = ABCDEFGHI$
					$= DEFGHJ$
					$= BCFGHK$
					$= ACEGHL$
					$= BDGHM$
14	$A-N$	256	V	1/64	$I = ABCDEFGHI$
					$= DEFGHJ$
					$= BCFGHK$
					$= ACEGHL$
					$= BDGHM$
					$= CEFHN$
15	$A-O$	256	V	1/128	$I = ABCDEFGHI$
					$= DEFGHJ$
					$= BCFGHK$
					$= ACEGHL$
					$= BDGHM$
					$= CEFHN$
					$= ADFHO$

APPENDIX 15

Resolution IV Two Level Fractional Factorial Designs

# factors	factors	# runs	resolution	fraction	generators
5	$A-E$	16	V	1/2	$I = ABCDE$
6	$A-F$	16	IV	1/4	$I = BCDE = ACDF$
7	$A-G$	16	IV	1/8	$I = BCDE = ACDF$ $= ABDG$
8	$A-H$	16	IV	1/16	$I = BCDE = ACDF$ $= ABDG = ABCH$
9	$A-I$	32	IV	1/16	$I = ABCDEF = CDEG$ $= BDEH = ADEI$
10	$A-J$	32	IV	1/32	$I = ABCDEF = CDEG$ $= BDEH = ADEI$ $= BCEJ$
11	$A-K$	32	IV	1/64	$I = ABCDEF = CDEG$ $= BDEH = ADEI$ $= BCEJ = ACEK$
12	$A-L$	32	IV	1/128	$I = ABCDEF = CDEG$ $= BDEH = ADEI$ $= BCEJ = ACEK$ $= ABEL$
13	$A-M$	32	IV	1/256	$I = ABCDEF = CDEG$ $= BDEH = ADEI$ $= BCEJ = ACEK$ $= ABEL = BCDM$
14	$A-N$	32	IV	1/512	$I = ABCDEF = CDEG$ $= BDEH = ADEI$ $= BCEJ = ACEK$ $= ABEL = BCDM$ $= ACDN$
15	$A-O$	32	IV	1/1024	$I = ABCDEF = CDEG$ $= BDEH = ADEI$ $= BCEJ = ACEK$ $= ABEL = BCDM$ $= ACDN = ABDO$

APPENDIX 16

Resolution III Two Level Fractional Factorial Designs

# factors	factors	# runs	resolution	fraction	generators
5	$A-E$	8	III	1/4	$I = ABCD = BCE$
6	$A-F$	8	III	1/8	$I = ABCD = BCE = ACF$
7	$A-G$	8	III	1/16	$I = ABCD = BCE = ACF = ABG$
8	$A-H$	16	IV	1/16	$I = BCDE = ACDF = ABDG = ABCH$
9	$A-I$	16	III	1/32	$I = ABCDE = BCDF = ACDG = CDH = ABDI$
10	$A-J$	16	III	1/64	$I = ABCDE = BCDF = ACDG = CDH = ABDI = BDJ$
11	$A-K$	16	III	1/128	$I = ABCDE = BCDF = ACDG = CDH = ABDI = BDJ = ADK$
12	$A-L$	16	III	1/256	$I = ABCDE = BCDF = ACDG = CDH = ABDI = BDJ = ADK = ABCL$
13	$A-M$	16	III	1/512	$I = ABCDE = BCDF = ACDG = CDH = ABDI = BDJ = ADK = ABCL = BCM$
14	$A-N$	16	III	1/1024	$I = ABCDE = BCDF = ACDG = CDH = ABDI = BDJ = ADK = ABCL = BCM = ACN$
15	$A-O$	16	III	1/2048	$I = ABCDE = BCDF = ACDG = CDH = ABDI = BDJ = ADK = ABCL = BCM = ACN = ABO$

APPENDIX 17

Orthogonal Main Effect Designs

		# Factors ≥ Level							
Name of Design	Factors	6	5	4	3	2	Size of Design	df Left	Page
2^3 OMED	2^3					3	4	0	382
2^7 OMED	2^7					7	8	0	382
2^7 OMED	4×2^4			1	1	5	8	0	382
3^4 OMED	3^4				4	4	9	0	382
2^{11} OMED	2^{11}					11	12	0	382
3×2^4 OMED	3×2^4				1	4	12	5	383
6×2^2 OMED	6×2^2	1	1	1	1	3	12	5	383
2^{15} OMED	2^{15}					15	16	0	383
2^{15} OMED	4×2^{12}			1	1	13	16	0	383
2^{15} OMED	$4^2 \times 2^9$			2	2	11	16	0	383
2^{15} OMED	$4^3 \times 2^6$			3	3	9	16	0	383
8×2^8 OMED	8×2^8	1	1	1	1	9	16	0	383
2^{15} OMED	$4^4 \times 2^3$			4	4	7	16	0	383
3^7 OMED	3^7				7	7	16	1	384
2^{15} OMED	4^5			5	5	5	16	0	383
$3^7 \times 2$ OMED	$3^7 \times 2$				7	8	18	2	384
$3^7 \times 2$ OMED	6×3^6	1	1	1	7	7	18	0	384
2^{19} OMED	2^{19}					19	20	0	385
2^{23} OMED	2^{23}					23	24	0	385
2^{23} OMED	4×2^{20}			1	1	21	24	0	385
6×2^{14} OMED	6×2^{14}	1	1	1	1	15	24	4	386
$4 \times 3 \times 2^{13}$ OMED	$4 \times 3 \times 2^{13}$			1	2	15	24	5	386
6×2^{14} OMED	$6 \times 4 \times 2^{11}$	1	1	2	2	13	24	4	386
5^6 OMED	5^6		6	6	6	6	25	0	387
3^{13} OMED	3^{13}				13	13	27	0	388
3^{13} OMED	9×3^9	1	1	1	10	10	27	0	388
2^{27} OMED	2^{27}					27	28	0	389
2^{31} OMED	2^{31}					31	32	0	390
2^{31} OMED	4×2^{28}			1	1	29	32	0	390
2^{31} OMED	$4^2 \times 2^{25}$			2	2	27	32	0	390
2^{31} OMED	$4^3 \times 2^{22}$			3	3	25	32	0	390
8×2^{24} OMED	8×2^{24}	1	1	1	1	25	32	0	391
2^{31} OMED	$4^4 \times 2^{19}$			4	4	23	32	0	390
8×2^{24} OMED	$8 \times 4 \times 2^{21}$	1	1	2	2	23	32	0	391
$3^7 \times 2^{16}$ OMED	$3^7 \times 2^{16}$				7	23	32	1	392
2^{31} OMED	$4^5 \times 2^{16}$			5	5	21	32	0	390
8×2^{24} OMED	$8 \times 4^2 \times 2^{18}$	1	1	3	3	21	32	0	391
2^{31} OMED	$4^6 \times 2^{13}$			6	6	19	32	0	390

Orthogonal Main Effect Designs—Continued

Name of Design	Factors	# Factors ≥ Level 6	5	4	3	2	Size of Design	df Left	Page
8×2^{24} OMED	$8 \times 4^3 \times 2^{15}$	1	1	4	4	19	32	0	391
2^{31} OMED	$4^7 \times 2^{10}$			7	7	17	32	0	390
8×2^{24} OMED	$8 \times 4^4 \times 2^{12}$	1	1	5	5	17	32	0	391
2^{31} OMED	$4^8 \times 2^7$			8	8	15	32	0	390
8×2^{24} OMED	$8 \times 4^5 \times 2^9$	1	1	6	6	15	32	0	391
2^{31} OMED	$4^9 \times 2^4$			9	9	13	32	0	390
8×2^{24} OMED	$8 \times 4^6 \times 2^6$	1	1	7	7	13	32	0	391
8×2^{24} OMED	$8 \times 4^7 \times 2^3$	1	1	8	8	11	32	0	391
8×2^{24} OMED	8×4^8	1	1	9	9	9	32	0	391
2^{35} OMED	2^{35}					35	36	0	393
$3^{12} \times 2^{11}$ OMED	$3^{12} \times 2^{11}$				12	23	36	0	394
$3^{13} \times 2^4$ OMED	$3^{13} \times 2^4$				13	17	36	5	395
$6 \times 3^{12} \times 2^2$ OMED	$6 \times 3^{12} \times 2^2$	1	1	1	13	15	36	4	395
4×3^{13} OMED	4×3^{13}			1	14	14	36	6	395
6^3 OMED	6^3	3	3	3	3	3	36	20	396
2^{39} OMED	2^{39}					39	40	0	397
2^{39} OMED	4×2^{36}			1	1	37	40	0	397
5×2^{28} OMED	5×2^{28}		1	1	1	33	40	7	398
5×2^{28} OMED	$5 \times 4 \times 2^{25}$		1	2	2	27	40	7	398
2^{43} OMED	2^{43}					43	44	0	399
2^{47} OMED	2^{47}					47	48	0	400
2^{47} OMED	4×2^{44}			1	1	45	48	0	400
2^{47} OMED	$4^2 \times 2^{41}$			2	2	43	48	0	400
2^{47} OMED	$4^3 \times 2^{38}$			3	3	41	48	0	400
8×2^{40} OMED	8×2^{40}	1	1	1	1	41	48	0	401
2^{47} OMED	$4^4 \times 2^{35}$			4	4	39	48	0	400
2^{47} OMED	$4^5 \times 2^{32}$			5	5	37	48	0	400
$6 \times 4 \times 2^{35}$ OMED	$6 \times 4 \times 2^{35}$	1	1	2	2	37	48	4	402
2^{47} OMED	$4^6 \times 2^{29}$			6	6	35	48	0	400
$6 \times 4 \times 2^{35}$ OMED	$6 \times 4^2 \times 2^{32}$	1	1	3	3	35	48	4	402
2^{47} OMED	$4^7 \times 2^{26}$			7	7	33	48	0	400
$6 \times 4 \times 2^{35}$ OMED	$6 \times 4^3 \times 2^{29}$	1	1	4	4	33	48	4	402
$8 \times 6 \times 2^{31}$ OMED	$8 \times 6 \times 2^{31}$	2	2	2	2	33	48	4	403
2^{47} OMED	$4^8 \times 2^{23}$			8	8	31	48	0	400
$6 \times 4 \times 2^{35}$ OMED	$6 \times 4^4 \times 2^{26}$	1	1	5	5	31	48	4	402
2^{47} OMED	$4^9 \times 2^{20}$			9	9	29	48	0	400
$6 \times 4 \times 2^{35}$ OMED	$6 \times 4^5 \times 2^{23}$	1	1	6	6	29	48	4	402
2^{47} OMED	$4^{10} \times 2^{17}$			10	10	27	48	0	400
$6 \times 4 \times 2^{35}$ OMED	$6 \times 4^6 \times 2^{20}$	1	1	7	7	27	48	4	402
2^{47} OMED	$4^{11} \times 2^{14}$			11	11	25	48	0	400

Orthogonal Main Effect Designs—Continued

Name of Design	Factors	# Factors ≥ Level 6	5	4	3	2	Size of Design	df Left	Page
$6 \times 4 \times 2^{35}$ OMED	$6 \times 4^{7} \times 2^{17}$	1	1	8	8	25	48	4	402
2^{47} OMED	$4^{12} \times 2^{11}$			12	12	23	48	0	400
$6 \times 4 \times 2^{35}$ OMED	$6 \times 4^{8} \times 2^{14}$	1	1	9	9	23	48	4	402
$6 \times 4 \times 2^{35}$ OMED	$6 \times 4^{9} \times 2^{11}$	1	1	10	10	21	48	4	402
$6 \times 4 \times 2^{35}$ OMED	$6 \times 4^{10} \times 2^{8}$	1	1	11	11	19	48	4	402
$6 \times 4 \times 2^{35}$ OMED	$6 \times 4^{11} \times 2^{5}$	1	1	12	12	17	48	4	402
$4^{12} \times 3 \times 2^{4}$ OMED	$4^{12} \times 3 \times 2^{4}$			12	13	17	48	5	404
$6 \times 4 \times 2^{35}$ OMED	$6 \times 4^{12} \times 2^{2}$	1	1	13	13	15	48	4	402
$4^{13} \times 3$ OMED	$4^{13} \times 3$			13	14	14	48	6	404
7^{8} OMED	7^{8}	8	8	8	8	8	49	0	405
$5^{11} \times 2$ OMED	$5^{11} \times 2$		11	11	11	12	50	4	406
$5^{11} \times 2$ OMED	10×5^{10}	1	11	11	11	11	50	0	406

2^3 Orthogonal Main Effect Design

2	2	2
0	0	1
0	1	0
1	0	0
1	1	1

2^7 Orthogonal Main Effect Design

4						
2	2	2	2	2	2	2
0	0	0	0	0	0	0
0	0	0	1	1	1	1
0	1	1	0	0	1	1
0	1	1	1	1	0	0
1	0	1	0	1	0	1
1	0	1	1	0	1	0
1	1	0	0	1	1	0
1	1	0	1	0	0	1

Association

2	2	2	4
0	0	0	0
0	1	1	1
1	0	1	2
1	1	0	3

3^4 Orthogonal Main Effect Design

3	3	3	3
0	0	0	0
0	1	2	2
0	2	1	1
1	0	2	1
1	1	1	0
1	2	0	2
2	0	1	2
2	1	0	1
2	2	2	0

2^{11} Orthogonal Main Effect Design

2	2	2	2	2	2	2	2	2	2	2
1	0	1	0	0	0	1	1	1	0	1
1	1	0	1	0	0	0	1	1	1	0
0	1	1	0	1	0	0	0	1	1	1
1	0	1	1	0	1	0	0	0	1	1
1	1	0	1	1	0	1	0	0	0	1
1	1	1	0	1	1	0	1	0	0	0
0	1	1	1	0	1	1	0	1	0	0
0	0	1	1	1	0	1	1	0	1	0
0	0	0	1	1	1	0	1	1	0	1
1	0	0	0	1	1	1	0	1	1	0
0	1	0	0	0	1	1	1	0	1	1
0	0	0	0	0	0	0	0	0	0	0

3×2^4 and 6×2^2 Orthogonal Main Effect Designs

3×2^4

3	2	2	2	2
0	0	0	0	0
0	0	1	0	1
0	1	0	1	1
0	1	1	1	0
1	0	0	1	1
1	0	1	1	0
1	1	0	0	1
1	1	1	0	0
2	0	0	1	0
2	0	1	0	1
2	1	0	0	0
2	1	1	1	1

6×2^2

6	2	2
0	0	0
1	0	0
2	0	0
3	0	1
4	0	1
5	0	1
0	1	0
1	1	0
2	1	0
3	1	1
4	1	1
5	1	1

2^{15} Orthogonal Main Effect Design

4			4			4			4			4		
2	2	2	2	2	2	2	2	2	2	2	2	2	2	2
0	0	0	0	0	0	1	1	0	1	0	1	0	0	0
0	0	0	0	1	1	1	0	1	1	1	0	0	1	1
0	0	0	1	0	1	0	0	0	0	1	1	1	1	0
0	0	0	1	1	0	0	1	1	0	0	0	1	0	1
0	1	1	0	0	0	1	0	1	0	0	0	1	1	0
0	1	1	0	1	1	1	1	0	0	1	1	1	0	1
0	1	1	1	0	1	0	1	1	1	1	0	0	0	0
0	1	1	1	1	0	0	0	0	1	0	1	0	1	1
1	0	1	0	0	0	0	0	0	1	1	0	1	0	1
1	0	1	0	1	1	0	1	1	1	0	1	1	1	0
1	0	1	1	0	1	1	1	0	0	0	0	0	1	1
1	0	1	1	1	0	1	0	1	0	1	1	0	0	0
1	1	0	0	0	0	0	1	1	0	1	1	0	1	1
1	1	0	0	1	1	0	0	0	0	0	0	0	0	0
1	1	0	1	0	1	1	0	1	1	0	1	1	0	1
1	1	0	1	1	0	1	1	0	1	1	0	1	1	0

Association

2	2	2	4
0	0	0	0
0	1	1	1
1	0	1	2
1	1	0	3

8×2^8 Orthogonal Main Effect Design

8	2	2	2	2	2	2	2	2
0	0	0	0	0	0	1	1	1
1	0	0	0	1	1	0	0	1
2	0	0	1	0	1	0	1	0
3	0	0	1	1	0	1	0	0
4	0	1	0	0	1	1	0	0
5	0	1	0	1	0	0	1	0
6	0	1	1	0	0	0	0	1
7	0	1	1	1	1	1	1	1
0	1	1	1	1	1	0	0	0
1	1	1	1	0	0	1	1	0
2	1	1	0	1	0	1	0	1
3	1	1	0	0	1	0	1	1
4	1	0	1	1	0	0	1	1
5	1	0	1	0	1	1	0	1
6	1	0	0	1	1	1	1	0
7	1	0	0	0	0	0	0	0

3^7 Orthogonal Main Effect Design

3	3	3	3	3	3	3
2	1	1	0	1	0	0
0	2	1	1	0	1	0
0	0	2	1	1	0	1
1	0	0	2	1	1	0
0	1	0	0	2	1	1
1	0	1	0	0	2	1
1	1	0	1	0	0	2
1	1	1	1	1	1	1
0	1	1	2	1	2	2
2	0	1	1	2	1	2
2	2	0	1	1	2	1
1	2	2	0	1	1	2
2	1	2	2	0	1	1
1	2	1	2	2	0	1
1	1	2	1	2	2	0
1	1	1	1	1	1	1

$3^7 \times 2$ Orthogonal Main Effect Design

						┌ 6	┐
3	3	3	3	3	3	3	2
0	0	0	0	0	0	0	0
1	1	2	1	1	1	0	0
2	2	1	2	2	2	0	0
0	1	1	1	2	0	0	1
1	2	0	2	0	1	0	1
2	0	2	0	1	2	0	1
0	2	2	1	0	2	1	0
1	0	1	2	1	0	1	0
2	1	0	0	2	1	1	0
0	2	1	0	1	1	1	1
1	0	0	1	2	2	1	1
2	1	2	2	0	0	1	1
0	0	2	2	2	1	2	0
1	1	1	0	0	2	2	0
2	2	0	1	1	0	2	0
0	1	0	2	1	2	2	1
1	2	2	0	2	0	2	1
2	0	1	1	0	1	2	1

Association

3	2	6
0	0	0
0	1	1
1	0	2
1	1	3
2	0	4
2	1	5

2^{19} Orthogonal Main Effect Design

2	2	2	2	2	2	2	2	2	2	2	2	2	2	2	2	2	2	2
1	0	1	1	0	0	0	0	1	0	1	0	1	1	1	1	0	0	1
1	1	0	1	1	0	0	0	0	1	0	1	0	1	1	1	1	0	0
0	1	1	0	1	1	0	0	0	0	1	0	1	0	1	1	1	1	0
0	0	1	1	0	1	1	0	0	0	0	1	0	1	0	1	1	1	1
1	0	0	1	1	0	1	1	0	0	0	0	1	0	1	0	1	1	1
1	1	0	0	1	1	0	1	1	0	0	0	0	1	0	1	0	1	1
1	1	1	0	0	1	1	0	1	1	0	0	0	0	1	0	1	0	1
1	1	1	1	0	0	1	1	0	1	1	0	0	0	0	1	0	1	0
0	1	1	1	1	0	0	1	1	0	1	1	0	0	0	0	1	0	1
1	0	1	1	1	1	0	0	1	1	0	1	1	0	0	0	0	1	0
0	1	0	1	1	1	1	0	0	1	1	0	1	1	0	0	0	0	1
1	0	1	0	1	1	1	1	0	0	1	1	0	1	1	0	0	0	0
0	1	0	1	0	1	1	1	1	0	0	1	1	0	1	1	0	0	0
0	0	1	0	1	0	1	1	1	1	0	0	1	1	0	1	1	0	0
0	0	0	1	0	1	0	1	1	1	1	0	0	1	1	0	1	1	0
0	0	0	0	1	0	1	0	1	1	1	1	0	0	1	1	0	1	1
1	0	0	0	0	1	0	1	0	1	1	1	1	0	0	1	1	0	1
1	1	0	0	0	0	1	0	1	0	1	1	1	1	0	0	1	1	0
0	1	1	0	0	0	0	1	0	1	0	1	1	1	1	0	0	1	1
0	0	0	0	0	0	0	0	0	0	0	0	0	0	0	0	0	0	0

2^{23} Orthogonal Main Effect Design

┌─	4	─┐																				
2	2	2	2	2	2	2	2	2	2	2	2	2	2	2	2	2	2	2	2	2	2	2
0	0	0	0	0	0	0	0	0	0	0	0	0	0	0	0	0	0	0	0	0	0	0
0	0	0	0	0	1	1	1	0	1	1	0	1	0	0	1	1	1	0	1	1	0	1
0	0	0	0	1	1	1	0	1	1	0	1	0	0	1	1	1	0	1	1	0	1	0
0	0	0	1	0	0	0	1	1	1	0	1	1	1	0	0	0	1	1	1	0	1	1
0	0	0	1	1	0	1	0	0	0	1	1	1	1	1	0	1	0	0	0	1	1	1
0	0	0	1	1	1	0	1	1	0	1	0	0	1	1	1	0	1	1	0	1	0	0
0	1	1	0	0	0	1	1	1	0	1	1	0	0	0	0	1	1	1	0	1	1	0
0	1	1	0	1	0	0	0	1	1	1	0	1	0	1	0	0	0	1	1	1	0	1
0	1	1	0	1	1	0	1	0	0	0	1	1	0	1	1	0	1	0	0	0	1	1
0	1	1	1	0	1	0	0	0	1	1	1	0	1	0	1	0	0	0	1	1	1	0
0	1	1	1	0	1	1	0	1	0	0	0	1	1	0	1	1	0	1	0	0	0	1
0	1	1	1	1	0	1	1	0	1	0	0	0	1	1	0	1	1	0	1	0	0	0
1	0	1	0	0	0	0	0	0	0	0	0	0	1	1	1	1	1	1	1	1	1	1
1	0	1	0	0	1	1	1	0	1	1	0	1	1	1	0	0	0	1	0	0	1	0
1	0	1	0	1	1	1	0	1	1	0	1	0	1	0	0	0	1	0	0	1	0	1
1	0	1	1	0	0	0	1	1	1	0	1	1	0	1	1	1	0	0	0	1	0	0
1	0	1	1	1	0	1	0	0	0	1	1	1	0	0	1	0	1	1	1	0	0	0
1	0	1	1	1	1	0	1	1	0	1	0	0	0	0	0	1	0	0	1	0	1	1
1	1	0	0	0	0	1	1	1	0	1	1	0	1	1	1	0	0	0	1	0	0	1
1	1	0	0	1	0	0	0	1	1	1	0	1	1	0	1	1	1	0	0	0	1	0
1	1	0	0	1	1	0	1	0	0	0	1	1	1	0	0	1	0	1	1	1	0	0
1	1	0	1	0	1	0	0	0	1	1	1	0	0	1	0	1	1	1	0	0	0	1
1	1	0	1	0	1	1	0	1	0	0	0	1	0	1	0	0	1	0	1	1	1	0
1	1	0	1	1	0	1	1	0	1	0	0	0	0	0	1	0	0	1	0	1	1	1

Association

2	2	2	4
0	0	0	0
0	1	1	1
1	0	1	2
1	1	0	3

6×2^{14} Orthogonal Main Effect Design

	4													
6	2	2	2	2	2	2	2	2	2	2	2	2	2	2
0	0	0	0	1	0	0	1	0	1	1	1	0	0	0
0	0	1	1	1	1	1	0	1	0	0	0	1	1	1
0	1	0	1	0	0	1	0	0	0	1	1	1	0	1
0	1	1	0	0	1	0	1	1	1	0	0	0	1	0
1	0	0	0	1	0	0	0	1	0	0	1	0	1	1
1	0	1	1	1	1	1	1	0	1	1	0	1	0	0
1	1	0	1	0	1	0	1	0	0	0	1	1	1	0
1	1	1	0	0	0	1	0	1	1	1	0	0	0	1
2	0	0	0	1	1	0	0	0	1	0	0	1	0	1
2	0	1	1	1	0	1	1	1	0	1	1	0	1	0
2	1	0	1	0	0	1	1	0	1	0	0	0	1	1
2	1	1	0	0	1	0	0	1	0	1	1	1	0	0
3	0	0	0	0	1	1	0	0	0	1	0	0	1	0
3	0	1	1	0	0	0	1	1	1	0	1	1	0	1
3	1	0	1	1	1	0	1	1	0	1	0	0	0	1
3	1	1	0	1	0	1	0	0	1	0	1	1	1	0
4	0	0	0	0	0	1	1	1	0	0	0	1	0	0
4	0	1	1	0	1	0	0	0	1	1	1	0	1	1
4	1	0	1	1	1	1	0	1	1	0	1	0	0	0
4	1	1	0	1	0	0	1	0	0	1	0	1	1	1
5	0	0	0	0	1	1	1	1	1	1	1	1	1	1
5	0	1	1	0	0	0	0	0	0	0	0	0	0	0
5	1	0	1	1	0	0	0	1	1	1	0	1	1	0
5	1	1	0	1	1	1	1	0	0	0	1	0	0	1

Association

2	2	2	4
0	0	0	0
0	1	1	1
1	0	1	2
1	1	0	3

$4 \times 3 \times 2^{13}$ Orthogonal Main Effect Design

4	3	2	2	2	2	2	2	2	2	2	2	2	2	2
0	0	0	0	0	1	1	0	1	0	0	0	1	1	1
0	0	1	0	1	1	1	1	0	1	1	0	1	0	0
0	1	0	1	1	0	1	1	1	0	1	1	0	1	0
0	1	1	1	0	0	0	1	1	1	0	1	1	0	1
0	2	0	1	0	1	0	0	0	1	1	1	0	1	1
0	2	1	0	1	0	0	0	0	0	0	0	0	0	0
1	0	0	1	1	0	1	0	0	0	1	1	1	0	1
1	0	1	1	0	1	0	1	0	0	0	1	1	1	0
1	1	0	0	1	0	1	1	0	1	0	0	0	1	1
1	1	1	0	0	1	0	1	1	0	1	0	0	0	1
1	2	0	0	0	1	1	0	1	1	0	1	0	0	0
1	2	1	1	1	0	0	0	1	1	1	0	1	1	0
2	0	0	0	0	0	0	1	0	1	1	1	0	0	0
2	0	1	0	1	0	0	0	1	0	0	1	0	1	1
2	1	0	1	1	1	0	0	0	1	0	0	1	0	1
2	1	1	1	0	1	1	0	0	0	1	0	0	1	0
2	2	0	1	0	0	1	1	1	0	0	0	1	0	0
2	2	1	0	1	1	1	1	1	1	1	1	1	1	1
3	0	0	1	1	1	0	1	1	1	0	0	0	1	0
3	0	1	1	0	0	1	0	1	1	1	0	0	0	1
3	1	0	0	1	1	0	0	1	0	1	1	1	0	0
3	1	1	0	0	0	1	0	0	1	0	1	1	1	0
3	2	0	0	0	0	0	1	0	0	1	0	1	1	1
3	2	1	1	1	1	1	1	0	0	0	1	0	0	1

5^6 Orthogonal Main Effect Design

5	5	5	5	5	5
0	0	0	0	0	0
0	1	1	2	3	4
0	2	2	4	1	3
0	3	3	1	4	2
0	4	4	3	2	1
1	0	1	1	1	1
1	1	2	3	4	0
1	2	3	0	2	4
1	3	4	2	0	3
1	4	0	4	3	2
2	0	2	2	2	2
2	1	3	4	0	1
2	2	4	1	3	0
2	3	0	3	1	4
2	4	1	0	4	3
3	0	3	3	3	3
3	1	4	0	1	2
3	2	0	2	4	1
3	3	1	4	2	0
3	4	2	1	0	4
4	0	4	4	4	4
4	1	0	1	2	3
4	2	1	3	0	2
4	3	2	0	3	1
4	4	3	2	1	0

3^{13} Orthogonal Main Effect Design

9												
3	3	3	3	3	3	3	3	3	3	3	3	3
0	0	0	0	0	0	0	0	0	0	0	0	0
0	0	0	0	1	1	2	2	1	1	2	2	1
0	0	0	0	2	2	1	1	2	2	1	1	2
0	1	1	2	0	1	1	1	1	0	0	2	2
0	1	1	2	1	2	0	0	2	1	2	1	0
0	1	1	2	2	0	2	2	0	2	1	0	1
0	2	2	1	0	2	2	2	2	0	0	1	1
0	2	2	1	1	0	1	1	0	1	2	0	2
0	2	2	1	2	1	0	0	1	2	1	2	0
1	0	1	1	0	2	1	0	0	1	1	2	1
1	0	1	1	1	0	0	2	1	2	0	1	2
1	0	1	1	2	1	2	1	2	0	2	0	0
1	1	2	0	0	0	2	1	1	1	1	1	0
1	1	2	0	1	1	1	0	2	2	0	0	1
1	1	2	0	2	2	0	2	0	0	2	2	2
1	2	0	2	0	1	0	2	2	1	1	0	2
1	2	0	2	1	2	2	1	0	2	0	2	0
1	2	0	2	2	0	1	0	1	0	2	1	1
2	0	2	2	0	1	2	0	0	2	2	1	2
2	0	2	2	1	2	1	2	1	0	1	0	0
2	0	2	2	2	0	0	1	2	1	0	2	1
2	1	0	1	0	2	0	1	1	2	2	0	1
2	1	0	1	1	0	2	0	2	0	1	2	2
2	1	0	1	2	1	1	2	0	1	0	1	0
2	2	1	0	0	0	1	2	2	2	2	2	0
2	2	1	0	1	1	0	1	0	0	1	1	1
2	2	1	0	2	2	2	0	1	1	0	0	2

Association

3	3	3	3	9
0	0	0	0	0
0	1	1	2	1
0	2	2	1	2
1	0	1	1	3
1	1	2	0	4
1	2	0	2	5
2	0	2	2	6
2	1	0	1	7
2	2	1	0	8

2^{27} Orthogonal Main Effect Design

2	2	2	2	2	2	2	2	2	2	2	2	2	2	2	2	2	2	2	2	2	2	2	2	2	2	2
0	1	1	1	1	1	1	1	1	1	1	1	1	1	1	1	1	1	1	1	1	1	1	1	1	1	1
1	0	1	0	1	0	1	0	1	0	1	0	1	0	1	0	1	0	1	0	1	0	1	0	1	0	1
1	1	0	1	1	0	0	1	1	1	1	0	0	0	0	0	0	0	0	1	1	1	1	0	0	1	1
0	0	0	1	0	0	1	1	0	1	0	0	1	0	1	0	1	0	1	1	0	1	0	0	1	1	0
1	1	1	1	0	1	1	0	0	1	1	1	1	0	0	0	0	0	0	0	0	1	1	1	1	0	0
0	1	0	0	0	1	0	0	1	1	0	1	0	0	1	0	1	0	1	0	1	1	0	1	0	0	1
1	0	0	1	1	1	0	1	1	0	0	1	1	1	1	0	0	0	0	0	0	0	0	1	1	1	1
0	0	1	1	0	0	0	1	0	0	1	1	0	1	0	0	1	0	1	0	1	0	1	1	0	1	0
1	1	1	0	0	1	1	1	0	1	1	0	0	1	1	1	1	0	0	0	0	0	0	0	0	1	1
0	1	0	0	1	1	0	0	0	1	0	0	1	1	0	1	0	0	1	0	1	0	1	0	1	1	0
1	1	1	1	1	0	0	1	1	1	0	1	1	0	0	1	1	1	1	0	0	0	0	0	0	0	0
0	1	0	1	0	0	1	1	0	0	0	1	0	0	1	1	0	1	0	0	1	0	1	0	1	0	1
1	0	0	1	1	1	1	0	0	1	1	1	0	1	1	0	0	1	1	1	1	0	0	0	0	0	0
0	0	1	1	0	1	0	0	1	1	0	0	0	1	0	0	1	1	0	1	0	0	1	0	1	0	1
1	0	0	0	0	1	1	1	1	0	0	1	1	1	0	1	1	0	0	1	1	1	1	0	0	0	0
0	0	1	0	1	1	0	1	0	0	1	1	0	0	0	1	0	0	1	1	0	1	0	0	1	0	1
1	0	0	0	0	0	0	1	1	1	1	0	0	1	1	1	0	1	1	0	0	1	1	1	1	0	0
0	0	1	0	1	0	1	1	0	1	0	0	1	1	0	0	0	1	0	0	1	1	0	1	0	0	1
1	0	0	0	0	0	0	0	0	1	1	1	1	0	0	1	1	1	0	1	1	0	0	1	1	1	1
0	0	1	0	1	0	1	0	1	1	0	1	0	0	1	1	0	0	0	1	0	0	1	1	0	1	0
1	1	1	0	0	0	0	0	0	0	0	1	1	1	1	0	0	1	1	1	0	1	1	0	0	1	1
0	1	0	0	1	0	1	0	1	0	1	1	0	1	0	0	1	1	0	0	0	1	0	0	1	1	0
1	1	1	1	1	0	0	0	0	0	0	0	0	1	1	1	1	0	0	1	1	1	0	1	1	0	0
0	1	0	1	0	0	1	0	1	0	1	0	1	1	0	1	0	0	1	1	0	0	0	1	0	0	1
1	0	0	1	1	1	1	0	0	0	0	0	0	0	0	1	1	1	1	0	0	1	1	1	0	1	1
0	0	1	1	0	1	0	0	1	0	1	0	1	0	1	1	0	1	0	0	1	1	0	0	0	1	0
1	1	1	0	0	1	1	1	1	0	0	0	0	0	0	0	0	1	1	1	1	0	0	1	1	1	0
0	1	0	0	1	1	0	1	0	0	1	0	1	0	1	0	1	1	0	1	0	0	1	1	0	0	0

2^{31} Orthogonal Main Effect Design

4			4			4			4			4			4			4			4			4						
2	2	2	2	2	2	2	2	2	2	2	2	2	2	2	2	2	2	2	2	2	2	2	2	2	2	2	2	2	2	2
0	0	0	0	0	0	0	0	0	1	1	0	0	1	1	1	0	1	1	0	1	0	1	1	0	0	0	1	1	1	1
0	0	0	0	0	0	1	1	0	0	0	0	1	0	1	0	1	1	0	1	1	1	0	1	1	1	0	0	0	1	1
0	0	0	0	1	1	0	1	1	0	1	1	1	1	0	1	0	1	1	1	0	0	0	0	1	0	1	0	0	1	1
0	0	0	0	1	1	1	0	1	1	0	1	0	0	0	0	1	1	0	0	0	1	1	0	0	1	1	1	1	1	1
0	0	0	1	0	1	0	1	1	1	1	0	0	1	1	0	0	0	0	0	0	1	1	0	1	1	0	0	0	0	0
0	0	0	1	0	1	1	0	1	0	0	0	1	0	1	1	1	0	1	1	0	0	0	0	0	0	0	1	1	0	0
0	0	0	1	1	0	0	0	0	0	1	1	1	1	0	0	0	0	0	1	1	1	0	1	0	1	1	1	1	0	0
0	0	0	1	1	0	1	1	0	1	0	1	0	0	0	1	1	0	1	0	1	0	1	1	1	0	1	0	0	0	0
0	1	1	0	0	0	0	1	1	1	0	1	1	1	0	1	1	0	0	0	0	1	0	1	0	0	0	0	1	1	0
0	1	1	0	0	0	1	0	1	0	1	1	0	0	0	0	0	0	1	1	0	0	1	1	1	1	0	1	0	1	0
0	1	1	0	1	1	0	0	0	0	0	0	0	1	1	1	1	0	0	1	1	1	1	0	1	0	1	1	0	1	0
0	1	1	0	1	1	1	1	0	1	1	0	1	0	1	0	0	0	1	0	1	0	0	0	0	1	1	0	1	1	0
0	1	1	1	0	1	0	0	0	1	0	1	1	1	0	0	1	1	1	0	1	0	0	0	1	1	0	1	0	0	1
0	1	1	1	0	1	1	1	0	0	1	1	0	0	0	1	0	1	0	1	1	1	1	0	0	0	0	0	1	0	1
0	1	1	1	1	0	0	1	1	0	0	0	0	1	1	0	1	1	1	1	0	0	1	1	0	1	1	0	1	0	1
0	1	1	1	1	0	1	0	1	1	1	0	1	0	1	1	0	1	0	0	0	1	0	1	1	0	1	1	0	0	1
1	0	1	0	0	0	0	1	1	1	1	0	0	0	0	1	1	0	0	1	1	0	0	0	0	1	1	1	0	0	1
1	0	1	0	0	0	1	0	1	0	0	0	1	1	0	0	0	0	1	0	1	1	1	0	1	0	1	0	1	0	1
1	0	1	0	1	1	0	0	0	0	1	1	1	0	1	1	1	0	0	0	0	0	1	1	1	1	0	0	1	0	1
1	0	1	0	1	1	1	1	0	1	0	1	0	1	1	0	0	0	1	1	0	1	0	1	0	0	0	1	0	0	1
1	0	1	1	0	1	0	0	0	1	1	0	0	0	0	0	1	1	1	1	0	1	0	1	1	0	1	0	1	1	0
1	0	1	1	0	1	1	1	0	0	0	0	1	1	0	1	0	1	0	0	0	0	1	1	0	1	1	1	0	1	0
1	0	1	1	1	0	0	1	1	0	1	1	1	0	1	0	1	1	1	0	1	1	1	0	0	0	0	1	0	1	0
1	0	1	1	1	0	1	0	1	1	0	1	0	1	1	1	0	1	0	1	1	0	0	0	1	1	0	0	1	1	0
1	1	0	0	0	0	0	0	0	1	0	1	1	0	1	1	0	1	1	1	0	1	1	0	0	1	1	0	0	0	0
1	1	0	0	0	0	1	1	0	0	1	1	0	1	1	0	1	1	0	0	0	0	0	0	1	0	1	1	1	0	0
1	1	0	0	1	1	0	1	1	0	0	0	0	0	0	1	0	1	1	0	1	1	0	1	1	1	0	1	1	0	0
1	1	0	0	1	1	1	0	1	1	1	0	1	1	0	0	1	1	0	1	1	0	1	1	0	0	0	0	0	0	0
1	1	0	1	0	1	0	1	1	1	0	1	1	0	1	0	0	0	0	1	1	0	1	1	1	0	1	1	1	1	1
1	1	0	1	0	1	1	0	1	0	1	1	0	1	1	1	1	0	1	0	1	1	0	1	0	1	1	0	0	1	1
1	1	0	1	1	0	0	0	0	0	0	0	0	0	0	0	0	0	0	0	0	0	0	0	0	0	0	0	0	1	1
1	1	0	1	1	0	1	1	0	1	1	0	1	1	0	1	1	0	1	1	0	1	1	0	1	1	0	1	1	1	1

Association

2	2	2	4
0	0	0	0
0	1	1	1
1	0	1	2
1	1	0	3

8×2^{24} Orthogonal Main Effect Design

	4			4			4			4			4			4			4			4		
8	2	2	2	2	2	2	2	2	2	2	2	2	2	2	2	2	2	2	2	2	2	2	2	2
0	0	0	0	1	0	1	0	1	1	1	1	0	0	1	1	1	0	1	1	1	0	0	1	1
0	0	1	1	0	0	0	1	0	1	0	1	1	1	0	1	1	1	0	0	1	1	1	1	0
0	1	0	1	0	1	1	1	1	0	1	0	1	1	1	0	0	1	1	1	0	1	1	0	1
0	1	1	0	1	1	0	0	0	0	0	0	0	0	0	0	0	0	0	0	0	0	0	0	0
1	0	0	0	0	1	1	1	0	1	1	0	1	0	0	0	0	0	0	1	1	0	1	1	0
1	0	1	1	1	1	0	0	1	1	0	0	0	1	1	0	0	1	1	0	1	1	0	1	1
1	1	0	1	1	0	1	0	0	0	1	1	0	1	0	1	1	1	0	1	0	1	0	0	0
1	1	1	0	0	0	0	1	1	0	0	1	1	0	1	1	1	0	1	0	0	0	1	0	1
2	0	0	0	1	1	0	0	0	0	0	1	1	0	1	1	0	1	1	1	0	1	1	1	0
2	0	1	1	0	1	1	1	1	0	1	1	0	1	0	1	0	0	0	0	0	0	0	1	1
2	1	0	1	0	0	0	1	0	1	0	0	0	1	1	0	1	0	1	1	1	0	0	0	0
2	1	1	0	1	0	1	0	1	1	1	0	1	0	0	0	1	1	0	0	1	1	1	0	1
3	0	0	0	0	0	0	1	1	0	0	0	0	0	0	0	1	1	0	1	0	1	0	1	1
3	0	1	1	1	0	1	0	0	0	1	0	1	1	1	0	1	0	1	0	0	0	1	1	0
3	1	0	1	1	1	0	0	1	1	0	1	1	1	0	1	0	0	0	1	1	0	1	0	1
3	1	1	0	0	1	1	1	0	1	1	1	0	0	1	1	0	1	1	0	1	1	0	0	0
4	0	0	0	0	0	0	0	0	0	1	1	0	1	1	0	0	0	0	0	1	1	1	0	1
4	0	1	1	1	0	1	1	1	0	0	1	1	0	0	0	0	1	1	1	1	0	0	0	0
4	1	0	1	1	1	0	1	0	1	1	0	1	0	1	1	1	1	0	0	0	0	0	1	1
4	1	1	0	0	1	1	0	1	1	0	0	0	1	0	1	1	0	1	1	0	1	1	1	0
5	0	0	0	1	1	0	1	1	0	1	0	1	1	0	1	1	0	1	0	1	1	0	0	0
5	0	1	1	0	1	1	0	0	0	0	0	0	0	1	1	1	1	0	1	1	0	1	0	1
5	1	0	1	0	0	0	0	1	1	1	1	0	0	0	0	0	1	1	0	0	0	1	1	0
5	1	1	0	1	0	1	1	0	1	0	1	1	1	1	0	0	0	0	1	0	1	0	1	1
6	0	0	0	0	1	1	0	1	1	0	1	1	1	1	0	1	1	0	0	0	0	0	0	0
6	0	1	1	1	1	0	1	0	1	1	1	0	0	0	0	1	0	1	1	0	1	1	0	1
6	1	0	1	1	0	1	1	1	0	0	0	0	0	1	1	0	0	0	0	1	1	1	1	0
6	1	1	0	0	0	0	0	0	0	1	0	1	1	0	1	0	1	1	1	1	0	0	1	1
7	0	0	0	1	0	1	1	0	1	0	0	0	1	0	1	0	1	1	0	0	0	1	0	1
7	0	1	1	0	0	0	0	1	1	1	0	1	0	1	1	0	0	0	1	0	1	0	0	0
7	1	0	1	0	1	1	0	0	0	0	1	1	0	0	0	1	0	1	0	1	1	0	1	1
7	1	1	0	1	1	0	1	1	0	1	1	0	1	1	0	1	1	0	1	1	0	1	1	0

Association

2	2	2	4
0	0	0	0
0	1	1	1
1	0	1	2
1	1	0	3

$3^7 \times 2^{16}$ Orthogonal Main Effect Design

3	3	3	3	3	3	3	2	2	2	2	2	2	2	2	2	2	2	2	2	2	2	2
2	1	1	0	1	0	0	0	0	0	0	0	0	1	1	1	0	0	0	0	1	1	1
0	2	1	1	0	1	0	0	0	0	0	1	1	0	0	1	0	0	1	1	0	0	1
0	0	2	1	1	0	1	0	0	0	1	0	1	0	1	0	0	1	0	1	0	1	0
1	0	0	2	1	1	0	0	0	0	1	1	0	1	0	0	0	1	1	0	1	0	0
0	1	0	0	2	1	1	0	0	1	0	0	1	1	0	0	1	0	0	1	1	0	0
1	0	1	0	0	2	1	0	0	1	0	1	0	0	1	0	1	0	1	0	0	1	0
1	1	0	1	0	0	2	0	0	1	1	0	0	0	0	1	1	1	0	0	0	0	1
1	1	1	1	1	1	1	0	0	1	1	1	1	1	1	1	1	1	1	1	1	1	1
0	1	1	2	1	2	2	0	1	1	1	1	1	0	0	0	0	0	0	0	1	1	1
2	0	1	1	2	1	2	0	1	1	1	0	0	1	1	0	0	0	1	1	0	0	1
2	2	0	1	1	2	1	0	1	1	0	1	0	1	0	1	0	1	0	1	0	1	0
1	2	2	0	1	1	2	0	1	1	0	0	1	0	1	1	0	1	1	0	1	0	0
2	1	2	2	0	1	1	0	1	0	1	1	0	0	1	1	1	0	0	1	1	0	0
1	2	1	2	2	0	1	0	1	0	1	0	1	1	0	1	1	0	1	0	0	1	0
1	1	2	1	2	2	0	0	1	0	0	1	1	1	1	0	1	1	0	0	0	0	1
1	1	1	1	1	1	1	0	1	0	0	0	0	0	0	0	1	1	1	1	1	1	1
2	1	1	0	1	0	0	1	0	0	0	0	0	1	1	1	1	1	1	1	0	0	0
0	2	1	1	0	1	0	1	0	0	0	1	1	0	0	1	1	1	0	0	1	1	0
0	0	2	1	1	0	1	1	0	0	1	0	1	0	1	0	1	0	1	0	1	0	1
1	0	0	2	1	1	0	1	0	0	1	1	0	1	0	0	1	0	0	1	0	1	1
0	1	0	0	2	1	1	1	0	1	0	0	1	1	0	0	0	1	1	0	0	1	1
1	0	1	0	0	2	1	1	0	1	0	1	0	0	1	0	0	1	0	1	1	0	1
1	1	0	1	0	0	2	1	0	1	1	0	0	0	0	1	0	0	1	1	1	1	0
1	1	1	1	1	1	1	1	0	1	1	1	1	1	1	1	0	0	0	0	0	0	0
0	1	1	2	1	2	2	1	1	1	1	1	1	0	0	0	1	1	1	1	0	0	0
2	0	1	1	2	1	2	1	1	1	1	0	0	1	1	0	1	1	0	0	1	1	0
2	2	0	1	1	2	1	1	1	1	0	1	0	1	0	1	1	0	1	0	1	0	1
1	2	2	0	1	1	2	1	1	1	0	0	1	0	1	1	1	0	0	1	0	1	1
2	1	2	2	0	1	1	1	1	0	1	1	0	0	1	1	0	1	1	0	0	1	1
1	2	1	2	2	0	1	1	1	0	1	0	1	1	0	1	0	1	0	1	1	0	1
1	1	2	1	2	2	0	1	1	0	0	1	1	1	1	0	0	0	1	1	1	1	0
1	1	1	1	1	1	1	1	1	0	0	0	0	0	0	0	0	0	0	0	0	0	0

2^{35} Orthogonal Main Effect Design

2	2	2	2	2	2	2	2	2	2	2	2	2	2	2	2	2	2	2	2	2	2	2	2	2	2	2	2	2	2	2	2	2	2	2
0	1	1	1	1	1	1	1	1	1	1	1	1	1	1	1	1	1	1	1	1	1	1	1	1	1	1	1	1	1	1	1	1	1	1
1	0	1	0	1	0	1	0	1	0	1	0	1	0	1	0	1	0	1	0	1	0	1	0	1	0	1	0	1	0	1	0	1	0	1
1	1	0	1	1	1	1	0	0	1	1	0	0	0	0	0	0	1	1	1	1	0	0	0	0	0	0	1	1	0	0	1	1	1	1
0	0	0	1	0	1	0	0	1	1	0	0	1	0	1	0	1	1	0	1	0	0	1	0	1	0	1	1	0	0	1	1	0	1	0
1	1	1	1	0	1	1	1	1	0	0	1	1	0	0	0	0	0	0	1	1	1	1	0	0	0	0	0	0	1	1	0	0	1	1
0	1	0	0	0	1	0	1	0	0	1	1	0	0	1	0	1	0	1	1	0	1	0	0	1	0	1	0	1	1	0	0	1	1	0
1	1	1	1	1	1	0	1	1	1	1	0	0	1	1	0	0	0	0	0	0	1	1	1	1	0	0	0	0	0	0	1	1	0	0
0	1	0	1	0	0	0	1	0	1	0	0	1	1	0	0	1	0	1	0	1	1	0	1	0	0	1	0	1	0	1	1	0	0	1
1	0	0	1	1	1	1	1	0	1	1	1	1	0	0	1	1	0	0	0	0	0	0	1	1	1	1	0	0	0	0	0	0	1	1
0	0	1	1	0	1	0	0	0	1	0	1	0	0	1	1	0	0	1	0	1	0	1	1	0	1	0	0	1	0	1	0	1	1	0
1	1	1	0	0	1	1	1	1	1	0	1	1	1	1	0	0	1	1	0	0	0	0	0	0	1	1	1	1	0	0	0	0	0	0
0	1	0	0	1	1	0	1	0	0	0	1	0	1	0	0	1	1	0	0	1	0	1	0	1	1	0	1	0	0	1	0	1	0	1
1	0	0	1	1	0	0	1	1	1	1	1	0	1	1	1	1	0	0	1	1	0	0	0	0	0	0	1	1	1	1	0	0	0	0
0	0	1	1	0	0	1	1	0	1	0	0	0	1	0	1	0	0	1	1	0	0	1	0	1	0	1	1	0	1	0	0	1	0	1
1	0	0	0	0	1	1	0	0	1	1	1	1	1	0	1	1	1	1	0	0	1	1	0	0	0	0	0	0	1	1	1	1	0	0
0	0	1	0	1	1	0	0	1	1	0	1	0	0	0	1	0	1	0	0	1	1	0	0	1	0	1	0	1	1	0	1	0	0	1
1	0	0	0	0	0	0	1	1	0	0	1	1	1	1	1	0	1	1	1	1	0	0	1	1	0	0	0	0	0	0	1	1	1	1
0	0	1	0	1	0	1	1	0	0	1	1	0	1	0	0	0	1	0	1	0	0	1	1	0	0	1	0	1	0	1	1	0	1	0
1	1	1	0	0	0	0	0	0	1	1	0	0	1	1	1	1	1	0	1	1	1	1	0	0	1	1	0	0	0	0	0	0	1	1
0	1	0	0	1	0	1	0	1	1	0	0	1	1	0	1	0	0	0	1	0	1	0	0	1	1	0	0	1	0	1	0	1	1	0
1	1	1	1	1	0	0	0	0	0	0	1	1	0	0	1	1	1	1	1	0	1	1	1	1	0	0	1	1	0	0	0	0	0	0
0	1	0	1	0	0	1	0	1	0	1	1	0	0	1	1	0	1	0	0	0	1	0	1	0	0	1	1	0	0	1	0	1	0	1
1	0	0	1	1	1	1	0	0	0	0	0	0	1	1	0	0	1	1	1	1	1	0	1	1	1	1	0	0	1	1	0	0	0	0
0	0	1	1	0	1	0	0	1	0	1	0	1	1	0	0	1	1	0	1	0	0	0	1	0	1	0	0	1	1	0	0	1	0	1
1	0	0	0	0	1	1	1	1	0	0	0	0	0	0	1	1	0	0	1	1	1	1	1	0	1	1	1	1	0	0	1	1	0	0
0	0	1	0	1	1	0	1	0	0	1	0	1	0	1	1	0	0	1	1	0	1	0	0	0	1	0	1	0	0	1	1	0	0	1
1	0	0	0	0	0	0	1	1	1	1	0	0	0	0	0	0	1	1	0	0	1	1	1	1	1	0	1	1	1	1	0	0	1	1
0	0	1	0	1	0	1	1	0	1	0	0	1	0	1	0	1	1	0	0	1	1	0	1	0	0	0	1	0	1	0	0	1	1	0
1	1	1	0	0	0	0	0	0	1	1	1	1	0	0	0	0	0	0	1	1	0	0	1	1	1	1	1	0	1	1	1	1	0	0
0	1	0	0	1	0	1	0	1	1	0	1	0	0	1	0	1	0	1	1	0	0	1	1	0	1	0	0	0	1	0	1	0	0	1
1	0	0	1	1	0	0	0	0	0	0	1	1	1	1	0	0	0	0	0	0	1	1	0	0	1	1	1	1	1	0	1	1	1	1
0	0	1	1	0	0	1	0	1	0	1	1	0	1	0	0	1	0	1	0	1	1	0	0	1	1	0	1	0	0	0	1	0	1	0
1	1	1	0	0	1	1	0	0	0	0	0	0	1	1	1	1	0	0	0	0	0	0	1	1	0	0	1	1	1	1	1	0	1	1
0	1	0	0	1	1	0	0	1	0	1	0	1	1	0	1	0	0	1	0	1	0	1	1	0	0	1	1	0	1	0	0	0	1	0
1	1	1	1	1	0	0	1	1	0	0	0	0	0	0	1	1	1	1	0	0	0	0	0	0	1	1	0	0	1	1	1	1	1	0
0	1	0	1	0	0	1	1	0	0	1	0	1	0	1	1	0	1	0	0	1	0	1	0	1	1	0	0	1	1	0	1	0	0	0

$3^{12} \times 2^{11}$ Orthogonal Main Effect Design

3	3	3	3	3	3	3	3	3	3	3	3	2	2	2	2	2	2	2	2	2	2	2
0	0	0	1	1	0	0	1	0	2	2	0	1	0	1	0	0	0	1	1	1	0	1
0	0	0	0	2	0	2	0	2	0	0	1	1	1	0	1	0	0	0	1	1	1	0
0	0	1	0	0	2	1	2	0	0	1	0	0	1	1	0	1	0	0	0	1	1	1
0	0	2	2	0	1	0	0	1	1	0	0	1	0	1	1	0	1	0	0	0	1	1
0	1	2	2	0	0	1	1	2	0	2	2	1	1	0	1	1	0	1	0	0	0	1
0	1	2	1	2	1	2	2	2	2	1	0	1	1	1	0	1	1	0	1	0	0	0
0	1	0	0	2	2	0	2	1	1	2	2	0	1	1	1	0	1	1	0	1	0	0
0	1	1	2	1	2	2	0	0	2	0	2	0	0	1	1	1	0	1	1	0	1	0
0	2	1	2	1	0	0	2	2	1	1	1	0	0	0	1	1	1	0	1	1	0	1
0	2	1	0	0	1	2	1	1	2	2	1	1	0	0	0	1	1	1	0	1	1	0
0	2	2	1	2	2	1	1	0	1	0	1	0	1	0	0	0	1	1	1	0	1	1
0	2	0	1	1	1	1	0	1	0	1	2	0	0	0	0	0	0	0	0	0	0	0
1	1	1	2	2	1	1	2	1	0	0	1	1	0	1	0	0	0	1	1	1	0	1
1	1	1	1	0	1	0	1	0	1	1	2	1	1	0	1	0	0	0	1	1	1	0
1	1	2	1	1	0	2	0	1	1	2	1	0	1	1	0	1	0	0	0	1	1	1
1	1	0	0	1	2	1	1	2	2	1	1	1	0	1	1	0	1	0	0	0	1	1
1	2	0	0	1	1	2	2	0	1	0	0	1	1	0	1	1	0	1	0	0	0	1
1	2	0	2	0	2	0	0	0	0	2	1	1	1	1	0	1	1	0	1	0	0	0
1	2	1	1	0	0	1	0	2	2	0	0	0	1	1	1	0	1	1	0	1	0	0
1	2	2	0	2	0	0	1	1	0	1	0	0	0	1	1	1	0	1	1	0	1	0
1	0	2	0	2	1	1	0	0	2	2	2	0	0	0	1	1	1	0	1	1	0	1
1	0	2	1	1	2	0	2	2	0	0	2	1	0	0	0	1	1	1	0	1	1	0
1	0	0	2	0	0	2	2	1	2	1	2	0	1	0	0	0	1	1	1	0	1	1
1	0	1	2	2	2	2	1	2	1	2	0	0	0	0	0	0	0	0	0	0	0	0
2	2	2	0	0	2	2	0	2	1	1	2	1	0	1	0	0	0	1	1	1	0	1
2	2	2	2	1	2	1	2	1	2	2	0	1	1	0	1	0	0	0	1	1	1	0
2	2	0	2	2	1	0	1	2	2	0	2	0	1	1	0	1	0	0	0	1	1	1
2	2	1	1	2	0	2	2	0	0	2	2	1	0	1	1	0	1	0	0	0	1	1
2	0	1	1	2	2	0	0	1	2	1	1	1	1	0	1	1	0	1	0	0	0	1
2	0	1	0	1	0	1	1	1	1	0	2	1	1	1	0	1	1	0	1	0	0	0
2	0	2	2	1	1	2	1	0	0	1	1	0	1	1	1	0	1	1	0	1	0	0
2	0	0	1	0	1	1	2	2	1	2	1	0	0	1	1	1	0	1	1	0	1	0
2	1	0	1	0	2	2	1	1	0	0	0	0	0	0	1	1	1	0	1	1	0	1
2	1	0	2	2	0	1	0	0	1	1	0	1	0	0	0	1	1	1	0	1	1	0
2	1	1	0	1	1	0	0	2	0	2	0	0	1	0	0	0	1	1	1	0	1	1
2	1	2	0	0	0	0	2	0	2	0	1	0	0	0	0	0	0	0	0	0	0	0

$3^{13} \times 2^4$, $6 \times 3^{12} \times 2^2$, and 4×3^{13} Orthogonal Main Effect Designs

												6 × 2²	4				Substitutions			
3	3	3	3	3	3	3	3	3	3	3	3	3	2	2	2	2	6	2	2	4
0	0	0	1	1	0	0	1	0	2	2	0	0	0	0	0	0	0	0	0	0
0	0	0	0	2	0	2	0	2	0	0	1	0	0	1	0	1	1	0	0	1
0	0	1	0	0	2	1	2	0	0	1	0	0	1	0	1	1	2	0	0	2
0	0	2	2	0	1	0	0	1	1	0	0	0	1	1	1	0	3	0	1	3
0	1	2	2	0	0	1	1	2	0	2	2	1	0	0	1	1	4	0	1	0
0	1	2	1	2	1	2	2	2	2	1	0	1	0	1	1	0	5	0	1	1
0	1	0	0	2	2	0	2	1	1	2	2	1	1	0	0	1	0	1	0	2
0	1	1	2	1	2	2	0	0	2	0	2	1	1	1	0	0	1	1	0	3
0	2	1	2	1	0	0	2	2	1	1	1	2	0	0	1	0	2	1	0	0
0	2	1	0	0	1	2	1	1	2	2	1	2	0	1	0	1	3	1	1	1
0	2	2	1	2	2	1	1	0	1	0	1	2	1	0	0	0	4	1	1	2
0	2	0	1	1	1	1	0	1	0	1	2	2	1	1	1	1	5	1	1	3
1	1	1	2	2	1	1	2	1	0	0	1	0	0	0	0	0	0	0	0	0
1	1	1	1	0	1	0	1	0	1	1	2	0	0	1	0	1	1	0	0	1
1	1	2	1	1	0	2	0	1	1	2	1	0	1	0	1	1	2	0	0	2
1	1	0	0	1	2	1	1	2	2	1	1	0	1	1	1	0	3	0	1	3
1	2	0	0	1	1	2	2	0	1	0	0	1	0	0	1	1	4	0	1	0
1	2	0	2	0	2	0	0	0	0	2	1	1	0	1	1	0	5	0	1	1
1	2	1	1	0	0	1	0	2	2	0	0	1	1	0	0	1	0	1	0	2
1	2	2	0	2	0	0	1	1	0	1	0	1	1	1	0	0	1	1	0	3
1	0	2	0	2	1	1	0	0	2	2	2	2	0	0	1	0	2	1	0	0
1	0	2	1	1	2	0	2	2	0	0	2	2	0	1	0	1	3	1	1	1
1	0	0	2	0	0	2	2	1	2	1	2	2	1	0	0	0	4	1	1	2
1	0	1	2	2	2	2	1	2	1	2	0	2	1	1	1	1	5	1	1	3
2	2	2	0	0	2	2	0	2	1	1	2	0	0	0	0	0	0	0	0	0
2	2	2	2	1	2	1	2	1	2	2	0	0	0	1	0	1	1	0	0	1
2	2	0	2	2	1	0	1	2	2	0	2	0	1	0	1	1	2	0	0	2
2	2	1	1	2	0	2	2	0	0	2	2	0	1	1	1	0	3	0	1	3
2	0	1	1	2	2	0	0	1	2	1	1	1	0	0	1	1	4	0	1	0
2	0	1	0	1	0	1	1	1	1	0	2	1	0	1	1	0	5	0	1	1
2	0	2	2	1	1	2	1	0	0	1	1	1	1	0	0	1	0	1	0	2
2	0	0	1	0	1	1	2	2	1	2	1	1	1	1	0	0	1	1	0	3
2	1	0	1	0	2	2	1	1	0	0	0	2	0	0	1	0	2	1	0	0
2	1	0	2	2	0	1	0	0	1	1	0	2	0	1	0	1	3	1	1	1
2	1	1	0	1	1	0	0	2	0	2	0	2	1	0	0	0	4	1	1	2
2	1	2	0	0	0	0	2	0	2	0	1	2	1	1	1	1	5	1	1	3

6^3 Orthogonal Main Effect Design

6	6	6
0	0	3
0	1	5
0	2	4
0	3	0
0	4	2
0	5	1
1	0	5
1	1	4
1	2	3
1	3	2
1	4	1
1	5	0
2	0	4
2	1	3
2	2	5
2	3	1
2	4	0
2	5	2
3	0	0
3	1	2
3	2	1
3	3	3
3	4	5
3	5	4
4	0	2
4	1	1
4	2	0
4	3	5
4	4	4
4	5	3
5	0	1
5	1	0
5	2	2
5	3	4
5	4	3
5	5	5

2^{39} Orthogonal Main Effect Design

┌ 4 ┐

2	2	2	2	2	2	2	2	2	2	2	2	2	2	2	2	2	2	2	2	2	2	2	2	2	2	2	2	2	2	2	2	2	2	2	2	2	2	2
0	0	0	0	0	0	0	0	0	0	0	0	0	0	0	0	0	0	0	0	0	0	0	0	0	0	0	0	0	0	0	0	0	0	0	0	0	0	0
0	0	0	0	0	0	1	0	1	0	1	1	1	1	0	0	1	1	0	1	1	0	0	0	1	0	1	0	1	1	1	1	0	0	1	1	0	1	1
0	0	0	0	0	1	0	1	0	1	1	1	1	0	0	1	1	0	1	1	0	0	0	1	0	1	0	1	1	1	1	0	0	1	1	0	1	1	0
0	0	0	0	1	0	1	0	1	1	1	1	0	0	1	1	0	1	1	0	0	0	1	0	1	0	1	1	1	1	0	0	1	1	0	1	1	0	0
0	0	0	0	1	1	0	1	1	0	0	0	0	1	0	1	0	1	1	1	1	0	1	1	0	1	1	0	0	0	0	1	0	1	0	1	1	1	1
0	0	0	1	0	1	0	1	1	1	1	0	0	1	1	0	1	1	0	0	0	1	0	1	0	1	1	1	1	0	0	1	1	0	1	1	0	0	0
0	0	0	1	0	1	1	1	1	0	0	1	1	0	1	1	0	0	0	0	1	1	0	1	1	1	1	0	0	1	1	0	1	1	0	0	0	0	1
0	0	0	1	1	0	0	0	0	1	0	1	0	1	1	1	1	0	0	1	1	1	1	0	0	0	0	1	0	1	0	1	1	1	1	0	0	1	1
0	0	0	1	1	0	1	1	0	0	0	0	1	0	1	0	1	1	1	1	0	1	1	0	1	1	0	0	0	0	1	0	1	0	1	1	1	1	0
0	0	0	1	1	1	1	0	0	1	1	0	1	1	0	0	0	0	1	0	1	1	1	1	1	0	0	1	1	0	1	1	0	0	0	0	1	0	1
0	1	1	0	0	0	0	1	0	1	0	1	1	1	1	0	0	1	1	0	1	0	0	0	0	1	0	1	0	1	1	1	1	0	0	1	1	0	1
0	1	1	0	0	1	1	0	1	1	0	0	0	0	1	0	1	0	1	1	1	0	0	1	1	0	1	1	0	0	0	0	1	0	1	0	1	1	1
0	1	1	0	1	0	1	1	1	1	0	0	1	1	0	1	1	0	0	0	0	0	1	0	1	1	1	1	0	0	1	1	0	1	1	0	0	0	0
0	1	1	0	1	1	0	0	0	0	1	0	1	0	1	1	1	1	0	0	1	0	1	1	0	0	0	0	1	0	1	0	1	1	1	1	0	0	1
0	1	1	0	1	1	1	1	0	0	1	1	0	1	1	0	0	0	0	1	0	0	1	1	1	1	0	0	1	1	0	1	1	0	0	0	0	1	0
0	1	1	1	0	0	0	0	1	0	1	0	1	1	1	1	0	0	1	1	0	1	0	0	0	0	1	0	1	0	1	1	1	1	0	0	1	1	0
0	1	1	1	0	0	1	1	0	1	1	0	0	0	0	1	0	1	0	1	1	1	0	0	1	1	0	1	1	0	0	0	0	1	0	1	0	1	1
0	1	1	1	0	1	1	0	0	0	0	1	0	1	0	1	1	1	1	0	0	1	0	1	1	0	0	0	0	1	0	1	0	1	1	1	1	0	0
0	1	1	1	1	0	0	1	1	0	1	1	0	0	0	0	1	0	1	0	1	1	1	0	0	1	1	0	1	1	0	0	0	0	1	0	1	0	1
0	1	1	1	1	1	0	0	1	1	0	1	1	0	0	0	0	1	0	1	0	1	1	1	0	0	1	1	0	1	1	0	0	0	0	1	0	1	0
1	0	1	0	0	0	0	0	0	0	0	0	0	0	0	0	0	0	0	0	0	2	2	2	2	2	2	2	2	2	2	2	2	2	2	2	2	2	2
1	0	1	0	0	0	1	0	1	0	1	1	1	1	0	0	1	1	0	1	1	2	2	2	1	2	1	2	1	1	1	1	2	2	1	1	2	1	1
1	0	1	0	0	1	0	1	0	1	1	1	1	0	0	1	1	0	1	1	0	2	2	1	2	1	2	1	1	1	1	2	2	1	1	2	1	1	2
1	0	1	0	1	0	1	0	1	1	1	1	0	0	1	1	0	1	1	0	0	2	1	2	1	2	1	1	1	1	2	2	1	1	2	1	1	2	2
1	0	1	0	1	1	0	1	1	0	0	0	0	1	0	1	0	1	1	1	1	2	1	1	2	1	1	2	2	2	2	1	2	1	2	1	1	1	1
1	0	1	1	0	1	0	1	1	1	1	0	0	1	1	0	1	1	0	0	0	1	2	1	2	1	1	1	1	2	2	1	1	2	1	1	2	2	2
1	0	1	1	0	1	1	1	1	0	0	1	1	0	1	1	0	0	0	0	1	1	2	1	1	1	1	2	2	1	1	2	1	1	2	2	2	2	1
1	0	1	1	1	0	0	0	0	1	0	1	0	1	1	1	1	0	0	1	1	1	1	2	2	2	2	1	2	1	2	1	1	1	1	2	2	1	1
1	0	1	1	1	0	1	1	0	0	0	0	1	0	1	0	1	1	1	1	0	1	1	2	1	1	2	2	2	2	1	2	1	2	1	1	1	1	2
1	0	1	1	1	1	1	0	0	1	1	0	1	1	0	0	0	0	1	0	1	1	1	1	1	2	2	1	1	2	1	1	2	2	2	2	1	2	1
1	1	0	0	0	0	0	1	0	1	0	1	1	1	1	0	0	1	1	0	1	2	2	2	2	1	2	1	2	1	1	1	1	2	2	1	1	2	1
1	1	0	0	0	1	1	0	1	1	0	0	0	0	1	0	1	0	1	1	1	2	2	1	1	2	1	1	2	2	2	2	1	2	1	2	1	1	1
1	1	0	0	1	0	1	1	1	1	0	0	1	1	0	1	1	0	0	0	0	2	1	2	1	1	1	1	2	2	1	1	2	1	1	2	2	2	2
1	1	0	0	1	1	0	0	0	0	1	0	1	0	1	1	1	1	0	0	1	2	1	1	2	2	2	2	1	2	1	2	1	1	1	1	2	2	1
1	1	0	0	1	1	1	1	0	0	1	1	0	1	1	0	0	0	0	1	0	2	1	1	1	1	2	2	1	1	2	1	1	2	2	2	2	1	2
1	1	0	1	0	0	0	0	1	0	1	0	1	1	1	1	0	0	1	1	0	1	2	2	2	2	1	2	1	2	1	1	1	1	2	2	1	1	2
1	1	0	1	0	0	1	1	0	1	1	0	0	0	0	1	0	1	0	1	1	1	2	2	1	1	2	1	1	2	2	2	2	1	2	1	2	1	1
1	1	0	1	0	1	1	0	0	0	0	1	0	1	0	1	1	1	1	0	0	1	2	1	1	2	2	2	2	1	2	1	2	1	1	1	1	2	2
1	1	0	1	1	0	0	1	1	0	1	1	0	0	0	0	1	0	1	0	1	1	1	2	2	1	1	2	1	1	2	2	2	2	1	2	1	2	1
1	1	0	1	1	1	0	0	1	1	0	1	1	0	0	0	0	1	0	1	0	1	1	1	2	2	1	1	2	1	1	2	2	2	2	1	2	1	2

Association

2	2	2	4
0	0	0	0
0	1	1	1
1	0	1	2
1	1	0	3

5×2^{28} Orthogonal Main Effect Design

	4																											
5	2	2	2	2	2	2	2	2	2	2	2	2	2	2	2	2	2	2	2	2	2	2	2	2	2	2	2	2
0	0	0	0	0	0	0	0	0	0	0	1	1	0	1	1	0	0	0	0	1	0	1	0	1	1	1	1	0
0	0	0	0	1	0	0	0	1	1	1	0	1	1	0	1	1	0	0	0	0	1	0	1	0	1	1	1	1
0	0	1	1	0	1	1	1	0	0	0	0	1	1	0	0	0	0	1	0	1	0	1	1	1	1	0	0	1
0	0	1	1	1	1	1	1	1	1	1	1	0	1	1	0	0	0	0	1	0	1	0	1	1	1	1	0	0
0	1	0	1	0	1	1	1	0	0	0	1	0	0	1	1	1	1	0	1	0	1	0	0	0	0	1	1	0
0	1	0	1	1	1	1	1	1	1	1	0	1	0	0	1	1	1	1	0	1	0	1	0	0	0	0	1	1
0	1	1	0	0	0	0	0	0	0	0	0	0	1	0	0	1	1	1	1	0	1	0	1	0	0	0	0	1
0	1	1	0	1	0	0	0	1	1	1	1	0	0	1	0	0	1	1	1	1	0	1	0	1	0	0	0	0
1	0	0	0	0	0	0	1	0	1	1	1	1	1	1	0	0	1	1	0	1	1	0	0	0	0	1	0	1
1	0	0	0	1	1	1	0	0	1	0	1	0	1	1	1	1	0	0	1	1	0	1	1	0	0	0	0	1
1	0	1	1	0	0	1	0	1	0	1	0	0	1	1	0	1	1	0	0	0	0	1	0	1	0	1	1	1
1	0	1	1	1	1	0	1	1	0	0	1	0	0	1	1	0	1	1	0	0	0	0	1	0	1	0	1	1
1	1	0	1	0	0	1	0	1	0	1	1	1	0	0	1	0	0	1	1	1	1	0	1	0	1	0	0	0
1	1	0	1	1	1	0	1	1	0	0	0	1	1	0	0	1	0	0	1	1	1	1	0	1	0	1	0	0
1	1	1	0	0	0	0	1	0	1	1	0	0	0	0	1	1	0	0	1	0	0	1	1	1	1	0	1	0
1	1	1	0	1	1	1	0	0	1	0	0	1	0	0	0	0	1	1	0	0	1	0	0	1	1	1	1	0
2	0	0	0	0	0	1	1	1	0	0	1	0	1	0	1	1	1	1	0	0	1	1	0	1	1	0	0	0
2	0	0	0	1	1	1	0	0	0	1	0	1	0	1	0	1	1	1	1	0	0	1	1	0	1	1	0	0
2	0	1	1	0	1	0	0	1	1	1	1	1	0	0	1	1	0	1	1	0	0	0	0	1	0	1	0	1
2	0	1	1	1	0	0	1	0	1	0	1	1	1	0	0	1	1	0	1	1	0	0	0	0	1	0	1	0
2	1	0	1	0	1	0	0	1	1	1	0	0	1	1	0	0	1	0	0	1	1	1	1	0	1	0	1	0
2	1	0	1	1	0	0	1	0	1	0	0	0	0	1	1	0	0	1	0	0	1	1	1	1	0	1	0	1
2	1	1	0	0	0	1	1	1	0	0	0	1	0	1	0	0	0	0	1	1	0	0	1	0	0	1	1	1
2	1	1	0	1	1	1	0	0	0	1	1	0	1	0	1	0	0	0	0	1	1	0	0	1	0	0	1	1
3	0	0	0	0	1	1	0	1	1	0	0	0	1	0	1	0	1	1	1	1	0	0	1	1	0	1	1	0
3	0	0	0	1	1	0	1	0	0	1	0	0	0	1	0	1	0	1	1	1	1	0	0	1	1	0	1	1
3	0	1	1	0	0	1	1	0	1	1	0	1	1	1	1	0	0	1	1	0	1	1	0	0	0	0	1	0
3	0	1	1	1	0	0	0	1	0	0	0	1	0	1	1	1	1	0	0	1	1	0	1	1	0	0	0	0
3	1	0	1	0	0	1	1	0	1	1	1	0	0	0	0	1	1	0	0	1	0	0	1	1	1	1	0	1
3	1	0	1	1	0	0	0	1	0	0	1	0	1	0	0	0	0	1	1	0	0	1	0	0	1	1	1	1
3	1	1	0	0	1	1	0	1	1	0	1	1	0	1	0	1	0	0	0	0	1	1	0	0	1	0	0	1
3	1	1	0	1	1	0	1	0	0	1	1	1	1	0	1	0	1	0	0	0	0	1	1	0	0	1	0	0
4	0	0	0	0	1	0	1	1	0	1	1	1	0	0	0	0	1	0	1	0	1	1	1	1	0	0	1	1
4	0	0	0	1	0	1	1	1	1	0	0	0	0	0	0	0	0	0	0	0	0	0	0	0	0	0	0	0
4	0	1	1	0	1	0	0	0	1	0	0	0	0	0	1	0	1	0	1	1	1	1	0	0	1	1	0	1
4	0	1	1	1	0	1	0	0	0	1	1	0	0	0	0	1	0	1	0	1	1	1	1	0	0	1	1	0
4	1	0	1	0	1	0	0	0	1	0	1	1	1	1	0	1	0	1	0	0	0	0	1	1	0	0	1	0
4	1	0	1	1	0	1	0	0	0	1	0	1	1	1	1	0	1	0	1	0	0	0	0	1	1	0	0	1
4	1	1	0	0	1	0	1	1	0	1	0	0	1	1	1	1	0	1	0	1	0	0	0	0	1	1	0	0
4	1	1	0	1	0	1	1	1	1	0	1	1	1	1	1	1	1	1	1	1	1	1	1	1	1	1	1	1

Association

2	2	2	4
0	0	0	0
0	1	1	1
1	0	1	2
1	1	0	3

2^{43} Orthogonal Main Effect Design

2	2	2	2	2	2	2	2	2	2	2	2	2	2	2	2	2	2	2	2	2	2	2	2	2	2	2	2	2	2	2	2	2	2	2	2	2	2	2	2	2	2	2
1	0	1	1	0	1	0	1	1	0	0	0	1	0	0	0	0	0	1	1	1	0	1	0	0	0	1	1	1	1	1	0	1	1	1	0	0	1	0	1	0	0	1
1	1	0	1	1	0	1	0	1	1	0	0	0	1	0	0	0	0	0	1	1	1	0	1	0	0	0	1	1	1	1	1	0	1	1	1	0	0	1	0	1	0	0
0	1	1	0	1	1	0	1	0	1	1	0	0	0	1	0	0	0	0	0	1	1	1	0	1	0	0	0	1	1	1	1	1	0	1	1	1	0	0	1	0	1	0
0	0	1	1	0	1	1	0	1	0	1	1	0	0	0	1	0	0	0	0	0	1	1	1	0	1	0	0	0	1	1	1	1	1	0	1	1	1	0	0	1	0	1
1	0	0	1	1	0	1	1	0	1	0	1	1	0	0	0	1	0	0	0	0	0	1	1	1	0	1	0	0	0	1	1	1	1	1	0	1	1	1	0	0	1	0
0	1	0	0	1	1	0	1	1	0	1	0	1	1	0	0	0	1	0	0	0	0	0	1	1	1	0	1	0	0	0	1	1	1	1	1	0	1	1	1	0	0	1
1	0	1	0	0	1	1	0	1	1	0	1	0	1	1	0	0	0	1	0	0	0	0	0	1	1	1	0	1	0	0	0	1	1	1	1	1	0	1	1	1	0	0
0	1	0	1	0	0	1	1	0	1	1	0	1	0	1	1	0	0	0	1	0	0	0	0	0	1	1	1	0	1	0	0	0	1	1	1	1	1	0	1	1	1	0
0	0	1	0	1	0	0	1	1	0	1	1	0	1	0	1	1	0	0	0	1	0	0	0	0	0	1	1	1	0	1	0	0	0	1	1	1	1	1	0	1	1	1
1	0	0	1	0	1	0	0	1	1	0	1	1	0	1	0	1	1	0	0	0	1	0	0	0	0	0	1	1	1	0	1	0	0	0	1	1	1	1	1	0	1	1
1	1	0	0	1	0	1	0	0	1	1	0	1	1	0	1	0	1	1	0	0	0	1	0	0	0	0	0	1	1	1	0	1	0	0	0	1	1	1	1	1	0	1
1	1	1	0	0	1	0	1	0	0	1	1	0	1	1	0	1	0	1	1	0	0	0	1	0	0	0	0	0	1	1	1	0	1	0	0	0	1	1	1	1	1	0
0	1	1	1	0	0	1	0	1	0	0	1	1	0	1	1	0	1	0	1	1	0	0	0	1	0	0	0	0	0	1	1	1	0	1	0	0	0	1	1	1	1	1
1	0	1	1	1	0	0	1	0	1	0	0	1	1	0	1	1	0	1	0	1	1	0	0	0	1	0	0	0	0	0	1	1	1	0	1	0	0	0	1	1	1	1
1	1	0	1	1	1	0	0	1	0	1	0	0	1	1	0	1	1	0	1	0	1	1	0	0	0	1	0	0	0	0	0	1	1	1	0	1	0	0	0	1	1	1
1	1	1	0	1	1	1	0	0	1	0	1	0	0	1	1	0	1	1	0	1	0	1	1	0	0	0	1	0	0	0	0	0	1	1	1	0	1	0	0	0	1	1
1	1	1	1	0	1	1	1	0	0	1	0	1	0	0	1	1	0	1	1	0	1	0	1	1	0	0	0	1	0	0	0	0	0	1	1	1	0	1	0	0	0	1
1	1	1	1	1	0	1	1	1	0	0	1	0	1	0	0	1	1	0	1	1	0	1	0	1	1	0	0	0	1	0	0	0	0	0	1	1	1	0	1	0	0	0
0	1	1	1	1	1	0	1	1	1	0	0	1	0	1	0	0	1	1	0	1	1	0	1	0	1	1	0	0	0	1	0	0	0	0	0	1	1	1	0	1	0	0
0	0	1	1	1	1	1	0	1	1	1	0	0	1	0	1	0	0	1	1	0	1	1	0	1	0	1	1	0	0	0	1	0	0	0	0	0	1	1	1	0	1	0
0	0	0	1	1	1	1	1	0	1	1	1	0	0	1	0	1	0	0	1	1	0	1	1	0	1	0	1	1	0	0	0	1	0	0	0	0	0	1	1	1	0	1
1	0	0	0	1	1	1	1	1	0	1	1	1	0	0	1	0	1	0	0	1	1	0	1	1	0	1	0	1	1	0	0	0	1	0	0	0	0	0	1	1	1	0
0	1	0	0	0	1	1	1	1	1	0	1	1	1	0	0	1	0	1	0	0	1	1	0	1	1	0	1	0	1	1	0	0	0	1	0	0	0	0	0	1	1	1
1	0	1	0	0	0	1	1	1	1	1	0	1	1	1	0	0	1	0	1	0	0	1	1	0	1	1	0	1	0	1	1	0	0	0	1	0	0	0	0	0	1	1
1	1	0	1	0	0	0	1	1	1	1	1	0	1	1	1	0	0	1	0	1	0	0	1	1	0	1	1	0	1	0	1	1	0	0	0	1	0	0	0	0	0	1
1	1	1	0	1	0	0	0	1	1	1	1	1	0	1	1	1	0	0	1	0	1	0	0	1	1	0	1	1	0	1	0	1	1	0	0	0	1	0	0	0	0	0
0	1	1	1	0	1	0	0	0	1	1	1	1	1	0	1	1	1	0	0	1	0	1	0	0	1	1	0	1	1	0	1	0	1	1	0	0	0	1	0	0	0	0
0	0	1	1	1	0	1	0	0	0	1	1	1	1	1	0	1	1	1	0	0	1	0	1	0	0	1	1	0	1	1	0	1	0	1	1	0	0	0	1	0	0	0
0	0	0	1	1	1	0	1	0	0	0	1	1	1	1	1	0	1	1	1	0	0	1	0	1	0	0	1	1	0	1	1	0	1	0	1	1	0	0	0	1	0	0
0	0	0	0	1	1	1	0	1	0	0	0	1	1	1	1	1	0	1	1	1	0	0	1	0	1	0	0	1	1	0	1	1	0	1	0	1	1	0	0	0	1	0
0	0	0	0	0	1	1	1	0	1	0	0	0	1	1	1	1	1	0	1	1	1	0	0	1	0	1	0	0	1	1	0	1	1	0	1	0	1	1	0	0	0	1
1	0	0	0	0	0	1	1	1	0	1	0	0	0	1	1	1	1	1	0	1	1	1	0	0	1	0	1	0	0	1	1	0	1	1	0	1	0	1	1	0	0	0
0	1	0	0	0	0	0	1	1	1	0	1	0	0	0	1	1	1	1	1	0	1	1	1	0	0	1	0	1	0	0	1	1	0	1	1	0	1	0	1	1	0	0
0	0	1	0	0	0	0	0	1	1	1	0	1	0	0	0	1	1	1	1	1	0	1	1	1	0	0	1	0	1	0	0	1	1	0	1	1	0	1	0	1	1	0
0	0	0	1	0	0	0	0	0	1	1	1	0	1	0	0	0	1	1	1	1	1	0	1	1	1	0	0	1	0	1	0	0	1	1	0	1	1	0	1	0	1	1
1	0	0	0	1	0	0	0	0	0	1	1	1	0	1	0	0	0	1	1	1	1	1	0	1	1	1	0	0	1	0	1	0	0	1	1	0	1	1	0	1	0	1
1	1	0	0	0	1	0	0	0	0	0	1	1	1	0	1	0	0	0	1	1	1	1	1	0	1	1	1	0	0	1	0	1	0	0	1	1	0	1	1	0	1	0
0	1	1	0	0	0	1	0	0	0	0	0	1	1	1	0	1	0	0	0	1	1	1	1	1	0	1	1	1	0	0	1	0	1	0	0	1	1	0	1	1	0	1
1	0	1	1	0	0	0	1	0	0	0	0	0	1	1	1	0	1	0	0	0	1	1	1	1	1	0	1	1	1	0	0	1	0	1	0	0	1	1	0	1	1	0
0	1	0	1	1	0	0	0	1	0	0	0	0	0	1	1	1	0	1	0	0	0	1	1	1	1	1	0	1	1	1	0	0	1	0	1	0	0	1	1	0	1	1
1	0	1	0	1	1	0	0	0	1	0	0	0	0	0	1	1	1	0	1	0	0	0	1	1	1	1	1	0	1	1	1	0	0	1	0	1	0	0	1	1	0	1
1	1	0	1	0	1	1	0	0	0	1	0	0	0	0	0	1	1	1	0	1	0	0	0	1	1	1	1	1	0	1	1	1	0	0	1	0	1	0	0	1	1	0
0	1	1	0	1	0	1	1	0	0	0	1	0	0	0	0	0	1	1	1	0	1	0	0	0	1	1	1	1	1	0	1	1	1	0	0	1	0	1	0	0	1	1
0	0	0	0	0	0	0	0	0	0	0	0	0	0	0	0	0	0	0	0	0	0	0	0	0	0	0	0	0	0	0	0	0	0	0	0	0	0	0	0	0	0	0

2^{47} Orthogonal Main Effect Design

4		4		4		4		4		4		4		4		4		4		4		4																								
2	2	2	2	2	2	2	2	2	2	2	2	2	2	2	2	2	2	2	2	2	2	2	2	2	2	2	2	2	2	2	2	2	2	2	2	2	2	2	2	2	2	2	2	2	2	2
0	0	0	0	0	0	0	0	0	0	0	0	0	0	0	0	0	0	0	0	0	0	0	0	0	0	0	0	0	0	0	0	0	0	0	0	0	0	1	0	0	0	1	1	1	0	1
0	0	0	0	0	0	0	0	0	0	1	1	0	1	1	0	1	1	1	1	0	1	1	0	1	1	0	1	0	1	1	0	1	1	0	1	1	1	0	1	0	0	0	1	1	1	0
0	0	0	0	0	0	0	0	0	1	1	0	1	1	0	1	1	0	1	0	1	1	0	1	1	0	1	0	1	1	0	1	1	0	1	1	1	1	1	0	1	0	0	0	1	1	1
0	0	0	1	1	0	0	1	1	1	0	1	0	1	1	1	1	0	0	1	1	1	0	1	0	0	0	1	1	0	0	0	0	1	0	1	0	0	1	1	0	1	0	0	0	1	1
0	0	0	1	1	0	0	1	1	1	1	0	1	0	1	0	1	1	0	0	0	0	1	1	1	0	1	1	0	1	1	1	0	0	0	0	1	1	0	1	1	0	1	0	0	0	1
0	0	0	1	1	0	0	1	1	0	1	1	1	1	0	1	0	1	1	0	1	0	0	0	0	1	1	0	0	0	1	0	1	1	1	0	1	1	1	0	1	1	0	1	0	0	0
0	0	0	0	1	1	1	0	1	1	1	0	0	0	0	1	0	1	0	1	1	0	0	0	1	1	0	0	1	1	1	1	0	1	0	1	1	1	1	1	0	1	1	0	1	0	0
0	0	0	0	1	1	1	0	1	1	0	1	1	1	0	0	0	0	1	1	0	0	1	1	0	0	0	1	0	1	0	1	1	1	1	0	0	0	1	1	1	0	1	1	0	1	0
0	0	0	0	1	1	1	0	1	0	0	0	1	0	1	1	1	0	0	0	0	1	1	0	0	1	1	1	1	0	1	0	1	0	1	1	0	0	0	1	1	1	0	1	1	0	1
0	0	0	1	0	1	1	1	0	0	1	1	1	0	1	0	0	0	0	1	1	1	1	0	1	0	1	0	1	1	0	0	0	1	1	0	0	0	0	0	1	1	1	0	1	1	0
0	0	0	1	0	1	1	1	0	0	0	0	0	1	1	1	0	1	1	0	1	0	1	1	1	1	0	1	1	0	0	1	1	0	0	0	1	1	0	0	0	1	1	1	0	1	1
0	0	0	1	0	1	1	1	0	1	0	1	0	0	0	0	1	1	1	1	0	1	0	1	0	1	1	0	0	0	1	1	0	0	1	1	0	0	0	0	0	0	0	0	0	0	0
0	1	1	0	1	1	0	1	1	0	1	1	0	1	1	0	1	1	0	1	1	0	1	1	0	1	1	0	1	1	0	1	1	0	1	1	0	0	1	0	0	0	1	1	1	0	1
0	1	1	0	1	1	0	1	1	1	0	1	1	0	1	1	0	1	0	0	0	0	0	0	0	0	0	1	1	0	1	1	0	1	1	0	1	1	0	1	0	0	0	1	1	1	0
0	1	1	0	1	1	0	1	1	0	0	0	0	0	0	0	0	0	1	1	0	1	1	0	1	1	0	1	0	1	1	0	1	1	0	1	1	1	1	0	1	0	0	0	1	1	1
0	1	1	0	0	0	1	0	1	1	1	0	1	0	1	0	0	0	1	0	1	1	1	0	0	1	1	0	0	0	0	1	1	1	1	0	0	0	1	1	0	1	0	0	0	1	1
0	1	1	0	0	0	1	0	1	0	0	0	1	1	0	1	0	1	0	1	1	1	0	1	1	1	0	1	1	0	0	0	0	0	1	1	1	1	0	1	1	0	1	0	0	0	1
0	1	1	0	0	0	1	0	1	1	0	1	0	0	0	1	1	0	1	1	0	0	1	1	1	0	1	0	1	1	1	1	0	0	0	0	1	1	1	0	1	1	0	1	0	0	0
0	1	1	1	0	1	1	1	0	0	0	0	0	1	1	1	1	0	1	0	1	0	1	1	0	0	0	1	0	1	0	0	0	1	1	0	1	1	1	1	0	1	1	0	1	0	0
0	1	1	1	0	1	1	1	0	1	1	0	0	0	0	0	1	1	0	0	0	1	0	1	0	1	1	1	1	0	1	0	1	0	0	0	0	0	1	1	1	0	1	1	0	1	0
0	1	1	1	0	1	1	1	0	0	1	1	1	1	0	0	0	0	0	1	1	0	0	0	1	0	1	0	0	0	1	1	0	1	0	1	0	0	0	1	1	1	0	1	1	0	1
0	1	1	1	1	0	0	0	0	1	0	1	1	1	0	0	1	1	1	0	1	0	0	0	1	1	0	1	0	1	0	1	1	0	0	0	0	0	0	0	1	1	1	0	1	1	0
0	1	1	1	1	0	0	0	0	0	1	1	1	0	1	1	1	0	1	1	0	1	0	1	0	0	0	0	0	0	1	0	1	0	1	1	1	1	0	0	0	1	1	1	0	1	1
0	1	1	1	1	0	0	0	0	1	1	0	0	1	1	1	0	1	0	0	0	1	1	0	1	0	1	0	1	1	0	0	0	1	0	1	0	0	0	0	0	0	0	0	0	0	0
1	0	1	1	0	1	1	0	1	1	0	1	1	0	1	1	0	1	1	0	1	1	0	1	1	0	1	1	0	1	1	0	1	1	0	1	0	0	1	0	0	0	1	1	1	0	1
1	0	1	1	0	1	1	0	1	1	1	0	1	1	0	1	1	0	0	1	1	0	1	1	0	1	1	0	0	0	0	0	0	0	0	0	1	1	0	1	0	0	0	1	1	1	0
1	0	1	1	0	1	1	0	1	0	1	1	0	1	1	0	1	1	0	0	0	0	0	0	0	0	0	1	1	0	1	1	0	1	1	1	0	1	0	0	0	1	1	1
1	0	1	0	1	1	1	1	0	0	0	0	1	1	0	0	1	1	1	1	0	0	0	0	1	0	1	0	1	1	1	0	1	0	0	0	0	0	1	1	0	1	0	0	0	1	1
1	0	1	0	1	1	1	1	0	0	1	1	0	0	0	1	1	0	1	0	1	1	1	0	0	0	0	0	0	0	0	1	1	1	0	1	1	1	0	1	1	0	1	0	0	0	1
1	0	1	0	1	1	1	1	0	1	1	0	0	1	1	0	0	0	0	0	0	1	0	1	1	1	0	1	0	1	0	0	0	0	1	1	1	1	1	0	1	1	0	1	0	0	0
1	0	1	1	1	0	0	0	0	0	1	1	1	0	1	0	0	0	1	1	0	1	0	1	0	1	1	1	1	0	0	1	1	0	0	0	1	1	1	1	0	1	1	0	1	0	0
1	0	1	1	1	0	0	0	0	0	0	0	0	1	1	1	0	1	0	1	1	1	1	0	1	0	1	0	0	0	1	1	0	0	1	1	0	0	1	1	1	0	1	1	0	1	0
1	0	1	1	1	0	0	0	0	1	0	1	0	0	0	0	1	1	1	0	1	0	1	1	1	1	0	0	1	1	0	0	0	1	1	0	0	0	0	1	1	1	0	1	1	0	1
1	0	1	0	0	0	0	1	1	1	1	0	0	0	0	0	1	0	1	1	1	0	0	1	1	0	0	0	1	1	0	1	0	1	0	1	1	0	0	0	1	1	1	0	1	1	0
1	0	1	0	0	0	0	1	1	1	0	1	1	1	0	0	0	0	0	0	0	1	1	0	0	1	1	0	1	1	1	1	0	1	0	1	1	1	0	0	0	1	1	1	0	1	1
1	0	1	0	0	0	0	1	1	0	0	0	1	0	1	1	1	0	0	1	1	0	0	0	1	1	0	1	0	1	0	1	1	1	1	0	0	0	0	0	0	0	0	0	0	0	0
1	1	0	1	1	0	1	1	0	1	1	0	1	1	0	1	1	0	1	1	0	1	1	0	1	1	0	1	1	0	1	1	0	1	1	0	0	0	1	0	0	0	1	1	1	0	1
1	1	0	1	1	0	1	1	0	0	0	0	0	0	0	0	0	0	1	0	1	1	0	1	1	0	1	0	1	1	0	1	1	0	1	1	1	1	0	1	0	0	0	1	1	1	0
1	1	0	1	1	0	1	1	0	1	0	1	1	0	1	1	0	1	0	1	1	0	1	1	0	1	1	0	0	0	0	0	0	0	0	0	1	1	1	0	1	0	0	0	1	1	1
1	1	0	1	0	1	0	0	0	0	1	1	0	0	0	1	0	1	0	0	0	0	1	1	1	1	0	1	0	1	1	1	0	0	1	1	0	0	1	1	0	1	0	0	0	1	1
1	1	0	1	0	1	0	0	0	1	0	1	0	1	1	0	0	0	1	1	0	0	0	0	0	1	1	0	1	1	1	0	1	1	1	0	1	1	0	1	1	0	1	0	0	0	1
1	1	0	1	0	1	0	0	0	0	0	0	1	0	1	0	1	1	0	1	1	1	1	0	0	0	0	1	1	0	0	1	1	1	0	1	1	1	1	0	1	1	0	1	0	0	0
1	1	0	0	0	0	0	1	1	1	0	1	1	1	0	0	1	1	0	0	0	1	1	0	1	0	1	0	0	0	1	0	1	0	1	1	1	1	1	1	0	1	1	0	1	0	0
1	1	0	0	0	0	0	1	1	0	1	1	1	0	1	1	1	0	1	0	1	0	0	0	1	1	0	0	1	1	0	0	0	1	0	1	0	0	1	1	1	0	1	1	0	1	0
1	1	0	0	0	0	0	1	1	1	1	0	0	1	1	1	0	1	1	1	0	1	0	1	0	0	0	1	0	1	0	1	1	0	0	0	0	0	0	1	1	1	0	1	1	0	1
1	1	0	0	1	1	1	0	1	0	0	0	0	1	1	1	1	0	0	0	0	0	1	0	1	0	1	1	0	0	0	1	1	0	1	0	1	0	0	0	1	1	1	0	1	1	0
1	1	0	0	1	1	1	0	1	1	1	0	0	0	0	0	1	1	0	1	1	0	0	0	1	0	1	1	0	1	0	0	0	1	1	0	1	1	0	0	0	1	1	1	0	1	1
1	1	0	0	1	1	1	0	1	0	1	1	1	1	0	0	0	0	1	0	1	0	1	1	0	0	0	1	1	0	1	0	1	0	0	0	0	0	0	0	0	0	0	0	0	0	0

Association

2	2	2	4
0	0	0	0
0	1	1	1
1	0	1	2
1	1	0	3

8×2^{40} Orthogonal Main Effect Design

8	2	2	2	2	2	2	2	2	2	2	2	2	2	2	2	2	2	2	2	2	2	2	2	2	2	2	2	2	2	2	2	2	2	2	2	2	2	2	2	2
0	0	1	0	0	0	1	1	1	0	1	0	1	0	0	0	1	1	1	0	1	0	1	0	0	0	1	1	1	0	1	0	1	0	0	0	1	1	1	0	1
1	1	0	1	0	0	0	1	1	1	0	1	0	1	0	0	0	1	1	1	0	1	0	1	0	0	0	1	1	1	0	1	0	1	0	0	0	1	1	1	0
1	1	1	0	1	0	0	0	1	1	1	1	1	0	1	0	0	0	1	1	1	1	1	0	1	0	0	0	1	1	1	1	1	0	1	0	0	0	1	1	1
0	0	1	1	0	1	0	0	0	1	1	0	1	1	0	1	0	0	0	1	1	0	1	1	0	1	0	0	0	1	1	0	1	1	0	1	0	0	0	1	1
1	1	0	1	1	0	1	0	0	0	1	1	0	1	1	0	1	0	0	0	1	1	0	1	1	0	1	0	0	0	1	1	0	1	1	0	1	0	0	0	1
1	1	1	0	1	1	0	1	0	0	0	1	1	0	1	1	0	1	0	0	0	1	1	0	1	1	0	1	0	0	0	1	1	0	1	1	0	1	0	0	0
1	1	1	1	0	1	1	0	1	0	0	1	1	1	0	1	1	0	1	0	0	1	1	1	0	1	1	0	1	0	0	1	1	1	0	1	1	0	1	0	0
0	0	1	1	1	0	1	1	0	1	0	0	1	1	1	0	1	1	0	1	0	0	1	1	1	0	1	1	0	1	0	0	1	1	1	0	1	1	0	1	0
0	0	0	1	1	1	0	1	1	0	1	0	0	1	1	1	0	1	1	0	1	0	0	1	1	1	0	1	1	0	1	0	0	1	1	1	0	1	1	0	1
0	0	0	0	1	1	1	0	1	1	0	0	0	0	1	1	1	0	1	1	0	0	0	0	1	1	1	0	1	1	0	0	0	0	1	1	1	0	1	1	0
1	1	0	0	0	1	1	1	0	1	1	1	0	0	0	1	1	1	0	1	1	1	0	0	0	1	1	1	0	1	1	1	0	0	0	1	1	1	0	1	1
0	0	0	0	0	0	0	0	0	0	0	0	0	0	0	0	0	0	0	0	0	0	0	0	0	0	0	0	0	0	0	0	0	0	0	0	0	0	0	0	0
2	0	1	0	0	0	1	1	1	0	1	0	1	0	0	0	1	1	1	0	1	1	0	1	1	1	0	0	0	1	0	1	0	1	1	1	0	0	0	1	0
3	1	0	1	0	0	0	1	1	1	0	1	0	1	0	0	0	1	1	1	0	0	1	0	1	1	1	0	0	0	1	0	1	0	1	1	1	0	0	0	1
3	1	1	0	1	0	0	0	1	1	1	1	1	0	1	0	0	0	1	1	1	0	0	1	0	1	1	1	0	0	0	0	0	1	0	1	1	1	0	0	0
2	0	1	1	0	1	0	0	0	1	1	0	1	1	0	1	0	0	0	1	1	1	0	0	1	0	1	1	1	0	0	1	0	0	1	0	1	1	1	0	0
3	1	0	1	1	0	1	0	0	0	1	1	0	1	1	0	1	0	0	0	1	0	1	0	0	1	0	1	1	1	0	0	1	0	0	1	0	1	1	1	0
3	1	1	0	1	1	0	1	0	0	0	1	1	0	1	1	0	1	0	0	0	0	0	1	0	0	1	0	1	1	1	0	0	1	0	0	1	0	1	1	1
3	1	1	1	0	1	1	0	1	0	0	1	1	1	0	1	1	0	1	0	0	0	0	0	1	0	0	1	0	1	1	0	0	0	1	0	0	1	0	1	1
2	0	1	1	1	0	1	1	0	1	0	0	1	1	1	0	1	1	0	1	0	1	0	0	0	1	0	0	1	0	1	1	0	0	0	1	0	0	1	0	1
2	0	0	1	1	1	0	1	1	0	1	0	0	1	1	1	0	1	1	0	1	1	1	0	0	0	1	0	0	1	0	1	1	0	0	0	1	0	0	1	0
2	0	0	0	1	1	1	0	1	1	0	0	0	0	1	1	1	0	1	1	0	1	1	1	0	0	0	1	0	0	1	1	1	1	0	0	0	1	0	0	1
3	1	0	0	0	1	1	1	0	1	1	1	0	0	0	1	1	1	0	1	1	0	1	1	1	0	0	0	1	0	0	0	1	1	1	0	0	0	1	0	0
2	0	0	0	0	0	0	0	0	0	0	0	0	0	0	0	0	0	0	0	0	1	1	1	1	1	1	1	1	1	1	1	1	1	1	1	1	1	1	1	1
4	0	1	0	0	0	1	1	1	0	1	1	0	1	1	1	0	0	0	1	0	0	1	0	0	0	1	1	1	0	1	1	0	1	1	1	0	0	0	1	0
5	1	0	1	0	0	0	1	1	1	0	0	1	0	1	1	1	0	0	0	1	1	0	1	0	0	0	1	1	1	0	0	1	0	1	1	1	0	0	0	1
5	1	1	0	1	0	0	0	1	1	1	0	0	1	0	1	1	1	0	0	0	1	1	0	1	0	0	0	1	1	1	0	0	1	0	1	1	1	0	0	0
4	0	1	1	0	1	0	0	0	1	1	1	0	0	1	0	1	1	1	0	0	0	1	1	0	1	0	0	0	1	1	1	0	0	1	0	1	1	1	0	0
5	1	0	1	1	0	1	0	0	0	1	0	1	0	0	1	0	1	1	1	0	1	0	1	1	0	1	0	0	0	1	0	1	0	0	1	0	1	1	1	0
5	1	1	0	1	1	0	1	0	0	0	0	0	1	0	0	1	0	1	1	1	1	1	0	1	1	0	1	0	0	0	0	0	1	0	0	1	0	1	1	1
5	1	1	1	0	1	1	0	1	0	0	0	0	0	1	0	0	1	0	1	1	1	1	1	0	1	1	0	1	0	0	0	0	0	1	0	0	1	0	1	1
4	0	1	1	1	0	1	1	0	1	0	1	0	0	0	1	0	0	1	0	1	0	1	1	1	0	1	1	0	1	0	1	0	0	0	1	0	0	1	0	1
4	0	0	1	1	1	0	1	1	0	1	1	1	0	0	0	1	0	0	1	0	0	0	1	1	1	0	1	1	0	1	1	1	0	0	0	1	0	0	1	0
4	0	0	0	1	1	1	0	1	1	0	1	1	1	0	0	0	1	0	0	1	0	0	0	1	1	1	0	1	1	0	1	1	1	0	0	0	1	0	0	1
5	1	0	0	0	1	1	1	0	1	1	0	1	1	1	0	0	0	1	0	0	1	0	0	0	1	1	1	0	1	1	0	1	1	1	0	0	0	1	0	0
4	0	0	0	0	0	0	0	0	0	0	1	1	1	1	1	1	1	1	1	1	0	0	0	0	0	0	0	0	0	0	1	1	1	1	1	1	1	1	1	1
6	0	1	0	0	0	1	1	1	0	1	1	0	1	1	1	0	0	0	1	0	1	0	1	1	1	0	0	0	1	0	0	1	0	0	0	1	1	1	0	1
7	1	0	1	0	0	0	1	1	1	0	0	1	0	1	1	1	0	0	0	1	0	1	0	1	1	1	0	0	0	1	1	0	1	0	0	0	1	1	1	0
7	1	1	0	1	0	0	0	1	1	1	0	0	1	0	1	1	1	0	0	0	0	0	1	0	1	1	1	0	0	0	1	1	0	1	0	0	0	1	1	1
6	0	1	1	0	1	0	0	0	1	1	1	0	0	1	0	1	1	1	0	0	1	0	0	1	0	1	1	1	0	0	0	1	1	0	1	0	0	0	1	1
7	1	0	1	1	0	1	0	0	0	1	0	1	0	0	1	0	1	1	1	0	0	1	0	0	1	0	1	1	1	0	1	0	1	1	0	1	0	0	0	1
7	1	1	0	1	1	0	1	0	0	0	0	0	1	0	0	1	0	1	1	1	0	0	1	0	0	1	0	1	1	1	1	1	0	1	1	0	1	0	0	0
7	1	1	1	0	1	1	0	1	0	0	0	0	0	1	0	0	1	0	1	1	0	0	0	1	0	0	1	0	1	1	1	1	1	0	1	1	0	1	0	0
6	0	1	1	1	0	1	1	0	1	0	1	0	0	0	1	0	0	1	0	1	1	0	0	0	1	0	0	1	0	1	0	1	1	1	0	1	1	0	1	0
6	0	0	1	1	1	0	1	1	0	1	1	1	0	0	0	1	0	0	1	0	1	1	0	0	0	1	0	0	1	0	0	0	1	1	1	0	1	1	0	1
6	0	0	0	1	1	1	0	1	1	0	1	1	1	0	0	0	1	0	0	1	1	1	1	0	0	0	1	0	0	1	0	0	0	1	1	1	0	1	1	0
7	1	0	0	0	1	1	1	0	1	1	0	1	1	1	0	0	0	1	0	0	0	1	1	1	0	0	0	1	0	0	1	0	0	0	1	1	1	0	1	1
6	0	0	0	0	0	0	0	0	0	0	1	1	1	1	1	1	1	1	1	1	1	1	1	1	1	1	1	1	1	1	0	0	0	0	0	0	0	0	0	0

6 × 4 × 2^{35} Orthogonal Main Effect Design

		4			4			4			4			4			4			4			4			4			4			4				
6	4	2	2	2	2	2	2	2	2	2	2	2	2	2	2	2	2	2	2	2	2	2	2	2	2	2	2	2	2	2	2	2	2	2	2	2
0	0	0	0	0	0	0	0	0	0	0	0	0	0	0	0	0	0	0	0	0	0	0	0	0	0	0	0	0	0	0	0	0	0	0	0	0
1	0	0	0	0	0	0	0	0	1	1	0	1	1	0	1	1	1	1	0	1	1	0	1	1	0	1	0	1	1	0	1	1	0	1	0	0
2	0	0	0	0	0	0	0	1	1	0	1	1	0	1	1	0	1	0	1	1	0	1	1	0	1	0	1	1	0	1	1	0	1	1	0	0
3	0	1	1	0	0	1	1	1	0	1	0	1	1	1	1	0	0	1	1	1	0	1	0	0	0	1	1	0	0	0	0	1	0	1	0	1
4	0	1	1	0	0	1	1	1	1	0	1	0	1	0	1	1	0	0	0	0	1	1	1	0	1	1	0	1	1	1	0	0	0	0	0	1
5	0	1	1	0	0	1	1	0	1	1	1	1	0	1	0	1	1	0	1	0	0	0	0	1	1	0	0	0	1	0	1	1	1	0	0	1
0	0	0	1	1	1	0	1	1	1	0	0	0	0	1	0	1	0	1	1	0	0	0	1	1	0	0	1	1	1	1	0	1	0	1	1	0
1	0	0	1	1	1	0	1	1	0	1	1	1	0	0	0	0	1	1	0	0	1	1	0	0	0	1	0	1	0	1	1	1	1	0	1	0
2	0	0	1	1	1	0	1	0	0	0	1	0	1	1	1	0	0	0	0	1	1	0	0	1	1	1	1	0	1	0	1	0	1	1	1	0
3	0	1	0	1	1	1	0	0	1	1	1	0	1	0	0	0	0	1	1	1	1	0	1	0	1	0	1	1	0	0	0	1	1	0	1	1
4	0	1	0	1	1	1	0	0	0	0	0	1	1	1	0	1	1	0	1	0	1	1	1	1	0	1	1	0	0	1	1	0	0	0	1	1
5	0	1	0	1	1	1	0	1	0	1	0	0	0	0	1	1	1	1	0	1	0	1	0	1	1	0	0	0	1	1	0	0	1	1	1	1
0	1	0	1	1	0	1	1	0	1	1	0	1	1	0	1	1	0	1	1	0	1	1	0	1	1	0	1	1	0	1	1	0	1	1	0	0
1	1	0	1	1	0	1	1	1	0	1	1	0	1	1	0	1	0	0	0	0	0	0	0	0	0	1	1	0	1	1	0	1	1	0	0	0
2	1	0	1	1	0	1	1	0	0	0	0	0	0	0	0	0	1	1	0	1	1	0	1	1	0	1	0	1	1	0	1	1	0	1	0	0
3	1	0	0	0	1	0	1	1	1	0	1	0	1	0	0	0	1	0	1	1	1	0	0	1	1	0	0	0	0	1	1	1	1	0	0	1
4	1	0	0	0	1	0	1	0	0	0	1	1	0	1	0	1	0	1	1	1	0	1	1	1	0	1	1	0	0	0	0	0	1	1	0	1
5	1	0	0	0	1	0	1	1	0	1	0	0	0	1	1	0	1	1	0	0	1	1	1	0	1	0	1	1	1	1	0	0	0	0	0	1
0	1	1	0	1	1	1	0	0	0	0	0	1	1	1	1	0	1	0	1	0	1	1	0	0	0	1	0	1	0	0	0	1	1	0	1	0
1	1	1	0	1	1	1	0	1	1	0	0	0	0	0	1	1	0	0	0	1	0	1	0	1	1	1	1	0	1	0	1	0	0	0	1	0
2	1	1	0	1	1	1	0	0	1	1	1	1	0	0	0	0	0	1	1	0	0	0	1	0	1	0	0	0	1	1	0	1	0	1	1	0
3	1	1	1	0	0	0	0	1	0	1	1	1	0	0	1	1	1	0	1	0	0	0	1	1	0	1	0	1	0	1	1	0	0	0	1	1
4	1	1	1	0	0	0	0	0	1	1	1	0	1	1	1	0	1	1	0	1	0	1	0	0	0	0	0	0	1	0	1	0	1	1	1	1
5	1	1	1	0	0	0	0	1	1	0	0	1	1	1	0	1	0	0	0	1	1	0	1	0	1	0	1	1	0	0	0	1	0	1	1	1
0	2	1	0	1	1	0	1	1	0	1	1	0	1	1	0	1	1	0	1	1	0	1	1	0	1	1	0	1	1	0	1	1	0	1	0	0
1	2	1	0	1	1	0	1	1	1	0	1	1	0	1	1	0	0	1	1	0	1	1	0	1	1	0	0	0	0	0	0	0	0	0	0	0
2	2	1	0	1	1	0	1	0	1	1	0	1	1	0	1	1	0	0	0	0	0	0	0	0	0	1	1	0	1	1	0	1	1	0	0	0
3	2	0	1	1	1	1	0	0	0	0	1	1	0	0	1	1	1	1	0	0	0	0	1	0	1	0	1	1	1	0	1	0	0	0	0	1
4	2	0	1	1	1	1	0	0	1	1	0	0	0	1	1	0	1	0	1	1	1	0	0	0	0	0	0	0	0	1	1	1	0	1	0	1
5	2	0	1	1	1	1	0	1	1	0	0	1	1	0	0	0	0	0	0	1	0	1	1	1	0	1	0	1	0	0	0	0	1	1	0	1
0	2	1	1	0	0	0	0	0	1	1	1	0	1	0	0	0	1	1	0	1	0	1	0	1	1	1	1	0	0	1	1	0	0	0	1	0
1	2	1	1	0	0	0	0	0	0	0	0	1	1	1	0	1	0	1	1	1	1	0	1	0	1	0	0	0	1	1	0	0	1	1	1	0
2	2	1	1	0	0	0	0	1	0	1	0	0	0	0	1	1	1	0	1	0	1	1	1	1	0	0	1	1	0	0	0	1	1	0	1	0
3	2	0	0	0	0	1	1	1	1	0	0	0	0	1	0	1	1	1	0	0	1	1	0	0	0	1	1	0	1	0	1	0	1	1	1	1
4	2	0	0	0	0	1	1	1	0	1	1	1	0	0	0	0	0	0	0	1	1	0	0	1	1	0	1	1	1	1	0	1	0	1	1	1
5	2	0	0	0	0	1	1	0	0	0	1	0	1	1	1	0	0	1	1	0	0	0	1	1	0	1	0	1	0	1	1	1	1	0	1	1
0	3	1	1	0	1	1	0	1	1	0	1	1	0	1	1	0	1	1	0	1	1	0	1	1	0	1	1	0	1	1	0	1	1	0	0	0
1	3	1	1	0	1	1	0	0	0	0	0	0	0	0	0	0	1	0	1	1	0	1	1	0	1	0	1	1	0	1	1	0	1	1	0	0
2	3	1	1	0	1	1	0	1	0	1	1	0	1	1	0	1	0	1	1	0	1	1	0	1	1	0	0	0	0	0	0	0	0	0	0	0
3	3	1	0	1	0	0	0	0	1	1	0	0	0	1	0	1	0	0	0	0	1	1	1	1	0	1	0	1	1	1	0	0	1	1	0	1
4	3	1	0	1	0	0	0	1	0	1	0	1	1	0	0	0	1	1	0	0	0	0	0	1	1	0	1	1	1	0	1	1	1	0	0	1
5	3	1	0	1	0	0	0	0	0	0	1	0	1	0	1	1	0	1	1	1	1	0	0	0	0	1	1	0	0	1	1	1	0	1	0	1
0	3	0	0	0	0	1	1	1	0	1	1	1	0	0	1	1	0	0	0	1	1	0	1	0	1	0	0	0	1	0	1	0	1	1	1	0
1	3	0	0	0	0	1	1	0	1	1	1	0	1	1	1	0	1	0	1	0	0	0	1	1	0	0	1	1	0	0	0	1	0	1	1	0
2	3	0	0	0	0	1	1	1	1	0	0	1	1	1	0	1	1	1	0	1	0	1	0	0	0	1	0	1	0	1	1	0	0	0	1	0
3	3	0	1	1	1	0	1	0	0	0	0	1	1	1	1	0	0	0	0	1	0	1	0	1	1	0	0	0	1	1	0	1	0	1	1	1
4	3	0	1	1	1	0	1	1	1	0	0	0	0	0	1	1	0	1	1	0	0	0	1	0	1	1	0	1	0	0	0	1	1	0	1	1
5	3	0	1	1	1	0	1	0	1	1	1	1	0	0	0	0	1	0	1	0	1	1	0	0	0	1	1	0	1	0	1	0	0	0	1	1

Association

2	2	2	4
0	0	0	0
0	1	1	1
1	0	1	2
1	1	0	3

$6^2 \times 2^{31}$ Orthogonal Main Effect Design

6	6	2	2	2	2	2	2	2	2	2	2	2	2	2	2	2	2	2	2	2	2	2	2	2	2	2	2	2	2	2	2	2
0	0	0	0	1	0	0	0	1	1	1	0	1	0	1	0	0	0	1	1	1	0	1	0	1	0	0	0	1	1	1	0	1
0	1	0	0	1	1	0	1	0	0	0	1	1	0	1	1	0	1	0	0	0	1	1	0	1	1	0	1	0	0	0	1	1
0	2	0	0	1	1	1	0	1	1	0	1	0	0	1	1	1	0	1	1	0	1	0	0	1	1	1	0	1	1	0	1	0
0	3	0	1	0	1	1	1	0	1	1	0	1	0	0	1	1	1	0	1	1	0	1	0	0	1	1	1	0	1	1	0	1
0	4	0	1	0	0	1	1	1	0	1	1	0	0	0	0	1	1	1	0	1	1	0	0	0	0	1	1	1	0	1	1	0
0	5	0	1	0	0	0	0	0	0	0	0	0	0	0	0	0	0	0	0	0	0	0	0	0	0	0	0	0	0	0	0	0
1	0	1	1	0	1	0	0	0	1	1	1	0	1	0	1	0	0	0	1	1	1	0	1	0	1	0	0	0	1	1	1	0
1	1	1	1	1	0	1	0	0	0	1	1	1	1	1	0	1	0	0	0	1	1	1	1	1	0	1	0	0	0	1	1	1
1	2	1	1	0	1	1	0	1	0	0	0	1	1	0	1	1	0	1	0	0	0	1	1	0	1	1	0	1	0	0	0	1
1	3	1	0	1	0	1	1	0	1	0	0	0	1	1	0	1	1	0	1	0	0	0	1	1	0	1	1	0	1	0	0	0
1	4	1	0	1	1	0	1	1	0	1	0	0	1	1	1	0	1	1	0	1	0	0	1	1	1	0	1	1	0	1	0	0
1	5	1	0	0	0	0	1	1	1	0	1	1	1	0	0	0	1	1	1	0	1	1	1	0	0	0	1	1	1	0	1	1
2	0	0	0	1	0	0	0	1	1	1	0	1	1	0	1	1	1	0	0	0	1	0	1	0	1	1	1	0	0	0	1	0
2	1	0	0	1	1	0	1	0	0	0	1	1	1	0	0	1	0	1	1	1	0	0	1	0	0	1	0	1	1	1	0	0
2	2	0	0	1	1	1	0	1	1	0	1	0	1	0	0	0	1	0	0	1	0	1	1	0	0	0	1	0	0	1	0	1
2	3	0	1	0	1	1	1	0	1	1	0	1	1	1	0	0	0	1	0	0	1	0	1	1	0	0	0	1	0	0	1	0
2	4	0	1	0	0	1	1	1	0	1	1	0	1	1	1	0	0	0	1	0	0	1	1	1	1	0	0	0	1	0	0	1
2	5	0	1	0	0	0	0	0	0	0	0	0	1	1	1	1	1	1	1	1	1	1	1	1	1	1	1	1	1	1	1	1
3	0	1	1	0	1	0	0	0	1	1	1	0	0	1	0	1	1	1	0	0	0	1	0	1	0	1	1	1	0	0	0	1
3	1	1	1	1	0	1	0	0	0	1	1	1	0	0	1	0	1	1	1	0	0	0	0	0	1	0	1	1	1	0	0	0
3	2	1	1	0	1	1	0	1	0	0	0	1	0	1	0	0	1	0	1	1	1	0	0	1	0	0	1	0	1	1	1	0
3	3	1	0	1	0	1	1	0	1	0	0	0	0	0	1	0	0	1	0	1	1	1	0	0	1	0	0	1	0	1	1	1
3	4	1	0	1	1	0	1	1	0	1	0	0	0	0	0	1	0	0	1	0	1	1	0	0	0	1	0	0	1	0	1	1
3	5	1	0	0	0	0	1	1	1	0	1	1	0	1	1	1	0	0	0	1	0	0	0	1	1	1	0	0	0	1	0	0
4	0	1	0	0	1	1	1	0	0	0	1	0	0	1	0	0	0	1	1	1	0	1	1	0	1	1	1	0	0	0	1	0
4	1	1	0	0	0	1	0	1	1	1	0	0	0	1	1	0	1	0	0	0	1	1	1	0	0	1	0	1	1	1	0	0
4	2	1	0	0	0	0	1	0	0	1	0	1	0	1	1	1	0	1	1	0	1	0	1	0	0	0	1	0	0	1	0	1
4	3	1	1	1	0	0	0	1	0	0	1	0	0	0	1	1	1	0	1	1	0	1	1	1	0	0	0	1	0	0	1	0
4	4	1	1	1	1	0	0	0	1	0	0	1	0	0	0	1	1	1	0	1	1	0	1	1	1	0	0	0	1	0	0	1
4	5	1	1	1	1	1	1	1	1	1	1	1	0	0	0	0	0	0	0	0	0	0	1	1	1	1	1	1	1	1	1	1
5	0	0	1	1	0	1	1	1	0	0	0	1	1	0	1	0	0	0	1	1	1	0	0	1	0	1	1	1	0	0	0	1
5	1	0	1	0	1	0	1	1	1	0	0	0	1	1	0	1	0	0	0	1	1	1	0	0	1	0	1	1	1	0	0	0
5	2	0	1	1	0	0	1	0	1	1	1	0	1	0	1	1	0	1	0	0	0	1	0	1	0	0	1	0	1	1	1	0
5	3	0	0	0	1	0	0	1	0	1	1	1	1	1	0	1	1	0	1	0	0	0	0	0	1	0	0	1	0	1	1	1
5	4	0	0	0	0	1	0	0	1	0	1	1	1	1	1	0	1	1	0	1	0	0	0	0	0	1	0	0	1	0	1	1
5	5	0	0	1	1	1	0	0	0	1	0	0	1	0	0	0	1	1	1	0	1	1	0	1	1	1	0	0	0	1	0	0
3	0	1	0	0	1	1	1	0	0	0	1	0	1	0	1	1	1	0	0	0	1	0	0	1	0	0	0	1	1	1	0	1
3	1	1	0	0	0	1	0	1	1	1	0	0	1	0	0	1	0	1	1	1	0	0	0	1	1	0	1	0	0	0	1	1
3	2	1	0	0	0	0	1	0	0	1	0	1	1	0	0	0	1	0	0	1	0	1	0	1	1	1	0	1	1	0	1	0
3	3	1	1	1	0	0	0	1	0	0	1	0	1	1	0	0	0	1	0	0	1	0	0	0	1	1	1	0	1	1	0	1
3	4	1	1	1	1	0	0	0	1	0	0	1	1	1	1	0	0	0	1	0	0	1	0	0	0	1	1	1	0	1	1	0
3	5	1	1	1	1	1	1	1	1	1	1	1	1	1	1	1	1	1	1	1	1	1	0	0	0	0	0	0	0	0	0	0
5	0	0	1	1	0	1	1	1	0	0	0	1	0	1	0	1	1	1	0	0	0	1	1	0	1	0	0	0	1	1	1	0
5	1	0	1	0	1	0	1	1	1	0	0	0	0	0	1	0	1	1	1	0	0	0	1	1	0	1	0	0	0	1	1	1
5	2	0	1	1	0	0	1	0	1	1	1	0	0	1	0	0	1	0	1	1	1	0	1	0	1	1	0	1	0	0	0	1
5	3	0	0	0	1	0	0	1	0	1	1	1	0	0	1	0	0	1	0	1	1	1	1	1	0	1	1	0	1	0	0	0
5	4	0	0	0	0	1	0	0	1	0	1	1	0	0	0	1	0	0	1	0	1	1	1	1	1	0	1	1	0	1	0	0
5	5	0	0	1	1	1	0	0	0	1	0	0	0	1	1	1	0	0	0	1	0	0	1	0	0	0	1	1	1	0	1	1

$4^{12} \times 3 \times 2^4$ and $4^{13} \times 3$ Orthogonal Main Effect Design

													4				Substitute
4	4	4	4	4	4	4	4	4	4	4	4	3	2	2	2	2	4
0	0	0	0	0	0	0	0	0	0	0	0	0	0	0	0	0	0
0	0	0	1	1	1	3	3	3	2	2	2	0	0	1	0	1	1
0	0	0	3	3	3	2	2	2	1	1	1	1	0	0	1	1	0
0	3	1	2	1	3	1	2	0	3	0	2	1	0	1	1	0	1
0	3	1	3	2	1	0	1	2	2	3	0	2	0	0	1	0	0
0	3	1	1	3	2	2	0	1	0	2	3	2	0	1	0	1	1
0	1	2	3	0	2	1	0	3	1	3	2	0	1	0	1	1	2
0	1	2	2	3	0	3	1	0	2	1	3	0	1	1	1	0	3
0	1	2	0	2	3	0	3	1	3	2	1	1	1	0	0	1	2
0	2	3	1	2	0	1	3	2	1	0	3	1	1	1	0	0	3
0	2	3	0	1	2	2	1	3	3	1	0	2	1	0	0	0	2
0	2	3	2	0	1	3	2	1	0	3	1	2	1	1	1	1	3
1	1	1	1	1	1	1	1	1	1	1	1	0	0	0	0	0	0
1	1	1	2	2	2	0	0	0	3	3	3	0	0	1	0	1	1
1	1	1	0	0	0	3	3	3	2	2	2	1	0	0	1	1	0
1	0	2	3	2	0	2	3	1	0	1	3	1	0	1	1	0	1
1	0	2	0	3	2	1	2	3	3	0	1	2	0	0	1	0	0
1	0	2	2	0	3	3	1	2	1	3	0	2	0	1	0	1	1
1	2	3	0	1	3	2	1	0	2	0	3	0	1	0	1	1	2
1	2	3	3	0	1	0	2	1	3	2	0	0	1	1	1	0	3
1	2	3	1	3	0	1	0	2	0	3	2	1	1	0	0	1	2
1	3	0	2	3	1	2	0	3	2	1	0	1	1	1	0	0	3
1	3	0	1	2	3	3	2	0	0	2	1	2	1	0	0	0	2
1	3	0	3	1	2	0	3	2	1	0	2	2	1	1	1	1	3
2	2	2	2	2	2	2	2	2	2	2	2	0	0	0	0	0	0
2	2	2	3	3	3	1	1	1	0	0	0	0	0	1	0	1	1
2	2	2	1	1	1	0	0	0	3	3	3	1	0	0	1	1	0
2	1	3	0	3	1	3	0	2	1	2	0	1	0	1	1	0	1
2	1	3	1	0	3	2	3	0	0	1	2	2	0	0	1	0	0
2	1	3	3	1	0	0	2	3	2	0	1	2	0	1	0	1	1
2	3	0	1	2	0	3	2	1	3	1	0	0	1	0	1	1	2
2	3	0	0	1	2	1	3	2	0	3	1	0	1	1	1	0	3
2	3	0	2	0	1	2	1	3	1	0	3	1	1	0	0	1	2
2	0	1	3	0	2	3	1	0	3	2	1	1	1	1	0	0	3
2	0	1	2	3	0	0	3	1	1	3	2	2	1	0	0	0	2
2	0	1	0	2	3	1	0	3	2	1	3	2	1	1	1	1	3
3	3	3	3	3	3	3	3	3	3	3	3	0	0	0	0	0	0
3	3	3	0	0	0	2	2	2	1	1	1	0	0	1	0	1	1
3	3	3	2	2	2	1	1	1	0	0	0	1	0	0	1	1	0
3	2	0	1	0	2	0	1	3	2	3	1	1	0	1	1	0	1
3	2	0	2	1	0	3	0	1	1	2	3	2	0	0	1	0	0
3	2	0	0	2	1	1	3	0	3	1	2	2	0	1	0	1	1
3	0	1	2	3	1	0	3	2	0	2	1	0	1	0	1	1	2
3	0	1	1	2	3	2	0	3	1	0	2	0	1	1	1	0	3
3	0	1	3	1	2	3	2	0	2	1	0	1	1	0	0	1	2
3	1	2	0	1	3	0	2	1	0	3	2	1	1	1	0	0	3
3	1	2	3	0	1	1	0	2	2	0	3	2	1	0	0	0	2
3	1	2	1	3	0	2	1	0	3	2	0	2	1	1	1	1	3

7^8 Orthogonal Main Effect Design

7	7	7	7	7	7	7	7
0	0	0	0	0	0	0	0
0	1	6	6	6	6	6	6
0	2	5	5	5	5	5	5
0	3	4	4	4	4	4	4
0	4	3	3	3	3	3	3
0	5	2	2	2	2	2	2
0	6	1	1	1	1	1	1
1	0	6	5	4	3	2	1
1	1	5	4	3	2	1	0
1	2	4	3	2	1	0	6
1	3	3	2	1	0	6	5
1	4	2	1	0	6	5	4
1	5	1	0	6	5	4	3
1	6	0	6	5	4	3	2
2	0	5	3	1	6	4	2
2	1	4	2	0	5	3	1
2	2	3	1	6	4	2	0
2	3	2	0	5	3	1	6
2	4	1	6	4	2	0	5
2	5	0	5	3	1	6	4
2	6	6	4	2	0	5	3
3	0	4	1	5	2	6	3
3	1	3	0	4	1	5	2
3	2	2	6	3	0	4	1
3	3	1	5	2	6	3	0
3	4	0	4	1	5	2	6
3	5	6	3	0	4	1	5
3	6	5	2	6	3	0	4
4	0	3	6	2	5	1	4
4	1	2	5	1	4	0	3
4	2	1	4	0	3	6	2
4	3	0	3	6	2	5	1
4	4	6	2	5	1	4	0
4	5	5	1	4	0	3	6
4	6	4	0	3	6	2	5
5	0	2	4	6	1	3	5
5	1	1	3	5	0	2	4
5	2	0	2	4	6	1	3
5	3	6	1	3	5	0	2
5	4	5	0	2	4	6	1
5	5	4	6	1	3	5	0
5	6	3	5	0	2	4	6
6	0	1	2	3	4	5	6
6	1	0	1	2	3	4	5
6	2	6	0	1	2	3	4
6	3	5	6	0	1	2	3
6	4	4	5	6	0	1	2
6	5	3	4	5	6	0	1
6	6	2	3	4	5	6	0

$5^{11} \times 2$ Orthogonal Main Effect Design

										10	
5	5	5	5	5	5	5	5	5	5	5	2
0	0	0	0	0	0	0	0	0	0	0	0
1	1	2	3	4	1	1	2	3	4	0	0
2	2	4	1	3	2	2	4	1	3	0	0
3	3	1	4	2	3	3	1	4	2	0	0
4	4	3	2	1	4	4	3	2	1	0	0
0	1	1	1	1	1	2	2	2	2	1	0
1	2	3	4	0	2	3	4	0	1	1	0
2	3	0	2	4	3	4	1	3	0	1	0
3	4	2	0	3	4	0	3	1	4	1	0
4	0	4	3	2	0	1	0	4	3	1	0
0	2	2	2	2	4	1	1	1	1	2	0
1	3	4	0	1	0	2	3	4	0	2	0
2	4	1	3	0	1	3	0	2	4	2	0
3	0	3	1	4	2	4	2	0	3	2	0
4	1	0	4	3	3	0	4	3	2	2	0
0	3	3	3	3	4	2	2	2	2	3	0
1	4	0	1	2	0	3	4	0	1	3	0
2	0	2	4	1	1	4	1	3	0	3	0
3	1	4	2	0	2	0	3	1	4	3	0
4	2	1	0	4	3	1	0	4	3	3	0
0	4	4	4	4	1	0	0	0	0	4	0
1	0	1	2	3	2	1	2	3	4	4	0
2	1	3	0	2	3	2	4	1	3	4	0
3	2	0	3	1	4	3	1	4	2	4	0
4	3	2	1	0	0	4	3	2	1	4	0
0	2	1	4	3	0	4	1	1	4	0	1
1	3	3	2	2	1	0	2	2	0	0	1
2	4	0	0	1	2	1	3	3	1	0	1
3	0	2	3	0	3	2	4	4	2	0	1
4	1	4	1	4	4	3	0	0	3	0	1
0	3	2	0	4	2	3	2	4	4	1	1
1	4	4	3	3	3	4	3	0	0	1	1
2	0	1	1	2	4	0	4	1	1	1	1
3	1	3	4	1	0	1	0	2	2	1	1
4	2	0	2	0	1	2	1	3	3	1	1
0	4	3	1	0	3	1	2	1	3	2	1
1	0	0	4	4	4	2	3	2	4	2	1
2	1	2	2	3	0	3	4	3	0	2	1
3	2	4	0	2	1	4	0	4	1	2	1
4	3	1	3	1	2	0	1	0	2	2	1
0	0	4	2	1	3	3	1	2	1	3	1
1	1	1	0	0	4	4	2	3	2	3	1
2	2	3	3	4	0	0	3	4	3	3	1
3	3	0	1	3	1	1	4	0	4	3	1
4	4	2	4	2	2	2	0	1	0	3	1
0	1	0	3	2	2	4	4	2	3	4	1
1	2	2	1	1	3	0	0	3	4	4	1
2	3	4	4	0	4	1	1	4	0	4	1
3	4	1	2	4	0	2	2	0	1	4	1
4	0	3	0	3	1	3	3	1	2	4	1

Association

5	2	10
0	0	0
1	0	1
2	0	2
3	0	3
4	0	4
0	1	5
1	1	6
2	1	7
3	1	8
4	1	9

Index

Q

R